T0312663

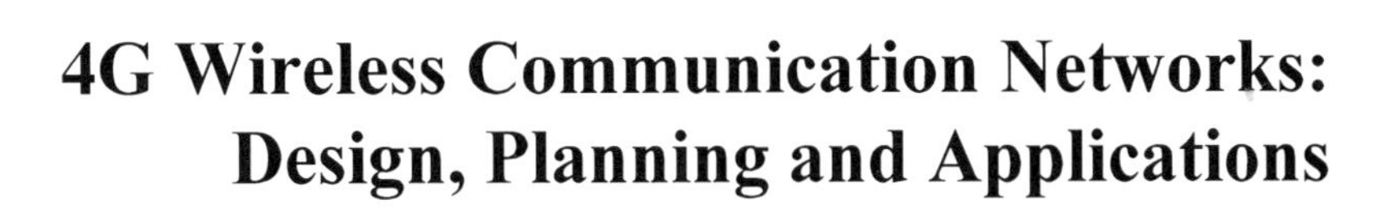

4G Wireless Communication Networks: Design, Planning and Applications

RIVER PUBLISHERS SERIES IN COMMUNICATIONS

This series focuses on communications science and technology. This includes the theory and use of systems involving all terminals, computers, and information processors; wired and wireless networks; and network layouts, procontentsols, architectures, and implementations.
Furthermore, developments toward newmarket demands in systems, products, and technologies such as personal communications services, multimedia systems, enterprise networks, and optical communications systems.

- Wireless Communications
- Networks
- Security
- Antennas & Propagation
- Microwaves
- Software Defined Radio

4G Wireless Communication Networks: Design, Planning and Applications

Editors

Johnson I Agbinya
Mari Carmen Aguayo-Torres
Ryszard Klempous
Jan Nikodem

ISBN 978-87-92982-71-1 (hardback)

Published, sold and distributed by:
River Publishers
P.O. Box 1657
Algade 42
9000 Aalborg
Denmark

Tel.: +4536995З197
www.riverpublishers.com

Dedicated to our families and friends

Table of Contents

Preface

The revolution in wireless communication technology since the nineties has led to rapid development of new mobile communication systems including GSM, GPRS, EDGE, WCDMA, WLAN, HSDPA, WiMAX and recently LTE-Advanced. These new technologies have revolutionized the way and manner modern communication is undertaken. The new technologies laid foundations in wireless and broadband network design, handsets and Internet enabled devices. These technologies were created with the support of other fields such as advanced signal processing techniques including OFDM, MIMO, coding, voice, image and video compression. These developments have also caused the migration in focus from fixed telephony to IP telephony with complimentary consequences in the use of social networking and online retail and sales. As welcome as the developments are, the overwhelming depth of technical expertise required to understand and follow the progress in technology advancement makes it harder and harder for the telecommunication engineer to follow without a logical and detailed compendium of the major concepts leading to the advancements.

This book is a detailed compendium of some of the major advancements focusing exclusively on the emerging broadband wireless communication technologies which support broadband wireless data rate transmissions. Several applications of wireless communication networks including health care, underground communication, biomedical and bio-telemetry systems are detailed in the book.

Several chapters in the book were written as lecture notes over a period of several years and have appeared in various forms with students that we have had the opportunity to teach mobile communication networks in several institutions around the world.

Acknowledgment

We acknowledge on behalf of all the authors the contributions of several research funding agencies which have enabled their chapters in this book. We also acknowledge the contributions of several authors who have contributed chapters in the book and of our host institutions for providing favourable research and writing environments.

Chapter 1.

Cellular Mobile Communication Concepts

Johnson I Agbinya

Department of Electronic Engineering, La Trobe University, Australia
J.Agbinya@latrobe.edu.au

1.1 CELLULAR TECHNOLOGY

Cellular communication concepts are used in the design of modern cellular communication systems such as GSM, UMTS, WiMAX and LTE. To understand how these technologies work, familiarity with the main concepts of the cellular networks is essential. In this respect, cellular technology is discussed in some details in this section. Nature also uses cellular concepts where possible. For example honeycombs are built by bees using the cellular concept and provide very rigid combs for storage of honey and development of colonies.

The growing number of subscribers requiring access to cellular communication networks in one hand and the limited available radio frequency spectrum on the other hand, implied that more system capacity is required. This has led to development cellular communication concepts. The concepts allow for reuse of allocated radio frequency in a small area and hence pave the way for improved system capacity as we shall soon illustrate. A cellular network refers to a radio network consisting of a number of cells. A cell is defined by the geographical coverage of a base station. Each base station serves a specific geographical area by means of a radio transceiver connection. In other words, a cell provides coverage for data transmission. Each cell is allocated either a set of frequencies as in GSM or a set of orthogonal codes as in UMTS, WiMAX and LTE-Advance. In GSM for example in other to minimize interference between transmissions using neighbouring frequencies, none of the adjacent cells can use the same set of frequencies. Cells far enough from each other can be allocated the

4G Wireless Communication Networks: Design, Planning and Applications, 1-36,

same set of frequencies [6]. The separation distance is called the reuse distance and the ability to use the same sets of frequencies far away is referred to as frequency reuse. These concepts are discussed in some details in the section 1.4.2.

The cost of spectrum contributes significantly to the need for efficient design of cellular mobile networks. As an example, Table 1.1 provides a snapshot on the price paid globally for LTE spectrum in the 700/800 MHz spectrum range. In most countries the spectrum price is offered per MHz per head of population and normalized for some number of years of usage. Table 1.1 shows a comparative normalized base price for Australian 15 year license.

For example the population of Germany was about 82 million persons in 2012. Hence assuming that an operator plans to cover all the 82 million Germans, then for s MHz of spectrum, an operator would have paid the price $Total\ price(\$) = price\ paid\ x\ p\ x\ s$.

If s=20 MHz, the price paid by the operator is 0.8x20x82,000,000 = AU$1.312billions. This is a huge initial capital which requires that such a

Table 1.1 LTE Spectrum Prices in the 700/800MHz band

Country	Reserve Price (AU$)	Price Paid
Australia	1.36	Due in 2013
Belgium	0.51	Due in 2013
Canada	0.69	Due in 2013
Croatia	0.34	0.34
Czech Republic	0.52	Due in 2013
Denmark	0.1	0.25
Finland	0.28	Due in 2013
France	0.41	0.60
Germany	0	0.87
India	0.79	Due in 2013
Ireland	0.21	Multi-band
Italy	0.58	0.8
Macedonia	0.65	Not sold
Netherlands	0.22	Multi-band
Portugal	0.5	0.5
Romania	0.19	Multi-band
Spain	0.36	0.46
Sweden	0.14	0.32
Switzerland	0.36	Multi-band
United Kingdom	0.41	Due in 2013

spectrum should be used efficiently so as to break even within the number of years the operator is permitted to use the spectrum. Assuming the operator is permitted to use the spectrum for 15 years, the net spectrum fee per year is about AU$87.5 million.

This kind of huge initial capital plus the costs of equipment and personnel and overhead motivates the need for efficient network design and planning to ensure that profit can still be made by the operator. In this chapter, some of the techniques which support efficient use of spectrum and network design are presented and explained. In general the performance of a particular mobile communication system in terms of how well it uses available spectrum is quanitified in terms of the spectrum efficiency. Spectrum efficiency is defined as the number of bits/second/Hz of spectrum used by the system. The spectrum efficiencies of 1G and 2G networks are typically very poor and typically around 0.17 for GSM and CDMA 2000. Modern mobile communication networks such as 3G and 4G networks provide better spectrum efficiency typically up to 4.8 for WiMAX and 16.32 for LTE. These improved values are due to use of advanced signal processing techniques such as OFDM, MIMO and quadrature amplitude modulation (QAM) schemes.

1.2 CELL SHAPE AND SIZE

A hexagon is usually assumed to be the shape of a cell theoretically. However, in practice the shape is amorphous due to the environmental impact on the signal propagation and also the technology in use such as the type of antenna applied at the base station and its location. Figure 1.1(a) illustrates an ideal cell shape. Other

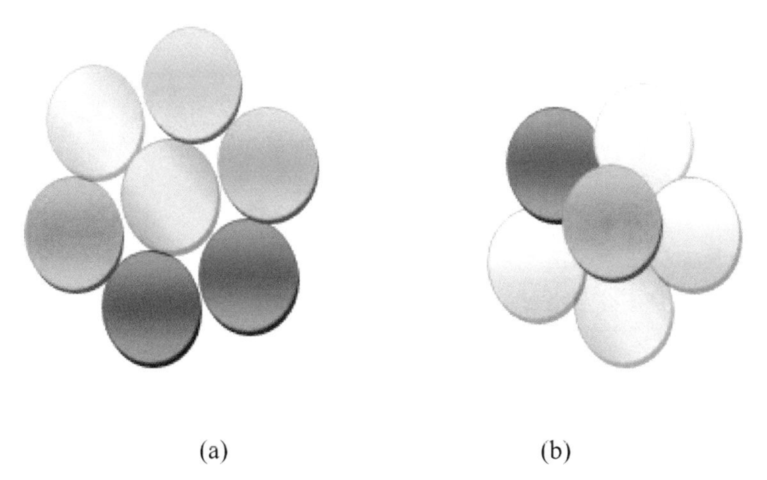

Figure 1.1: (a) Circular Cells with Blind Spots; (b) overlapping circular cells.

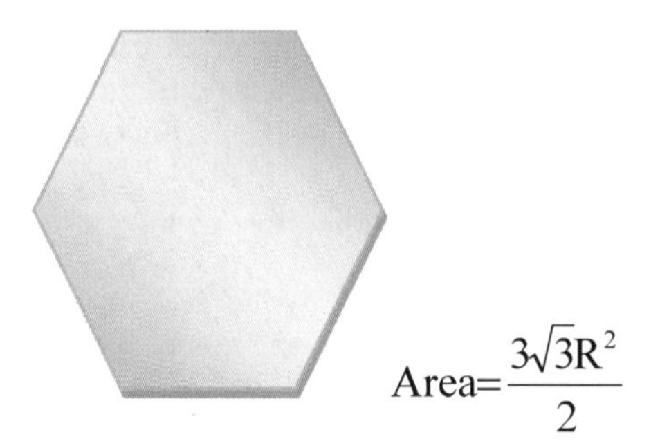

Figure 1.2 cell shape.

possible cell shapes are circle and square. The use of circles to define the area of coverage of a cell is not optimum. For example if the geographical area is divided into a number of circles, gaps appear between them as in Figure 1.1(a). The gaps represent blind spots. Therefore the use of circular cell templates is not advised. In Figure 1.1(a) there are blind spots between the circles. Including the purple circle in the middle has not completely removed the blind spots. To remove the blind spots, the cell footprints must overlap as in Figure 1.1(b) which results to wasting of resources and coverage. Figure 1.2 shows a hexagonal cell of radius R and its area.

A rectangle is also inappropriate because a rectangular cell has unequal distances with its neighbors. Consider instead the use of a square template for cells. A square width of *d* has four neighbors (Figure 1.3) at distance *d* and four at distance $\sqrt{2}d$ therefore it requires complicated computations for some required algorithms. Designing antennas with square radiation patterns is almost impossible.

A hexagonal cell, having equal distance with all adjacent cells therefore provides the best coverage model with the least number of required base stations

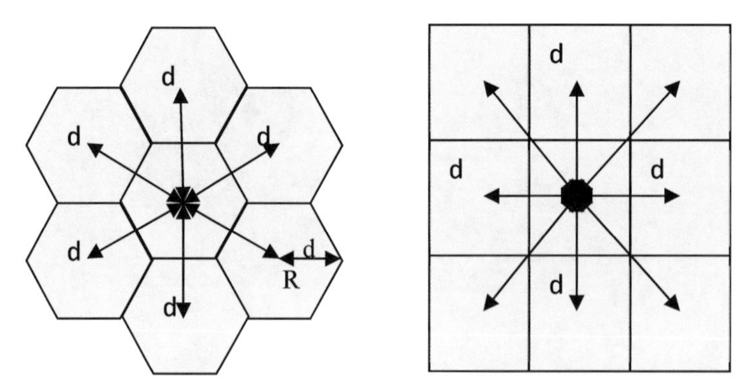

Figure 1.3 Comparison between the cell shape and computations/square and hexagon.

to serve a particular area. The rest of this book is developed using hexagonal cell shapes.

Assume R as the cell radius. The distance between two adjacent cells is $d=\sqrt{3}R$ and the coverage area of a cell is approximately $2.598R^2$.

The physical size of cells is often used to describe them. The popular descriptions of cells are macro cell, micro cell and pico cells. A macro cell is large and has radius between 100m to 10km or more. It is suitable for rural coverage which requires large cell range normally enhanced with very high towers. Micro cells are the medium size cells and have radius of about several 100m to a couple of kilometers. Micro cells are efficient for crowded urban areas such as small suburbs and shopping malls. The smallest cells are referred to as pico cells, which have size of a few dozen meters. They are usually deployed to serve very small sized environments such as offices and lifts. They are also often used to fill up where ever there are coverage holes.
As the size of the cell decreases, the required transmission power also decreases and need less transmission power. The heights of the base stations antenna are lower for the smaller cells.

1.3 CLUSTER AND FREQUENCY REUSE

To increase the capacity of a cellular communication network, while using a limited number of radio channels, cellular systems use the frequency reuse technology. It means that the same number of frequencies is used as many times as required in a network by imposing a careful frequency plan. Frequencies are planned for example around the concept of clusters. A cellular network consists of at least one cluster. A number of cells, each one using a different group of frequencies from the other cells, form a cluster. The coverage range of a cluster is known as a footprint. Different cells associated with a cluster must be allocated a

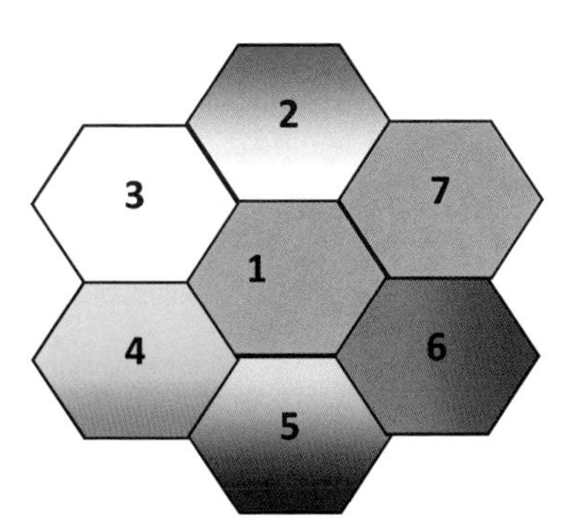

Figure 1.4 Cluster with K=1/7.

specific and unique subset of all the available duplex channels to prevent interference between the same frequencies. Assuming *S* as all the available pairs of channels and *N* as the number of channels used by each cell, the number of cells in a cluster is deduced by $K=S/N$. The inverse of this number of the cells is referred to as frequency reuse factor *(1/K)*. The possible values for *K* are 1, 3, 4, 7, 9, 12, 13, 16, 19, 21. Figure 1.4 shows a cluster with a reuse factor K=1/7.
A cluster can be repeated in the network as many times as needed. In fact networks are planned using clusters of cells. If we assume the cluster to be repeated *M* times, the total capacity of the network will be $C=M*S$, which is the total number of the radio channels available to be assigned to the subscribers at the same time.

To prevent interference between co-channel cells, which are the cells using the same group of radio frequencies, they need to be far enough apart from each other. The separation takes advantage of the expected power loss or power decay of radio wave transmissions with distance. The minimum required distance between co-channel cells is calculated in this section as an illustration

- Assuming the base stations in the co-channel cells are located in $C(u_1,v_1)$ and $F(u_2,v_2)$ and distance between them is D derived by:

$$D=\left\{(u_2-u_1)^2\left(\cos 30^0\right)^2+\left[(v_2-v_1)+(u_2-u_1)\sin 30^0\right]^2\right\}^{1/2}$$

$$D=\left\{(u_2-u_1)^2+(v_2-v_1)^2+(v_2-v_1)(u_2-u_1)\right\}^{1/2} \quad (1.1)$$

When we restrict (u_1,v_1) to integer values (i,j) and set $(u_1,v_1)=(0,0)$ the result is given by the expression

$$D^2=i^2+ij+j^2 \quad (1.2)$$

In order therefore to connect without gaps between adjacent cells, the number of cells per cluster (cluster size), D, must satisfy the equation:

$$D^2=i^2+ij+j^2 \quad (1.3)$$

Where i, j are non-negative integers. To find the nearest co-channel neighbours of a particular cell the steps required are:

(1) Move i cells along any chain of hexagons (Figure 1.5) and
(2) Turn 60 degrees counter-clockwise and move j cells.

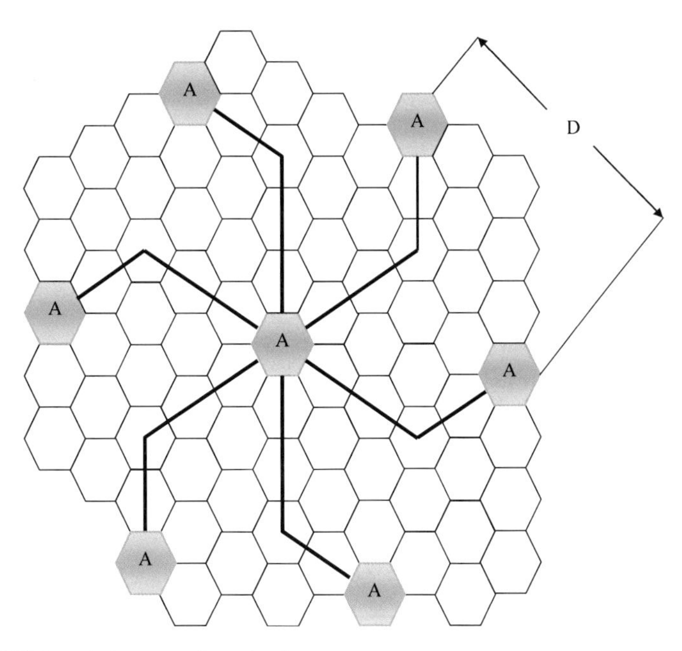

Figure 1.5 Distance between co-channel cells.

Table 1.2 Reuse Distance and Signal-to-Interference Ratio for Cluster Size 7 (n=2)

Path loss exponent	N	2	
Cluster size	N	7	
Co-channel reuse ratio	Q	$\sqrt{(3N)}$	4.582576
Radius of cell	R		meter
Distance to co-channel cells	D		meter
Ratio of D to R	Q=D/R	$\sqrt{(3N)}$	4.582576
Number of first tier cells	i_0	6	One per side of hexagon
Signal to noise ratio	S/I	$\left(\frac{D}{R}\right)^n / i_0$	3.5
(S/I) in decibels	S/I	$10\log_{10}(S/I)$	5.441dB

An application of reuse distance is given in the following tables. At the point of system design on the drawing board, there is actually no recorded RF signals to fall back on. The use of reuse distance ratio fills this gap because it allows the designer to estimate the signal to noise ratio per reuse ratio to ascertain the efficiency in the choices of system parameters.

Consider these three tables for cases shown when the cluster sizes are 7 and 4 and the path loss exponent varies from 2 to 4. The analysis is for the first tier sources of interference (base stations located on a first-tier circular locus around a target transmitting station) to the desired signal. The values of Q and hence the signal-to-noise ratio (SNR) for the selected range of the desired base stations are given and used as guide to show whether good reception exists at range R with

Table 1.3 Reuse Distance and Signal-to-Interference Ratio for Cluster Size 7 (n=4)

Path loss exponent	n	4	
Cluster size	N	7	
Co-channel reuse ratio	Q	$\sqrt{(3N)}$	4.582576
Radius of cell	R		meter
Distance to co-channel cells	D		meter
Ratio of D to R	Q=D/R	$\sqrt{(3N)}$	4.582576
Number of first tier cells	i_0	6	One per side of hexagon
Signal to noise ratio	S/I	$\left(\frac{D}{R}\right)^n / i_0$	73.5
(S/I) in decibels	S/I	$10\log_{10}(S/I)$	18.663dB

Table 1.4 Reuse Distance and Signal-to-Interference Ratio for Cluster Size 4 (n=4)

Path loss exponent	n	4	
Cluster size	N	4	
Co-channel reuse ratio	Q	$\sqrt{(3N)}$	3.464
Radius of cell	R		meter
Distance to co-channel cells	D		meter
Ratio of D to R	Q=D/R	$\sqrt{(3N)}$	3.464
Number of first tier cells	i_0	6	One per side of hexagon
Signal to noise ratio	S/I	$\left(\frac{D}{R}\right)^n / i_0$	24
(S/I) in decibels	S/I	$10\log_{10}(S/I)$	13.80211dB

the selected reuse distances. Assume that good reception is possible only when the SNR is equal to or greater than 15dB.

We find from Table 1.2 that the signal to noise ratio is actually too small for reception because we need at least about 15dB for that to happen. In the next set of parameters, the selected cluster size and hence the reuse distance ratio to the cell radius leads to at least 18dB which is enough for good reception (Table 1.3).

In the next set of parameters in Table 1.4, the obtained signal to noise ratio is about 14dB which is still below the 15dB required for good reception.

In a nutshell, we can estimate the desired SNR for adequate reception at a point before the actual base stations are deployed to the sites.

1.3.1 Adjacent Channel Interference

Interference resulting from signals that are adjacent in frequency to the desired signal is called *adjacent channel interference (ACI)*. ACI is the interference from a cell using a frequency or code that is adjacent to the one being used by a handset of interest. Adjacent channel interference results from imperfect receiver filters that allow nearby frequencies to leak into the adjacent pass bands. Adjacent channel interference can be minimised through careful filtering and channel assignments.

Isolation filters between channels are usually not ideal. As a result, signal from one channel can leak into adjacent channels. ACI is a function of the distance between the two cells under consideration. For omni-directional cells, ACI is given by the expression:

$$ACI = -10\,Log\left[\left(\frac{d_i}{d_c}\right)^n\right] + Filter\ isolation \tag{1.4}$$

1.3.2 Near End Far End Interference

In the uplink, signals are attenuated differently because they take different paths. Signals to and from mobiles nearest to the base station are stronger than signals from mobiles located much farther away. This is a fact from the normal nature of radio wave transmission decaying with distance. In the downlink however, mobiles at the cell edge experience larger degradation and interference compared to mobiles close to the base station, because by the time the transmitted signal reaches the edge of the cell, it has decayed significantly. Therefore mobiles at the edge of cells (as in Figure 1.6) experience a higher level of interference from own base station compared to other base stations. Mobiles very close to base stations therefore often cause more interference, particularly in terms of adjacent cell interference. Let us consider the worst case scenario.

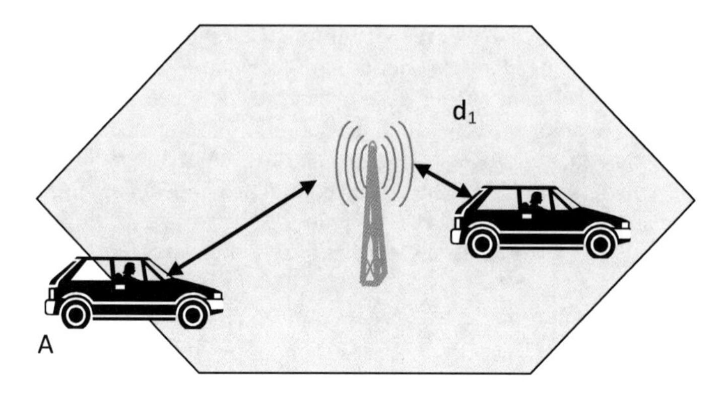

Figure 1.6 Worst Case Adjacent Cell Interference between Two Mobiles.

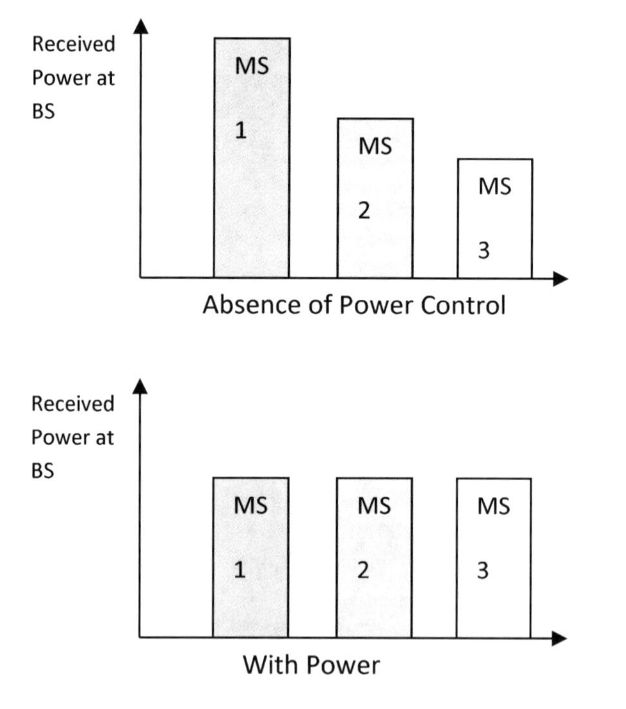

Figure 1.7 Power Control of Mobile Stations.

Two mobiles use the same base station. The mobile A is at the edge of the cell and is transmitting at frequency 1. The mobile B is very close to the base station and is transmitting at a channel y. Because of the large signal level from the mobile close to the base station the base station receiver filter for channel 1 need to attenuate the signal level. If the cell radius R = 10 km and the distance $d_1 = 250$ metres, the carrier to interference ratio seen in channel 1 is given by the expression:

$$C/I = \left(R/d_1 \right)^{-n} \tag{1.5}$$

When n=4, this interference level is 64 dB above the desired signal. A typical filter to use for attenuating this undesired interference is a Butterworth filter. Consider its degree is m. Its attenuation as a function of frequency ω is $10\log_{10}\left(1+\omega^{2m}\right)$, and for a large value of ω, the expression is $20 * m\log_{10}(\omega)$.

It can be shown that the channel separation of the mobiles should be higher than 4 ($\omega > 4$) for the interference to be tolerable. Power control is used to mitigate near far effect as shown in the Figure 1.7.

With power control, the powers received from all mobiles at the base station are equalised.

1.4 POWER CONTROL FOR REDUCING INTERFERENCE

In practical cellular radio and personal communication systems the power levels transmitted by every subscriber unit are constant controlled by the serving base stations. This is done to ensure that each mobile transmits the smallest power necessary to maintain a good quality link on the reverse channel. Power control not only helps prolong battery life for the subscriber unit, but also dramatically reduces the reverse channel S/I in the system.

1.5 TECHNIQUES FOR IMPROVING CAPACITY IN CELLULAR SYSTEMS

As the demand for wireless services increases, the number of channels assigned to a cell eventually becomes insufficient to support the required number of users. At this point, cellular design techniques are needed to provide more channels per unit coverage area. Several techniques for increased capacity can be identified. Any technique which increases the system bandwidth and/or signal to noise ratio is always going to increase capacity. This is predicted by Shannon's equation stating the system capacity as

$$C = B\log_2\left(1+\frac{S}{N}\right) \tag{1.6}$$

Where C is the system capacity in bits per second, B is bandwidth (Hertz), S is signal power in watt and N is noise power in watts. There are significant number of techniques for improving the capacity of the system. Some of them are listed below:

1) Techniques common to all cellular systems
   a) cell splitting and
   b) sectoring of cells
2) Diversity techniques including
   c) allocating multiple time slots as in GPRS
   d) multiple-input-multiple-output systems (HSDPA, WiMAX, LTE-A, WLAN)
3) Bandwidth increase methods including the use of
   e) orthogonal frequency division multiplexing (OFDM) which uses many sub-carriers (WiMAX, LTE-A, WLAN)
   f) larger system bandwidth networks (eg., WCDMA, LTE-A and WiMAX) using 5MHz carriers
4) Signal processing techniques
   g) digital modulation techniques which allows many bits to be concurrently transmitted as symbols (BPSK, QPSK, QAM)
   h) signal compression methods (voice, image, video and text compression)
5) Noise reduction techniques
   i) Clusters and reuse distance
   j) Co-channel interference reduction
   k) Reduction of fading and use of fading and diffraction margins

Techniques common to all cellular networks such as *cell splitting, sectoring are discussed in this section.* Other methods listed above will be discussed in chapters where they are used.

1.5.1 Cell Splitting

Overwhelming increased demand or subscribers within the coverage area of a cell above its possible capacity can be alleviated with cell splitting. Cell splitting is the process of subdividing larger cells into smaller cells for the purpose of reducing congestion and increasing capacity. Cell splitting allows for an orderly growth of the cellular system.

Rapidly growing traffic within a cell often means capacity will be used up and consideration should be given to splitting the cell into smaller ones. Further increase in capacity in the cell is possible by increasing the number of voice channels in the cell or reviewing the cell boundaries so that the area formerly regarded as a single cell is divided among several smaller cells. Decreasing the area of each cell helps in meeting growing traffic in a cell without having to allocate more spectrum. Cell splitting not only helps to serve higher demand areas by smaller cells but also to serve lower demand areas by larger cells. A theoretical cell splitting scenario is shown in Figure 1.8. Each small cell has its own base station and a corresponding reduction in antenna height and transmitter power. The price paid is new base stations but with shorter antenna rigs and less transmission power. Antenna feeder losses are also reduced because short cables are used. Cell splitting is used mostly in high traffic density areas or any place where added capacity is needed. Sectors use directional antennas to further control the interference and frequency reuse of channels. How does cell splitting help to increase capacity? Consider the case when the distance between adjacent cell sites is reduced by a half, or reduce the cell radius by half. Therefore, the new cell radius = 0.5 x old cell radius. If the new cell carries the same traffic load as the old parent cell, then the new traffic load is 4 times the old traffic load. This increase in capacity can be seen by the expression:

$$\text{New traffic load} = \frac{\pi R^2}{\pi\left(\frac{1}{2}R\right)^4} \text{ x old traffic capacity} = 4 \text{ x old traffic capacity} \tag{1.7}$$

Let us consider the theoretical cell splitting case in Figure 1.8. Since the radius of the new cells are reduced, the power transmitted need to be reduced as well to cover just the range of the new cell.

Let the power transmitted by the old big cell of Radius R be P_0; the power transmitted by any of the smaller cells of radius 0.5xR be P_n. We need to determine what power is to be transmitted by any of the smaller cells so it does not spill beyond its boundary. Ideally, the power received by a mobile at the

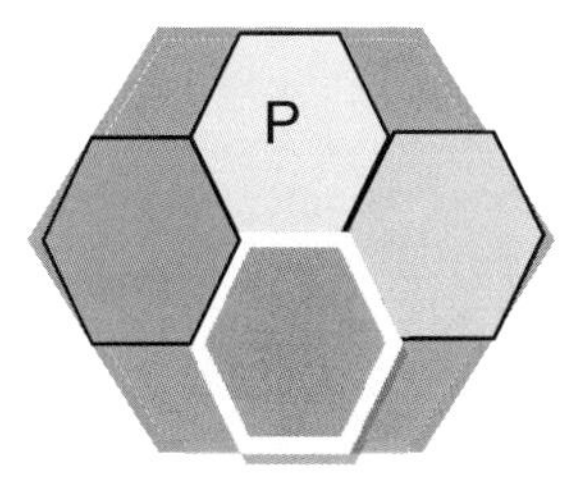

Figure 1.8 Cell Splitting Into 4 smaller Cells.

boundary of any of the new smaller cells should be equal to the power at the boundary of the old big cell. This power relationship is given by the expression $P_0 R^{-n} = P_n \left(R/2\right)^{-n}$ and 'n' is the propagation path loss exponent, determined by the mobile environment. Considering n=4, this relationship becomes $P_n/P_0 = 2^{-4} = 1/16$. This means the power transmitted by the base stations of the new cells is one-sixteenth of the power transmitted by the old base station covering the larger cell. This power rating of the new base stations is 12 dB less than P_0. Hence a lot smaller base stations could be used at lower costs.

What has happened to the ratio D/R for the split cells? Since the radius is now R/2, D is also reduced to D/2. Hence D/R remains constant. We can prove this by the following analysis. For the old cell plans, the carrier-to-interference ratio is

$$\frac{C}{I} = \frac{R^{-n}}{\left(D_1 + D_2 + D_3 + D_4 + D_5 + D_6\right)^{-n}} \tag{1.8}$$

By splitting the cell, the carrier-to-interference ratio for the new plan is

$$\frac{C}{I} = \frac{\left(R/2\right)^{-n}}{\left(D_1/2 + D_2/2 + D_3/2 + D_4/2 + D_5/2 + D_6/2\right)^{-n}} \tag{1.9}$$

One of the issues that often arise in practice is finding the best locations of the split cells inside the area covered by the old cell. The ideal locations for the new cell sites are at a point midway between neighbouring existing cells. In practice, as long as the split cells are anywhere within a distance of a quarter of the smaller cell radius, they are adequately located.

1.5.2 Sectoring

Cell splitting achieves capacity improvement by essentially re-scaling the system. By decreasing the cell radius R and keeping the co-channel reuse ratio D / R unchanged, cell splitting increases the number of channels per unit area. However, another way to increase capacity is to keep the cell radius unchanged and seek methods to decrease the D / R ratio. In this approach, capacity improvement is achieved by reducing the number of cells in a cluster and thus increasing the frequency reuse. However, in order to do this, it is necessary to reduce the relative interference without decreasing the transmit power.

The co-channel interference in a cellular system may be decreased by replacing a single omni-directional antenna at the base station by several directional antennas, each radiating within a specified sector. By using directional antennas, a given cell will receive interference and transmit with only a fraction of the available co-channel cells. The technique for decreasing co-channel interference and thus increasing system capacity by using direction antennas is called *sectoring.* The factor by which the co-channel interference is reduced depends on the amount of sectors used. A cell is normally partitioned into three 120^0 sectors or six 60^0 sectors. Consider the network in Figure 1.9 with 120^0 sectors, and a frequency reuse ratio 7.

Of the 6 co-channel cells shown in the first tier in Figure 1.3, only 2 of them interfere with the centre cell. If omni-directional antennas were used at each base station, all 6 co-channel cells would interfere with the centre cell.

When sectoring is employed, the channels used in a particular cell are broken down into sectored groups and are used only within a particular sector. Assuming 7-cell reuse, for the case of 120° sectors, the number of interferers in the first tier is reduced from 6 to 2. This is because only 2 of the 6 co-channel cells receive interference with a particular sectored channel group. The *S/I* is increased to about 24.2 dB, which is a significant improvement over the omni-directional worst case *S/I=17dB*.

In practice, the reduction offered by sectoring enables planners to reduce the cluster size N. The cost for improved $S\,/\,I$ and capacity is an increased number of

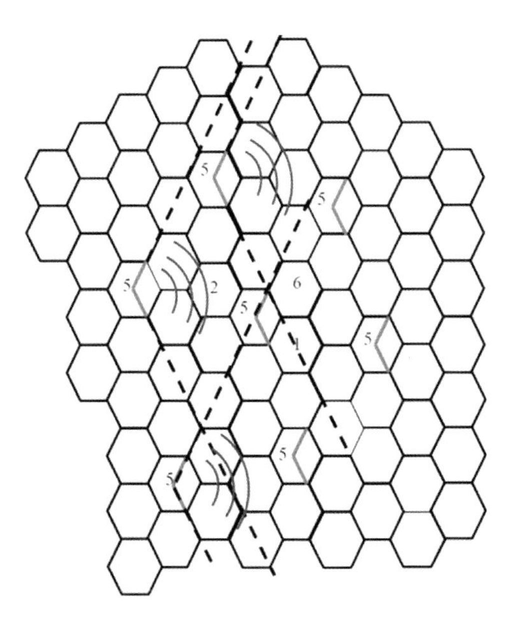

Figure 1.9 120^0 Sectoring in a wireless cellular mobile system.

antennas at each base station, and a decrease in trunking efficiency due to channel sectoring at the base station. Since sectoring reduces the coverage area of a particular group of channels, the number of handovers increases. Fortunately, base stations allow mobiles to be handed off from sector to sector within the same cell without intervention from the MSC, therefore the handover problem is often not a major concern.

1.6 CHANNEL ASSIGNMENT IN CELLULAR SYSTEMS

Each cell can use a specific number of channels for allocation to its users. Channel assignment, which means designation of the available channels to the base stations or to the user's calls, requires a precise planning to minimize the possible interference to enhance the capacity of the network. To achieve that there are three main strategies, static (fixed), dynamic and a hybrid form of both. In the static channel assignment, which suites networks with uniform traffic and high total traffic load in all cells, all the available channels are allocated to the cells and each base station has its own number of channels [7]. In this pattern, if all the channels are in use, new call request will be blocked. However, in dynamic channel assignment, the channel will be allocated to a call at the time that the call is made. In other words, no base station has its own specific and constant number of channels. A combination of the two methods is used in networks with multiform traffic load but slightly the same traffic ratio, which means the traffic load have not many changes over time.

To implement a hybrid channel assignment, all the available channels are divided into two sets of static and dynamic channels. In the case that the cells have different traffic load, three main strategies can be used to prevent calls being blocked. Borrowing available channels from adjacent cells is a solution for serving users, while all channels in a cell are occupied by other users. In this case the borrowed channels are released to the original cell when the calls are terminated. Another way to overcome this problem is to allocate the channels to the cells based on their needs. In this case, all the available channels are maintained in a pool and are assigned to the cells when needed. Channels can also be allocated by a central decision element, which decides allocation of the channels; however it can slow down the network.

As the number of users grows the capacity available to a user decreases. Alongside, frequency borrowing from adjacent cells, as a method of increasing the network's capacity, cell splitting (Figure 1.10(b)) also can be used to achieve this goal. Cell splitting refers to minimizing the cell size. In this scheme, a cell is divided into smaller cells, therefore the number of cell increases consequently the number of required base stations is also boosted and he achieved capacity is much higher. This scheme operates properly in crowded areas such as big cities, particularly in busy spots such as shopping malls.

Another method of increasing the capacity is cell sectoring. Using directional antennas at the base station within a cell, instead of using Omni-directional antenna, is referred to as cell sectoring (Figure 1.10(a)). Three 120° or six 180° directional antennas are usually deployed in a base station to sector a cell. This leads to less co-channel interference but more intra cell hand over.

These schemes are used in WiMAX networks with some unique variations which are discussed in latter sections of the chapter.

1.6.1 Handoff (Handover)

One of the most important operations in a cellular system is the concept of handover or handoff. When a mobile user changes the base station in use, which means moving from one cell to another, handover occurs. As it is described in the previous sections, different cells use different frequency channels, therefore when a user changes a cell and consequently its assigned channel the call would be dropped. Dropping a call is undesirable for a communication; therefore handoff solves this problem and maintains the connectivity while changing the cell. However, the handoff is unrecognizable to the mobile phone user.

The reason for handoff is to achieve a stronger signal for a desirable communication quality. When the received signal is not strong enough, the mobile station requests a handoff and the call will be switched to the new base station with higher signal level.

There are two groups of handoff processes called horizontal and vertical handoff. Vertical handoff occurs when a mobile device changes its point of attachment from a network of one type to another (eg. from GPRS network to UMTS or from WLAN to UMTS etc). Handoff can occur between networks owned by different operators if commercial agreements exist between them to permit it.

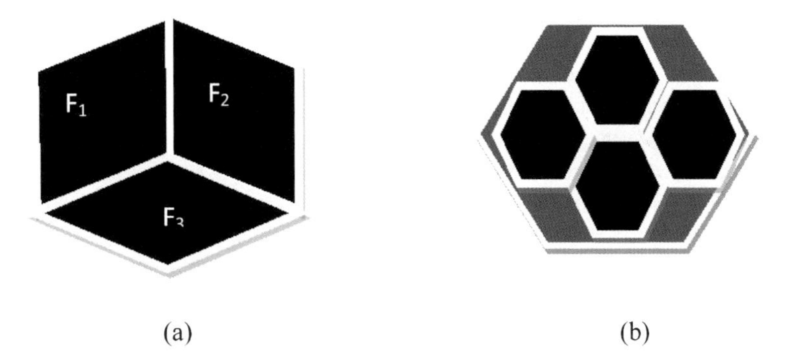

Figure 1.10 cells sectoring (a) and splitting (b).

Horizontal handoff occurs within the same network type (eg. within a UMTS network). There are several forms of handoff depending on the network conditions including

a) Handoff between neighboring base stations (BTS) or cells
b) Handoff between two RNCs
c) Handoff between SGSNs

This is important to perform the handover within less time as possible. To do that, three important factors should be identified, which are:

- Minimum usable signal level for required quality.
- A threshold signal level
- An optimum signal level for a handoff to be performed.

Having the above factors handover margin can be defined as:

$$\Delta = P_{r(handover)} - P_{r(\min imumusable)} \tag{1.10}$$

The handoff margin should be small enough to obtain optimum handoff. Because unnecessary handovers occur if Δ is too large, whereas too small handover margin leads to call disconnection, as there is not enough time for switching between the cells and weak signal can lead to dropped call.

To decide on handoff, different performance metrics can be used. Some of the metrics are defined below:

- *Cell blocking probability:* the probability of blocking a new call, when the traffic load is very high and the base station cannot serve the call.
- *Cell dropping probability:* the probability of disconnecting a call in order to perform the handover.
- *Cell completion probability:* the probability of having an active call terminated successfully.
- *Probability of a successful handover:* the probability of performing a handover successfully
- *Handover blocking probability:* the probability of an unsuccessful handover
- *Handover probability:* the probability of a handover to occur before the call is terminated
- *Rate of handover:* the number of handovers occurring per a period of time

There are two types of handovers, soft handover and hard handover. Hard handoffs are operated when the connection is broken with the current base station before connecting to the target base station. However, if the connection to the new base station is performed before breaking with the current base station, soft handoffs occurs [6]. Softer handover occurs when the handover is between two sectors of the same base station.

1.7 TRAFFIC ENGINEERING

Traffic engineering refers to techniques for assessing and planning the traffic and traffic behaviours of a network. It also refers to optimizing the network performance by determination of the traffic patterns, predicting the growth rate and data analysis. In this section, traffic engineering is discussed based on the network capacity such as predicting the required capacity and data analysis on capacity issues.

As explained earlier, the capacity of a cell is determined by the number of available channels within a cell which allows the same number of simultaneous users per cell. However, it is unrealistic to have capacity to serve all the subscribers at the same time. In practice, an expected number of simultaneous users will be identified for frequency planning and channel allocation purposes, because not all subscribers request for calls at the same time. Even if they do, priorities used in the allocation of channels help to maintain same network access. Assuming the call capacity to be N with S subscribers, the network is described as a blocking system if $S>N$ and a non-blocking system if $S<N$. In practice systems are usually blocking systems. In case of a call being blocked two scenarios may happen based on the network configuration [6]. The call might be waiting in a queue to be served and the call might be dropped and rejected. In the second scenario, a cellular system assumes that the user hangs up and replicates the call after a certain amount of time (LLC model); however, another assumption is that user requests to call repeatedly (LCH model) [6].

For a blocking system to perform efficiently, there are some factors to be considered such as the call blocking probability and average delay, if the blocked calls are waiting to be served by the system.

Traffic is determined as traffic density (A), which is stated in terms of a dimensionless unit in *Erlang*. The traffic density is the result of an average holding time per successful calls (h) multiplied by the average rate of call request per unit of time (λ), or it can be defined as the average number of call requests during an average holding time.

$$A = \lambda * h \tag{1.11}$$

A network should be deployed in such a way that it can support the highest possible traffic. The highest possible traffic occurs during the busy hour of the day. To estimate the highest possible traffic load, according to the International Telecommunication Unit-Telecommunication (ITU-T), the average traffic load of transmitted during the busy hour of 30 busiest days of the year should be identified.

In traffic engineering, the number of users may be considered as either finite or infinite. If an infinite number of users is assumed, the arrival rate will be dependent on the number of users which are using the channels at the time. The arrival rate at time t is expressed as:

$$ArrivalRate = \lambda(S-K)/S \tag{1.12}$$

Where:

- S is the total number of users
- Each user average rate of call is λ/S
- K is the simultaneous number of user at time t

However, based on the assumption that the number of users is infinite and the network follows a LLC pattern, the important parameter, the Grade of Service (GoS) Probability or call blocking probability (P) should be considered. GoS determines the probability of a call to be blocked during the busy hour. Low grade of service values between 0.01 and 0.001 are good values. High grades of service in fact means that the network is performing poorly. Assuming traffic density of A and N number of channels, the grade of service is determined by the expression:

$$P = \frac{\frac{A^N}{N!}}{\sum_{x=0}^{N} \frac{A^X}{N!}} \tag{1.13}$$

From this equation, given the capacity of the system, the number of traffic can be determined for a given value of grade of service. Also given the amount of traffic, the capacity is identified for a certain grade of service using Erlang B table. Equation (1.13) is often provided in terms of Erlang B tables which allows fast determination of the required number of channels for assumed GOS and traffic A.

1.8 BENEFITS AND DRAWBACKS OF CELLULAR TECHNOLOGY

To conclude the overview of cellular system, some advantages and disadvantages of cellular technology are summarized in this section:

Advantages

- It is normally possible to increase the system capacity through creating new cells, sectoring, cell splitting and or by simply scaling the network up progressively
- Lower transmission power
- Increasing the coverage area
- Robustness

Disadvantages

The major disadvantages include

- The need to engineer necessary handover within (and between) the networks
- Handoff is necessary
- Inter cell interference occurs due to co-channel interference and adjacent channel interference
- Fixed base stations are required. Newer versions of networks on platforms or vehicular networks seek to solve this limitation
- Modern cellular networks require the use of good network planning software and network optimization. Network optimization is often not as straightforward as expected. It is almost an art on its own

1.9 DIMENSIONING FOR SERVICE

In the process of service dimensioning, different types of services offered to different customer profiles is determined and also how it affects the air interface is analyzed. For example, typically, there are three important services offered by WiMAX which are VoIP, broadband data and guaranteed bandwidth [1]. According to the type of the service, a portion of the total available bandwidth in each sector is allocated to that particular type of service to deliver the QoS required by WiMAX. There are three different categories of applications required divergent QoS known as Unsolicited Grant Service (UGS), Polling Service (PS), Best Effort service (BE) [2]. VoIP is categorized as a PS while broadband data which corresponds to the Internet access is an application and requires BE service and guaranteed bandwidth is UGS.

Usually a fixed portion of the bandwidth identified as 20% [1] of the total bandwidth is assigned to the UGS during the primary configuration of the network. However, the remaining channel bandwidth is allocated dynamically to the BE and PS, each one a specific portion according to their needs. The impact of three different services on the WiMAX network is analyzed in the following sections.

1.9.1 Voice Over IP Service

VoIP refers to the vocal communication delivered in a packet-switched network. For example in WiMAX technology, the voice stream is broken up into a number of data packets and transmitted via the air interface. One significant task in WiMAX planning is to estimate the capacity needed for delivering the VoIP service to subscribers, since any defeat to a desirable resource allocation results in degeneracy of the grade of service and subsequently degradation of several calls in the sector. To achieve that, some important characteristics of the voice service needs to be identified within a sector, such as total lines, active lines, grade of service, traffic activity and required data rate per call [1].

Mostly the required capacity per sector is needed to be calculated for a given number of simultaneous calls during the busy hour. Using Erlang B equation or table the number of active call will be identified for a given GoS:

$$P_b = \frac{\frac{A^N}{N!}}{\sum_{x=0}^{N} \frac{A^X}{X!}} \tag{1.14}$$

Where

- A: Sector traffic activity
- N: number of active calls
- P_b: blocking probability or grade of service (GoS)

1.9.2 Broadband data service

This service offers access to data at high rates. It is evaluated by the peak information ratio *(PIR)* in Mbps, which is defined as the maximum data rate that a network can offer to the subscribers if the network is not congested [1]. However, since the broadband data service is a BE service, data rate available per user will be lower when the number of simultaneous subscribers increases during the busy hour known as data over-subscription rate *(O)*. The ratio of peak information ratio P_{pir} to data over-subscription ratio is a performance indicator that expresses the committed information ratio, P_{cir} , which is the required data rate during the busy hour.

$$P_{cir} = \frac{P_{pir}}{O} \tag{1.15}$$

$$P_{total} = \sum \left(P_{cir,i} \times N_{s,i}\right) \tag{1.16}$$

- $N_{s,i}$: corresponding number of subscribers
- P_{total}: Summation of the per profile rates
- P_{pir}: profile's CIR

The rate for FDD is derived from the *Max (DL, UL)* and for TDD as the sum of DL and UL rates [1].

1.10 DIMENSIONING FOR COVERAGE

Through the coverage dimensioning process, the number of base stations required for coverage of a given area in km^2 is identified to meet the requirements of KPIs [1] (See chapter 10 for examples). The footprint of each cell needs to be calculated for the estimation of the number of base stations,

$$Number of\ required base stations = \frac{Service Area\left(km^2\right)}{Cell\,Footprint\left(km^2\right)} \tag{1.17}$$

Cell footprint varies for different scenarios and equipment configuration. It refers to the maximum range that a base station supports to radiate the required electromagnetic waves to achieve a SNR performance threshold based on the signal power at the receiver. The received signal strength can be calculated using the following form of link budgets and for example the formula [1]:

$$S(dBm) = P + G + G_{sp} - L_d - L_p - M \tag{1.18}$$

Where:

- S: Signal strength at receiver
- P: power of transmission
- G: antenna gain
- G_{sp}: signal processing gain
- L_d: system loss (shadowing, path, fading)
- L_p: penetration loss

- *M*: design margin (to address mobility, interference, reliability and implementation)

For different terminal profiles, the gains and losses and also design margins are different. Different terminal profiles [1] are defined as follows:

a) Fixed outdoor
    - Located on the rooftop or outside the building walls.
    - Usually for delay sensitive services such as VoIP also Ethernet
    - Connection to the indoor terminal via cable
    - High data rate is possible at a large range with low impact on the channel resources
    - This type of terminals are mostly suited to small-to-medium enterprises (SMEs)
    - Requires high cost of equipment and implementation

b) Fixed portable indoor unit
    - Located indoor close to the window or outer wall
    - Unit is portable within the indoor space and has smaller coverage range
    - Requires power supply
    - Suitable for residential access
    - Self-installation
    - Lower cost compared with the fixed outdoor terminals

c) Nomadic mobile unit
    - Mobile units can be used within the outdoor and indoor spaces
    - Suitable for individual customers requiring particular services

However, customers usually prefer to have a combination of different networks and services.

As mentioned earlier, received signal strength needs to meet at least a signal strength threshold which itself depends on SNR threshold and noise floor (N_{th}). Noise floor refers to the sum of all noise sources and SNR threshold has to have a particular level for the system to work acceptably. The SNR threshold for a received signal is typically defined to have less than 10^{-6} bit error rate when decoding at receiver [1]. However, SNR threshold varies with the modulation and coding scheme.

To estimate the cell footprint we consider factors such as cell shape, network layout, number of sectors per cell to identify the operating system range. Operating system range depends on the deployment of different scenarios. If the network is deployed in a rural outdoor area the base station should be one that has footprint at its maximum range in use while if the same base station is used in urban area or for mobile users, the footprint is a percentage of the maximum footprint as overlapping is essential for handoff and a margin of about *10%* is considered for this reason.

In terms of cell shape, although square cells are more suitable for the fixed networks, hexagonal cells are employed for implementation of the mobile network [1]. Assuming r as the operating range and F the footprint of a cell, the relationship between them is defined as:

$$F_{sq} = 2r^2 \tag{1.19}$$

$$F_{hex} = \frac{3}{2}\sqrt{3}r^2 \tag{1.20}$$

Using hexagonal cells, larger footprint are supported compared to the square cells. Therefore the number of required base stations is less than deployment by square cells. However, for achieving high capacity, using square cells is more suitable because they provide more capacity per unit area unit. For mobile network deployment hexagonal cells result in lower cost according to the number of required base stations. In some cases hybrid scenarios are in use, and deployment should be based on the worst scenario [1]. For instance if indoor and outdoor deployments are required, the planning should be based on indoor terminals because they operate in the shorter range and if the planning only meet the outdoor requirement, indoor coverage faces the problem of blind spots and out of coverage areas, but if the plan is for indoor terminals with shorter range the required range for the outdoor terminals has already been met.

The number of cells required is a function of the modulation scheme used. The more agile the modulation scheme is the higher the bit rate and the smaller the cell coverage. Hence, more base stations are required for modulation schemes which use more bits per subcarrier (modulation schemes which provide better spectral efficiency). BPSK uses 1 bit per subcarrier, QPSK 2 bits, 16 QAM 4 bits and 64 QAM uses 8 bits per carrier.

The size of the cell is a function of the modulation type and the scheme is shown in Figure 1.11. Different modulation types can reach different cell range. From Figure 1.11, the largest cell range is achieved using BPSK ½. The smallest cell range is provided by 64QAM ¾. The cell area of coverage is inversely proportional to the capacity obtainable with a given modulation type. Thus 64QAM ¾ provides the largest cell capacity. The area of coverage A for each modulation type is given by the expression:

$$A = \frac{3\sqrt{3}}{2}\left(d_k^2 - d_{k-1}^2\right) \tag{1.21}$$

Each modulation scheme covers a percentage of the total area of the cell and provides some known capacity. Therefore the capacity of the cell can be estimated to be

$$C = \sum_{k=1}^{M} C_{\text{modulation}\,k} \cdot \left(\frac{A_{\text{modulation}\,k}}{A_{Total}} \right) \tag{1.22}$$

In Figure 1.11 using WiMAX as an example, M=7.

1.11 WIMAX CELL CAPACITY

The use of over subscription ratio (OSR) as a criterion for design is a popular method when starting to develop the characteristics and parameters of a WiMAX cell (base station). OSR is the ratio of the total subscribers traffic demand over the reference capacity of the base station when taking into account adaptive

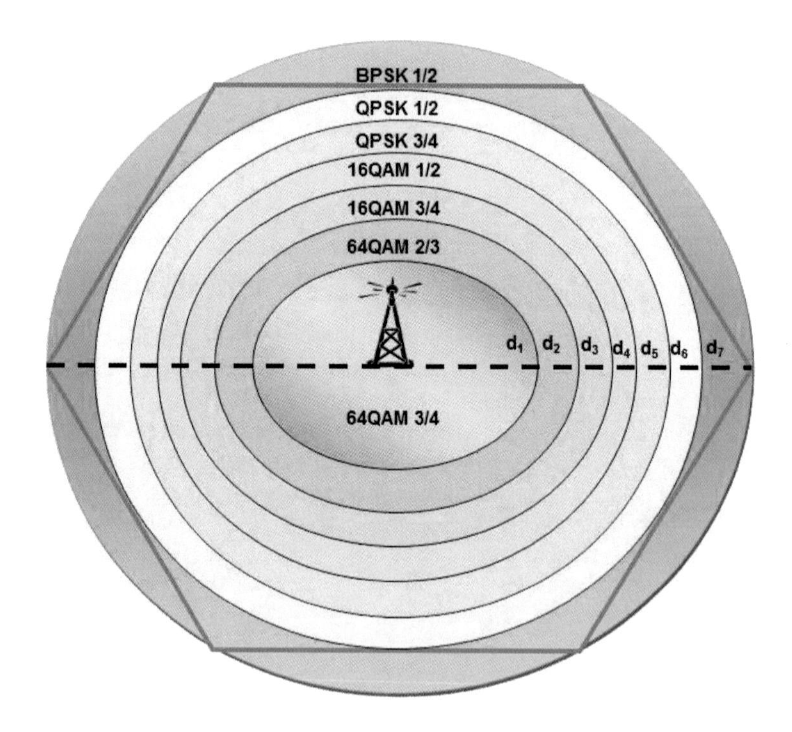

Figure 1.11 Size of WiMAX Cell as a Function of Modulation Type.

modulation. The reference capacity of the base station is the available bit rate obtainable with the lowest modulation scheme. In the WiMAX case, this reference is provided by BPSK ½ which is the lowest modulation scheme served by the base station. This reference capacity is given by the expression

$$C_{ref} = \frac{FFT_{used}}{2T_S} \tag{1.23}$$

With $T_S = 40\mu s$ and 256 FFT (7MHz bandwidth and subcarrier spacing of 31.25kHz), the reference capacity is 3.2Mbps and this value depends on the cyclic prefix value and T_S. The subcarrier spacing used here is fairly large and better reference capacities can be obtained with more fine grain subcarrier spacing. The following steps could be followed in the design specifications for the base station.

Step 1: Define the service area of interest
This is the area where the different types of services desired by the subscribers are provided. This means for example that the city of Sydney is taken first and divided down into urban, high density urban, suburban, and rural areas. Taking for example the high density urban (CBD) area, repartition the area according to the types of subscribers who are likely going to use services at different bit rates from this base station. For example

- 100 subscribers at 1Mbps
- 200 subscribers at 512kbps
- 300 subscribers at 256kbps and
- 400 subscribers at 128kbps

Thus this base station will handle 1000 subscribers at a time at the different bit rates specified. With this number of subscribers it is necessary to also estimate the total subscriber traffic required.

Step 2: Create Sites According to the OSR
For the sake of illustration, we assume that the OSR is between 1 and 2.

Table 1.5 Subscriber distribution

Modulation scheme	Threshold (dBm)	Bit rate	C/(N+I) dB
QPSK 1/2	-97	2, 820	3
16QAM 1/2	-91	5, 640	9
16QAM ¾	-88	8, 545	12
64QAM 2/3	-83	11, 280	17
64QAM 3/4	-82	12, 818	18

Therefore the total traffic demand is either equal to the reference capacity or twice the reference capacity. The average traffic demand based on the subscriber distribution in Table 1.5 the average traffic demand is:

$$\begin{aligned} C_{avg} &= (100\times1000000+200\times512000+300\times256000+400\times128000)/1000 \\ &= 330.4kbps \end{aligned}$$

Thus at a minimum bit rate of 2,820, each base station need to be able to connect 2820/330.4 = 8 (8.54) subscribers. At the high OSR of 2, the base station can handle at most 17 subscribers. At the modulation rate of 64QAM ¾, the base station can connect 38 subscribers.

Consider a practical example where a WiMAX operator has the following service classes to attend to in the form of

- Platinum subscribers at a VBR of 1 Mbps minimum reserved (OSR 10), 3 Mbps maximum sustained (OSR 20), SME users, 5% of total number of subscribers
- Gold subscribers at a VBR, 540 kbps minimum reserved (OSR 10), 1 Mbps maximum sustained (OSR 20), SOHO users, 15% of total number of subscribers
- Silver subscribers at BE, 1 Mbps maximum sustained (OSR 20), residential users, 80% of total number of subscribers

The average capacity per subscriber for this case follows the same method used earlier and is

$$\begin{aligned} C_{avg} &= \left(\frac{5}{100}\times\left[\frac{1000}{10}+\frac{(3000-1000)}{20}\right]+\frac{15}{100}\times\left[\frac{540}{15}+\frac{(1000-540)}{20}\right]+\frac{85}{100}\times\frac{1000}{20}\right) \\ &= 61.35kbps/subscriber \end{aligned}$$

At a raw bit rate of 9.6Mbps for a base station in Table 1.5, the base station can therefore service (9.6 Mbps/61.35 kbps) 156 subscribers using QPSK with two bits per symbol. This number of subscribers would be reduced to a half (78 subscribers) if a FEC of ½ to three quarters (117 subscribers) if the FEC is ¾. The range of this base station is 2.4km. Overheads in the PHY and MAC layers will reduce this subscriber base.

Step 3: Determine the number of base stations

If the total geographical area to be covered is A km square, what is the number of base stations to deploy? We assume hexagonal cells.

$$N_{BaseStations} = \frac{Total\ Area}{2.598R^2} \tag{1.24}$$

For a 30, 000 km square total area with base stations of radius 2.4km, 2000 base stations are required.

1.12 DIMENSIONING FOR CAPACITY AND FREQUENCY PLANNING

After accomplishing the coverage planning, capacity dimensioning and planning needs to be completed. Capacity dimensioning is performed to ensure that the capacity offered to the customers satisfies the required capacity according to the number of subscribers, type of service using by them and data traffic transmitted throughout the network [1]. The outcome of this process is determination of the number of the sectors per cell to address the need of required data rate for the subscribers using different services [1]. By estimation of the number of the subscribers per sector, total capacity of a cell can be identified, subsequently the capacity of whole system. To calculate the number of the subscriber per sector, the capacity of the sector needs to be identified.

Average throughput of a sector is prescribed by the vendor, but it mainly depends on the deployment scenario. Different modulation schemes are used by IEEE802.16e standard. Based on the modulation scheme used in each sector, the throughput varies. Table 1.6 suggests the different modulation technique and coding and corresponding throughput in both UL and DL according to the bandwidth.

According to the Table 1.6, the throughput will be maximized using 64QAM-1/2 in the 10MHz bandwidth in the interference free environment. The equipment specification as usually explained in a table shows the modulation schemes and required SINR and Ethernet throughput for UL and DL using different bandwidth [1]. The data rate achieved through a sector is calculated using different

Table 1.6 Ethernet rate per modulation scheme based on 802.16e (Adapted from [1]).

PHY Mode (CTC)	SINR(dB)	DL(Mbps) 5MHz	UL(Mbps) 5MHz	DL(Mbps) 10MHz	UL(Mbps) 10MHz
QPSK1/2	2.9	1.4	0.9	2.7	1.7
QPSK3/4	6.3	2.2	1.3	4.3	2.7
16QAM1/2	8.6	2.8	1.7	5.8	3.6
16QAM3/4	12.7	4.3	2.6	8.6	5.4
64QAM1/2	16.9	5.8	3.5	11.6	7.1
64QAM3/4	18.0	6.5	3.9	12.9	8.2

parameters according to the modulation scheme and coding [3]:

$$R_b = R_s \frac{MC}{R_r} \tag{1.25}$$

where:

- M: modulation gain(2 for QPSK, 4 for 16-QAM, 6 for 64-QAM)
- C: coding rate (1/2, 3/4, …)
- R_r : repetition rate of 1,2,4,6
- R_b: bit rate
- R_s : symbol rate

Although different modulation schemes are used in different sectors, because different terminals experience divergent interference due to frequency reuse, IEEE802.16, recommends a default assumption for the average sector throughput. Based on this assumption, the average sector throughput is defined as the throughput corresponding to the16-QAM modulation scheme and 1/2 coding rate; however, the throughput will be higher if deployment conditions are desirable.

Since the average sector throughput is determined, customer data should be examined in terms of type of services, average data rate and VoIP Committed Information Ratio. After that a graph is required to show the subscribers per sector versus throughput.

The next step is to determine the number of required sectors identified by dividing the number of customers per area with the number of customers per sector. Assume the number of customers per unit area is S, the uplink bit rate to be $R_{b,UL}$ and the number of customers per sector is S_S , then the number of sectors is

$$N_S = S/S_S \tag{1.26}$$

The bit rate per sector can be determined based on system parameters and is given by the expression:

$$R_{b,DL} = B.n.\frac{N_{dataDL}}{N_{FFT}} \times bits\ per\ symbol\ x \frac{1-overhead}{1+T_g} xTDD_{down/up\ ratio} \tag{1.27}$$

The TDD down/up ratio between the UL and DL times may be 3:1 and the overhead time includes times used for synchronisation, initialisation and headers.

However, adaptive modulation is one of the important factors that affect the capacity of the network [1]. As described earlier , WiMAX is able to adopt different modulation schemes for data transmission based on the link quality. For example if the SINR is high, transmission can be done using higher order of the modulation schemes such as 64-QAM to achieve higher throughput while in the case of weak SINR lower orders of modulation schemes such as QPSK can be employed.

Alongside the adoption of different modulation schemes, to increase the capacity of the network and minimize the interference between the co-channel cells, WiMAX uses powerful sub-channelization modes which are PUSC (Partial Usage of Sub-Channels) and FUSC (Full Usage of Sub-Channels) [1], [4], [5]. FUSC refers to the usage of all sub-carriers within a cell, while PUSC is defined as allocation of specific group of sub-carriers to each sector within a cell. Using PUSC mode allows the network to use the frequency reuse factor of one when the spectrum is limited without interference between adjacent sectors, because they use divergent sub-carriers. This scheme is very efficient as there is no need for guard band and also provides more reliability of links in both uplink and downlink by using different sub-carriers in the neighbouring cells subsequently reducing the chance of interference [1], [4]. While FUSC is used only in downlink PUSC can be used in both uplink and downlink. FUSC cannot be used in uplink because of the nature of uplink which is more unpredictable than downlink [5]. To plan the frequency channels to achieve higher capacity and optimum use of the channels, there are factors that need to be considered in of uplink and downlink.

Planning for downlink is easier in comparison with uplink because the major source of interference in downlink is the neighbouring base station using the same frequency channel. Therefore it is possible to calculate the SINR in each spot and prepare an interference matrix for each base station in the service area and the proper modulation scheme for each spot [5]. For example in a location that has suitable condition higher modulation order can be used to enhance the capacity of the network. However, most of the downlink interference is expected at the borders of cells where the SINR is low and signal is more likely to collide with other signals on the same carrier frequency [5]. To overcome this problem, WiMAX divides the TDD frame, which consists of two portions of uplink and downlink, into 48 sub-carriers [5]. At the first part of the downlink frame, the base station adopts the PUSC mode to transmit data to the subscribers located at the edge of the cell to reduce the possible interference while using FUSC mode in the second part of the downlink frame, to transmit data to the subscribers near to the base station with higher SINR for increasing the capacity and finally efficient use of the spectrum [5].

In spite of downlink, uplink planning is more complicated as it is less predictable. The subscriber station can be anywhere within the service area or anywhere within a particular cell. Therefore it is difficult to determine the received level of SINR and consequently the level of interference. Interference can vary not only by changing the location of the user but also by the number of the slot that it uses to transmit the data. Using more slots by a subscriber leads to more number of sub-carriers and subscribers it collides with [5]. For these reasons the uplink can use only PUSC scheme for data transmission to achieve a robust network. To allocate an optimum modulation scheme to the link, base station estimates the SINR according to the feedback that it receives from the subscriber station through the previous received frame. Since in mobile WiMAX system, the user changes the channel frequently because of mobility, the base station needs to be aware of the current channel that user is using. Therefore to allocate the best possible modulation scheme to the connection, the base station has to use any of the average SINR received previously, the best SINR or the worst SINR. The average SINR is more suitable since the selection of the worst SINR results in capacity reduction due to data transmission on lower modulation order, while assigning the modulation scheme based on the best amount of SINR causes less robust links due to high error occurrence in data retransmission [1], [5].

1.13 BACKHAUL DIMENSIONING

The backhaul network links a number of base stations and ASN-gateways, via a radio interface or fibre rings [1]. Base stations can be connected together through a point to point, if the number of the required base stations are high, or point to multipoint radio interface, if the base stations are close to each other. Also the base stations need to be connected to a gateway within the backhaul network using an IP transport or IP link, which carries the IP packet between the base station and the gateway. To evaluate or plan a backhaul network, the information rate of the backhaul, the number of the base station and gateways also their positions needs to be determined.

The backhaul information rate increases with the number of the sectors per cell and the sector's throughput in a linear manner. However, the sector's throughput varies with the deployment scenarios. Total TDD rate that can be delivered by the backhaul is the product of the number of the sectors and each sector's throughput. To calculate the FDD rate, the total throughput needs to be multiplied by DL/(DL+UL) rate [1].

According to WiMAX forum, using MIMO system and a channel bandwidth of 10MHz, the uplink air interface can support 28Mbps per sector peak data rate and 63Mbps at downlink [1]. Employing three sectors per base station, total peak data rate that a base station can deliver in uplink and downlink is 84Mbps and 189Mbps respectively [1]. To calculate the total data rate supported by a base station, two other factors need to be determined, which are the multiplexing technique used and the uplink and downlink rate. Assuming FDD technique and 3:1 downlink and uplink rate, the peak downlink data rate is 46Mbps/sector and 8Mbps/sector for uplink, which leads to total peak data rate of 46Mbps per sector for each base station, as the total data rate in a FDD system is equivalent to the maximum rate between uplink and downlink [1]. Therefore by multiplying the total throughput for each sector by three, total peak data rate per a base station is calculated, which is approximately 150Mbps in this case. It means that the link between the base station and gateway needs to be able to have the capacity equal to 150Mbps [1].

In contrast, if TDD technique is used in the same scenario, the peak data rate per sector in uplink and downlink is 1.83Mbps and 13.60Mbps respectively which results in total data rate of about 16Mbps per sector, as the sum of uplink and downlink data rate [4]. Tripling the throughput of each sector results in peak data rate of 48Mbps per base station.

Choosing either FDD or TDD, also the downlink and uplink rate depends on the deployment scenario and the type of application due to be delivered in the network. However comparison between TDD and FDD assists to select the best technique that suites the providers. Different metrics can be considered for the evaluation of the backhaul planning and the techniques that need to be deployed for achieving the better performance, such as different service type to be offered, the available bandwidth, data rate, complexity, cost and also the number of subscribers that can be served by the base station.

In terms of bandwidth availability, if the frequency band is limited, TDD is more likely to have the better performance than FDD; since FDD needs different frequencies for downlink and uplink, while TDD send the uplink and downlink data traffic on the same frequency band but different time slots. However, planning for different service types, FDD is more suitable for the applications which require more bandwidth such as video, whereas TDD which is proper for application such as voice requires lower bandwidth. Furthermore, FDD provides higher data rate than TDD but TDD is far less complex therefore lower deployment cost [4]. However, operators can benefit from the high number of subscribers supported by FDD technique when they demand applications with lower bandwidth requirements.

To achieve an efficient backhaul dimensioning, it is recommended by [1] to dimension the backhaul network and anticipate the total required capacity of the WiMAX backhaul network based on the predicted demand for service by dividing the total available spectrum into three different parts, each part serving a portion of the users with the same required quality of service applications. It is suggested in [1] that this method of dimensioning requires less backhaul deployment cost initially and scale well if the number of subscribers increases by adding more backhaul. To deploy that, assume three different types of applications are in demand defined as video calls with required high QoS guaranteed, voice calls which requires less QoS guaranteed compare to video calls and finally best effort service, which provides services such as the Internet access with no guarantee of QoS requirement. In this case, the aim is to identify the total capacity needed for the backhaul link between two base stations or between the base station and gateway to support all different service applications efficiently.

The first step is to divide the applications into three groups known as EF class (Expedited Forwarding), AF class (Assured Forwarding) and BE class (Best Effort) [1]. The applications such as video calls are placed in the EF category, which require absolute amount of delay and very sensitive to delay. AF class includes the applications with average delay and less sensitive to delay compared to the EF class applications, such as voice calls. BE class consists of the applications with no sensitivity to delay. The next step is to determine how many subscribers are using the spectrum simultaneously at each class, which is assumed to be *N* voice calls, *M* video calls and few BE type applications to be served by a base station. Based on the assumption that a portion of the spectrum is used by n EF subclasses and other portion of the spectrum is divided to be used by m AF subclasses and few BE. Also using a weighted fair queuing scheduler with h as a vector defining the weights used in the scheduling to determine the distribution of the spectrum for the AF and BE class, the capacity required by different classes are calculated by [1]:

$$c_i^{EF} = \frac{h_i^{EF} \quad c}{h^{BE} + h^{AF} + \sum_{i=1}^{n} h_i^{EF}} \tag{1.28}$$

$$c_i^{AF} = \frac{h_i^{AF} \quad c}{h^{BE} + h^{AF} + \sum_{i=1}^{n} h_i^{EF}} \tag{1.29}$$

$$c^{BE} = \frac{h_i^{BF} \quad c}{h^{BE} + h^{AF} + \sum_{i=1}^{n} h_i^{EF}} \tag{1.30}$$

Where:

- c_i^{EF} : each EF subclass needs to have at least this amount of capacity
- c_i^{AF} : total bandwidth available for AF class
- c^{BE} : total bandwidth available for BE class
- c : capacity of the backhaul link
- h^{EF} : weight for EF class
- h^{AF} : weight for AF class
- h^{BE} : weight for BE class

Now the backhaul link capacity can be determined by accumulating the data result from the calculation of capacity needed for each subclass.

1.14 CONCLUSIONS

This chapter has presented the basic concepts required for planning and designing of modern cellular communication networks. Most of the concepts apply in the design of GSM, UMTS, LTE-Advanced and other forms of cellular communication systems. The chapter ends with some illustration of requirements for WiMAX networks, filling a gap which is often missing in most modern communication books because they focus on more popular network types.

References

[1] Yan Zhang, WiMAX Network Planning and Optimization, USA: CRC Press, 2009.

[2] Johnson I. Agbinya, IP Communications and Services for NGN, New York: Taylor and Francis, 2009.

[3] Bharathi Upase, Mythri Hunukumbure and Sunil Vadgama, "Radio Network Dimentioning and Planning for WiMAX Networks" 43.4, pp.435-450 Available: http://www.fujitsu.com/downloads/MAG/vol43-4/paper09.pdf.

[4] WiMAX Forum. "Mobile WiMAX-Part I: A Technical Overview and Performance Evaluation," WiMAX Forum, 2006, pp. 1-53.

[5] Kai Dietze and Ted Hicks, "WiMAX Uplink and Downlink Design Consideration", Internet:http://www.edx.com/files/wimax_paper_2v4-1.pdf , [Sep.19, 2009].

[6] Steve Wisniewski, Wireless and Cellular Networks, USA: Pearson Prentice Hall, 2005.

[7] I. Katzela and M. Naghshineh, "Channel Assignment Schemes for Cellular Mobile Telecommunication Systems: A Comprehensive Survey," IEEE Personal Communications, pp. 10-31, June 1996.

[8] Jeffrey G. Andrews, Arunabha Ghosh and Rias Muhamed, Fundamentals of WiMAX, USA: Prentice Hall, 2007.

[9] Rhode and Schwarz, WiMAX General Information about the standard 802.16 (application note), 2006, pp. 1 – 34.

[10] WiMAX Capacity White Paper, SR Telecom Inc., Canada, 2006, pp. 1 – 34.

Chapter 2.

Radio Wave Propagation Models for Cellular Communications

Johnson I Agbinya

Department of Electronic Engineering, La Trobe University, Australia
J.Agbinya@latrobe.edu.au

2.1 PROPAGATION REGIMES AND MODELS

Based on different environments different path losses are experienced by radio waves. Consequently different radio propagation modelling should be employed to estimate the electric field strength [1]. Channel models can be seen from different angles; in terms of existence of a direct line of sight between the transmitter and receiver, channel modelling methods can be categorized as free space propagation model and modelling for the land propagation when the LOS does not exist [3]. However, according to the required statistics for channel modelling, two main groups of Deterministic and Empirical modelling can be considered [1]. Deterministic models are used when the channel modelling requires detailed information about the geometric environment such as the location and the surrounding area. In contrast, empirical models are based on the measured mean path loss for different environments; this method is less complex and costly but less accurate compared to the deterministic modelling.

Some of the empirical channels modelling used by WiMAX technology are Stanford University Interim (USI) model, Cost-231 Hata model, Macro Model and Ericsson model 9999 [1]. In this section the USI and free space channel modelling are introduced in some detail.

Terrestrial propagation (propagation in natural settings) is much different from free space propagation. Propagation of cellular signals in urban environment is difficult to predict because of man-made structures (tall buildings,

4G Wireless Communication Networks: Design, Planning and Applications, 37-66.

signs and other obstacles). On earth, a signal propagates through air and reflects, diffracts and scatters.

2.1.1 Deterministic Propagation Models

This section discusses three deterministic propagation models including Friis free-space model, two-ray model and the Egli model. The two-ray model is an improvement of the Friis model. The Egli model is also an improvement of the two-ray model.

2.1.1.1 Friis Model: Free Space Propagation

Historically, a great deal is known in relation to propagation of radio waves in various types of terrains and man-inhabited regions. Understandably only a few of the propagation models that have been published are recommended for use in standards. Radio wave propagation is affected by several factors including the radiation frequency, antenna height, distance, the curvature of the earth, atmospheric conditions (such as ionisation in the ionosphere, rain, gas and dust particles), and the presence of obstacles blocking or affecting the wave such as hills, tall buildings and vegetation.

The medium separating the receiver from the transmitter plays an important role in the propagation of the signal. Friis modelled free space propagation by assuming that the space between the transmitter and receiver is free of objects (free space) with the formula [3]:

$$P_r = P_t \frac{G_t G_r \lambda^2}{(4\pi d)^2} \tag{2.1}$$

Where P_r, P_t, G_r and G_t are the receiver power, transmitter power, receiver antenna and transmitter antenna gains respectively. It has been assumed in this expression that radiation is into a spherical space of radius d surrounding the antenna. Thus radiation is into an area equal to the surface area of a sphere ($4\pi d^2$). The medium between the transmitter and receiver is often a dielectric, air, a piece of wire, fibre or some liquid including water. The medium varies from application to application and from terrain to terrain. In equation (2.1) the effect of the terrain is not obviously apparent. In later propagation models, this will be made apparent. The distance between the transmitter and receiver introduces a path loss as shown in Figure 2.1. The loss is modelled with the expression:

$$L_p = 10xLog\left(\frac{4\pi d}{\lambda}\right)^2 \tag{2.2}$$

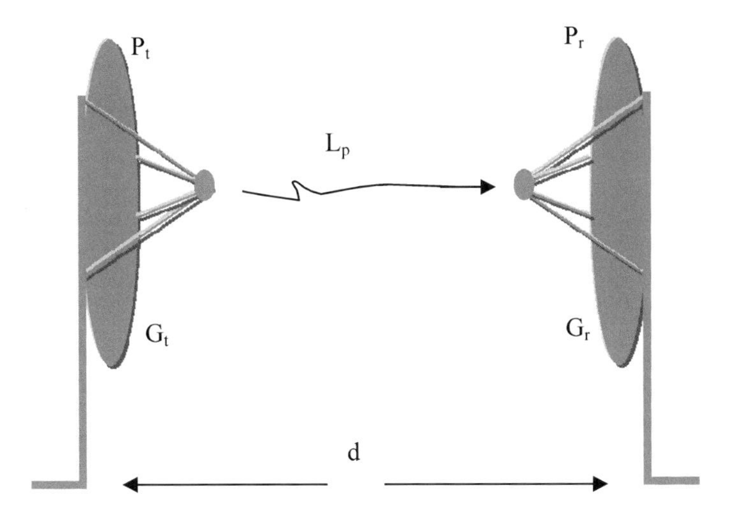

Figure 2.1 Free Space Propagation.

The propagation-exponent is a function of the terrain between the transmitter and receiver and has values in the range 2 to 5 in urban areas. A value 2 is used for free space. The free space propagation loss is often expressed in decibels and is:

$$L_p = 32.44 + 20Log(f) + 20Log(d) \tag{2.3}$$

The frequency f is measured in MHz and the distance d in kilometres. As the equation (3) implies, using higher carrier frequencies, the free space loss will be higher. For instance, if two receivers have the same sensitivity, using 450MHz carrier frequency the cell radius will be 7.78 times longer than when 3.5GHz carrier frequency is used [4].

Radio wave propagation can be discussed either in terms of the received power or the received field intensity. In free space the received field intensity at distance d is given by the expression

$$E_0 = \sqrt{\frac{\eta G_T P_T}{4\pi d^2}} = \frac{\sqrt{30 G_T P_T}}{d} volts / meter \tag{2.3a}$$

P_T is the radiated transmitter power (watts), η is intrinsic impedance of the medium in the terrain between the transmitter and receiver. The intrinsic impedance of free space is approximately 377 Ohms. G_T is the transmitter antenna power-gain ratio relative to an isotropic antenna. For an isotropic antenna

Table 2.1 Path loss exponent

Environment	Path loss exponent, n
Free space	2
Ideal specular reflection	4
Urban cells	2.7 - 3.5
Urban cells (shadowed)	3 - 5
In building (line of sight)	1.6 - 1.8
In building (obstructed path)	4 - 6
In factory (obstructed path)	2 - 3

$G_T = 1$. The gain is 1.5 for a doublet and 1.64 for a half-wave dipole antenna. Practical antennas normally have gains higher than that of isotropic antenna.

The path loss exponent varies from terrain to terrain as shown in Table 2.1.

2.1.1.2 Two Ray Model

The two ray model is an improvement over the free-space model. We need to prove this shortly. Theoretically, propagation can be predicted with the help of a 2-ray model and Fresnel Zones. A two-ray model is shown in Figure 2.2. The signal reaches the receiver through two paths, the direct path and a reflected ray In other words, the signal is approximated using ray-optics as having an optical ray linking the transmitter and receiver. This model assumes a flat earth.

The path difference between the direct and reflected rays can be shown to be approximated by the expression:

$$\Delta = \left[\sqrt{(h_r + h_t)^2 + d^2} - \sqrt{(h_t - h_r)^2 + d^2} \right] \approx \frac{2h_t h_r}{d} \tag{2.4}$$

In practice, destructive signals resulting from reflections from the earth are avoided by imposing the criteria:

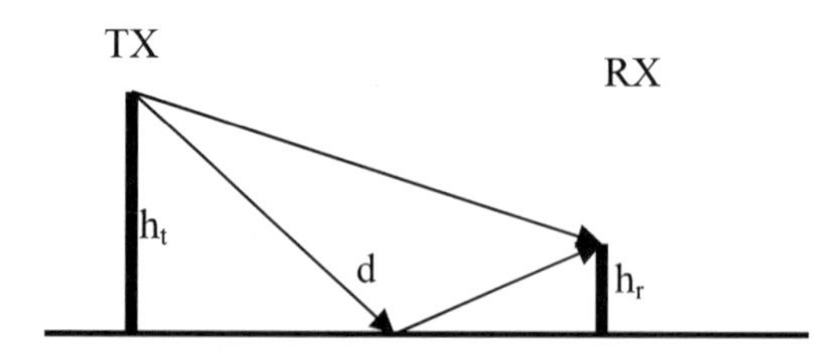

Figure2.2 Two-Ray Propagation.

$$\Delta > (2n-1)\lambda / 2, \quad n = 1, 2, 3, \cdots \tag{2.5}$$

Electromagnetic wave fronts are often divided into zones of concentric circles (separated by half a wavelength) called Fresnel zones. The zones define propagation break points. Waves reflected from the edges of each Fresnel zone arrive at the receiver constructively. In the first Fresnel zone n=1 so that $h_r h_t > \lambda d / 4$. The first breakpoint, d_o is called the first Fresnel zone and occurs at a distance $d_o = 4h_r h_t / \lambda$. Fresnel zones provide evidence of how far the wavefront or signal is able to clear or propagate above obstacles in the path linking a receiver and its transmitter. Until this point, the propagation is assumed to be free space. These two propagation models are useful for system design. The two models are used to predict microcell and indoor coverage. In many applications the distance between the transmitter and receiver is smaller than the first breakpoint, and for such cases, the Fresnel point does not help with the design.

Over ideal ground or the so-called specular ground, the received power is given by the modified free-space power:

$$P_r = P_{FS}\left[(1 - \exp(j2\pi\Delta / \lambda))\right]^2 \approx P_{FS}\left(\frac{4\pi h_t h_r}{\lambda d}\right)^2 \tag{2.6}$$

The modified free-space model shows the influences of the transmitter and receiver antenna heights. The next time you see a microwave tower, be appreciative of the need for good antenna heights as shown in equation (2.6). Since the received free-space (FS) power is

$$P_{FS} = \frac{P_r}{P_t} = G_t G_r \left(\frac{\lambda}{4\pi d}\right)^2 \tag{2.7}$$

Hence the two ray model leads to the expression:

$$P_r = P_t G_t G_r \frac{h_r^2 h_t^2}{d^4} \tag{2.8}$$

As a result of the distance dependence in this expression, every time we double the distance, we lose 12 dB of signal energy. This shows that frequency reuse should be done at shorter distances. Equation (2.8) is a very practical expression and relates only the system parameters to the communication range. Usually all the variables in this equation are known except the communication range d which is a variable. The received power in this equation should be more than the receiver sensitivity for good communication.

2.1.1.2.1 Estimation of Received Signal in Two-Ray Model

The two ray model assumes that there are two paths through which the radio wave reaches the receiver, the direct path (line-of-sight) and the ground reflected wave path. In most communication systems on earth, both the transmitting and receiving antennas are located on the ground. Even when they are mobile, we can reference the heights of the antennas to the ground level by adding the height of the object carrying the transceiver to the height of the actual antenna. That being the case, it is assumed that there is no impedance mis-match between the media in which the antennas are located. Hence no reflected waves back to the transmitter are expected. However, the earth's surface has different dielectric properties to the dielectric properties of ordinary space. Therefore wave reflection and absorption are expected. This architectural structure of the location of the antennas means that the receiver should receive at least two components of the transmitted signal, the line-of-sight component and the ground reflected component as shown in Figure 2.3.

Normally the distance d between the transmitter and receiver antennas is many times greater than their dimensions. Hence the phase difference between the two signals received can be quite significant and leads to losses in the received signal compared to the transmitted signal. This section provides the means to quantify the received signal. Define the ratio of the complex valued reflected electric field

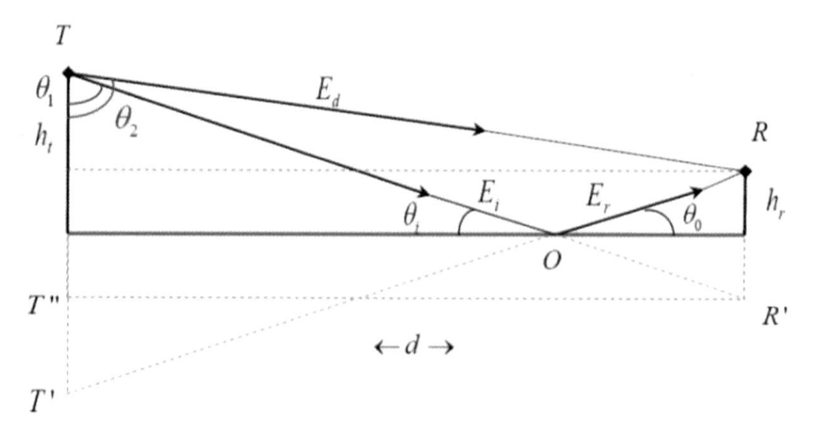

Figure 2.3 Two Ray Model.

intensity E_r at the receiver to the complex valued incident electric field intensity on the ground surface E_{0i} as the reflection coefficient with the expression:

$$\rho = \frac{E_r}{E_i} = |\rho| e^{j\psi} \tag{2.9}$$

Since the two field intensities are vectors, there is a phase difference due to reflection between them given by angle ψ. This information is used shortly in quantifying the magnitude of the received signal. The magnitude and phase of the reflection coefficient are functions of the dielectric constant of the ground and the polarisation of the antennas. The polarisation of an electromagnetic wave is the direction or orientation of the electric field vector. Formally, accounting for polarisation, the reflection coefficients for grazing angles θ_i are:

For vertical polarisation

$$\rho_v = \frac{\varepsilon_r \sin\theta_i - \sqrt{\varepsilon_r - \cos^2\theta_i}}{\varepsilon_r \sin\theta_i + \sqrt{\varepsilon_r - \cos^2\theta_i}} \tag{2.10}$$

For horizontal polarisation

$$\rho_h = \frac{\sin\theta_i - \sqrt{\varepsilon_r - \cos^2\theta_i}}{\sin\theta_i + \sqrt{\varepsilon_r - \cos^2\theta_i}} \tag{2.11}$$

Noticeably, for media with complex relative permittivity $\varepsilon_r > 1$, horizontally polarised transmissions could lead to larger reflection coefficients. This may mean in certain instances, reduced transmission range. Obviously, the two ray propagation model is the simplest example of multipath propagation and leads to fading. To estimate this well, we should first find the phase difference ϕ between the line-of-sight signal and the ground reflected signal. This phase difference is given by the difference in path lengths which is TO+OR – TR which will be expressed as a phase. Using Pythagoras theorem

$$TR = \sqrt{d^2 + (h_t - h_r)^2} \tag{2.12}$$

Also, the distance T'O+OR is given by the expression

$$T'OR = T'O + OR = \sqrt{d^2 + (h_t + h_r)^2} \tag{2.13}$$

Note that TO=T'O. The path difference is therefore given by

$$\begin{aligned}\Delta d &= \left[\sqrt{d^2 + (h_t + h_r)^2}\right] - \left[\sqrt{d^2 + (h_t - h_r)^2}\right] \\ &= d\sqrt{1 + \left(\frac{h_t + h_r}{d}\right)^2} - d\sqrt{1 + \left(\frac{h_t - h_r}{d}\right)^2}\end{aligned} \tag{2.14}$$

Assume that the $d >> (h_t - h_r)$ and $d >> (h_t + h_r)$. Also let

$$\mathrm{x} = \left(\frac{\mathrm{h_t} + \mathrm{h_r}}{\mathrm{d}}\right)^2 \quad \text{or} \quad x' = \left(\frac{h_t - h_r}{d}\right)^2 \tag{2.15a}$$

Therefore we can write equation (2.14) as

$$\Delta d = d(1 + x)^{\frac{1}{2}} - d(1 + x')^{\frac{1}{2}} \tag{2.15b}$$

By expanding this in Taylor series and retaining only the first two terms in each case we can write

$$\Delta d = d\left[1 + \frac{1}{2}\left(\frac{h_t + h_r}{d}\right)^2 - \left(1 + \frac{1}{2}\left(\frac{h_t - h_r}{d}\right)^2\right)\right] = \frac{2h_t h_r}{d} \tag{2.15c}$$

The phase difference is the product of the wave number and the path difference and is

$$\phi = \frac{2\pi}{\lambda}\Delta d = \frac{4\pi h_t h_r}{\lambda d} \tag{2.16}$$

The phase difference is a function of the transmitter and receiver antenna heights, the transmission wavelength and the range of communication. This phase difference is due to the path difference between the two rays. This is in addition to the phase difference due to reflection of waves on the ground. In both cases of vertical horizontal polarisation, provided the distance between the transmitter and receiver is very large, the incident angles $\theta_i \to 0$; $TR \cong TR'$. Hence

$$\sin\theta_1 = \frac{d}{TR'} \approx \sin\theta_2 = \frac{d}{TR} \tag{2.17}$$

When $\theta_i \to 0; \psi = 180^0$, the reflected wave is out of phase with the incident wave $(|\rho| = 1)$ but of the same magnitude. This means that the total electric field intensity at the receiver is

$$E_{total} = E_d + E_r \approx E_d\left(1 - \rho e^{j\phi}\right) = E_d\left(1 - e^{j\phi}\right) \tag{2.18}$$

To quantify the effect of the reflected ray on the received signal strength, we use Euler's formula to expand this equation as

$$E_{total} \approx E_d\left(1 - (\cos\phi + j\sin\phi)\right) \tag{2.19}$$

Therefore the magnitude of the received signal strength is

$$|E_{total}|^2 \approx |E_d|^2\left[(1-\cos\phi)^2 + \sin^2\phi\right] \approx 2|E_d|^2[1-\cos\phi] \tag{2.20}$$

Or

$$|E_{total}|^2 \approx 4|E_d|^2 \sin^2\frac{\phi}{2} \tag{2.21}$$

The total electric field intensity at the receiver therefore is

$$|E_{total}| \approx 2|E_d|\sin\frac{\phi}{2} \tag{2.22}$$

By substituting for ϕ in this expression the total electric field at the receiver is

$$|E_{total}| \approx 2|E_d|\sin\left(\frac{2\pi h_t h_r}{\lambda d}\right) \tag{2.23}$$

This signal strength is a function of the amplitude of the direct ray signal, frequency, separation between the transmitter and receiver and the heights of the transmitting and receiving antennas. This equation also implies that when there are obstructions which block the direct ray E_d, the received signal will be greatly affected. For this reason, a significant level of research and publications modelling diffractions (blocking and scattering of waves), some based on measurements and others on deterministic expressions have been proposed to tackle the problem of diffraction. Diffraction models are discussed in another chapter of this book.

The free space model given by Friis may now be compared with the two-ray model.

The Friis model gave us

$$E_d = \sqrt{\frac{\eta G_T P_T}{4\pi d^2}} = \frac{\sqrt{30 G_T P_T}}{d} volts/meter \tag{2.24}$$

By substituting for E_d the Two Ray Model showing the effect of the ground reflected wave is therefore quantified by the expression

$$|E_{total}| \approx 2\frac{\sqrt{30 G_T P_T}}{d}\sin\left(\frac{2\pi h_t h_r}{\lambda d}\right) \tag{2.25}$$

A more informative expression is obtained by using the series expansion of the sine function. Since

$$\sin\left(\frac{2\pi h_t h_r}{\lambda d}\right) \approx \frac{2\pi h_t h_r}{\lambda d} \tag{2.26}$$

Then

$$|E_{total}| \approx 4\pi\sqrt{30 G_T P_T}\frac{h_t h_r}{\lambda d^2} \tag{2.27}$$

The power received is also a function of the antenna aperture and the impedance of free space. Therefore

$$P_r = \frac{|E_{total}|^2}{\eta} A_{er} \approx \frac{(4\pi)^2 30 G_{T^2} P_T}{120\pi}\left(\frac{h_t h_r}{\lambda d^2}\right)^2 . \frac{G_R \lambda^2}{4\pi} \tag{2.28}$$

Upon simplification this becomes

$$P_r \approx G_T G_R P_T \frac{h_t^2 h_r^2}{d^4} \tag{2.29}$$

The power received is proportional to the fourth power of the distance between the transmitter and receiver and the product of the squares of their antenna heights. This simplified expression holds in most cellular communication systems including GSM, UMTS and LTE. The recourse to propagation models and diffraction models is to ensure more accurate estimate of the received signal which also accounts for the effects of the terrain separating the base station and receiving terminals.

Since practical communication takes place in none-free space situations, the actual field intensity or power at distance d must be estimated using some other expressions or approximations. Several propagation models are introduced in the next section.

2.1.1.3 Egli Model

The Egli model [13] is a simple improvement of the Two Ray model. It was obtained from measurements and takes into account the gains of the base station and terminal antennas as well as their heights. It is given by the expression

$$L_e = G_b G_m \left(\frac{h_b h_m}{d^2}\right)^2 \left(\frac{40}{f}\right)^2 \tag{2.30}$$

or

$$L_e(dB) = 32.04 + 10\log G_b + 10\log G_m + 20\log h_b + 20\log h_m - 40\log d - 20\log f \tag{2.31}$$

This model is also a function of distance and frequency as the free space and two ray path loss models. The frequency is in MHz. The model however fails to account explicitly for the presence of obstacles and the environmental degradation. The correction term $(40/f)^2$ is not explicit as to which type of terrain it is accounting for.

2.2 MORE ELABORATE PROPAGATION MODELS

Efficient propagation losses are useful for computing link budgets. Link budgets provide the maximum allowable path losses per link and an indication of the link that will likely be a limiting factor to the system. The maximum allowable path loss also sets a limit on the maximum cell size. We will use path losses to compute link budgets in this chapter.

Although the plane earth model is more appropriate for cellular channels, it ignores the curvature of the earth's surface and considers a two-path model of direct line of sight and a ground reflected paths. In this model the heights of the transmitting and receiving antennas feature prominently in the propagation loss expression. Provided the heights of the antennas are less than the separation between the transmitter and receiver (h_b and h_r), the propagation loss expression can be shown to be

$$L_p = -10\log_{10} G_t - 10\log_{10} G_r - 20xLog\,h_b - 20xLog\,h_r + 40xLog\,(d) \quad dB \tag{2.32}$$

Propagation out door is difficult to predict and as such, empirical models, without real analytical basis are applied. Most of the models used are accurate to within 10 to 14 decibels in urban and suburban areas. They tend to be less accurate in rural areas because most of the data used may have been collected in the urban and suburban areas. In practice there are huge variations in the types of terrain and environment to cover. The heights of antenna, clutter, tree density, beamwidth, wind speed, season (time of the year) and multipath, vary widely and affect mobile phone waves. Hence complex models are required for such situations. They are used to predict propagation loss. One of the other popular models is the Okumura model. Beyond the Okumura model, there are the Hata and Walfisch-Ikegami models and others by Egli, Lee, Carey, Longley-Rice, COST 231, Ibrahim-Parsons, Erceg, SUI and many more. Although most of them are beyond the scope of this chapter, we will examine a few briefly.

In a study of channel models for fixed wireless application by the Institution of Electrical and Electronic Engineers (IEEE) [5], a set of "propagation models applicable to the multi-cell architecture" was presented. They assumed a cell radius of less than 10 km, variety of terrain and tree density types, directional antenna (2-10m) installed under-the-eaves/window or rooftop, 15-40 m BTS antennas and high cell coverage requirements of between 80-90%. This section is derived from the analysis.

2.2.1 Okumura Model)

All the propagation models discussed so far here are highly idealised. They have focused mostly on the parameters of the transmitter and receivers with little to no consideration on the effect of the system topography and morphology. The Okumura model represents an improvement on the previous models.

The Okumura model was derived from extensive RF measurement reports in and around the city of Tokyo. The measurements were over frequencies up to 1920 MHz. The measurements were used to derive correction terms which match various types of urban, suburban and rural channels. Corrections were made to the free space path loss model using the expression

$$L(dB) = L_F + A_{mu} - H_{tu} - H_{ru} \qquad (2.33)$$

Where the correction terms shown are:

$L_F = 32.45 + 20\log_{10} f_{c(MHz)} + 20\log_{10} d$ is the free space path loss with d in km and frequency in MHz unit.

$A_{mu}(f,d)$ is the median path loss relative to free space losses in urban area as function of frequency f and distance d.

H_{tu} is a correction for the transmitter antenna height in an urban area and

H_{tu} is a correction term for the receiver antenna height in an urban area

While the free space loss is calculated, the correction terms are read off Okumura charts.

2.2.2 Suburban Path Loss (Hata-Okumura Model)

The use of Okumura model is strongly based on the use of the graphical representations given. Hence the model does not lend itself to the use of software for planning. Hata provided an improvement which makes it possible to not only accommodate for the effects of the environment during design but also to be able to calculate the required enhancements.

The Hata-Okumura model [6] is widely used below 1500MHz. The model is valid in the 150-1500 MHz frequency range (NMT and GSM), with receiver distances greater than a kilometre from the base station and base station antenna heights greater than 30m. As such, the model is applicable to mobile phone applications below 1500 MHz. The modified Hata-Okumura model extends this range to around 2 GHz [7]. This is the so-called COST 231 model. Although the model targeted 2G systems in the 900 and 1800MHz range, it has application to systems around 2GHz (eg., DCS1800) provided lower base station antenna heights, hilly or moderate-to-heavy wooded terrains are not involved. Corrections for these limitations were applied to cover most terrain conditions applicable to the US. The body of study for savannah, and dense forest regions in Africa and other regions need to be understood. Path loss regimes are divided into three broad categories A, B and C:

- Category A (maximum path loss): hilly terrain with moderate-to-heavy tree densities
- Category B (intermediate path loss): terrain conditions between category A and C
- Category C (minimum path loss): mostly flat terrain with light tree densities

The median path loss at 1.9 GHz for a distance d_o from a base station is given by:

Table 2.2 Path Loss (Terrain) Correction Variables

Model Parameter	Terrain Type A	Terrain Type B	Terrain Type C
A	4.6	4	4.6
B	0.0075	0.0065	0.005
C	12.6	17.1	20

$$L_p = A + 10 * n * Log_{10}(d/d_o) + s; \quad d > d_o \tag{2.34}$$

where, $A = 20xLog_{10}(4\pi d_o/\lambda)$ and λ is the wavelength in metres, n is the path loss exponent and

$$n = (a - b\,h_b + c/h_b) \tag{2.35}$$

The height of the base station h_b is between 10m and 80m, and d_o =100m, a, b, and c are constants that depend on the terrain category which is reproduced from [5] Table 2.2.

Shadowing effect is represented by s and follows a log-normal distribution with typical standard deviation between 8.2 and 10.6 dB. This too depends on the terrain and tree density type. Correction terms are used to account for antenna height and frequency region. For the model to apply to frequencies outside the range of specification (2GHz), and for receive antenna heights between 2m and 10m, correction terms are specified. The coarse form of the path loss model (in dB) has three correction terms as:

$$L_{pc} = L_p + \Delta L_f + \Delta L_h \tag{2.36}$$

and ΔL_f (in dB) is the frequency correction term given by the expression. The frequency correction term is given by the expression

$$\Delta L_f = 6\,Log(f/2000) \tag{2.37}$$

The frequency f is MHz, and is positive for frequencies higher than 2 GHz. The correction term for antenna height is:

$\Delta L_h = -10.8\,Log(h/2)$; for categories A and B and

$\Delta L_h = -20\,Log(h/2)$; for category C

The height of the receive antenna is in the range 2m < h < 10m.

2.2.3 COST 231 Hata Model (Alternative Flat Suburban)

The COST 231 Hata model is used for modelling for propagation in an urban environment in the PCS range and is calculated from the expression:

$$L_p = 46.3 + 33.9 \times Log(f) - 13.82 \times Log(h_b) - a(h_m) + [44.9 - 6.55 \times Log(h_b) * Log(d) + C_m] \tag{2.38}$$

where, $a(h_m)$ is the mobile antenna correlation factor given as:

$$a(h_m) = \begin{cases} [1.1 \times Log(f) - 0.7]h_m - [1.56 \times Log(f) - 0.8]dB & \text{for suburban/urban} \\ 3.2 \times [Log(11.75 h_m)]^2 - 4.97 dB & \text{for dense urban} \end{cases} \tag{2.39}$$

$$C_m = \begin{cases} 0\,dB \text{ for suburban / urban} \\ 3\,dB \text{ for dense urban} \end{cases} \tag{2.40}$$

$$L_{p(suburban)} = L_{p(urban)} - 2 \times [Log(f/28)]^2 - 5.4 \tag{2.41}$$

and
h_b = base station height (m), 30-200m
h_m = mobile height (m), 1-10 m
f = frequency (MHz), 1500 - 2000 MHz
d = distance (km), 1-20 km
It has been shown that the COST 231 Walfish-Ikegami (W-I) model provides a close match for extensive experimental data from suburban and urban areas and that the Category C in the Hata-Okamura model is in good agreement with the cost 231 W-I model. It also provides continuity between the two models. The COST 231 W-I model agrees well with measured data from urban areas, provided appropriate rooftop heights and building spaces are used.

2.2.4 Walfisch-Ikegami Model

The Walfisch-Ikagami Model is valid between 800 and 2000 MHz and over distances of 20 km to 5 km and was recommended by the WiMAX Forum for modelling of WiMAX singal propagation. It is useful for dense urban and

canyon-like environments where average building height is larger than the receiving antenna height. This means wireless signals undergo diffraction and are guided along the street like an urban canyon (Figure 2.4). The model applies with the following propagation parameters:
h_b = base station height (m), 3-50m
h_m = mobile height (m), 1-3 m
f = frequency (MHz), 800 - 2000 MHz
d = distance (km), 0.2-5 km

This model applies therefore fairly well when WiMAX is deployed in densely populated urban areas with significant number of semi-high rise buildings (eg. Sydney CBD, Figure 2.5). The model distinguishes between the LOS and the NLOS cases.

Line of Sight: For the line of sight case the loss equation is

$$\widetilde{L}_P = 42.6 + 26\log(d) + 20\log(f) \tag{2.42}$$

There are no correction terms in this case.

For an urban canyon with line of sight the suggested propagation model is

$$\widetilde{L}_P = -31.4 + 26.\log_{10}(d) + 20.\log_{10}(f) \tag{2.43}$$

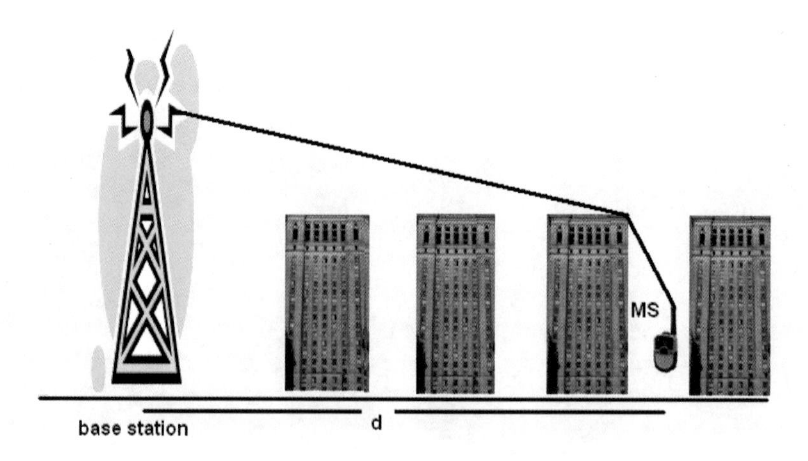

Figure 2.4 Communication from a MS in a Canyon.

None Line of Sight: For the none line of sight case,

$$\widetilde{L}_P = L_r + L_f + L_{ms} \tag{2.44}$$

The median path is given by the expression:

$$\widetilde{L}_P = \begin{cases} L_r + L_f + L_{ms} ; L_r + L_{ms} > 0 \\ L_f \qquad\qquad ; L_r + L_{ms} \leq 0 \end{cases} \tag{2.45}$$

where Lr, Lf and Lms are the roof top, free space and multiscreening path losses. The free space loss is given by the expression

$$L_f = 32.44 + 20.\log_{10}(f) + 20.\log_{10}(d) \tag{2.46}$$

The frequency f is measured in MHz and the distance in km. The rooftop diffraction model is given by the expression

$$L_{rt} = -6.9 + 10 \log(w) + 10 \log(f) + 20 \log(d_{hm}) + L_{ori} \tag{2.47}$$

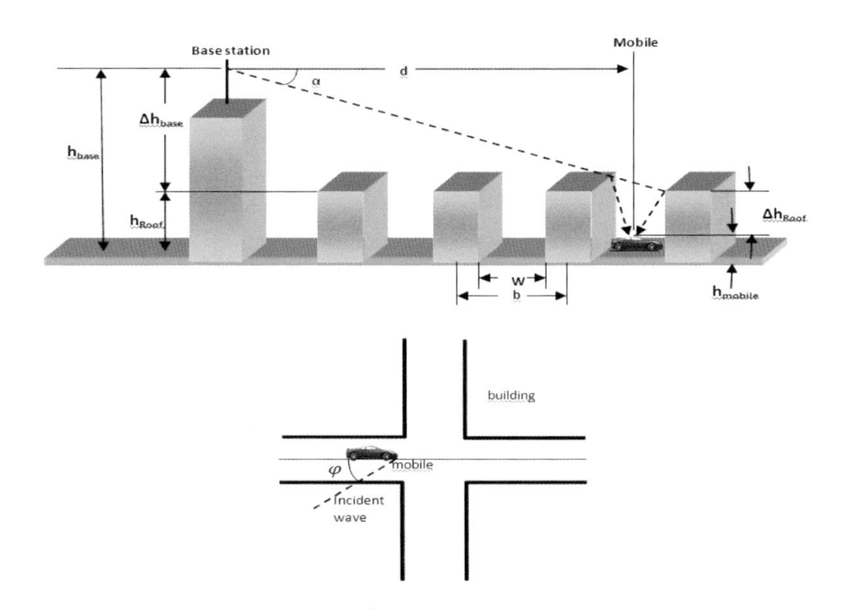

Figure 2.5 Walfisch-Ikegami Model.

Where

$$\begin{aligned} L_{ori} &= -10 + 0.354\varphi && for\ 0 \le \varphi < 35 \\ &= 2.5 + 0.075(\varphi - 35) && for\ 35 \le \varphi < 55 \\ &= 4 - 0.114(\varphi - 55) && for\ 55 \le \varphi < 90 \end{aligned} \tag{2.48a}$$

$$L_{(mult)} = k_0 + k_a + k_d \log(d) + k_f \log(f) - 9\log(W) \tag{2.48b}$$

$$\begin{aligned} &k_0 = 0, \\ &k_d = 18 - 15\left(d_{hb}/h_{roof}\right) \\ &k_a = 54 - 0.8d_{hb}\ and \\ &k_f = -4 + 1.5 \end{aligned} \tag{2.48c}$$

For the non line of sight case in a metropolitan area with the following parameters, the median loss model becomes

$$\widetilde{L}_P = -65.9 + 38\log_{10}(d) + \left(24.5 + \frac{1.5f}{925}\right)\log_{10}(f) \tag{2.49}$$

A 10dB fading margin is suggested by the WiMAX Forum with this formula.

2.2.5 Erceg Model

The Erceg model was adopted by the 802.16 group for fixed WiMAX. The model is mostly suited to fixed wireless network applications. The Erceg model consists of a base model and an extended model. The data leading to the model were collected in Dallas, Chicago, New Jersey, Atlanta and Seattle in the USA, at 1.9GHz and over 95 macrocells.

The base model has three models in one and each one accounts for specified terrain type.

Table 2.3 Walfisch – Ikegami Propagation Model

Parameter	Variable	Value
Height of base station	h_b	12.5m
Building height	h_{bd}	12m
Building-to-building distance	d	50m
Building width	w	25m
Height of mobile station	h_m	1.5m
Orientation of all paths	φ	30^0

i) Erceg A model is applicable to hilly terrain with moderate to heavy density of trees

ii) Erceg B model is also applicable to hilly terrain with light tree density. It applies also to flat terrain with moderate to heavy density of trees

iii) Erceg C model is applicable to flat terrain with light density of trees

The Erceg loss formula models instantaneous path loss as a sum of a median loss term and a shadow fade value given by the expression

$$L_P = \widetilde{L}_P + X = A + 10n\log_{10}\left(\frac{d}{d_0}\right) + X \tag{2.50}$$

Where X is the shadow fades and A is the free space path loss over a distance $d_0 = 100m$ and at a frequency f.

$$A = 20\log_{10}\left(\frac{4\pi f d_0}{C}\right) \tag{2.51}$$

The associated path loss exponent n is modelled as a Gaussian random variable with a mean value given by the expression

$$n = A - Bh_b + Ch_b^{-1} \tag{2.52}$$

Hence the instantaneous value of the path loss exponent is

$$n = A - Bh_b + Ch_b^{-1} + x\sigma_\alpha \tag{2.53}$$

Where x is a Gaussian random variable with zero mean and unit variance. The standard deviation of the distribution of the path loss exponent is σ_α. The parameters of the Erceg model are given below for the different terrain types.

Table 2.4 Parameters of the Erceg Model

Parameters	Erceg Model A	Erceg Model B	Erceg Model C
a	4.6	4	3.6
b	0.0075	0.0065	0.005
c	12.6	17.1	20
S_a	0.57	0.75	0.59
μ_s	10.6	9.6	8.2
σ_s	2.3	3	1.6

Notice the similarity between rows 1, 2 and 3 of Table 2.2 and Table 2.4. Erceg's model is an improvement on the Hata-Okumura model.

The base model applies only at 1.9Hz and for MS with omnidirectional antenna at a height of 2 meters and for base stations of heights 10 to 80m. The extended model modifies the base model to enable it work over larger frequency range and with the following parameters:
h_b = base station height (m), 10-80m
h_m = mobile height (m), 2-10 m
f = frequency (MHz), 1900 - 3500 MHz
d = distance (km), 0.1- 8 km
The median path loss model for the extended Erceg model is

$$\widetilde{L}_P = A + 10\gamma \log_{10}\left(\frac{d}{d_0}\right) + \Delta P.L_f + \Delta P.L_{hMS} + \Delta P.L_{\theta MS} \tag{2.54}$$

The correction terms in this equation are:

$$\Delta P.L_f = 6\log\left(\frac{f}{1900}\right) \tag{2.55a}$$

$$\Delta P.L_{hMS} = -10.8\log\left(\frac{h_m}{2}\right); \textit{ for Erceg A and B} \tag{2.55b}$$

$$\Delta P.L_{hMS} = -20\log\left(\frac{h_m}{2}\right); \textit{ for Erceg C} \tag{2.55c}$$

$$\Delta P.L_{\theta MS} = 0.64\ln\left(\frac{\theta}{360}\right) + 0.54\left(\ln\left(\frac{\theta}{360}\right)\right)^2 \tag{2.55d}$$

The correction term $\Delta P.L_{\theta MS}$ is called the antenna gain reduction factor because it accounts for the fact that the angular scattering is reduced due to the directivity of the antenna. This correction can be significant and is about 7dB at an antenna angle of 20^0.

2.2.6 Stanford University Interim (SUI) Model

To calculate the path loss using SUI model, the environment is categorized in three different groups with different characteristics, known as A, B, and C [1]. A is referred to hilly environment and moderate to very dense vegetation which results in highest path loss, while B refers to hilly environment but rare vegetation or high vegetation but flat terrain. However C is referred to flat area with rare vegetation which leads to lowest path loss.

SUI model is a suitable channel modelling for WiMAX implementation, using frequency band at 3.5GHz, which can support for cell radius in range of 0.1km and 8km, the base station antenna height between 10m and 80m and receiver antenna height in the range of 2m and 10m [1]. In SUI model the path loss is calculated using formula:

$$L_p = A + 10n\log_{10}\left(d/d_0\right) + X_f + X_h + s \qquad \textit{for } d > d_0 \tag{2.56}$$

In this expression, $d_0=100m$ and d is the distance between the transmitter and the receiver, s is standard deviation and is a random variable, X_f is a correction for frequency above 2GHz and X_h is a correction for transmitter antenna height [1].

$$X_f = 6.0\log_{10}\left(f/2000\right) \tag{2.57}$$

For A and B environment

$$X_h = -10.8\log_{10}\left(h_r/2000\right) \tag{2.58}$$

$$X_h = -20.0\log_{10}\left(h_r/2000\right) \qquad \textit{for C environment} \tag{2.59}$$

Where f is the frequency and h_r is the height of the antenna at the receiver. However, in equation (2.56), A is calculated by:

$$A = 20\log_{10}\left(\frac{4\pi d_0}{\lambda}\right) \tag{2.60}$$

Where λ is the wavelength of the signal in meters. Also γ is the path loss exponent which has different values between 2 and 5 for different environment and depends on the height of the base station antenna h_b and three constants of a, b and c which vary with different types of environment of A, B and C [1].

$$n = a - bh_b + \frac{c}{h_b} \tag{2.61}$$

The path loss exponent in urban areas when the LOS exists is approximately 2, while it is between 3 and 5 in the urban area with absence of a LOS. However the path loss exponent could be more than 5 if the signal propagation in unusual environments such as tunnels [1]. SUI model is used for planning for WiMAX in rural, urban and suburban areas.

2.2.7 Lee Model

The Lee model [9,10] is used for prediction of point-to-point propagation. It divides the signal propagation regime into two types:

1. Environments with man-made structures and their impacts on radio waves. Man-made structures of interests include tall buildings, street light poles, concrete structures, bridges and tunnels. Hence the propagation model for one city is different from the model for other cities.
2. Environments where the natural terrain varies. Natural terrains include mountains, valleys, water pools (ocean, rivers, lakes), forests, grasslands, deserts etc. Effective antenna gain and shadow loss are critical to the accuracy of the model.

Both types of environments are clearly varying from city to city and place to place hence the Lee model is given to address or estimate this variability.

For distances larger than one mile equivalent in metres, the received power in decibels for the Lee model for macrocells is described by the expression [9]

$$P_r = P_0 + \gamma \log \frac{R}{R_0} + G_{eh} + L + \alpha \quad (dB) \tag{2.62}$$

P_0 is the received power at the intercept R_0 (1 mile equivalent in metres); $R_0 < R$

R is the distance from the transmitter in miles

R_0 is the equivalent length in metres to 1 mile

γ is the path loss exponent

G_{eh} is the effective height gain of the antenna where $G_{eh} = 20\log(h_m / h_b)\, dB$. It is zero for shadow loss.

L is shadow loss in dB when the receiver is blocked. When the receiver is in line of sight, L=0

α is a standard adjustment in dB

h_m is the height of the mobile station antenna in meters

h_b is the height of the base station in meters

The Lee model consists of two sections: the human structure related part $P_r = P_0 - \gamma \log \frac{R}{R_0}$ and the natural terrain related part is obtained with the antenna (base station and mobile device) part $G_{eh} + L + \alpha \quad (dB)$[11]. Distances in the Lee model and optimized Lee model are given in miles and should be converted to the metric units for comparison with other models.

2.2.7.1 Optimized Lee Model

In [12] Agilent provides an optimised version of the Lee propagation model which accounts in addition to the traditional Lee model, knife-edge diffraction losses, antenna pattern effects and additional losses and gains in the link. The expression used is given in equation (2):

$$P_r = P_0 + \gamma \log \frac{R}{R_0} + 15 \log\left(\frac{h_{beff}}{h_{bref}}\right) + 10 \log\left(\frac{P_{tx}}{P_{txref}}\right) + 10 \log\left(\frac{h_m}{h_{mref}}\right) + \alpha \quad (dB) \tag{2.63}$$

where $R_0 < R$. The reference values are defined as follows [12]:

P_r is the mean signal level received at distance R from the transmitter

P_0 is the signal level expected with the reference parameters $(h_{breff}, h_{mref}, R_0, P_{txref})$

R is the distance from the transmitter

R_0 is the reference distance and for the Lee model it is taken as 1 mile (1609m)

γ is the path loss exponent

h_{beff} is the effective height of the base station in feet

h_m is the height of the mobile antenna (in feet)

P_{tx} is the transmit ERP in watts

P_{txref} is the effective transmit ERP in watts

α is a factor added to account for knife-edge diffraction losses, antenna pattern effects and any other losses in the link.

2.3 PATH LOSS EXPONENTS FOR DIFFERENT TERRAINS

2.3.1 Path Loss Exponents

Path loss exponent varies widely across propagation environments [8]. Therefore the bound on the hop distance and number is different for different types of propagation domains. For long-distance coverage the exponent for out door environments is around 4 except in none line-of-sight situations when it could be bigger than 4. The value of the path loss exponent is an indicator of how fast energy is lost between the transmitter and receiver. $\alpha < 2$ is a measure of the guiding effect of the channel and when $\alpha > 2$ the channel is considered to be scattering energy.

Table 2.5 Path Loss Exponent for Out Door Environments

Path Loss Exponents for Out Door Environments			
No	Location	Path loss exponent n	Frequency range
1	Urban	4.2	
2	Free space	2	Micro cellular
3	Log-normally shadowing area	2 to 4	Micro cellular
4	UWB LOS up to breakpoint	2	UWB range
	UWB LOS after breakpoint	4	UWB range
	LOS urban (antenna ht = 4m)	1.4	5.3 GHz range
	LOS urban (antenna ht = 12m)	2.5	5.3 GHz range
	NLOS urban (antenna ht = 4m)	2.8	5.3 GHz range
	NLOS urban (antenna ht = 12m)	4.5	5.3 GHz range
5	LOS rural (antenna ht = 55m)	3.3	5.3 GHz range
	NLOS rural (antenna ht = 55m)	5.9	5.3 GHz range
	LOS suburban (antenna ht = 5m)	2.5	5.3 GHz range
	NLOS suburban (antenna ht = 12m)	3.4	5.3 GHz range
6	Highway micro-cells	2.3	900 MHz
	Dual Carriage Highway	7.7	1.7 GHz
7	BFWA/directional antenna (5.5-6.5m)	1.6	3.5 GHz
	BFWA/directional antenna (6.5-7.5m)	2.2	
	BFWA/directional antenna (7.5-8.5m)	2.7	
	BFWA/directional antenna (8.5-9.5m)	2.6	
	BFWA/directional antenna (9.5-10.5m)	3.6	

The following tables provide typical values of α and also show how different structures guide radio waves and which ones scatter them. They also provide a database of α for design of wireless networks in different environmental situations.

2.3.2 Out Door Environments

The value of path loss exponent outdoors is a function of the terrain (free space, urban, suburban, rural and foliage type), if communication is line-of-sight (LOS) or non-LOS (NLOS), height of the antenna and the channel frequencies. Table 2.5 summarises these effects [8].

The dynamic range of alpha in Table 2.5 is 6.3. None line-of-sight communication often means higher path loss exponents. Similarly the higher the height of the antenna the higher the expected pass loss exponent. The value of the path loss exponent after the break point is normally greater than the value before the break point. The break point distance can be approximated with the expression:

$$d_b = \frac{4h_T h_R}{\lambda} \tag{2.64}$$

h_T and h_R are the heights of the transmitting and receiving antennas and λ is the wavelength of transmission. The high value of exponent for dual carriage highway is due to ground reflections from the road surface.

2.3.3 Indoor Environments

In Table 2.6, the path loss exponent for in door communications across a wide variation of frequencies is shown. The unpredictability of the path loss exponent is demonstrated by the range of values shown [8].

Communications indoors at various frequencies affect the path loss exponent and the predominant sources of effects are the height of the building (or height of antenna), antenna directivity, LOS or NLOS communication, the channel frequencies, the types of materials used in the construction of the buildings and the location of measurements in the building. Omni-directional antennas often result to lower path loss exponents compared to directional antennas. This is because, the omni-directional antennas collects signals from many more multipath sources. Building materials of different types lead to different path loss exponents. The dynamic range of α in Table 2.6 is 8.8. Therefore the optimum hop index will vary widely indoors. Measuring α is therefore required prior to establishing the relay nodes.

Table 2.6 Path Loss Exponent for Indoor Communications

Path Loss Exponents for Indoor Environments			
No	Location	Path loss exponent	Frequency range
1	LOS	1.83	802.11a (5.4 GHz)
2	LOS	1.91	802.11b (2.4 GHz)
3	NLOS	4.7	802.11a
4	NLOS	3.73	802.11b
5	Omni/Omnidirectional antennas	1.55	UWB
	Omni/directional antennas	1.65	UWB
	Directional/Directional–shadow	1.72	UWB
6	Indoor CDMA	1.8 ~ 2.2	20 GHz – 30 GHz
7	LOS	1.73	900 MHz
	NLOS	0.48 ~ 1.12	900 MHz
	LOS	2.23	1.89 GHz
	NLOS	-1.43 ~ 1.47	1.89 GHz
8	LOS (millimetre wave)	1.2 ~ 1.8	94 GHz
	Obstructed channel	3.6 ~ 4.1	94 GHz
	LOS	1.8 ~ 2.0	11.5 GHz
	LOS	1.2	37.2 GHz
9	Inside room of a building	0.77	900 MHz
	Inside room of a building	0.44	1.35 GHz
	LOS DECT picocells	-1.55	1.8 GHz
	NLOS DECT picocells	-3.76	1.8 GHz
10	Corridor Ground Floor	0.70	450 MHz
		0.48	900 MHz
		0.02	1.35 GHz
		-1.43	1.89 GHz
11	Corridor Floor 1 of building	1.12	450 MHz
		1.02	900 MHz
		0.07	1.35 GHz
		1.46	1.89 GHz
12	Corridor Floor 2 of building	1.79	450 MHz
		1.72	900 MHz
		0.44	1.35 GHz
		2.22	1.89 GHz
13	Indoor 3rd floor of a laboratory	1.3	2.45 GHz
		1.8	5.25 GHz
		1.7	10 GHz
		1.8	17 GHz
		1.7	24 GHz

Table 2.7 Path Loss Exponent for Underground Communications

Path Loss Exponents for Underground Communications			
No	Location	Path loss exponent	Frequency range
1	Underground (train) – front	12.45	465 MHz
2	Underground (train) – rear	9.72	
3	Underground (train) - front	8.58	820 MHz
4	Underground (train) – rear	8.17	
5	Train yard (parallel to track) Train yard (cross-track)	2.7 3.4	
6	Underground Mine Moving train – 140 km track sites	2.13 ~ 2.33 1.5 ~ 7.7	2.4 GHz 320 MHz

2.3.4 Underground Environments

Communications underground such as in tunnels and mines forms a vital component of the overall wireless communication industry. In many countries, tunnels form significant sections of roads and railways. Similarly communication inside mines is also a vital support for mining and mineral exploration. Table 2.7 records typical path loss exponents reported for underground communications. Understandably, low frequency applications are prevalent.

Path loss exponent in underground communications is normally predominantly very high as seen from Table 2.7 [8]. This is due to the terrain, the materials used for construction of the tunnels and to some extent the channel frequencies used. The scattering properties of the terrain also affect the path loss exponent. The dynamic range of α in this table is 10.95. Path losses underground are therefore very high and hop distances must be chosen with this in mind.

2.3.5 Unspecified Environments

Path loss exponent in other terrains of interest are summarised in Table 2.8 [8].

Table 2.8 demonstrates the varying nature of path loss exponent when different types of materials and the terrain types that affect communications are considered. These tables show that there is no universally accepted path loss model for indoor, outdoor or underground channels. The path loss model varies, from building to building and from terrain to terrain. All path loss models in use are approximations for only a few conditions.

Table 2.8 Effect of Materials on Path Loss Exponent

Path Loss Exponents for Different Environmental Structures			
No	Location	Path loss exponent	Frequency range
1	Engineering	1.4 ~ 2.2	0.8 GHz – 1.0 GHz
2	Apartment Hallway	1.9 ~ 2.2	
3	Parking structure	2.7 ~ 3.4	
4	One-sided corridor	1.4 ~ 2.4	
5	One-sided Patio	2.8 ~ 3.8	
6	Concrete Canyon	2.1 ~ 3.0	
7	Plant fence	4.6 ~ 5.1	
8	Small Boulders	3.3 ~ 3.7	
9	Sandy Flat beach	3.8 ~ 4.6	
10	Dense Bamboo	4.5 ~ 5.4	
11	Dry Tall Underbrush	3.0 ~ 3.9	

2.4 RECEIVER SENSITIVITY

The receiver power is limited by its design (implementation margin), thermal noise in the receiver, its noise figure and signal-to-noise ratio. Thus the receiver sensitivity is given by the expression:

$$R_{SS} = SNR_R + NF_R + L_{implementation} + N_{Thermal} \tag{2.65}$$

These factors are given by the expressions

$$N_{Thermal} = -147 + 10\log_{10}(\Delta f) = -147 + 10\log_{10}\left(B.n.\frac{N_{Used}}{N_{FFT}} \right) \tag{2.66}$$

Where Δf is the subcarrier spacing. The thermal noise is affected by its thermal noise density which is

$$N_0 = K.T.B \approx -174dB \tag{2.67}$$

K is Boltzman constant. The SNR is a function of the modulation scheme used and these have been provided in the standard and are given in Table 2.9.

Table 2.9 SNR Parameters as Function of Modulation Schemes

Modulation scheme	SNR CC (AWGN, BER 10-6)	SNR CTC (AWGN, BER 10-6)	Data bit per symbol
QPSK 1/2	5 dB	2.5 dB	1
QPSK 3/4	8 dB	6.3 dB	1.5
16-QAM 1/2	10.5 dB	8.6 dB	2
16-QAM 3/4	14 dB	12.7 dB	3
64-QAM 1/2	16 dB	13.8 dB	3
64-QAM 2/3	18 dB	16.9 dB	4
64-QAM 3/4	20 dB	18 dB	4.5

The receiver noise factor is caused by the electronics in its RF chain. It is a ratio of the input SNR to a device to its output SNR. This is normally measured at 290Kelvin. It provides a measure of the performance of a device. It is given by the expression:

$$NF = \frac{SNR_{in}}{SNR_{out}} \tag{2.68}$$

The receiver noise figure in equation (2.65) is defined as ten times the logarithm to base ten of the noise factor. Or NF=10.log (Noise factor). A detailed review and analysis of all published propagation model is not only unnecessary but also impossible as there have been significant number of models proposed. This chapter has selected some of the most widely used models in addition to the theoretical models.

References

[1] Josip Milanovic, Snjezana Rimac-Drlje, Krunoslav Bejuk, "Comparison of Propagation Models Accuracy for WiMAX on 3.5GHz" IEEE International Conference, 2007, pp:111-114.
[2] Mobile WiMAX Group, "Coverage of mobile WiMAX", pp. 1 – 18.
[3] Rana Ezzine, Ala Al Fuqaha, Rafik Braham, Abdelfettah Belghith. "A New Generic Model for Signal Propagation in WiFi and WiMAX Environment". Wireless Days, 2008, pp:1-5.
[4] Tamaz Javornik, Gorazd Kandus, Andrej Hrovat, Igor Ozimek. Software in Telecommunication and Computer Networks, 2006, pp: 71-75.
[5] Yan Zhang. WiMAX Network Planning and Optimization. USA: CRC Press, 2009.
[6] Johnson I Agbinya. IP Communications and Services for NGN. New York: Taylor and Francis, 2009.
[7] Kejie Lu, Yi Qian, Hsiao-Hwa Chen, Shengli Fu. "WiMAx Networks: From

Access to Service Platform", IEEE Computer Society, 22, pp. 38-45, May/June 2008.
[8] Johnson I Agbinya, "Design Consideration of Mohots and Wireless Chain Networks", Wireless Personal Communication", © Springer 2006, Vol. 40, pp. 91 -106.
[9] David J. Y. Lee and William C. Y. Lee, "Enhanced Lee Model from Rough Terrain Data Sampling Aspect", Proc. 72nd IEEE Conference on Vehicular Technology Conference Fall (VTC 2010-Fall), Ontario, Canada, Sept. 6-9, 2010, pp. 1-5.
[10] Lee's Model, Appendix VI, Propagation Special Issue, IEEE Transactions on Vehicular Technology, Vol. 37, No. 1, Feb. 1988, pp. 68–70.
[11] William C. Y. Leeand David J. Y. Lee, "The propagation characteristics in a cell coverage area", in Proc. IEEE 47th Conference on Vehicular Technology, 1997, Vol. 3, pp. 2238 – 2242.
[12] Propagation Model Optimisation (Sample Report), Envision Wireless, June 21, 2005, www.evwi.com.
[13] J. J. Egli, "Radio Propagation above 40 Mc Over Irregular Terrain", Proc. IRE, 1957, pp.1383-1391.

Chapter 3.

Diffraction Models

Johnson I Agbinya

Department of Electronic Engineering, La Trobe University, Australia
J.Agbinya@latrobe.edu.au

3.1 INTRODUCTION

Historically, a great deal of research has been undertaken on the propagation of radio waves in various types of terrains and man-inhabited regions. Understandably only a few of the propagation models that have been published are recommended for use in standards as shown in the previous chapter. A great deal of signal degradation in addition to path losses can take place in the channel to reduce the quality of communication. A major source of signal degradation is diffraction losses.

3.2 KNIFE EDGES

Practical telecommunication system design and planning need to take into account the presence of obstacles which are in the direct line of sight path. It is however impossible to account for all the obstacles which impact on the propagation of the signal. Normally the nature of the obstacle, its material components, size and features all affect the way and manner the signal propagates. In practice a simplified model is used and a popular approach is the use of a knife-edge to represent an obstruction to the signal. An obstacle in the line of sight path is often represented by a knife edge as shown in Figure 3.1. The obstacle in Figure 3.1 is a hill. For the sake of modelling the RF signal ray, the hill is replaced with a knife edge of height h above the line of sight. The position of the knife-edge from the transmitter is d_1 and it is d_2 from the receiving terminal in the car.

4G Wireless Communication Networks: Design, Planning and Applications, 67-76.

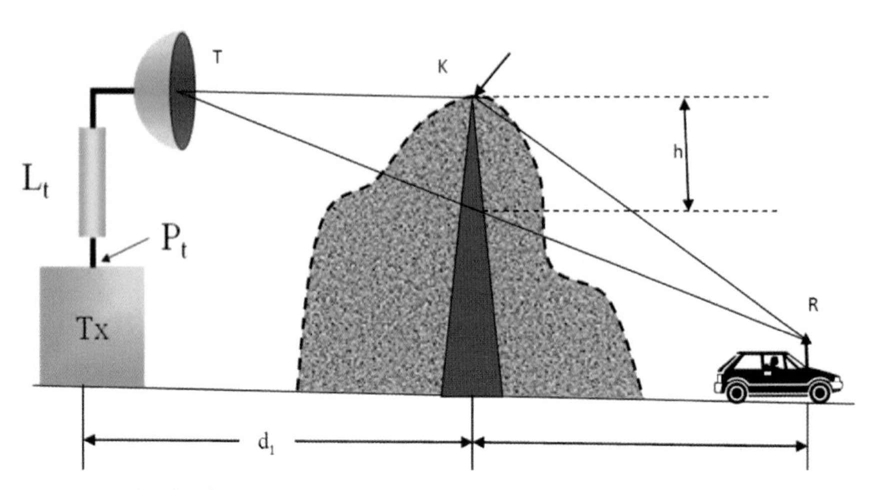

Figure 3.1 Refraction from a Knife-Edge.

To estimate the effect of the diffraction knife-edge, the extra path length of the ray TK-KR compared with the direct ray TR need to be found. To do so, consider that the triangle TKR is redrawn as in Figure 3.2.

By using Pythagoras theorem the following expressions can be derived from the triangle TKP:

$$\begin{aligned} d_1 + \Delta_1 &= \sqrt{h^2 + d_1^2} \\ (d_1 + \Delta_1)^2 &= h^2 + d_1^2 \\ &\textit{and} \\ \Delta_1 &= \frac{h^2 - \Delta_1^2}{2d_1} \end{aligned} \tag{3.1}$$

Following the same procedure for triangle KPR the following expression is obtained:

$$\Delta_2 = \frac{h^2 - \Delta_2^2}{2d_2} \tag{3.2}$$

By assuming that in practice

$(d_1 \text{ and } d_2) >> h$

Δ_1^2 and Δ_2^2 are too small and ignored

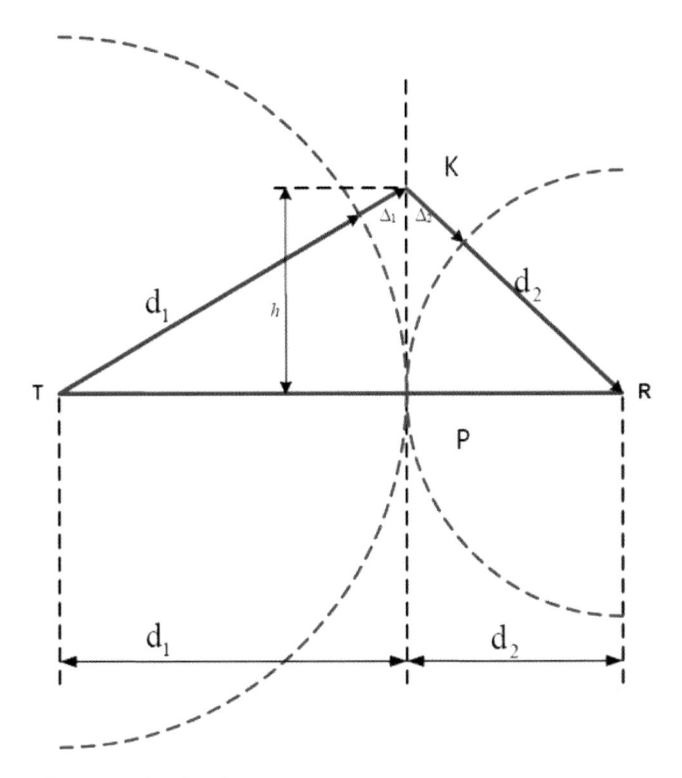

Figure 3.2 Diffraction Path Triangle.

Then the extra path length compared to the direct path is

$$\Delta = \Delta_1 + \Delta_2 = \frac{h^2 - \Delta_1^2}{2d_1} + \frac{h^2 - \Delta_2^2}{2d_2} \approx \frac{h^2}{2}\frac{(d_1 + d_2)}{d_1 d_2} \tag{3.3}$$

This extra path length leads to a phase difference between the direct ray and the refracted ray given by the expression:

$$\varphi = \frac{2\pi}{\lambda}\Delta = \frac{2\pi}{\lambda}\frac{h^2}{2}\frac{(d_1 + d_2)}{d_1 d_2} \tag{3.4}$$

Let us define the Fresnel-Kirchhoff parameter by the expression:

$$\upsilon = 2h\sqrt{\frac{(d_1 + d_2)}{\lambda d_1 d_2}} \tag{3.5}$$

The Fresnel-Kirchhoff parameter defines the radii of ellipsoids where the phase differences are multiples of π. This means that for n being an integer we can re-write the above expression as

$$n\pi = \frac{\pi}{2}\upsilon_n^2 \qquad (3.6)$$

$$\upsilon_n = \sqrt{2n}$$

Therefore the height of the knife-edge above the line of sight at the position of the knife edge is

$$h = \upsilon_n \sqrt{\frac{\lambda d_1 d_2}{2(d_1 + d_2)}} = \sqrt{\frac{n\lambda d_1 d_2}{(d_1 + d_2)}} \qquad (3.7)$$

The value n defines the individual Fresnel zones and the radius of the ellipsoids are

$$h = \upsilon_n \sqrt{\frac{\lambda d_1 d_2}{2(d_1 + d_2)}} = \sqrt{\frac{n\lambda d_1 d_2}{(d_1 + d_2)}} \qquad (3.8)$$

where

$$R_n = \sqrt{\frac{n\lambda d_1 d_2}{(d_1 + d_2)}}$$

$$= 17.32\sqrt{\frac{n d_1 d_2}{f_{GHz}(d_1 + d_2)}} \qquad (3.9)$$

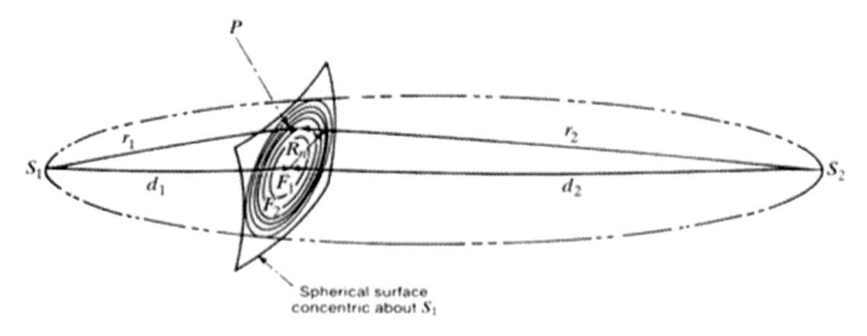

Figure 3.3 Fresnel Zones.

The value for n=1 defines the first Fresnel zone (Figure 3.3). In propagation, the required height clearance of the obstacle by the transmitter must be at least equal to the radius of the first Fresnel zone. This is often enough to ensure direct line of sight reception at the position of the receiver (R).

By obtaining the Fresnel – Kirchhoff diffraction parameter estimation of the extra power loss introduced by the obstruction may be made. This power loss is defined by the expressions:

$$G_d(dB) = 10\log_{10}|F(\upsilon)|^2 \tag{3.10}$$

where

$$F(\upsilon) = \begin{cases} 0, & \upsilon \le -1 \quad (i) \\ 0.5 - 0.62\upsilon, & -1 \le \upsilon \le 0 \quad (ii) \\ 0.5e^{-0.95\upsilon}, & 0 \le \upsilon \le 1 \quad (iii) \\ 0.4 - \sqrt{0.1184 - (0.38 - 0.1\upsilon)^2}, & 1 \le \upsilon \le 2.4 \quad (iv) \\ \dfrac{0.225}{\upsilon}, & \upsilon > 2.4 \quad (v) \end{cases} \tag{3.11}$$

3.2.1 Types of Single Knife - Edges

Three types of single knife edges are identified in Figures 3.4 a), b) and c). Figure 3.4 a) shows a knife-edge in which there is only the direct ray from the transmitter (T) to the receiver (R). It grazes the top of the obstacle. Since there is no obstacle in the path of the signal, h=0. Therefore: $\alpha = 0$ and $\upsilon = 0$.

Figure 3.4 b) is a knife-edge in which the height of the obstacle is negative (h<0). The direct ray to the receiver is well above the obstacle because the tip of the knife – edge is well below the direct ray at a height h less than zero. Therefore $\alpha < 0$ and $\upsilon < 0$ are as in equations (i) and (ii) above. This situation can occur when the transmitting and receiving antennas heights are well above the height of the obstacle.

In Figure 3.4 c) we have positive diffraction in which the diffraction coefficient, diffraction angle, height are all positive. This situation refers to most normal propagation settings in areas with varied obstacles (given as equations (iii) to (v) above).

3.2.2 Multiple Knife-Edge Diffraction (Epstein-Peterson Method)

In many practical propagation situations like in mountainous regions and city areas with many high rise buildings and structures, it is not unusually for the existence of multiple knife – edges in the path of the signal. The multiple knife – edges lead to multiple diffractions. Epstein and Peterson provided a recursive method of estimating the overall diffraction loss due to multiple knife – edges. The method recursively re-use the algorithm for a single knife – edge. Figure 3.5 illustrates the case for three knife edges.

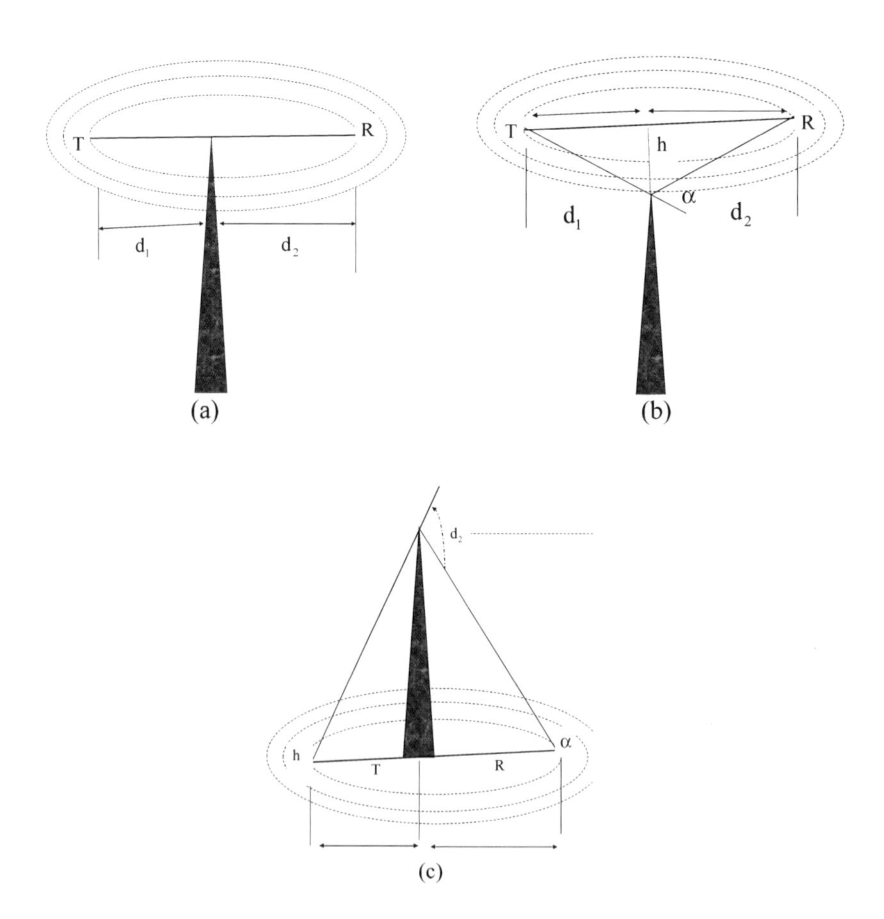

Figure3.4 a) Knife-edge without diffraction b) Knife-edge with negative diffraction (h<0) c) Knife-edge with negative diffraction (h<0)

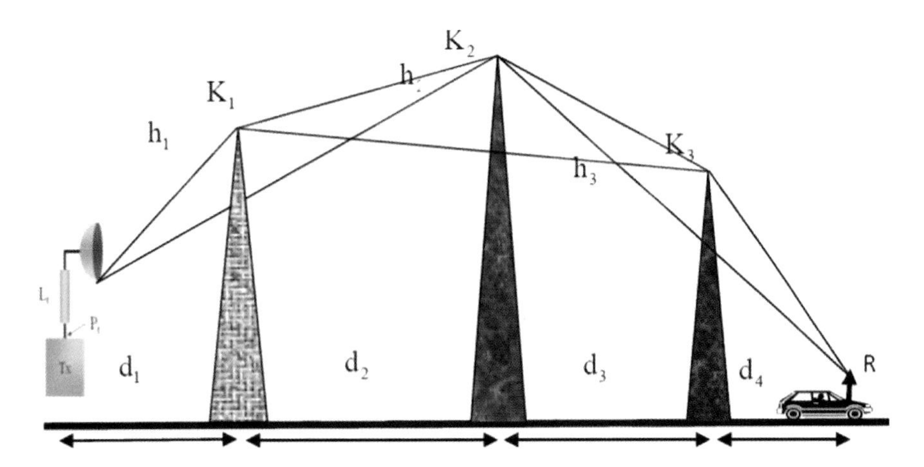

Figure 3.5 Diffraction From Multiple Knife – Edges.

Three triangles are defined in this case by drawing the line from the transmitter to the top of K_2 and from K_1 to K_3 and from K_2 to the receiver R. The diffraction loss is first computed for the knife edge K_1. This involves the distances d_1 and d_2. The diffraction loss computation is done for the knife-edge K_2 as if the transmitter is at K_1 and the receiver is at K_3. This computation involves the knife-edge distances d_2 and d_3. Lastly the diffraction loss computation is repeated for the knife-edge K_3 by assuming that the transmitter is now located at K_2. This last portion of the diffraction loss uses distances d_3 and d_4. The total diffraction loss is the sum of the three partial diffraction losses given by the general expression;

$$L_{sum} = \sum_{k=1}^{N} L_k \tag{3.12}$$

In the example of Figure 3.5, N=3. The Epstein-Peterson method provides a conservative estimate of the overall diffraction loss. For this reason corrections need to be made. One approach for doing this is called the Foose correction method. The Foose correction is based on the difference between the predicted received signal level and a measured received signal level using the expression:

$$S_{measured} - S_{predicted} = n_{obs} x \beta_F \tag{3.13}$$

Where $n_{obs} > 1$ is the number of obstructions and β_F $(dB / obstructio\ n)$ is the Foose correction term. The Foose correction term is a function of the nature of the obstruction and the type of environment. For suburban terrain with low foliage

β_F is about 3.91. In general the diffraction loss when the Foose correction is considered is given by the expression:

$$L_{diffraction} = L_{sum} + n_{obs}.\beta_F = \sum_{k=1}^{N} L_k + n_{obs}.\beta_F \tag{3.14}$$

3.2.3 The Deygout (principle edge) Method

This method is more involved; it splits the path into segments (Figure 3.6). Firstly we need to find the edge with largest value of parameter v, ignoring all other edges. This is called the "Principle Edge" (Figure 3.7) and its v parameter is saved.

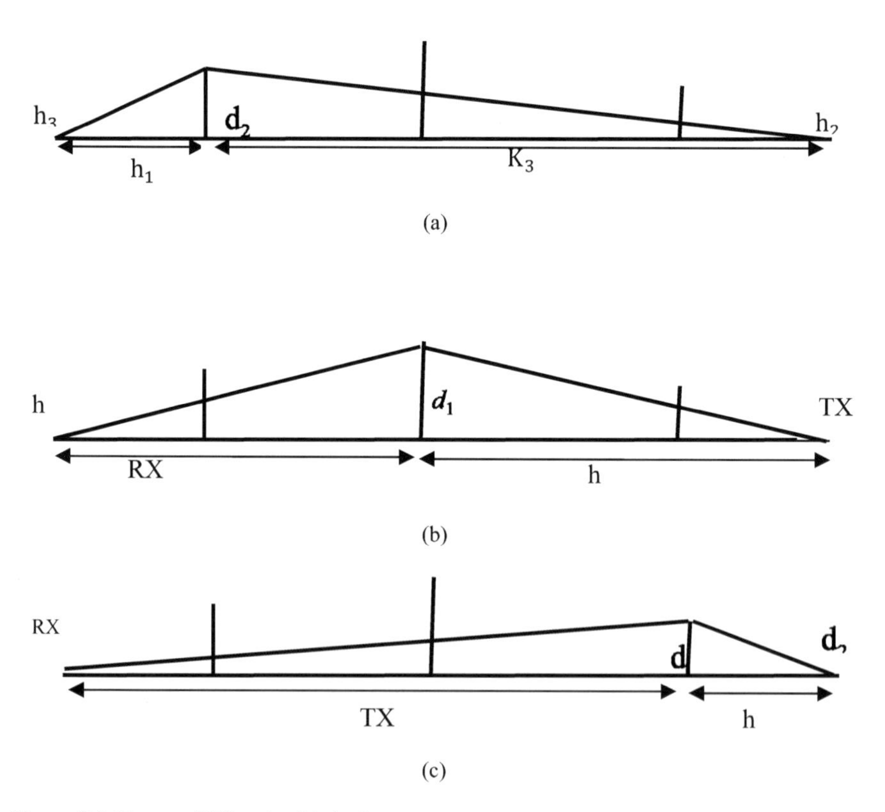

Figure 3.6: Deygout Diffraction Method.

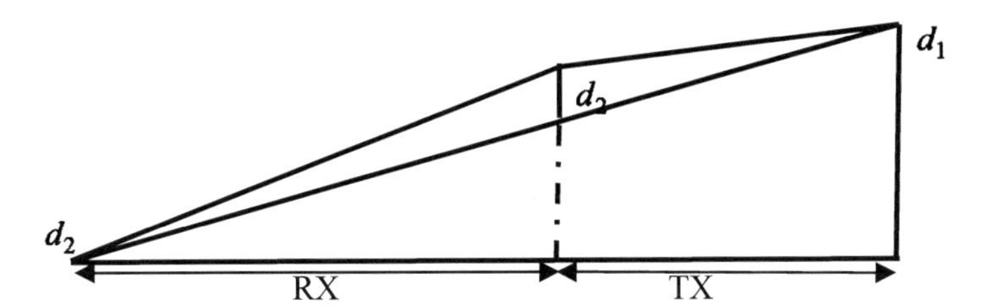

Figure 3.7 Deygout Principle Edge Diffraction.

Working from the principle edge P, we treat the computation process as if there is a new path between the TX and the principle edge and create a new reference plane and calculate v for the intermediate edge, if there is one, based on the height above the reference plane.

This edge will have a lower value of v and becomes the principle edge for the path from Tx to P. The process is recursive for multiple intermediate edges and can be repeated until all edges are considered. The method ignores any edges with 1st Fresnel zone clearance. The same process is used along the path from P to the receiver.

At the end of the procedure we will have a set of J(v) losses for each edge - the method simply adds these up. So for 3 edges:

$$L = J(vp) + J(vtp) + J(vpr) \tag{3.15}$$

Generally, modified Deygout methods are used with Foose factors and scaling factors to further improve the accuracy compared to measurements.

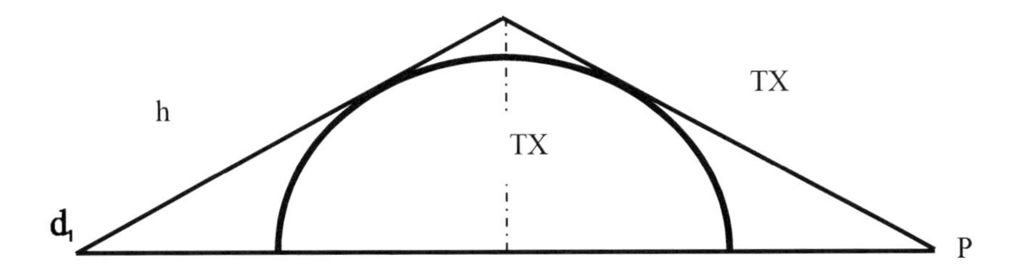

Figure 3.8 Natural Terrain Diffraction.

3.3 REAL TERRAIN

Normal terrain, for example hills do not really look like knife edges and are often better represented by half-cylinders, which have a higher loss. The half-cylinder is shown in Figure 3.8. The diffraction of height h lines graze the half-cylinder as two tangents.

Fortunately, we can approximate the additional loss, **L = J(v) + T** where T is an additional loss that accounts for diffraction at the tangents to the cylinder.

Most real-world obstructions are not like knife edges and it is only possible to solve the equations for idealised cases. Solutions for many objects and including reflection effects, loss from trees etc. rapidly become impractical. Usually we do not really know enough detail about the exact nature of the terrain. For example mobile systems would need re-analysing every 0.1 wavelengths and for 3G systems that would require a terrain map with points every 1.5cm. To overcome this and make a best guess, path loss prediction models are used.

References

[1] David J. Y. Lee and William C. Y. Lee, "Enhanced Lee Model from Rough Terrain Data Sampling Aspect", Proc. IEEE, 72nd Vehicular Technology Conference Fall (VTC 2010-Fall), 2010, 6-9 Sept. 2010
[2] Lee's Model, Appendix VI, Propagation Special Issue, IEEE Transactions on Vehicular Technology, Vol. 37, No. 1, Feb. 1988, pp. 68 – 70.
[3] William C. Y. Lee and David J. Y. Lee, "The propagation characteristics in a cell coverage area", in Proc. IEEE 47th Conference on Vehicular Technology, 1997, Vol. 3, pp. 2238 – 2242.
[4] Propagation Model Optimisation (Sample Report), Envision Wireless, June 21, 2005, www.evwi.com

Chapter 4.

Fading in Cellular Networks

Johnson I Agbinya

Department of Electronic Engineering, La Trobe University, Australia
J.Agbinya@latrobe.edu.au

4.1 PROPAGATION REGIMES

So far in previous chapters emphasis has been placed on path and diffraction losses. Some of the propagation models that were discussed made provisions for fading in the channel. This chapter aims to build upon those discussions. Network planning provides information on the feasibility of network role out. Specifically some of the information is on the estimate of the optimum number of base stations, the location of the base stations and antennas, determining the type of antenna and also the received power at receiver and the environment characteristics of the propagation environment. These optimum values are dependent on the services and the number of users.

In wireless systems, the transmitted signal is propagated in an open environment and losses some of its power as a result of phenomenon such as scattering, diffraction, reflection and fading when received at the receiver; therefore it is important to calculate the received signal power to determine factors such as the radius of a cell or the type of the antenna [1]. The degradation and reduction of the transmitted signal power through interaction with the environment is known as fading. Fading adds to the difference between the transmitted and received signal power. In general, path loss should include factors for fading and is calculated as:

$$Path\ Loss = P_T + G_T + G_R - P_R - L_T - L_R \quad [dB] \qquad (4.1)$$

4G Wireless Communication Networks: Design, Planning and Applications, 77-90.

Figure 4.1 Link Budget.

where, P_T shows the power at transmitter and P_R is power at the receiver, G_T and G_R is the transmitter and receiver antenna gain respectively, L_T and L_R express the feeder losses [1]. This equation describes the link budget. A link budget describes the extent to which the transmitted signal weakens in the link before it is received at the receiver. The link budget therefore accounts for all the gains and losses in the path the signal takes to the receiver including fading.

As shown in Figure 4.1, a link is created by three related communication entities:

a. the transmitter
b. the receiver
c. the channel (medium) between them. The medium introduces losses causing a reduction in the received power.

The link budget (Figure 4.1) equation therefore can be written either as in equation (4.2) to show the path loss or in terms of the receiver power to be:

$$P_R = P_T + G_T - L_P + G_R - A_M \tag{4.2}$$

This equation assumes that all the signal gains and losses are expressed in decibels. The units for these are as follows:

$A_M\,(dB)$, (L_T + L_R *in equation (4.2)*) represents all the attenuation losses such as feeder loss, link margin, diffraction losses, losses due to mobility (Doppler), and the effects of rain, trees and obstacles in the signal path.

$G_T(dBi)$ is the transmitter antenna gain

$G_R(dBi)$ is the receiver antenna gain

$P_R(dBm)$ is the received power at the receiver

$P_T(dBm)$ is the transmitted power

$L_P(dB)$ is the path loss in the physical medium between the transmitter and receiver.

Antennas are often modelled as isotropic sources. The source radiates microwave energy uniformly in all directions, the so-called isotropic source (into a spherical volume). Isotropic antennas are idealised sources which do not exist in practice. A more practical source is the half-wave dipole that is used to model an effective radiated power (ERP). Effective isotropic radiated power (EIRP)) is measured in terms of a half-wave dipole model and: EIRP = ERP + 2.15 dB. In the rest of this section we enumerate some causes of signal fading in the channel.

Interference margin (usually around 1 dB) is used to account for interference during the busy hour and depends on traffic load, frequency reuse plan and other factors.

Penetration of microwave into buildings varies and can be very severe. *Building penetration* is usually around 5 to 20 dB and accounts for penetration into different types of building materials for indoor coverage. Penetration is a function of the type of building. In-building losses can be quite dramatic in areas such as lifts and underground shelters.

Vehicle penetration is less severe compared to building penetration. The estimate for this is around 6 dB and accounts for the attenuation of signal by the frame of a car or truck. Other margins normally considered include:

Body loss (human body) of around 3 dB: This accounts for the absorption of the signal by a mobile user's head and torso. It is sometimes called head loss. This varies depending on hair structure and the proximity of the mobile terminal to the head.

Fade margin of between 4 to 10 dB accounts for multipath fading. Fading dips occur for slow moving mobiles. Fast moving mobiles tend to overcome this because they move faster out of dips before the fading affects the signal. Fade margin varies with the environmental conditions. In the next sections we describe

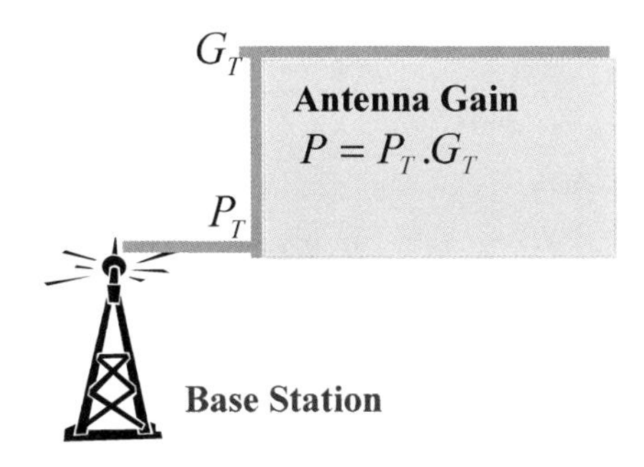

Figure 4.2 Base Station Antenna Power Gain.

the sequence of signal reducing events in the

i. transmitter
ii. channel (medium) and
iii. receiver

The following discussions explain what happens at the transmitter, receiver and channel respectively.

4.2 TRANSMITTER

Consider a typical RF emitter, a WiMAX signal emitter as an example. Although we use WiMAX, the discussions are applicable to other mobile telephone technologies. A WiMAX base station normally employs some form of MIMO system and the early implementations used 2x2 MIMO antennas. The base station also used adaptive antenna systems. Hence power is radiated to many paths linking the transmitter to the receiver. Hence several sources of signal fading are created. The transmitter antenna boosts the data signal power before launching it into the channel.

The antenna creates an effective isotropically radiated power (EIRP) and outputs (radiates) it into the channel (medium), where EIRP = ERP + 2.15 dBi. Notice that the gain of the antenna is in dBi, while ERP is in dB. This conversion is essential. Due to the connectors and cables (Figure 4.3) used in the transmitter circuit, power losses are made as in Figure 4.4

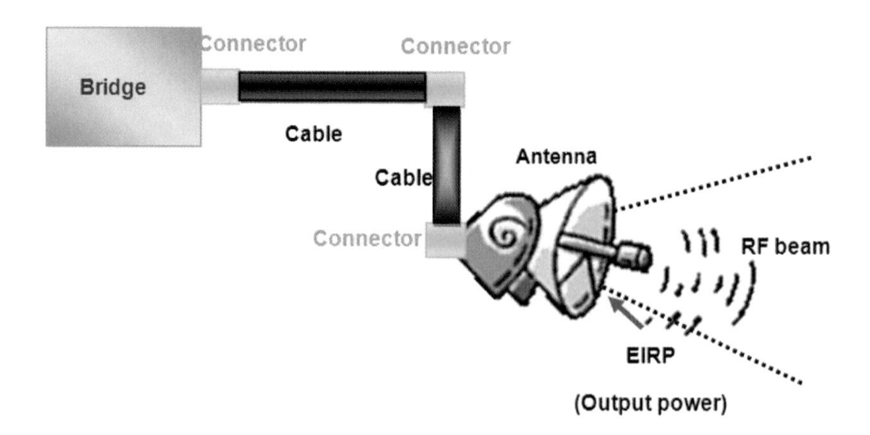

Figure 4.3 Sources of Power Losses in Transmitter Receiver).

The antenna EIRP is reduced as follows

$$EIRP(dB) = P_T - L_{connector,cable} + G_T \tag{4.3}$$

For a 2x2 MIMO base station case, the following power losses were recorded in [2] (Table 4.1).

The parameters for a set of customer premises equipment for both fixed and mobile cases are given in Table 4.2.

4.3 SIGNAL FADING

The WiMAX channel like all other communication channels introduces signal degradation (fading) and reduces the output signal launched to it by the transmitter (Figure 4.4).

The received signal power irrespective of the path loss and shadowing must be greater than the WiMAX receiver sensitivity (R_{SS}). The signal to noise ratio at the transmitter for a 2x2 MIMO base station is

$$SNR = P_R + 102 + 10\log_{10}\left(\frac{F_S N_{used}}{N_{FFT}}\right) \tag{4.4}$$

Table 4.1 Parameters of Base Station [2]

	Standard BS	BS with 2x2 MIMO	BS with 2x2 MIMO and 2 element AAS
DL TX power	35 dB_m	35 dB_m	35 dB_m
DL TX antenna gain	16 dB_i	16 dB_i	16 dB_i
Other DL TX gain	0 dB	9 dB	15 dB
UL RX antenna gain	16 dB_i	16 dB_i	16 dB_i
Other UL RX gain	0 dB	3 dB	6 dB
UL RX noise figure	5 dB	5 dB	5 dB

Table 4.2 Parameters for Customer Premises Equipment.

	Portable CPE	Mobile CPE
UL TX power	27 dB_m	27 dB_m
UL TX antenna gain	6 dB_i	2 dB_i
Other UL TX gain	0 dB	0 dB
DL RX antenna gain	6 dB_i	2 dB_i
Other DL RX gain	0 dB	0 dB
DL RX noise figure	6 dB	6 dB

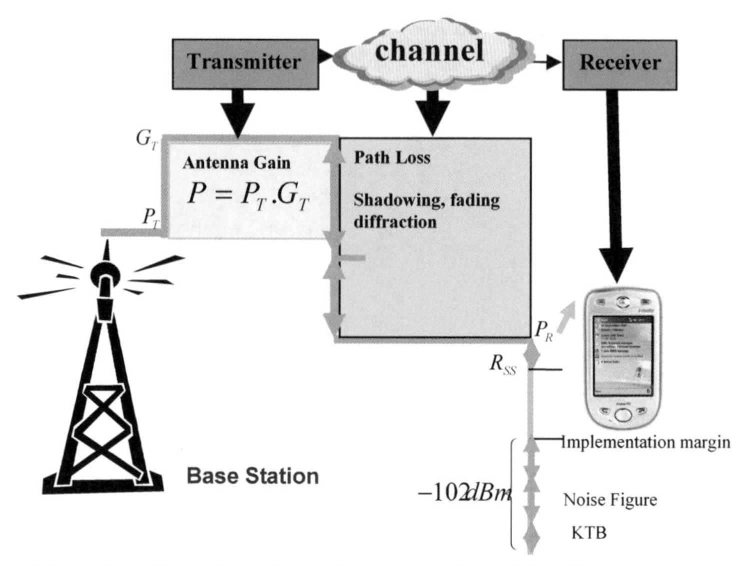

Figure 4.4 Received Signal Power Level after Passing Through the Channel.

This expression assumes that the implementation margin is 7dB and noise figure is 5dB. The effects of the channel and all other degradation sources must add up to the required minimum received signal power level or at least equal to or greater than the receiver sensitivity $P_R \geq R_{SS}$.

Signal degradation has always been a disturbing factor in telecommunications, and much more so in cellular communications. There are three principal sources of degradation in a cellular environment: noise (inter-modulation noise, additive white Gaussian noise (AWGN)), multiple access interference (MAI) and fading. Noise is mainly contributed by the environment and the equipment in use. MAI results from sharing bandwidth and communication channel. Multiple access interference includes inter-cell interference, intra-cell interference, co-channel interference and adjacent channel interference. Inter cell interference exists between two or more cells because of shared frequencies and frequency reuse. Intra-cell interference exists within a cell. Co-channel interference occurs between two users or stations using the same channel. The present chapter is dedicated to the third type of impairment in cellular systems - *fading*.

Fading is of two forms, large-scale and small-scale fading. Large-scale fading is mostly mean signal attenuation as a function of distance and signal variation around its mean value. Small-scale fading is of two forms, time spreading (which is composed of flat fading and frequency selective fading) and time variance of channel (which is composed of fast fading and slow fading). Figure 4.5 categorises these degradation sources.

Ideally, we would like communication signals to propagate (travel) without obstructions or disturbances. This can only happen in free space. Unfortunately, free-space is not a practical proposition in telecommunication and normal channels reflect, scatter and diffract radio waves. Scattering, reflection and diffraction of waves affect propagation of microwaves in many ways. Also multipath propagation, Doppler spread and coherence time, delay spread and coherence bandwidth all conspire against the transmitted signal. These effects reduce the system signal to noise ratio, and could lead to incoherent calls, dropped calls, and noisy channels.

Reflection occurs when the *path of a wave is obstructed* by a smooth surface. As the wave impinges on the surface, its direction of travel is changed. Usually the dimensions of the surface where reflection takes place are large relative to the wavelength of the wave.

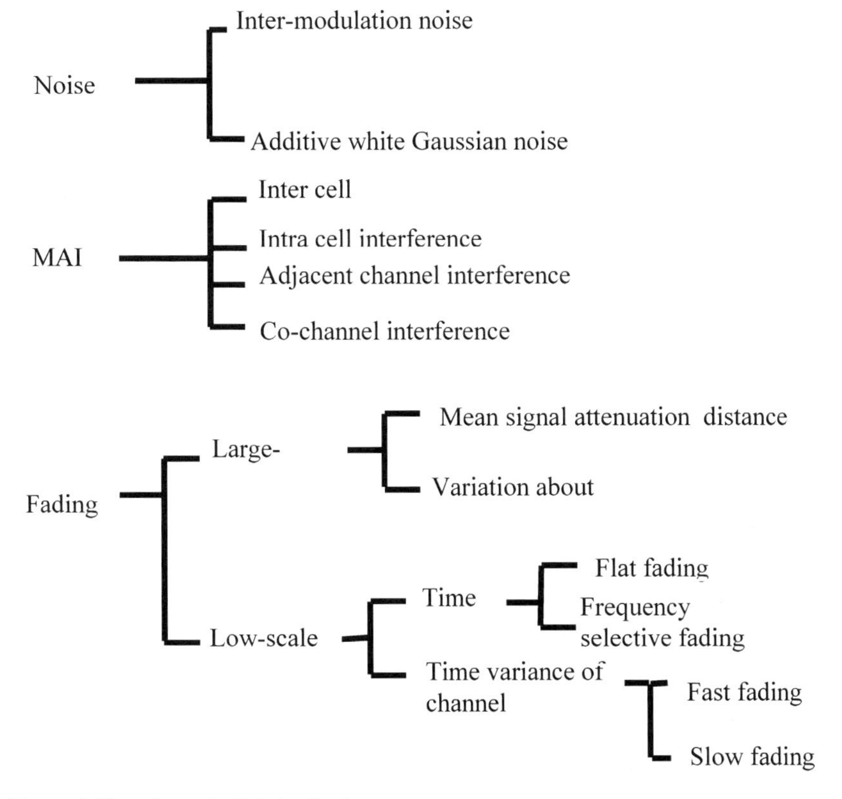

Figure 4.5 Impairmnt in Cellular Environment.

When an object with large dimensions relative to the wavelength of a signal *blocks the path of a wave, diffraction* takes place. The object's sharp edges lead to diffraction or shadowing. The obstacle causes a change in the forward direction of the wave either away or towards the receiver.

Scattering however occurs when a wave impinges on an object with dimensions comparable to the wavelength of the signal. The object causes the wave to spread out or scatter to different directions. As the wave scatters, the energy in the original wave is distributed into many components so that the components reaching the receiver carry a lot less power than the intended transmission. In urban applications street lights, signs and foliage are the worst culprits. Scattering is a loss of useful signal strength because the receiver is unable to collect all its energy as it is tuned to a narrow view, and collects signals only in the narrow view.

Microwave signals may therefore arrive at a receiver from many paths (multipath). The line of sight (LOS) signal is usually the preferred one in most applications as it travels the shortest path and thus arrives at a receiver with the strongest amplitude or power.

4.3.1 Multipath Propagation

Multipath propagation causes large variations in signal strength. The major effects are threefold:

- Time variations due to multipath delays;
- Random frequency modulation due to *Doppler shifts* from different multipath signals;
- Random changes in signal strength over short time periods;

In practice, multipath delays lead to time dispersion or 'fading effects' which is small-scale in nature.

A mobile communication multipath channel can be modelled as a linear time-varying filter with impulse response $h(t,\tau)$, where τ is the multipath delay in the channel for a fixed time t. In practice a low-pass model of the channel is easier to use than the actual complex model. The model allows a description of all other signal components relative to the signal which arrives first with delay $\tau_0 = 0$. All the delayed signals arriving latter than this are discretised in terms of delays in N equally spaced time intervals of width $\Delta\tau$. All the multipath wave components in bin i are represented in terms of one component with delay $\tau_i = i\,\Delta\tau$.

Multipath fading is measured by using channel sounding through direct pulse measurements, spread spectrum sliding correlator or swept-frequency channel analyser. These techniques provide time dispersion parameters (mean excess delay, maximum excess delay at some given signal to noise ratio and rms delay spread), coherence bandwidth and Doppler spread or spectral broadening.

We have shown in Chapter 2 for a two ray propagation model that the total received electric field intensity at the receiver is approximately

$$E_{total} = E_d + E_r \approx E_d\left(1 + \rho e^{j\psi} e^{j\phi}\right) \quad (4.5)$$

Where ψ the reflection angle and ϕ is phase change due to path length difference between the direct line of sight ray and the ground reflected ray. In a multipath situation, there will be N different paths contributing to the total electric field. Hence if we assume that for each path reflection of the field takes place once, then the total electric field at the receiver is given by the expression

$$E_{total} = E_d + E_d \sum_{k=1}^{N} \rho_k e^{j(\psi_k + \phi_k)} \quad (4.6)$$

Where

The reflection coefficient for path k is ρ_k

The phase change due to reflection in path k is ψ_k

The phase change proportional to the path length difference Δ_k is $\phi_k = \frac{2\pi\Delta_k}{\lambda}$

4.3.2 Effects of the Morphology

The channel in general is highly dynamic and variable due to changing contents of the environment and their variations. In windy wooded areas with tall trees, the effect of the wind is to vary the channel as the trees sway in the wind.

Wind causes trees to move and cause dynamic fading. Assume that the average speed of the swaying trees is v. If the trees move on average by distance d, the fading time due to this mobility is given by the expression:

$$\tau_t = \frac{d}{\upsilon} \quad (4.7a)$$

This time variation of tree branches and their leaves from the mean results to corresponding phase changes and fading.

Correspondingly, a mobile receiver in a car also experiences fading due also to the dynamics of the car. Vehicles reflect RF signals and this is worsened by their speed. This is estimated with a term

$$\tau_v = \frac{d}{\upsilon} \tag{4.7b}$$

Where d is the length of the vehicle and v is the speed of the vehicle.

4.3.3 Doppler Effect

A moving object experiences Doppler frequency shifts. Mobile phone signals in fast moving cars, air plane and ships also experience Doppler shifts. The measured frequency increases as the mobile moves towards a base station. As it moves away from the base station, the frequency decreases. Doppler effects therefore leads to a variation of the signal bandwidth and is governed by the expression:

$$f_d = \frac{v}{\lambda}\cos\theta \tag{4.8}$$

where θ is the angle made by the signal path to the base station and the ground plane as shown in Figure 4.6. In this expression, velocity v is measured in meters/second, wavelength λ in meters and frequency f_d in Hertz. In the time domain, Doppler frequency shift leads to coherence time. *Coherence time* is the time duration over which two signals have strong potential for amplitude correlation. Coherence time can be approximated by the expression:

$$T_c = \sqrt{\frac{9}{16\pi f_d^2}} \tag{4.9}$$

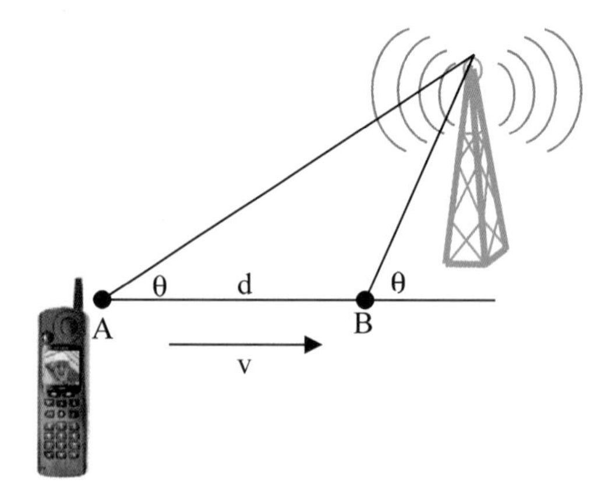

Figure4.6 Doppler Effect In Cellular Communications.

where f_d is the maximum Doppler shift, which occurs when $\theta = 0$ degrees. To avoid distortion due to motion in the channel, the symbol rate must be greater than the inverse of coherence time ($1/T_c$).

4.3.4 Delay Spread and Coherence Bandwidth

Multipath delay causes the signal to appear noise-like in amplitude. We can compute its statistical averages and parameters. The standard deviation of the distribution of multipath signal amplitudes is called delay spread, σ_τ. Delay spread varies with the terrain with typical values for rural, urban and suburban areas: $\sigma_t \approx 0.2\mu s\,(rural);\ \ \sigma_t \approx 0.5\mu s\,(suburban); \sigma_t \approx 3.0\mu s\,(urban)$. Since the signal bandwidth varies due to delays, what then is the best measure of the bandwidth in practice? To answer this question, we use coherence bandwidth. It is defined as the statistical measure of the range of frequencies over which the channel is considered constant or flat. It is the bandwidth over which two frequencies have a strong potential for amplitude correlation. Coherence bandwidth estimated for both strong and weak correlation are:

$B_c \approx \frac{0.02}{\sigma_t}$ for correlation greater than 0.9 and $B_c \approx \frac{0.2}{\sigma_t}$ for correlation greater than 0.5. Table 4.3 shows typical rms delay spreads for various types of terrain.

4.3.5 Categories of Fading

There are two major categories of fading, small-scale and large-scale fading. Large-scale fading is dependent on the distance between the transmitter and receiver. It is generally called path loss or 'large-scale path loss', *'log-normal* fading' or 'shadowing'. Small-scale fading is caused by the superposition of multipath signals, the speed of the receiver or transmitter and the bandwidth of the transmitted signal. Therefore, small-scale fading is a result of constructive and

Table 4.3 Delay Spread

Delay spread figures at 900 MHz	Delay in microseconds
Urban	1.3
Urban (worst-case)	10 – 25
Suburban (typical)	0.2 - 0.31
Suburban (extreme)	1.96 - 2.11
Indoor (maximum)	0.27
Delay Spread at 1900 MHz	
Buildings (average)	0.07 - 0.094
Buildings (worst – case)	1.47

destructive interference between several versions of the same signal causing attenuation of the average signal power. This type of fading is over a fraction of a wavelength and of the order of 20 to 30 dB. Small-scale fading is also known by other names such as 'fading', 'multipath' and '*Rayleigh*' fading. Rayleigh fading is a statistical variation of the received envelope of a flat fading signal. Multipath fading manifests as time spreading of the signal and a time variant behaviour. A time variant behaviour of the channel may be due to motion of the mobile and or changing environment (movement of foliage, reflectors and scatters).

If the impulse response of the mobile radio channel is $h(\tau, t)$ and time invariant and if it has zero mean, then the envelope of the impulse response has a Rayleigh distribution given by the expression:

$$p(r) = \frac{r}{\sigma^2} \exp\left(-\frac{r^2}{2\sigma^2}\right) \tag{4.10}$$

where σ^2 is the total power in the multipath signal. If however the impulse response has a non zero mean, then there is a component of the direct path (line of sight or specular component) signal in the channel and the magnitude of the impulse response has a Ricean distribution. (*Rice fading* is therefore the combination of Rayleigh fading with a significant non-fading (line of sight) component). Ricean distribution is given by the expression

$$p(r) = \frac{r}{\sigma^2} \exp\left(-\frac{r^2 + s^2}{2\sigma^2}\right) I_0\left(\frac{rs}{\sigma^2}\right) \tag{4.11}$$

The power of the line of sight signal is s^2 and I_0 is a Bessel function of the first kind.

The distance between either the dips or troughs in Rayleigh fading is of the order of half a wavelength. Small-scale fading occurs as either of four types:

- frequency selective fading in which the bandwidth of the signal is greater than the coherence bandwidth and the delay spread is greater than the symbol rate; Signals at some frequency components experience more fading than others
- flat fading when the bandwidth of the signal is less than the coherence bandwidth and the delay spread is less than the symbol rate
- fast fading when the Doppler spread is high and the coherence time is less than the symbol period and
- slow fading with a low Doppler spread and coherence time is greater than the symbol period

Table 4.4 Fading Effects.

Type of fading	Frequency Effects	Time Effects
Effects of Multipath Delay Spread		
Frequency Selective fading	BW of signal > coherence BW	Delay spread > symbol period
Flat fading	BW of signal < coherence BW	Delay spread < symbol period
Effects of Doppler Spread		
Slow fading	Low Doppler spread	Coherence time > symbol period
Fast fading	High Doppler spread	Coherence time < symbol period

The first two fading are caused by multipath delay spread and the last two by Doppler spread. Table 4.4 is a summary of the conditions that exist with each type of fading.

Small-scale fading may be corrected by using adaptive equalisers or through the use of modulation techniques such as spread spectrum and error correction.

References

[1] Josip Milanovic, Snjezana Rimac-Drlje, Krunoslav Bejuk, "Comparison of Propagation Models Accuracy for WiMAX on 3.5GHz" IEEE International Conference, 2007, pp:111-114

[2] Mobile WiMAX Group, "Coverage of mobile WiMAX", pp. 1 - 18

[3] Rana Ezzine, Ala Al Fuqaha, Rafik Braham, Abdelfettah Belghith. "A New Generic Model for Signal Propagation in WiFi and WiMAX Environment". *Wireless Days*, 2008, pp:1-5

[4] Tamaz Javornik, Gorazd Kandus, Andrej Hrovat, Igor Ozimek. *Software in Telecommunication and Computer Networks*, 2006, pp: 71-75

[5] Yan Zhang. WiMAX Network Planning and Optimization. USA: CRC Press, 2009

[6] Johnson I Agbinya. *IP Communications and Services for NGN*. New York: Taylor and Francis, 2009

[7] Kejie Lu, Yi Qian, Hsiao-Hwa Chen, Shengli Fu. " WiMAx Networks: From Access to Service Platform". *IEEE Computer Society,* 22.,pp.38-45, May/June 2008

[8] Johnson I Agbinya, "Design Consideration of Mohots and Wireless Chain Networks", Wireless Personal Communication", © Springer 2006, Vol. 40, pp. 91 -106

[9] David J. Y. Lee and William C. Y. Lee, "Enhanced Lee Model from Rough

Terrain Data Sampling Aspect",

[10] Lee's Model, Appendix VI, Propagation Special Issue, IEEE Transactions on Vehicular Technology, Vol. 37, No. 1, Feb. 1988, pp. 68 – 70.

[11] William **C. Y.** Lee and David **J. Y.** Lee, "The propagation characteristics in a cell coverage area", in Proc. IEEE 47th Conference on Vehicular Technology, 1997, Vol. 3, pp. 2238 – 2242.

[12] Propagation Model Optimisation (Sample Report), Envision Wireless, June 21, 2005, www.evwi.com

[13] J. J. Egli, "Radio Propagation above 40 Mc Over Irregular Terrain", *Proc. IRE*, 1957, pp.1383-1391

Chapter 5.

OFDMA and SC-FDMA BER Performance for Rayleigh and Nakagami-*m* Fading Channels

Juan Jesús Sánchez-Sánchez[1], Unai Fernández-Plazaola[2] and Mari Carmen Aguayo-Torres[2]

[1]*Ericsson Málaga*
[2]*Departamento de Ingeniería de Comunicaciones, Universidad de Málaga*
juan.jesus.sanchez.sanchez@ericsson.com, unai@ic.uma.es, aguayo@ic.uma.es

5.1 INTRODUCTION

In this chapter we present the performance analysis, in terms of BER (Bit Error Rate), for OFDMA (Orthogonal Frequency-Division Multiple Access) and SC-FDMA (Single Carrier Frequency-Division Multiple Access) transmissions over Rayleigh and Nakagami-*m* fading channels. With this purpose, we first provide the reader with a brief description of those transmission schemes that were chosen for downlink (OFDMA) and uplink (SC-FDMA) transmission in the Long Term Evolution (LTE) for the Universal Mobile Telecommunications System (UMTS). We also discuss the main features of the Nakagami-*m* fading channels and BER analysis technique applied in this chapter. Concretely, we address the case in which zero-forcing frequency domain equalization (ZF-FDE) is applied to compensate the effects of the channel on the received signal. Our analysis is focused on the effects of equalization on the noise term before the decision stage. Obtained results allow us to derive closed-form BER expressions as we showed in [1] and [2].

4G Wireless Communication Networks: Design, Planning and Applications, 91-116.

5.2 OVERVIEW OF OFDMA AND SC-FDMA

In OFDM (Orthogonal Frequency-Division Multiplexing), a large number of orthogonal sub-carriers are used to transmit information from several parallel streams modulated with a digital modulation scheme [3]. If these data streams belong to different terminals or users, OFDM becomes OFDMA in which different data signals are transmitted through a common physical media that is divided into frequency resources units. Both OFDM and OFDMA suffer from power distortion that may be particularly troublesome in uplink transmissions where excessive complexity in user terminal is an issue. With the purpose of overcoming that limitation in 3GPP LTE [4], SC-FDMA was chosen as the uplink multiple access scheme due to its lower power distortion [5].

5.2.1 Orthogonal Frequency-Division Multiplexing

In OFDM, the available bandwidth is divided into sub-carriers that are orthogonal in the sense that the peak of one sub-carrier coincides with the nulls of the other sub-carriers, thereby avoiding the use of frequency guard bands and increasing spectral efficiency. OFDM multiplexes the data on these orthogonal sub-carriers and transmits them in parallel. This allows dividing the high-speed digital signal to be transmitted into several slower signals that are sent in parallel through separated narrower frequency bands.

The block diagram for an OFDM system (transmitter and receiver) is depicted in Figure 5.1. The sequence of bits to be transmitted is mapped into a sequence of complex symbols according to the modulation scheme used, typically Quadrature Amplitude Modulation (QAM) or Phase Shift Keying (PSK). These complex symbols are allocated in orthogonal sub-carriers and transformed by means of an Inverse Discrete Fourier Transform of size M (M-IDFT) that produces an OFDM symbol in the time domain. A Cyclic Prefix (CP) of length

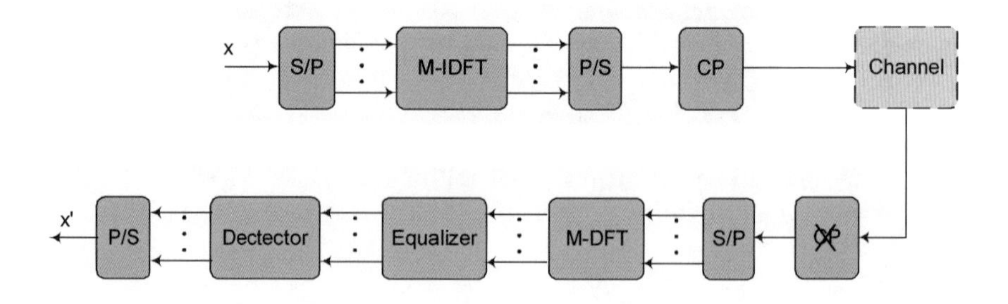

Figure 5.1 Transmitter and receiver for OFDM.

greater than the channel response is added as a guard period, to reduce the temporal dispersion and to eliminate the Inter-Symbolic Interference (ISI) [6]. Additionally, the prefix also helps to preserve the orthogonality among sub-carriers, thereby avoiding Inter-Carrier Interference (ICI) as well. The addition of the cyclic prefix also transforms the time domain convolution between the OFDM symbols and the channel response in a circular convolution. In the frequency domain, this circular convolution becomes a point-wise multiplication between the complex symbol allocated in each sub-carrier and the corresponding channel frequency response. That makes possible to perform the equalization in the frequency domain.

The main advantage of OFDM is its behavior against bad channel conditions, such as the fading caused by multi-path propagation, without using complex equalization filters. Normally, as the bandwidth of the transmitted signal is less than the coherence bandwidth, the channel response in each carrier can be considered flat. Thus, the channel can be modeled as a set of narrowband fading channels, one for each single sub-carrier [3]. Besides, thanks to its inherent immunity to multipath effects, OFDM provides multipath and interference tolerance in non-Line Of Sight (non-LOS) conditions.

5.2.2 Orthogonal Frequency-Division Multiple Access

In OFDMA all the available sub-carriers are grouped into different sub-channels that are assigned to distinct users. It takes advantage of the orthogonality among sub-carriers to avoid interference between users, thereby achieving greater flexibility and efficiency in the allocation of system resources. Each sub-channel may be compounded of adjacent localized sub-carriers (Localized FDMA, LFDMA) or by sub-carriers distributed across the total bandwidth (Distributed FDMA, DFDMA). A particular case of DFDMA is Interleaved FDMA (IFDMA) in which the sub-carriers are equally spaced over the entire system bandwidth. In Figure 5.2 four users are multiplexed following the localized scheme and the

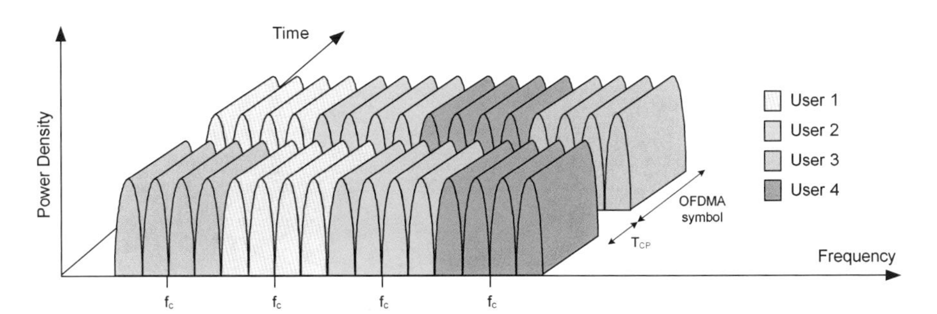

Figure 5.2 User allocation example in OFDMA (LFDMA).

entire system bandwidth is divided into four different sub-bands.

Due to the fact that it combines scalability, multi-path robustness, and Multiple-Input Multiple-Output (MIMO) compatibility [3], OFDMA has been adopted in LTE [4] specifications. However, the OFDMA (and OFDM) waveform exhibits very pronounced envelope fluctuations, resulting in a high Peak to Average Power Ratio (PAPR) [5]. Because of this drawback, OFDMA requires expensive linear power amplifiers, which in turn increases the cost of the terminal and reduces its autonomy, as the battery drains faster.

5.2.3 Single Carrier with Frequency Domain Equalization

A SC-FDE system shares some common elements with OFDM as depicted in Figure 5.3. These similarities allow the coexistence of SC-FDE and OFDM modems, that is, an equipment may be able to operate with any of these transmission techniques. In an SC-FDE system, information bits are grouped and mapped into a complex symbol belonging to a given complex constellation. The sequence of modulated symbols is divided into data blocks and transmitted sequentially. Each of those data blocks is cyclically extended with a copy of the last part of the block (i.e., a CP), which is transmitted as a guard interval. The insertion of the CP has similar advantages to those already described in OFDM. First, if the length of the CP is longer than the length of the channel impulse response, the Inter-Block Interference (IBI) due to multipath propagation is avoided. Second, the transmitted data propagating through the channel can be modeled as a circular convolution between the channel impulse response and the transmitted data block. This becomes a point-wise multiplication of the DFT samples in the frequency domain. Thus, channel distortion can be removed dividing the DFT of the received signal by the DFT of the channel impulse [5].

This frequency domain equalization is performed at the receiver after

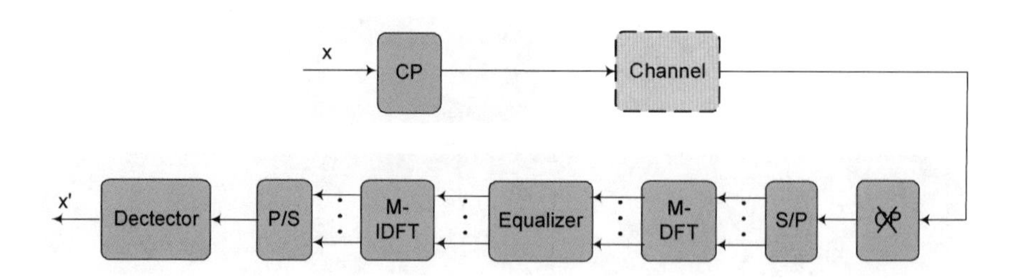

Figure 5.3: Transmitter and receiver for SC-FDE.

transforming the received signal to the frequency domain by applying a DFT. After the equalization, an IDFT transforms the single carrier signal back to the time domain. Unlike OFDM, where a separate detector is employed for each subcarrier, a single detector recovers the original modulation symbols in this case.

In the overall, SC-FDE performance is similar to that of OFDM with essentially the same complexity, even for long channel delay. However, there are several advantages when it is compared to OFDM [5]:

- it is less sensitive to nonlinear distortion and hence, it allows the use of low-cost power amplifiers [7],
- it offers a greater robustness against spectral nulls and,
- it is less sensitive to carrier frequency offset.

On the other hand, its main disadvantage with respect to OFDM is that neither channel-adaptive subcarrier bit nor power loading are possible with this transmission technique.

5.3 SINGLE CARRIER FREQUENCY DIVISION MULTIPLE ACCESS

SC-FDMA is an extension of SC-FDE that allows multiple access with a complexity similar to that of OFDMA. Both technologies use almost the same transceiver blocks, being the DFT pre-coding and inverse pre-coding stages, which are added in SC-FDMA at transmitter and receiver ends, the main difference between them (see Figure 5.4). Thanks to those blocks, SC-FDMA has better capabilities in terms of envelope fluctuations of the transmitted signal and, therefore, its PAPR is lower (up to 2 dB) than in OFDMA [5]. That leads to greater efficiency in power consumption, a desirable feature in user equipments. in addition, SC-FDMA complexity focuses on the receiver-end, hence, it is an appropriate technology for uplink transmission since complexity at the base station is not an issue. Due to these features, SC-FDMA was chosen as the medium access technique for 3GPP LTE uplink [4].

After computing the DFT of the input data, the result is either distributed over the entire bandwidth (DFDMA) or placed in consecutive sub-carriers (LFDMA). In both cases, unoccupied carriers are set to zero. When the sub-carriers are equally spaced over the entire system bandwidth, it is called interleaved mode (IFDMA) [5]. Regardless of the manner in which the sub-carriers are mapped, the result is a sequence X_l (l=0,1,2,M-1) of M complex amplitudes that is transformed by means of an M-IDFT to a signal in the time domain which is transmitted sequentially. As in SC-FDE, a cyclic prefix is also added as a guard interval between blocks to avoid IBI caused by multi-path propagation and to make possible frequency domain equalization.

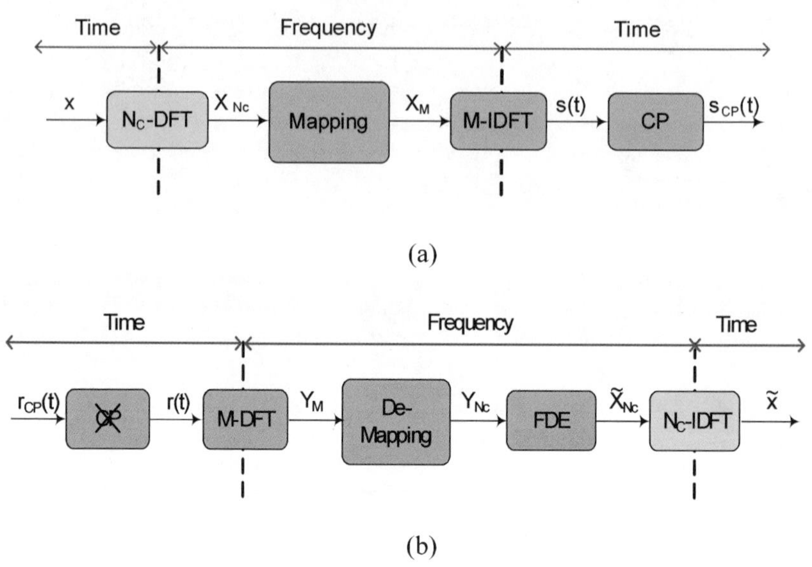

Figure 5.4 SC-FDMA transmitter and receiver schemes (a) Transmitter (b) Receiver.

In Figure 5.5 four users are multiplexed following the localized scheme, thereby the entire system bandwidth is used by the users in different instants. Note that, in OFDMA, each data symbol modulates a single sub-carrier whereas in SC-FDMA it modulates the whole allocated wideband carrier. Thus, the modulated symbols are transmitted sequentially over air and the final transmitted signal is a single carrier one unlike OFDMA where the final signal is compounded of the superposition of the allocated sub-carriers. Both techniques transmit the same amount of data symbols in the same time period using the same bandwidth. However SC-FDMA symbols are transmitted sequentially over a single carrier

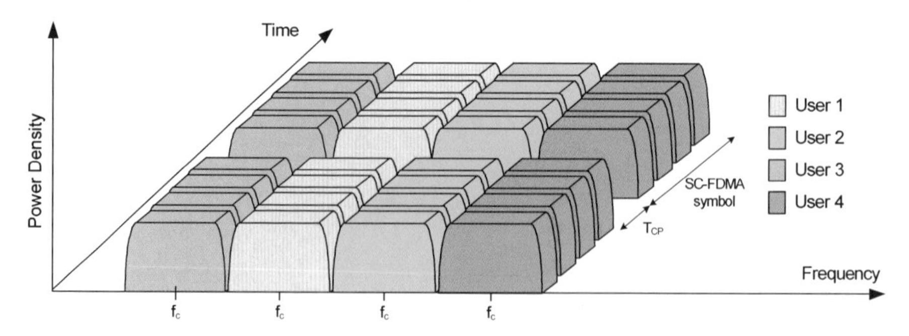

Figure 5. 5 User allocation example in SC-FDMA (LFDMA).

as opposed to the parallel transmission of OFDM/OFDMA over multiple carriers. Also, the users are orthogonally multiplexed and de-multiplexed in the frequency domain, which gives SC-FDMA the aspect of FDMA [5].

5.4 BER ANALYSIS TECHNIQUES DEVICES

In this chapter, the BER analysis is performed by applying the framework proposed in [10] that allows obtaining a closed-form expression for the BER for multilevel QAM (M-QAM). This approach is based in the assumption of independent bit-mapping for in-phase and quadrature components (e.g., Gray mapping). The resulting general expression can be simplified if the complex noise affecting the symbol before the decision stage, the enhanced noise, is circularly symmetric as, in that case, only one of its marginal distributions is needed. In this case, we can use the following expression for the BER

$$BER = \sum_{n=1}^{L-1} w(n) I(n) \tag{5.1}$$

where $I(n)$ are the so-called Components of Error Probability (CEP), $w(n)$ are coefficients dependent on constellation mapping, $L=2$ for BPSK, and $L = \sqrt{M}$ for M-QAM. The CEPs can be expressed in function of the CDF $F_{\eta_r}()$ corresponding to the real part of the noise η affecting the symbol before the decision stage $\eta_r = \Re\{\eta\}$ as follows

$$I(n) = \Pr\{\Re\{\eta\} > (2n-1)d\} = 1 - F_{\eta_r}((2n-1)d) \tag{5.2}$$

where d is the minimum distance between each symbol and the decision boundary, e.g., $d = \sqrt{E_s}$ for BPSK and $d = \sqrt{\frac{3E_s}{2(M-1)}}$ for M-QAM, and the coefficients $w(n)$ are directly computed as in [10].

As stated before, the resulting closed-from expression is written as a function of the CDF of the real part of the noise affecting the symbol before the decision stage. Thus, in the following, we study the stochastic nature of this complex noise and provide functions to perform the evaluation of this distribution required in (2.2) for different fading channels when FDE is applied. Note that, we only consider circularly symmetric enhanced noise and, therefore, we only need to obtain one of its marginal distributions in order to apply (5.2).

5.4.1 Channel Model

In the system under analysis, the received signal is the sum of the line-of-sight (LOS) paths and all resolvable multipath components. In our study we consider only the multipath fading due to the constructive and destructive combination of randomly delayed, reflected, scattered, and diffracted signal components. This type of fading is relatively fast and is, therefore, responsible for the short-term signal variations. In the following sections we analyze two different scenarios in which the fast fading follows different distributions, that is, Rayleigh and Nakagami-*m* distributions.

Rayleigh distribution is frequently used to model multipath fading when there is no direct LOS path [9]. That distribution is obtained by using mathematics to capture the underlying physical properties of the channel models. However, some experimental data does not fit well into that distribution. Thus, the more general Nakagami-*m* fading distribution was developed [11]; its parameters can be adjusted to fit a variety of empirical measurements [12]. Usually, Nakagami-*m* fading occurs for multi-path scattering with relatively large delay-time spreads and with different clusters of reflected waves. The parameter *m* allows modeling signal fading conditions that range from severe to moderate or light fading. Note that, in frequency-selective Nakagami-*m* fading channels, the magnitude of the channel frequency response can be approximated with Nakagami-*m* distributed random variables [13].

In the following, the channel frequency response for each system sub-carrier is represented by the $M \times M$ diagonal matrix $\mathbf{H}$ whose entries are assumed to be i.i.d. complex circularly symmetric random variables. In this point, we must keep in mind that the Fourier transform is a linear transformation and therefore, it preserves normality [14]. Thus, as the underlying random variables for the considered impulse response are Gaussian, the respective underlying random variables for the channel frequency response follow the same distribution. For instance, for Rayleigh fading channels, the channel frequency response has as underlying variable a zero-mean complex Gaussians with unitary-variance.

Kang et al. show in [15] how, for frequency-selective Nakagami-*m* fading channels, the magnitudes of the channel frequency response can be also approximated as Nakagami-*m* distributed random variables. Their fading and mean power parameters can be expressed as explicit functions of the fading and mean power parameters of the channel impulse response.

5.4.2 OFDM BER Analysis

The purpose of this section is to study the statistical distribution of the enhanced noise after linear FDE in an OFDM system. In the following sections we find the density for the resulting noise term after equalization for Rayleigh and Nakagami-

m fast fading channels. Those densities will serve us to derive an expression for the BER in an OFDM un-coded transmission.

5.4.3 System Model

The block diagram for a simple OFDM system compounded by transmitter and receiver-ends is depicted in Figure 5.1. We showed previously how the available bandwidth in OFDM is divided into orthogonal sub-carriers. Thus, a sequence of modulated complex symbols $\mathbf{x}$ is mapped in parallel into the allocated sub-carriers whereas the remaining sub-carriers are forced to zero. Thanks to the dual IDFT and DFT operations at transmitter and receiver-ends, and to the addition of the cyclic prefix, the received signal in the frequency domain can be expressed as a version of the transmitted one transformed by channel response and contaminated with thermal noise. In a more formal way,

$$\mathbf{r} = \mathbf{H}\mathbf{x} + \eta \tag{5.3}$$

where $\mathbf{H}$ is the channel frequency response diagonal matrix and η is a vector with the complex AWGN component for each sub-carrier η_k. The complex random variable corresponding to the channel frequency response for a given sub-carrier k, h_k, is assumed to have zero mean and to be normalized in power, whereas each noise term has zero mean and variances N_0. Both η_k, and h_k are circularly symmetric random variables.

Throughout this dissertation, perfect synchronization and ideal estimation of the channel frequency response are assumed at the receiver end.

5.4.4 Noise Analysis

After applying zero-forcing equalization and under the assumption of ideal estimation of the channel frequency response, the expression for the received symbol can be derived from (5.3) as

$$\hat{\mathbf{x}} = \mathbf{x} + (\mathbf{H}^H\mathbf{H})^{-1}\mathbf{H}^H\eta \tag{5.4}$$

Thus, the expression for the k-th received symbol is

$$\hat{x}_k = x_k + \frac{\eta_k}{h_k} = x_k + \hat{\eta}_k \tag{5.5}$$

where h_k is the channel frequency response at sub-carrier k.

The enhanced noise term $\hat{\eta}_k$ is the ratio of two i.i.d. complex circularly symmetric random variables; one of them corresponds to the Additive White Gaussian Noise (AWGN), at the receiver and the other to the channel frequency response. This second random variable depends on the nature of the aforementioned multipath fading.

In the following, we derive the PDF (Probability Density Function) corresponding to those quotients, that is, the Rayleigh/Rayleigh and the Rayleigh/Nakagami-*m* ratios. Note that, for a given subcarrier, the *k* index will be suppressed for the sake of simplicity.

5.4.4.1 Enhanced Noise for Rayleigh Fading Channels

In this case, the enhanced noise term $\hat{\eta}$ is the ratio of two i.i.d. circularly symmetric Gaussian random variables (i.e., the noise at the receiver and the channel frequency response for the considered sub-carrier). Both variables have zero-mean but different variances.

The ratio or quotient distribution is the distribution of the ratio of random variables which have known distributions. Translated into densities, it can be defined as follows.

Proposition 1

Let *X* and *Y* be continuous random variables with a joint density $f_{X,Y}(x,y)$. Let $V = X/Y$. Then the density of V is given by

$$f_v(v) = \int_{-\infty}^{\infty} f_{X,Y}(vy, y)|y|dy, \tag{5.6}$$

Note that we first apply a transformation based on a simple change of variable and then, we calculate the marginal distribution.

In the case under study, the ratio can be derived formally as follows.

Proposition 2

Let *X* and *Y* be two independent complex Gaussian random variables with zero-mean and variance σ_X^2 and σ_Y^2 respectively. Let R_X and R_Y be the random variables for their respective modulus, both of them following a Rayleigh distribution, and let θ_X and θ_Y be random variables with a uniform distribution in $[0, 2\pi]$ for their respective phases. Let R_X be the ratio X/Y and let R_Z be its modulus. Then, the density of R_X is calculated as

$$f_{R_Z}(r)=\frac{2\sigma_X^2\sigma_Y^2 r}{\left(\sigma_X^2+\sigma_Y^2 r^2\right)^2} \qquad r\in\mathbb{R}, r\geq 0 \tag{5.7}$$

As θ_Z is a uniform random variable defined in $[0,2\pi]$ and R_Z and θ_Z are independent random variables, the density of the ratio $Z = X / Y$ yields

$$f_Z(r,\theta)=\frac{1}{2\pi}\frac{2\sigma_X^2\sigma_Y^2 r}{\left(\sigma_X^2+\sigma_Y^2 r^2\right)^2} \qquad r\in\mathbb{R},\ r\geq 0 \tag{5.8}$$

The variables X and Y are circularly symmetric and they belong to the class of elliptical distributions [17]. Its ratio Z is also circularly symmetric and its modulus and phase are independent. In this case, as the channel frequency response is normalized $\sigma_Y^2=1$ and $\sigma_X^2=\sigma^2=N_0/E_S$ where E_S and N_0 are the signal and noise powers respectively.

Hence, the distribution of the enhanced noise $\hat{\eta}$ can be written as

$$f_{\hat{\eta}}(r,\theta)=\frac{1}{\pi}\frac{\sigma^2 r}{\left(\sigma^2+r^2\right)^2} \qquad r\in\mathbb{R}, r\geq 0 \tag{5.9}$$

With a change of variables, equation (5.9) can be transformed into the bivariate joint PDF for the real and imaginary components of the resulting complex random variable

$$f_{\hat{\eta}_r,\hat{\eta}_i}(x,y)=\frac{1}{\pi}\frac{\sigma^2}{\left(\sigma^2+\left(x^2+y^2\right)\right)^2} \qquad x,y\in\mathbb{R} \tag{5.10}$$

where $\hat{\eta}_r$ and $\hat{\eta}_i$ are the real and imaginary components.

The expression in equation (5.10) belongs to the Pearson type VII family of distributions. In fact, it is the probability density function of a bivariate Student-t distribution with 2 degrees of freedom, mean $(0,0)$, scale matrix $\frac{\sigma^2}{2}I_2$, where I_2 is the identity matrix of size 2. The marginal distributions are also Student-t distributions and then, it is obtained that $\hat{\eta}_R \stackrel{d}{=} \hat{\eta}_I \sim t\left(0,\frac{\sigma^2}{2},2\right)$, and the PDF of the real and the imaginary components are equal and given by

$$f_{\hat{\eta}_r}(x) = f_{\hat{\eta}_i}(x) = \frac{1}{2}\frac{\sigma^2}{\left(\sigma^2 + x^2\right)^{3/2}} \quad x \in \mathbb{R} \tag{5.11}$$

It is a heavy-tailed distribution whose tails decay following a power law x^{-3} and it has no finite variance.

The CDF of $\hat{\eta}_r$ results in the following expression

$$F_{\hat{\eta}_r}(x) = \frac{1}{2} + \frac{x}{2\sqrt{\left(\sigma^2 + x^2\right)}} \quad x \in \mathbb{R} \tag{5.12}$$

Note that, from equation (2.12) with $\sigma^2 = N_0 / E_S$, it is possible to derive the well known analytical expression of BER for BPSK in OFDM over a Rayleigh fading channel.

$$p_e = F_{\hat{\eta}_r}(-1) = \frac{1}{2}\left(1 - \sqrt{\frac{E_s}{E_s + N_0}}\right). \tag{5.13}$$

In [15] we provide the derivation for the corresponding Characteristic Function (CHF) that results

$$\Psi_{\hat{\eta}_r}(\omega) = \sigma|\omega|K_1(\sigma|\omega|) \quad \omega \in \mathbb{R} \tag{5.14}$$

where $|\omega|$ means the absolute value of $|\omega|$ and $K_1(\,)$ is the modified first order Bessel function of the second kind [8].

5.4.4.2 Enhanced Noise for Nakagami-*m* fading channels

In this case we assume that the modulus of the complex random variable corresponding to the channel frequency response follows a Nakagami-*m* distribution [15].

Although, in general, it is not possible to assume the phase to be a random uniform distributed random variable, we know that, as in the previous analysis, the numerator X is a complex circular random variable, thereby rotation invariant [16]. That means that it preserves the circular symmetry when it is multiplied by $e^{\cdot\theta}$ for any given α. That implies that the resulting random variable Z is also a circularly symmetric random variable and, consequently, its phase Φ_Z follows a uniform phase distribution between 0 and 2π and it is independent from the modulus [17].

$$f_{R_Z,\Phi_Z}(r,\varphi)=\frac{1}{2\pi}\frac{m}{\sigma^2}\left(\frac{m}{\Omega}\right)^m r\left(\frac{r^2}{2\sigma^2}+\frac{m}{\Omega}\right)^{m-1} \quad r\in\mathbb{R} \text{ and } r\geq 0 \tag{5.15}$$

Assuming that the channel is normalized in power, i.e., $\Omega=1$, expression (5.15) can be particularized to obtain the density of the enhanced noise $\hat{\eta}$. Its joint distribution in polar coordinates yields

$$f_{R_{\hat{\eta}},\Phi_{\hat{\eta}}}(r,\varphi)=\frac{1}{2\pi}\frac{m^{m+1}}{\sigma^2}r\left(\frac{r^2}{2\sigma^2}+m\right)^{-m-1} \quad r\in\mathbb{R} \text{ and } r\geq 0 \tag{5.16}$$

where $\sigma=\frac{N_0}{E_s}$.

For $m=1$ the Nakagami-*m* distribution becomes a Rayleigh distribution.

The joint PDF in Cartesian coordinates is

$$f_{\hat{\eta}_r,\hat{\eta}_i}(x,y)=\frac{m^{m+1}}{2\pi\sigma^2}r\left(\frac{x^2+y^2}{2\sigma^2}+m\right)^{-m-1} \quad x,y\in\mathbb{R} \tag{5.17}$$

where $\hat{\eta}_r$ and $\hat{\eta}_i$ are the real and imaginary components of the enhanced noise. The resulting random variable follows a bivariate Pearson type VII distribution. An example of this circularly symmetric density is provided in Figure 5.6.

In this case it is possible to compute each corresponding marginal PDF

$$f_{\hat{\eta}_r}(x)=f_{\hat{\eta}_i}(x)=\sqrt{\frac{1}{\pi}}\frac{\Gamma\left(m+\frac{1}{2}\right)2^m m^m\sigma^{2m}}{\Gamma(m)\left(2m\sigma^2+x^2\right)^{m-\frac{1}{2}}} \quad x\in\mathbb{R} \tag{5.18}$$

where $\Gamma(n)$ is the Gamma function [8].

For this marginal PDF the mean is zero and the variance yields $\sigma^2_{\hat{\eta}_r}=\frac{m\sigma^2}{m-1}$ provided $m>1$. For $m\leq 1$ it has neither a finite variance nor higher order statistics and, therefore, has no MGF. Nonetheless, there is no restriction regarding the CHF which is derived as

$$\Psi_{\hat{\eta}_r}(\omega)=\frac{\sigma^m|\omega|^m}{2^{1-m/2}\Gamma(m)}m^{m/2}K_m\left(\sigma\sqrt{2m}|\omega|\right) \quad \omega\in\mathbb{R} \tag{5.19}$$

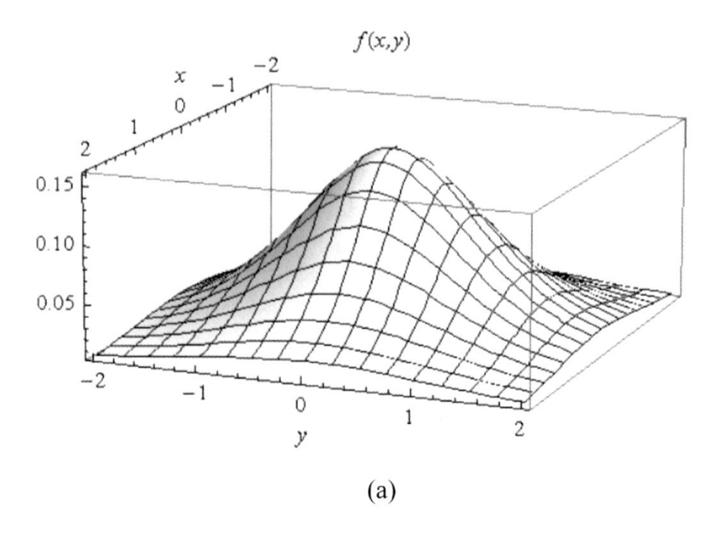

(a)

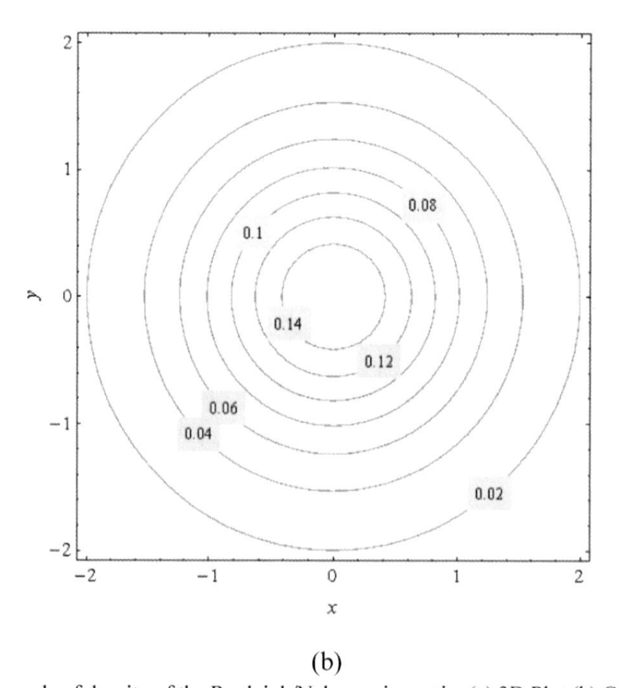

(b)

Figure 5.6 Example of density of the Rayleigh/Nakagami-m ratio. (a) 3D Plot (b) Contour Plot.

where $|\omega|$ is the absolute value of ω and $K_m()$ is the modified m-th order Bessel function of the second kind [8].

The corresponding CDF of $\hat{\eta}_r$ results in the following expression

$$F_{\hat{\eta}_r}(x) = \frac{(-1)^{-m}\Gamma\left(m+\frac{1}{2}\right)}{2\sqrt{\pi}\Gamma(m)} B_{-\frac{2m\sigma^2}{x^2}}\left(m, \tfrac{1}{2}-m\right) \quad x \in \mathbb{R} \tag{5.20}$$

where $B()$ is the incomplete Beta function [8].

5.4.4.3 BER Analysis for OFDM

As described in section 3, the expressions of the CDF for the enhanced noise after equalization, that is, of the noise term in the detection stage, allow us to compute the BER for the different channels considered. More specifically, as the enhanced noise considered is circularly symmetric in all the cases, we only need it marginal CDF $F_{\hat{\eta}_r}()$. Thus, plugging equation (5.12) into expression (5.2) we can compute the CEPs for the Rayleigh fading channel as follows;

$$BER = \sum_{n=1}^{L-1} w(n)\left(\frac{1}{2} - \frac{(2n-1)d}{2\sqrt{((2n-1)d)^2+\sigma^2}}\right). \tag{5.21}$$

In a similar fashion, it is possible to derive the expression for Nakagami-m fading channels using equation (5.20). The resulting BER expression is given by

$$BER = \sum_{n=1}^{L-1} w(n)\left(1 - \frac{(-1)^{-m}\Gamma\left(m+\frac{1}{2}\right)}{2\sqrt{\pi}\Gamma(m)}\right) B_{-\frac{2m\sigma^2}{((2n-1)d)^2}}\left(m, \tfrac{1}{2}-m\right) \tag{5.22}$$

5.5 SIMULATIONS AND NUMERICAL RESULTS

In this section, we validate the results obtained for the considered fading channels. With this purpose, BER values are obtained for SNR ranging from 0 to 30 dB and compared with values obtained by means of simulations. In order to simplify the validation process in the case of Nakagami fading channels, we directly generate channel frequency response whose magnitudes are distributed according to a Nakagami-m distribution with some integers and half-integer values of m. In Figure 5.7 we show the numerical evaluation of the BER for a BPSK

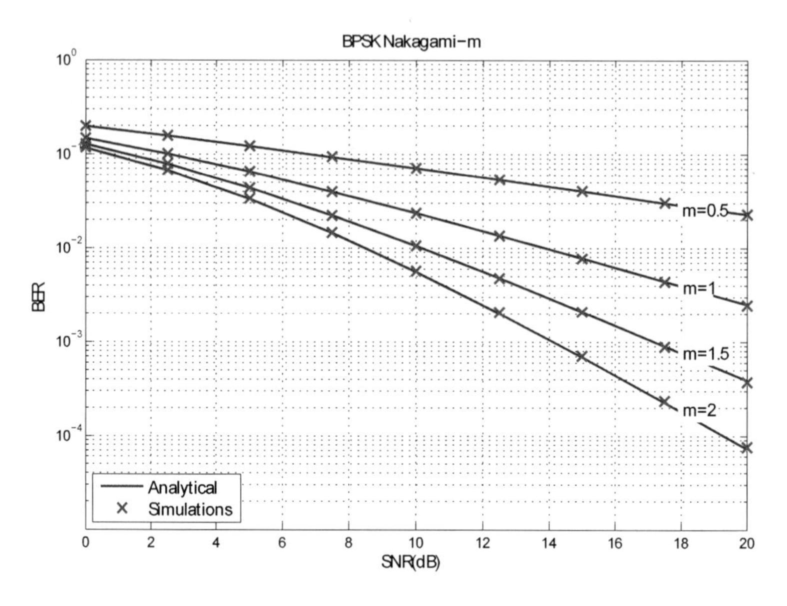

Figure 5.7 BER values for several Nakagami-m fading channels, including Rayleigh (m = 1), with BPSK.

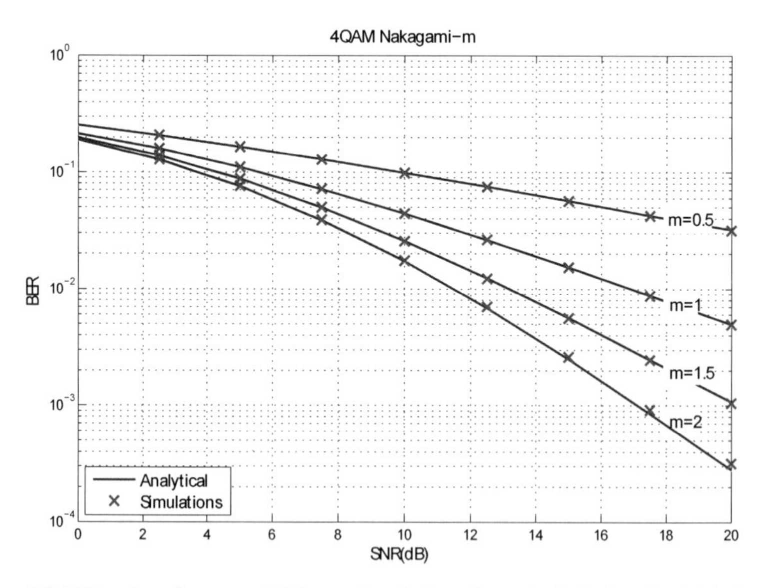

Figure 5.8 BER values for several Nakagami-m fading channels, including Rayleigh (m = 1), with 4QAM.

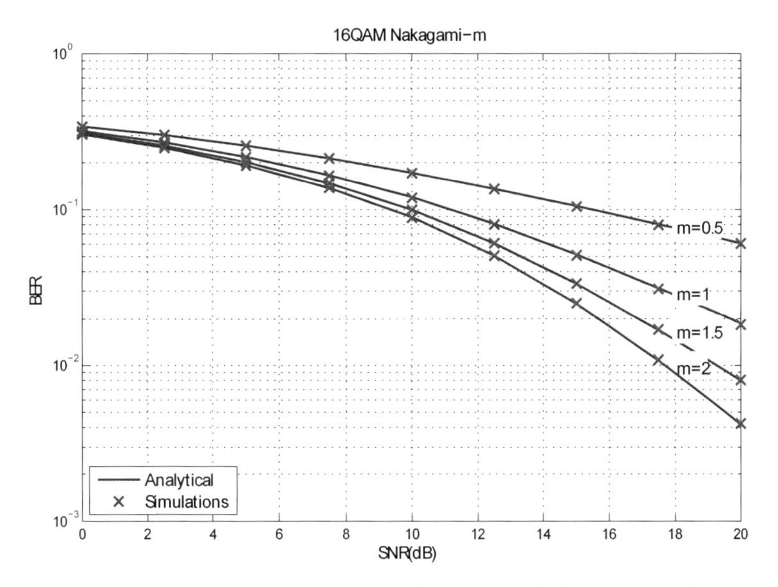

Figure 5.9 BER values for several Nakagami-m fading channels, including Rayleigh (m = 1), with 16QAM.

transmission for different values of m. Note that for $m=1$ the channel has a Rayleigh fading. If $m<1$ BER values worsen (i.e., it is worse than Rayleigh scenario) whereas for $m>1$ we obtain lower values of BER. A similar behavior can be observed in Figure 5.8 and 5.9 where results for 4QAM and 16QAM are shown.

5.6 SC-FDMA BER ANALYSIS

5.6.1 System Model

The transmission process in SC-FDMA is similar to that in OFDMA as depicted in Figure 5.4. For a given user, a sequence of transmitted bits is mapped into a constellation of complex symbols (e.g., BPSK or M-QAM). The resulting complex sequence $\mathbf{x}$ of length N_c is pre-coded by means of a DFT operation before being mapped onto the subset of N_c allocated sub-carriers. For a transmitted complex information vector of N_c symbols $\mathbf{x}$, the pre-coded complex symbol $\mathbf{X}_{\mathrm{N_c}}$ is obtained by multiplying by the unitary Fourier matrix $\mathbf{F}$ to perform the aforementioned N_c-DFT operation. Each element in $\mathbf{F}$ is defined as $F_{j,k} = exp\{\frac{2\pi i}{N_c} jk\} / \sqrt{N_c}$ and, therefore, $\mathbf{F}^{\mathrm{H}}\mathbf{F}=\mathbf{I}$.

The pre-coded sequence, represented by the column vector $\mathbf{X}_{\mathrm{N_c}}$, is then mapped onto a different subset of allocated sub-carriers per user, i.e., N_c out of the M sub-carriers in which the total system bandwidth is divided. The subset may consist of a group of adjacent (LFDMA) or interleaved sub-carriers (IFDMA) and it is determined by the $M \times N_c$ mapping matrix $\mathbf{L}$ for which $L_{i,j} = 1$ if the pre-coded symbol j is transmitted over the sub-carrier i and zero otherwise.

The frequency-domain symbol is defined as $\mathbf{X}_{\mathrm{M}} = \mathbf{L}\mathbf{X}_{\mathrm{N_c}}$ and, consequently, non-allocated sub-carriers are forced to zero. From this point on, transmission is similar to that of OFDMA: an *M*-IDFT operation converts each frequency-domain symbol $\mathbf{X}_{\mathrm{M}}$ into a time-domain symbol and a cyclic prefix, whose length must be greater than the channel impulse response, is added in order to avoid ISI.

At the receiver (see Figure 5.4), perfect channel estimation and synchronization are assumed, thereby avoiding interference from other users. The cyclic prefix is suppressed and an *M*-DFT operation converts each time-domain symbol into a frequency-domain symbol $\mathbf{Y}_{\mathrm{M}}$. After applying $\mathbf{L}^H$ for de-mapping, the N_c allocated sub-carriers $\mathbf{Y}_{\mathrm{N_c}}$ can be expressed as

$$\mathbf{Y}_{\mathrm{N_c}} = \mathbf{L}^H \mathbf{H}_{\mathrm{M}} \mathbf{L}\mathbf{F}\mathbf{x} + \mathbf{L}^H \eta_{\mathrm{M}} \tag{5.23}$$

where η_{M} is the noise vector whose entries are i.i.d. complex Gaussian $CN(0, N_0)$ and the channel frequency response for each sub-carrier is represented by the $M \times M$ diagonal matrix $\mathbf{H}_{\mathrm{M}}$ whose entries are complex circularly symmetric random variables [15]. The magnitude of the frequency response can be well approximated with a Nakagami-*m* random variables with new parameters m and Ω, which, in generally, are different from their counterparts in the time-domain [11]. In fact, for a flat fading channel $m = m_t$, whereas if $m_t \geq 1$ then $1 \leq m < m_t$, and if $1/2 \leq m_t < 1$ then $1 > m > m_t$ [15]. For a Rayleigh fading channel (i.e., $m = 1$), it is exact for any value of $\mathbf{L}$.

The expression for the recovered symbol in frequency after FDE yields

$$\widetilde{\mathbf{X}}_{\mathrm{N_c}} = \mathbf{W}\mathbf{Y}_{\mathrm{N_c}} = \mathbf{W}\mathbf{H}\mathbf{F}_{\mathbf{X}} + \mathbf{W}\mathbf{L}^{\mathbf{H}}\boldsymbol{\eta}_{\mathbf{M}} \tag{5.24}$$

where $\mathbf{H} = \mathbf{L}^H \mathbf{H}_{\mathrm{M}} \mathbf{L}$ is a $N_c \times N_c$ diagonal matrix whose entries $\mathbf{h} = diag(\mathbf{H})$ are the channel frequency responses for each allocated sub-carrier and $\boldsymbol{\eta} = \mathbf{L}^H \eta_{\mathrm{M}}$ is a vector whose elements are the corresponding complex noise values. The elements of $\mathbf{h}$ are assumed to be i.i.d. random variables; this working assumption makes sense when the spacing between consecutive allocated sub-carriers is bigger than the coherence bandwidth of the considered fading channel. The $N_c \times N_c$ equalization matrix $\mathbf{W}$ is defined as

$$\mathbf{W} = (\mathbf{H}^H\mathbf{H})^{-1}\mathbf{H}^H \tag{5.25}$$

Before detection, an inverse pre-coding is performed by means of an N_c-IDFT. Thus, for ZF the equalized symbol in time domain $\tilde{\mathbf{x}}$ yields

$$\tilde{\mathbf{x}} = \underbrace{\mathbf{F}^H}_{\text{Inv. Precoding}} \underbrace{(\mathbf{H}^H\mathbf{H})^{-1}\mathbf{H}^H}_{\text{FDE}} \mathbf{HFx} + \underbrace{\mathbf{F}^H}_{\text{Inv. Precoding}} \underbrace{(\mathbf{H}^H\mathbf{H})^{-1}\mathbf{H}^H}_{\text{FDE}} \eta \tag{5.26}$$

Hence, the expression for the k-th received symbol after FDE is given by

$$\tilde{x}_k = x_k + \sum_{j=1}^{N_c} \frac{F_{j,k}^*}{h_j}\eta_j = x_k + \sum_{j=1}^{N_c} \hat{\eta}_{j,k} = x_k + \tilde{\eta}_k \tag{5.27}$$

where $\hat{\eta}_{j,k} = \dfrac{F_{j,k}^* \eta_j}{h_j}$.

As it can been seen, each received symbol is the result of adding an effective noise term $\tilde{\eta}_k$ to the original transmitted symbol as shown in (5.27). Hence, when ZF-FDE is applied, the effective noise is the result of adding an elementary noise term $\hat{\eta}_{j,k}$ (enhanced noise) for each allocated sub-carrier. Each elementary noise term is equivalent to the enhanced noise in OFDM reception with ZF-FDE.

5.6.2 Noise Characterization

In order to derive a mathematical model for the effective noise in SC-FDMA after ZF-FDE, we must first consider the elementary noise. Any given elementary noise term $\hat{\eta}_{j,k}$ is a random variable resulting from the ratio of two complex circularly symmetric random variables. In the remainder of the dissertation, indexes are dropped when possible for the sake of readability. Each complex random variable $\hat{\eta}$ has a Pearson type VII distribution and the expression in Cartesian coordinates of its PDF (5.17) yields [15]

$$f_{\hat{\eta}_r,\hat{\eta}_i}(x,y)=\frac{m^{m+1}}{2\pi\sigma^2}\left(\frac{x^2+y^2}{2\sigma^2}+m\right)^{-m-1} \qquad x,y\in\mathbb{R} \tag{5.28}$$

where $\hat{\eta}_r$ and $\hat{\eta}_i$ are the real and imaginary components of the elementary noise term and $\sigma=\frac{N_0}{N_c E_S}$.

The effective noise is the sum of N_c elementary noise random variables as described in equation (5.27). As all the terms in the sum are circularly symmetric random variables, the effective noise has also circular symmetry and its marginal distributions are described by the same even function. Thus, effective noise is also a circularly symmetric random variable and just one of its marginal distributions is needed to characterize it. This marginal distribution can be expressed as

$$f_{\hat{\eta}_r}(x)=f_{\hat{\eta}_i}(x)=\sqrt{\frac{1}{\pi}}\frac{\Gamma\left(m+\frac{1}{2}\right)2^m m^m\sigma^{2m}}{\Gamma(m)\left(2m\sigma^2+x^2\right)^{m-\frac{1}{2}}} \qquad x\in\mathbb{R} \tag{5.29}$$

5.6.2.1 Effective Noise Characterization

As stated before, the effective noise after ZF-FDE is the sum of several elementary complex noise terms. The density function for the sum of independent Pearson type VII random variables is given by

$$f_{\hat{\eta}_r}(x)=\frac{1}{2\pi}\int_{-\infty}^{\infty}\left(\frac{\sigma^m|\omega|^m}{2^{1-m/2}\Gamma(m)}m^{m/2}K_m\left(\sigma\sqrt{2m}|\omega|\right)\right)^{N_c} e^{-j\omega x}\,d\omega. \tag{5.30}$$

Usually, it is not possible to obtain an analytical expression for the CDF of the effective noise $F_{\tilde{\eta}_r}(x)$ for $N_c\geq 2$,it is possible to use the inversion theorem as proposed by Gil-Peláez to evaluate it as

$$F_{\hat{\eta}_r}(x)=\frac{1}{2}+\frac{1}{2\pi}\int_0^{\infty}\frac{e^{jx\omega}\Psi_{\tilde{\eta}_r}(-\omega)-e^{-jx\omega}\Psi_{\tilde{\eta}_r}(\omega)}{j\omega}d\omega. \tag{5.31}$$

where $\Psi_{\tilde{\eta}_r}(\omega)$ corresponds here to the CHF for the real marginal of the enhanced noise. For a Rayleigh fading channel, we have to particularize that expression using the CHF for a univariate Student-t (5.14); the resulting expression follows

$$f_{\hat{\eta}_r}(x)=\frac{1}{2\pi}\int_{-\infty}^{\infty}\left(\sigma|\omega|K_1(\sigma|\omega|)\right)^{N_c}e^{-jx\omega}d\omega. \tag{5.32}$$

For a Nakagami-m fading channel, each elementary complex noise term follows a bivariate Pearson type VII distribution. The density function for the sum of independent Pearson type VII random variables is given by

$$f_{\hat{\eta}_r}(x)=\frac{1}{2\pi}\int_{-\infty}^{\infty}\left(\frac{\sigma^m|\omega|^m}{2^{1-m/2}\Gamma(m)}m^{m/2}K_m\left(\sigma\sqrt{2m}|\omega|\right)\right)^{N_c}e^{-j\omega x}d\omega. \tag{5.33}$$

5.6.3 SC-FDMA BER Analysis

The BER analysis presented in this subsection follows the same premises that the one performed in section 5.6.1: we consider square M-QAM with independent bit-mapping for in-phase and quadrature components, e.g., Gray mapping. As before, the effective noise term $\tilde{\eta}$ is circularly symmetric and the BER can be expressed as a sum of CEPs (see equation (5.1)) which depend on the CDF of the effective noise.

In the case of ZF-FDE, the integral in (5.31) can be accurately solved in a quite simple manner by means of the composite trapezoidal rule. Thus, the CDF can be calculated as

$$F_{\tilde{\eta}_r}(x)\approx\frac{1}{2}+\frac{1}{2\pi}\left(\frac{\omega_{\max}-\omega_{\min}}{n}\left[\frac{Y(x,\omega_{\min})+Y(x,\omega_{\max})}{2}+\sum_{k=1}^{n-1}Y\left(x,\omega_{\min}+k\frac{\omega_{\max}-\omega_{\min}}{n}\right)\right]\right) \tag{5.34}$$

where the auxiliary function $Y(x,\omega)$ is defined as

$$Y(x,\omega)=\frac{e^{\phi x\omega}\Psi_{\tilde{\eta}_r}(-\omega)-e^{-\phi x\omega}\Psi_{\tilde{\eta}_r}(\omega)}{\phi\omega}. \tag{5.35}$$

and $\Psi_{\tilde{\eta}_r}()$ depends on the fading channel considered.

In equation (5.33) the minimum value in the integration interval is ω_{min}, which is set to zero, whereas ω_{max} should be suitably chosen as a function of the shape of the corresponding CHF. The number of points to evaluate inside the interval is denoted as n. In the cases under study, $\Psi_{\tilde{\eta}_r}(\omega)$ is a symmetric, exponentially decaying function with a decay rate that depends on the argument of $K_1()$. In fact, its decay is inversely proportional to the values of SNR and directly proportional to the values of N_c values. By making a reasonable choice for the value of ω_{max}, one can achieve a sufficient accuracy at low computational cost. By keeping this in mind and using equation (5.34), it is possible to obtain an approximate closed-form expression for the BER as

$$BER = \sum_{n=1}^{L-1} w(n)\left(\frac{1}{2} - \frac{\omega_{\max}}{2\pi n}\left(\frac{1+Y((2n-1)d,\omega_{\max})}{2} + \sum_{k=1}^{n-1} Y\left((2n-1)d, k\frac{\omega_{\max}}{n}\right)\right)\right), \quad (5.36)$$

where $Y(\ )$ is defined in 5.35.

5.6.4 Validation of Closed-form Expression

In order to validate the closed-form expression presented in the previous section, results from the evaluation of equations (5.36) for different numbers of sub-carriers are compared to values obtained from simulations. A channel model, in which the channel frequency response for the N_c allocated sub-carriers are modeled as i.i.d. random variables which follow a Nakagami-*m* distribution. Two different Nakagami-*m* fading channels are considered with *m* values of 1 and 2, respectively. BER results for each channel are shown for BPSK and 16QAM modulations in Figure 5.10 and 5.11.

In the first case, the Nakagami-*m* fading channel with $m = 1$ is equivalent to a Rayleigh fading channel and obtained results presented in Figure 5.10 are consistent with those presented in [2]. For low SNR values, BER curves are above the OFDM curve with a difference that increases with the number of allocated sub-carriers. In this case, the probability of having values close to zero for the channel frequency response can not be neglected. When the number of terms in the sum N_c increases, the probability of having a term close to zero increments, the term becomes dominant and the BER worsens.

However, as SNR values increase, all the curves converge to the OFDM curve independently of N_c. This behavior is a direct consequence of the expansion in the ω axis of the characteristic function: as SNR increases, all CHFs tend to converge to a common shape and, subsequently, their corresponding PDFs and BER values also converge. These results are consistent with the fact that, for ZF-FDE, OFDM determines the lower bound for SC-FDMA BER values.

For greater values of m (e.g., $m = 3$), effective noise has a finite variance and BER curves for SC-FDMA have a significatively different behavior (see Figure 5.11). For low values of SNR, SC-FDMA BER curves are very close to

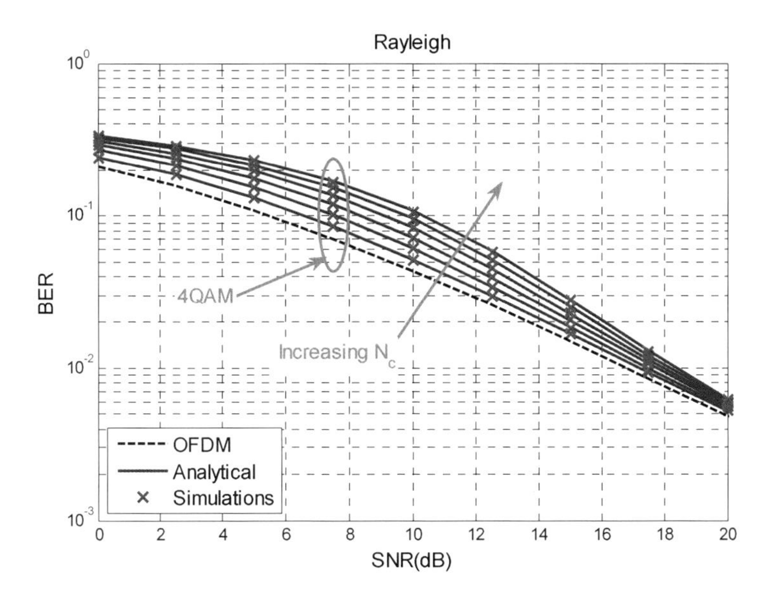

Figure 5.10: Analytical vs. Simulation BER values for 4QAM transmission over a Rayleigh fading channel with $N_c = 2, 4, 8, 16, 32$ and 64.

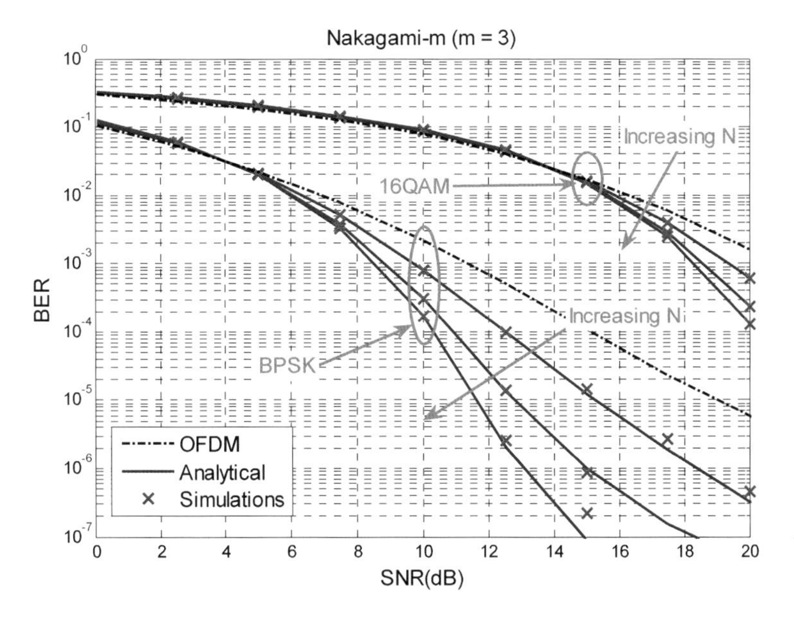

Figure 5.11: Analytical vs. Simulation BER values for BPSK and 16QAM transmissions over a Nakagami-*m* fading channel ($m = 3$) with $N_c = 4, 16$ and 64.

those for OFDM but, as the SNR increases, SC-FDMA seems to perform better that OFDM. In this case, the magnitude of the channel frequency responses have a non-zero minimum and, therefore, there is not nulls, thus the effective SNR always increases with the number of allocated sub-carriers N_c.

5.7 CONCLUSIONS

In this chapter we first study the enhanced noise after ZF-FDE for an OFDM signal transmitted over different fading channels (i.e., Rayleigh and Nakagami-*m*). We characterize the enhanced noise after equalization and apply derived expressions for the CDF to compute analytically the BER for Rayleigh and Nakagami-*m* fading channels. Additionally, we present our study of the stochastic nature of noise for SC-FDMA transmissions over a Nakagami-*m* fading channel. We show how, under the assumption of independent sub-carriers, the effective noise is also related to the Pearson type VII family of distributions. Closed-form expressions for BER values in SC-FDMA transmissions over a Nakagami-*m* fading channel with BPSK and M-QAM modulations were presented and validated by means of simulations.

References

[1] J. J. Sánchez-Sánchez, U. Fernández-Plazaola, and M. C. Aguayo-Torres, BER analysis for OFDM with ZF-FDE in Nakagami-*m* fading channels, in Fifth International Conference on Broadband and Biomedical Communications (IB2Com), pp. 1-5, dec. 2010.
[2] J. J. Sánchez-Sánchez, M. C. Aguayo-Torres, and U. Fernández-Plazaola, BER Analysis for Zero-Forcing SC-FDMA over Nakagami-m Fading Channels, IEEE Transactions on Vehicular Technology, 2011.
[3] R. v. Nee and R. Prasad, OFDM for Wireless Multimedia Communications. Norwood, MA, USA: Artech House, Inc., 2000.
[4] 3GPP, 36series: Evolved Universal Terrestrial Radio Access (E-UTRA), ts, 3rd Generation Partnership Project (3GPP), http://www.3gpp.org/ftp/Specs/html-info/36-series.htm, 2008.
[5] H. G. Myung, Single Carrier Orthogonal Multiple Access Technique for Broadband Wireless Communications. PhD thesis, Polytechnic University, Brooklyn, NY, USA, January 2007.
[6] J. van de Beek, P. Ödling, S. Wilson, and P. Börjesson, Review of Radio Science, 1996-1999, ch. Orthogonal Frequency Division Multiplexing (OFDM). Wiley, 2002.

[7] D. Falconer, S. L. Ariyavisitakul, A. Benyamin-Seeyar, and B. Eidson, Frequency Domain Equalization for Single-Carrier Broadband Wireless Systems, IEEE Commun. Mag, vol. 40, pp. 58-66, 2002.
[8] M. Abramowitz and I. A. Stegun, Handbook of Mathematical Functions with Formulas, Graphs, and Mathematical Tables. New York: Dover Publications, 1964.
[9] M. K. Simon and M.-S. Alouini, Digital communication over fading channels. Wiley series in telecommunications and signal processing, John Wiley & Sons, 2005.
[10] F. López-Martínez, E. Martos-Naya, J. Paris, and U. Fernández-Plazaola, Generalized BER Analysis of QAM and Its Application to MRC Under Imperfect CSI and Interference in Ricean Fading Channels, IEEE Transactions on Vehicular Technology, vol. 59, pp. 2598-2604, jun 2010.
[11] M. Nakagami, The m-Distribution, a general formula of intensity of rapid fading, Statistical Methods in Radio Wave Propagation: Proceedings of a Symposium held at the University of Californica, June 18-20, pp. 3-36, 1958.
[12] A. Goldsmith, Wireless Communications. New York, USA: Cambridge University Press, 2005.
[13] Z. Kang, K. Yao, and F. Lorenzelli, Nakagami-m fading modeling in the frequency domain for OFDM system analysis, IEEE Communications Letters, vol. 7, pp. 484 - 486, oct. 2003.
[14] J. K. Ghosh, Only linear transformations preserve normality, Sankhya: The Indian Journal of Statistics, Series A, vol. 31, no. 3, pp. 309-312, 1969.
[15] J. J. Sánchez-Sánchez, OFDMA and SC-FDMA BER Performance for Rayleigh and Nakagami-*m* Fading Channels. PhD thesis, Universidad de Málaga, 2011.
[16] B. Picinbono, On circularity, IEEE Transactions on Signal Processing, vol. 42, pp. 3473 -3482, dec. 1994.
[17] K. Fang, Encyclopedia of Statistical Sciences, vol. 3, ch. Elliptically contoured distributions, pp. 1910-1918. John Wiley and Sons, 2nd ed., 2006

Chapter 6.

SC-FDMA BER Performance for MMSE Equalized Fading Channels

Juan Jesús Sánchez-Sánchez, Mari Carmen Aguayo-Torres, Jyoti Gangane, and Unai Fernández-Plazaola

Department of Ingeniería de Comunicaciones, University of Malaga, Spain
aguayo@ic.uma.es

6.1 INTRODUCTION

In previous chapter we presented the performance analysis of Zero-Forcing (ZF) Frequency Domain Equalization (FDE) for OFDMA (Orthogonal Frequency-Division Multiple Access) and SC-FDMA (Single Carrier Frequency-Division Multiple Access) transmissions over Rayleigh and Nakagami-*m* fading channels. In order to obtain a closed form expression for the Bit Error Rate (BER), the statistical distribution of the enhanced noise after linear FDE was analyzed. Only the Cumulative Distribution Function (CDF) of the effective noise was necessary to evaluate the BER for SC-FDMA.

In this chapter a complete different approach is used in order to evaluate the BER for Minimum Mean Squared Error (MMSE) FDE SC-FDMA systems. The main difference between ZF and MMSE equalized SC-FDMA is that for MMSE-FDE there exists, in addition to the term corresponding to noise, a component of interference due to the other symbols. In this case, we use a more usual approach of evaluating the noise probability density conditioned to the channel frequency response and applying a Gaussian approximation through the Central Limit Theorem (CLT).

In order to be self-contained, this chapter briefly introduces the system and channel models, which were thoroughly described in the previous chapter.

4G Wireless Communication Networks: Design, Planning and Applications, 117-140.

6.2 SC-FDMA SYSTEM MODEL

In this section, a brief description of the system model is given. Interested reader may refer to previous chapter for those details omitted here.

6.2.1 Channel Model

In a wireless communication system, the transmitted signal always suffers from weakening, which can take place both in large and small scale. Large-scale weakening is caused by path loss and shadowing, and small-scale fading is due to the productive and unhelpful interference of multipath signals. The mobile environment radio wave propagation channel can be described by multiple paths which arise from elimination and spreading. If there are L distinct paths from the transmitter to the receiver, the impulse response for this channel will be [1]:

$$h(t,\tau)=\sum_{i=0}^{L-1} a_i(t)\delta(\tau-\tau_i) \tag{6.1}$$

It is the well known tapped-delay line model. Channel models are characterized by the number of taps, the time delay relative to the first tap, the average power relative to the strongest tap (Power Delay Profile, PDP) and the Doppler spectrum of each tap.

As a result of the scattering, each path will be the result of the addition of a large number of scattered waves with approximately the same delay that gives rise to time-varying fading of the path amplitudes $|a_i|$, a fading which is well described by Rayleigh distributed amplitudes varying according to a classical Doppler. Moreover, in a radio propagation environment a Line of Sight (LOS) component can also exist between the transmitter and the receiver. The Rice factor K measures [2] the relative strength of the LOS compared to that of the whole varying amplitudes. It is a measure of the severity of the fading, being K=0 the most severe fading case (Rayleigh fading, i.e. no LOS), and K=∞ the usual AWGN channel.

It is well known that Nakagami-*m* distribution can approximately gather the Rayleigh and Rice distributions [3] while is more mathematically tractable. Thus, magnitudes of the taps of the channel impulse response $|a_i|$, *i*=0,…,*L*-1 will be modelled as i.i.d. Nakagami-*m* random variables whose distribution yields;

$$f_{|a_i|}(x)=\frac{2}{\Gamma(\mu_n)}\left(\frac{\mu_n}{\Omega_n}\right)^{\mu_n} x^{2\mu_n-1}e^{-\frac{x^2\mu_n}{\Omega_n}} \qquad x\in R \; and \; x\geq 0, \tag{6.2}$$

where $\mu_n > 1/2$ is the fading parameter and Ω_n the power of the n-th tap. $\Gamma()$ is the well known gamma function.

The frequency response of the channel $H(t,\upsilon)$ can be evaluated through Fourier transform (in variable τ) of the impulse response $h(t,\tau)$. Being usually L>1, there will exist certain correlation among $H(t,\upsilon)$ at different frequencies. Coherence Bandwidth (CB) [1] is a measurement of how close two frequencies have to be so that correlation among their responses cannot be dismissed.

For Nakagami-m distributed taps, the magnitude of the channel frequency response at each frequency can be well approximated with a Nakagami-*m* random variable with new parameters μ and Ω, which, in general, are different from their counterparts in the time-domain [4]. In particular, for Rayleigh or Rice fading distributed taps, approximation is exact, i.e. frequency response is also Rayleigh or Rice distributed.

The analytical expressions for the BER presented in the next section are derived for Nakagami-m channels under the assumption of independence among allocated sub-carriers. This assumption makes sense when coherence bandwidth of selective channels is sufficiently small in comparison to the distance between assigned sub-carriers. Later, we present simulation results for actual channel models and evaluate the effect of frequency correlation on SC-FDMA performance.

6.2.2 SC-FDMA Description

The transmission process in SC-FDMA is similar to that in OFDMA and is depicted in Figure 6.1. For a given user, a sequence of transmitted bits is mapped into a constellation of complex symbols (e.g., BPSK or M-QAM). The resulting complex sequence $\mathbf{x}$ of length N_c is pre-coded by means of a Discrete Fourier Transform (DFT) operation, that is, $\mathbf{X}_{\mathbf{N_c}}$ is obtained by multiplying $\mathbf{x}$ by the unitary Fourier matrix $\mathbf{F}$ described as $\mathrm{F}_{j,k} = exp\left\{\frac{2\pi i}{N_c} jk\right\} / \sqrt{N_c}$. X_{N_c} is then mapped onto a different subset of allocated sub-carriers for each user, i.e., N_c out of the M sub-carriers in which the total system bandwidth is divided. The subset may consist of a group of adjacent (LFDMA) or interleaved sub-carriers (IFDMA) [5].

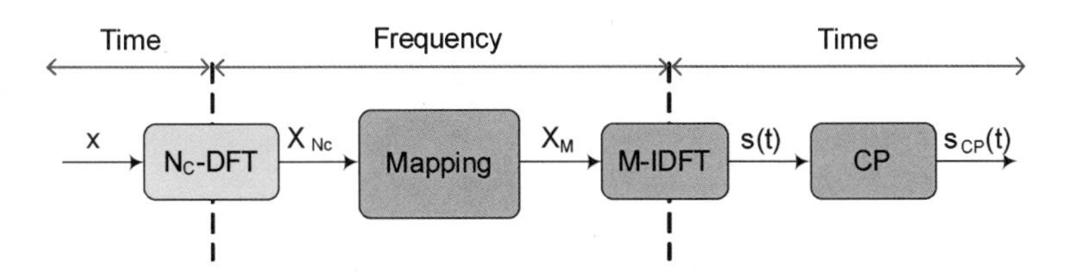

Figure 6.1 SC-FDMA Transmitter.

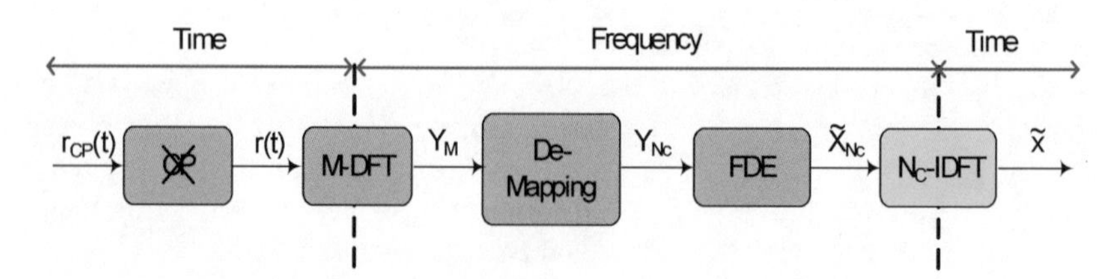

Figure 6.2 SC-FDMA Receiver.

In the frequency-domain, non-allocated sub-carriers are forced to zero. From this point on, transmission is similar to that of OFDMA: an M-IDFT operation converts each frequency-domain symbol $\mathbf{X}_{\mathbf{N_c}}$ into a time-domain symbol and a cyclic prefix, whose length must be greater than the channel impulse response, is added in order to avoid ISI.

At the receiver (see Figure 6.2), perfect channel estimation and synchronization are assumed, thereby avoiding interference from other users. The cyclic prefix is suppressed and an M-DFT operation converts each time-domain symbol into a frequency-domain symbol $\mathbf{Y_M}$. After de-mapping, the $\boldsymbol{N_C}$. allocated sub-carriers $\mathbf{Y}_{\mathbf{N_C}}$ can be expressed as

$$\mathbf{Y}_{\mathrm{N_c}} = \mathbf{Hx} + \boldsymbol{\eta}, \tag{6.3}$$

where $\mathbf{H}$ is a $\boldsymbol{N_C} \times \boldsymbol{N_C}$ diagonal matrix whose entries $\mathbf{h} = diag(\mathbf{H})$ are the circularly symmetric [6] channel frequency responses for each allocated sub-carrier. As previously described, those variables are assumed to be Nakagami-m i.i.d. in our analysis. For simulations, they are evaluated as the Fourier transform $H(t,\upsilon)$ of the channel impulse response given by (6.1), resulting in complex Gaussian correlated variables. $\boldsymbol{\eta}$ is a noise vector whose entries are i.i.d. complex Gaussian.

The expression for the recovered symbol in the frequency domain after FDE [7] $\tilde{\mathbf{X}}_{\mathrm{N_c}}$ is

$$\tilde{\mathbf{X}}_{\mathrm{N_c}} = \mathbf{WY}_{\mathrm{N_c}} = \mathbf{WHX}_{\mathrm{N_c}} + \mathbf{W}\boldsymbol{\eta}, \tag{6.4}$$

After the inverse precoding stage $\mathbf{F}^H$, the received symbol can be expressed as

$$\tilde{\mathbf{X}} = \mathbf{F}^H \tilde{\mathbf{X}}_{\mathrm{N_c}} = \mathbf{F}^H \mathbf{WHFx} + \mathbf{F}^H \mathbf{W} \eta, \tag{6.5}$$

where the $\boldsymbol{N_C} \times \boldsymbol{N_C}$ equalization diagonal matrix $\mathbf{W}$ is defined for MMSE-FDE as

$$\mathbf{W}_{MMSE} = \left(\frac{N_0}{E_S}\mathbf{I} + \mathbf{H}^H\mathbf{H}\right)^{-1}\mathbf{H}^H, \tag{6.6}$$

where $\boldsymbol{E_S}$ and $\boldsymbol{N_0}$ are the signal and noise powers respectively.

Note that under the ZF criterion used in the previous chapter, the equalization matrix **W** was

$$\mathbf{W}_{ZF} = (\mathbf{H}^H\mathbf{H})^{-1}\mathbf{H}^H, \tag{6.7}$$

Both matrices become similar for high SNR, while present quite different behavior for bad channel conditions.

The k -th received symbol after MMSE-FDE can be described as

$$\tilde{x}_k = x_k T_{k,k} + \sum_{l\neq kl=1}^{N_c} x_l T_{k,l} + \sum_{j=1}^{N_c} \frac{F_{j,k}^* h_j^*}{|h_j|^2 + N_0/E_S}\eta_j \tag{6.8}$$

where the elements $\boldsymbol{T_{k,l}}$ of $\mathbf{T} = \boldsymbol{F^H W_{MMSE}}\mathrm{HF}$ are expressed as

$$T_{k,l} = \frac{1}{N_c}\sum_{j=1}^{N_c} \frac{F_{k,j}^* F_{l,j}|h_j|^2}{|h_j|^2 + N_0/E_S}. \tag{6.9}$$

For MMSE-FDE there exists, in addition to the term corresponding to the noise, a component of interference due to the other symbols. Note that the receiver symbol after FDE $\tilde{\boldsymbol{x}}_{\boldsymbol{k}}$ is attenuated by the term $\boldsymbol{T_{k,k}}$. When the channel frequency response is such that $\frac{N_0}{E_S} << |h_j|^2\ \frac{N_0}{E_S} << |h_j|^2$, the effect of the interference term decreases and the noise term tends to that of ZF. After compensating the received symbol, we obtain the transmitted symbol contaminated by equivalent noise $\tilde{\eta}_k$ and interference δ_k terms, that is,

$$\hat{x}_k = x_k + \underbrace{T_{k,k}^{-1}\sum_{l\neq kl=1}^{N_c} x_l T_{k,l}}_{\text{Effective Interference}} + \underbrace{T_{k,k}^{-1}\sum_{j=1}^{N_c}\frac{F_{j,k}^* h_j^*}{|h_j|^2 + N_0/E_S}\eta_j}_{\text{Effective Noise}} = x_k + \delta_k + \tilde{\eta}_k = x_k + \xi_k, \tag{6.10}$$

where the term ξ_k gathers the effects of noise and interference, which are inherently independent [8].

When not ambiguous, index k is omitted in the remainder of the chapter for the sake of simplicity.

6.3 NOISE & INTERFERENCE CHARACTERIZATION FOR MMSE-FDE

In this section, the effective noise after MMSE-FDE is characterized. To the best of our knowledge, it is not possible to derive an analytical expression for its density function. However, it is possible to approximate its variance in function of a random variable β^M whose PDF can be approximated as Gaussian density function by means of the CLT.

From (6.10), we know that the recovered symbol before detection results:

$$\hat{x} = x + \delta + \tilde{\eta} = x + \xi, \tag{6.11}$$

where the effects of noise $\tilde{\eta}$ and interference δ are gathered in the effective noise term ξ. The density of the noise $\tilde{\eta}$ conditioned to the channel frequency response $\mathbf{h}=[h_1 \ldots h_{N_c}]$ can be expressed as

$$p(\tilde{\eta}/h) \sim CN(0, \sigma_{\tilde{\eta}}^2), \tag{6.12}$$

where $\sigma_{\tilde{\eta}}^2$ is obtained from (6.10) as

$$\sigma_{\tilde{\eta}}^2 = N_0 \frac{\frac{1}{N_c}\sum_{j=1}^{N_c}\left(\frac{|h_j|}{|h_j|^2 + N_0/E_S}\right)^2}{\left(\frac{1}{N_c}\sum_{j=1}^{N_c}\frac{|h_j|^2}{|h_j|^2 + N_0/E_S}\right)^2}. \tag{6.13}$$

In a similar fashion and as in [8], the interference component conditioned to the channel frequency response can be assumed to follow a complex circularly symmetric Gaussian distribution with zero-mean and variance σ_δ^2 given by

$$\sigma_\delta^2 = E_S \frac{\frac{1}{N_c}\sum_{j=1}^{N_c}\left(\frac{|h_j|^2}{|h_j|^2 + N_0/E_S}\right)^2 - \left(\frac{1}{N_c}\sum_{j=1}^{N_c}\frac{|h_j|^2}{|h_j|^2 + N_0/E_S}\right)^2}{\left(\frac{1}{N_c}\sum_{j=1}^{N_c}\frac{|h_j|^2}{|h_j|^2 + N_0/E_S}\right)^2}. \tag{6.14}$$

As noise and interference components are independent random variables, their sum can be modeled with another zero-mean complex circularly symmetric Gaussian random variable whose variance results:

$$\sigma_{\xi}^{2}=\frac{\frac{N_0}{N_c}\sum_{j=1}^{N_c}\left(\frac{|h_j|}{|h_j|^2+N_0/E_S}\right)^2+\frac{E_S}{N_c}\sum_{j=1}^{N_c}\left(\frac{|h_j|^2}{|h_j|^2+N_0/E_S}\right)^2-E_S\left(\frac{1}{N_c}\sum_{j=1}^{N_c}\frac{|h_j|^2}{|h_j|^2+N_0/E_S}\right)^2}{\left(\frac{1}{N_c}\sum_{j=1}^{N_c}\frac{|h_j|^2}{|h_j|^2+N_0/E_S}\right)^2} \tag{6.15}$$

We can rewrite expression (6.15) in a more elegant way as

$$\sigma_{\xi}^{2}=\sigma_{\tilde{\eta}}^{2}+\sigma_{\tilde{\delta}}^{2}=E_S\frac{\frac{N_0}{E_S}\beta^M}{1-\frac{N_0}{E_S}\beta^M}, \tag{6.16}$$

with

$$\beta^M=\frac{1}{N_c}\sum_{j=1}^{N_c}\frac{1}{|h_j|^2+N_0/E_S}. \tag{6.17}$$

and the density of the term ξ conditioned to the frequency channel yields;

$$f(\xi/h)=f(\xi/\beta^M)\sim CN(0,\sigma_{\xi}^{2}) \tag{6.18}$$

As we assume independent bit-mapping and the variable is circularly symmetric, it is possible to work only with the marginal distribution corresponding to its real component

$$f(\xi_r/h)=f(\xi_r/\beta^M)\sim N(0,\sigma_{\xi_r}^{2}/2) \tag{6.19}$$

Thus, the density of the real part of the effective noise term in MMSE can be expressed as

$$f_{\xi_r}(x)=\int_{-\infty}^{\infty}\frac{e^{-\frac{x^2}{2\sigma_{\xi_r}^2}}}{\sqrt{2\pi\sigma_{\xi_r}^2}}f_{\sigma_{\xi_r}^2}(\sigma_{\xi_r}^2)d\sigma_{\xi_r}^2=\int_0^{E_S/N_0}\sqrt{\frac{\frac{1}{E_S}\left(\frac{E_S}{\beta^M N_0}-1\right)}{\pi}}e^{-x^2\left(\frac{E_S}{\beta^M N_0}-1\right)}f_{\beta^M}(\beta^M)d\beta^M. \tag{6.20}$$

Analytical evaluation of $f_{\beta^M}(\beta^M)$ has not been possible. We are now trying to obtain an approximation by first evaluating the j-th term in the sum (6.17):

$$Z_j=\frac{1}{|h_j|^2+N_0/E_S}. \tag{6.21}$$

Its probability density can be derived as

$$f_{Z_j}(z) = f_{|h_j|^2}\left(\frac{1-z\frac{E_S}{N_0}}{z}\right)\frac{1}{z^2} = \frac{1}{z^2\Gamma(\mu)}\left(\frac{\mu}{\Omega}\right)^{\mu}\left(\frac{1}{z}-\frac{E_S}{N_0}\right)^{\mu-1} e^{\frac{\mu\left(\frac{E_S}{N_0}z-1\right)}{\Omega z}}. \tag{6.22}$$

It is not possible to obtain analytical expressions for the mean $\overline{Z}$ and variance σ_Z^2 of Z_j for arbitrary values of Ω and μ. However, those expressions are available provided μ is an integer. Specifically, for the case in which the magnitude of the channel frequency response follows a Rayleigh distribution (i.e., $\mu = 1$) and it is normalized ($\Omega = 1$), the density function of Z_j can be reduced to

$$f_{Z_j}(z) = \frac{e^{\frac{E_S}{N_0}-\frac{1}{z}}}{z^2} \quad \forall z \in [0, E_S/N_0]. \tag{6.23}$$

The corresponding expression for the mean yields;

$$\overline{Z} = e^{\frac{E_S}{N_0}}\Gamma(0,1), \tag{6.24}$$

where $\Gamma_{(z,x)}$ is the incomplete Gamma function defined as [9]

$$\Gamma(z,x) = \int_x^{\infty} t^{z-1}e^{-t}dt. \tag{6.25}$$

The variance is given by

$$\sigma_Z^2 = \frac{E_S}{N_0} - e^{\frac{E_S}{N_0}}\Gamma\left(0,\frac{E_S}{N_0}\right) - e^{2N_0}\Gamma\left(0,\frac{E_S}{N_0}\right)^2. \tag{6.26}$$

In other cases, it would be still possible to use numerical integration to make the computation of the mean $\overline{Z}$ and variance σ_Z^2.

Once those parameters are computed, the application of the central limit theorem to approximate the probability density function of β^M for high N_C is straightforward. Thus, as (6.17) is the average of N_C i.i.d. random variables each having finite values of expectation $\overline{Z}$ and variance σ_Z^2, the density of β^M can be approximated by the following Gaussian distribution:

$$f_{\beta^M}(\beta^M) \approx \hat{f}_{\beta^M}(\beta^M) = \frac{1}{\sqrt{2\pi\sigma_Z^2 / N_c}} e^{-\frac{\left(\beta^M - \overline{Z}_j\right)^2}{2\sigma_Z^2/N_c}}, \tag{6.27}$$

There is an additional restriction that should not be neglected: the variable β^M is limited by definition to the interval $[0,E_S/N_0]$. In order to apply this restriction, the density is now approximated by a bounded version of (6.23), that is,

$$f_{\beta^M}(\beta^M) \approx \tilde{f}_{\beta^M}(\beta^M) = \frac{1}{\widehat{F}(\frac{E_S}{N_0}) - \widehat{F}(0)} \frac{1}{\sqrt{2\pi\sigma_Z^2 / N_c}} e^{-\frac{(\beta^M - \bar{Z})^2}{2\sigma_Z^2/N_c}} \quad \beta^M \in [0, E_S / N_0], \tag{6.28}$$

where $\widehat{F}(x)$ is the CDF obtained through integration from (6.23).

Finally, probability density function of ξ_r is approximated as

$$\tilde{f}_{\xi_r}(x) = \int_0^{E_S/N_0} \sqrt{\frac{\frac{1}{E_S}\left(\frac{E_S}{\beta^M N_0} - 1\right)}{\pi}} e^{-x^2\left(\frac{E_S}{\beta^M N_0} - 1\right)} \tilde{f}_{\beta^M}(\beta^M) d\beta^M. \tag{6.29}$$

This approximation will be used to compute the BER for arbitrary Nakagami-m fading channels.

6.3.1 Effective Noise Characterization for MMSE-FDE

The expression for the CDF of the effective noise after MMSE-FDE can be derived from expression (6.29) through integration as follows;

$$\tilde{F}_{\xi_r}(x) = \int_{-\infty}^{x} \left(\int_0^{E_S/N_0} \sqrt{\frac{\frac{1}{E_S}\left(\frac{E_S}{\beta^M N_0} - 1\right)}{\pi}} e^{-x^2 \frac{1}{E_S}\left(\frac{E_S}{\beta^M N_0} - 1\right)} \tilde{f}_{\beta^M}(\beta^M) d\beta^M \right) dt$$

$$= \int_0^{E_S/N_0} \tilde{f}_{\beta^M}(\beta^M) \left(\int_{-\infty}^{x} \sqrt{\frac{\frac{1}{E_S}\left(\frac{E_S}{\beta^M N_0} - 1\right)}{\pi}} e^{-x^2 \frac{1}{E_S}\left(\frac{E_S}{\beta^M N_0} - 1\right)} dt \right) d\beta^M. \tag{6.30}$$

The inner integral can be rewritten using the $Q(\,)$ function:

$$\int_{-\infty}^{x} \sqrt{\frac{\frac{1}{E_S}\left(\frac{E_S}{\beta^M N_0} - 1\right)}{\pi}} e^{-x^2 \frac{1}{E_S}\left(1/\beta^M \frac{N_0}{E_S} - 1\right)} dt = \left\{ \begin{array}{l} u = \sqrt{\frac{2}{E_S}\left(\frac{E_S}{\beta^M N_0} - 1\right)} t (47) \\ du = \sqrt{\frac{2}{E_S}\left(\frac{E_S}{\beta^M N_0} - 1\right)} dt (48) \end{array} \right\}$$

$$= \frac{1}{\sqrt{2\pi}} \int_{-\infty}^{xA} e^{-\frac{u^2}{2}} du = \frac{1}{\sqrt{2\pi}} \int_{-xA}^{\infty} e^{-\frac{u^2}{2}} du = Q(-xA) = 1 - Q(xA),$$

with

$$A = \sqrt{\frac{2}{E_S}\left(\frac{E_S}{\beta^M N_0} - 1\right)}.$$

Hence, the CDF of the effective noise after MMSE-FDE can be expressed as

$$\tilde{F}_{\xi_r}(x) = 1 - \int_0^{E_S/N_0} \tilde{f}_{\beta^M}(\beta^M) Q\left(x\sqrt{\frac{2}{E_S}\left(\frac{E_S}{\beta^M N_0} - 1\right)}\right) d\beta^M. \quad (6.31)$$

6.3.2 Conditioned Effective SNR in MMSE-FDE

We showed above that, for MMSE-FDE, the effective noise ξ conditioned to the channel frequency response $\mathbf{h} = [h_1 \dots h_{N_c}]$ follows a complex normal distribution with zero-mean and variance σ_ξ^2 , that is,

$$\xi|\mathrm{h} \sim CN(0, \sigma_\xi^2)$$

where σ_ξ^2 is described in (6.15).

In this case, the expression for the instantaneous SNR yields;

$$\gamma = \frac{E_S}{\beta^M N_0} - 1, \quad (6.32)$$

with β^M being the random variable:

$$\beta^M = \frac{1}{N_c}\sum_{j=1}^{N_c} \frac{1}{|h_j|^2 + N_0/E_S}, \quad (6.33)$$

for which we derived above an approximation of its PDF (6.28).

6.4 SC-FDMA BER ANALYSIS

The BER analysis presented in this subsection follows the same premises that the one performed in previous chapter: we consider square M-QAM with independent bit-mapping for in-phase and quadrature components, e.g., Gray mapping. The effective noise plus interference term ξ is circularly symmetric and the BER can be expressed as [10]

$$BER = \sum_{n=1}^{L-1} w(n) I(n),$$

where $I(n)$ are the so-called Components of Error Probability (CEP), $w(n)$ are coefficients dependent on constellation mapping, $L=2$ for BPSK, and $L=\sqrt{M}$ for M-QAM. The CEPs can be expressed in function of the CDF $F_{\eta_r}()$ corresponding to the real part of the noise η affecting the symbol before the decision stage $\eta_r = \Re\{\eta\}$ as follows

$$I(n) = Pr\{\Re\{\xi\} > (2n-1)d\} = 1 - F_{\xi_r}((2n-1)d)$$

where d is the minimum distance between each symbol and the decision boundary, e.g., $d = \sqrt{E_s}$ for BPSK and $d = \sqrt{\frac{3E_s}{2(M-1)}}$ for M-QAM, and the coefficients $w(n)$ are directly computed as in [10].

For MMSE FDE SC-FDMA CEPs are computed directly using (6.36) and plugging those into the sum (6.35) we can compute BER values as follows;

$$BER_M = \sum_{n=1}^{L-1} w(n) \int_0^{E_S/N_0} \tilde{f}_{\beta^M}(\beta^M)\, Q\left((2n-1)d\sqrt{\frac{2}{E_S}\left(\frac{E_S}{\beta^M N_0}-1\right)}\right) d\beta^M .$$

BER values can be numerically evaluated using different numerical integration methods. Results presented in the following section are calculated by means of the double exponential integration [11].

6.5 NUMERICAL RESULTS

In order to validate the presented closed-form expressions, results from the evaluation of equation (6.37) for different numbers of sub-carriers are compared to values obtained from simulations. The channel frequency responses for the N_c allocated sub-carriers are modeled as i.i.d. random variables which follow a Nakagami-m distribution. For Figures. 6.3 to 6.5, we consider the expression (6.37) for BER when MMSE-FDE is applied to compensate the effects of a Rayleigh fading channel; values obtained for BPSK, 4QAM and 16QAM are compared to those obtained through simulations for different values of N_C. Moreover, results for OFDM are also presented.

By a simple inspection, it can be concluded that the validity of the provided expression depends highly on the number of sub-carriers and on the value of the power density N_0. For small values of SNR the values obtained with the aforementioned expression are consistent with those obtained through simulation. For higher values of SNR the expression values diverge with the smallest values of N_c considered. However, the approximation is still valid for greater values of N_c (e.g., 64 or greater) which are closer to practical values for a real system. The

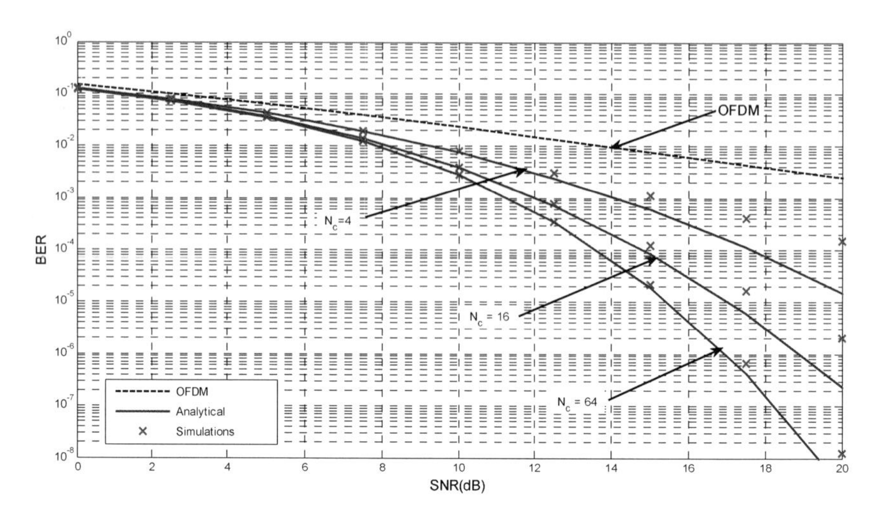

Figure 6.3 Analytical vs. simulation results for independent Rayleigh sub-carriers (BPSK) MMSE-FDE with N_C =4, 16 and 64.

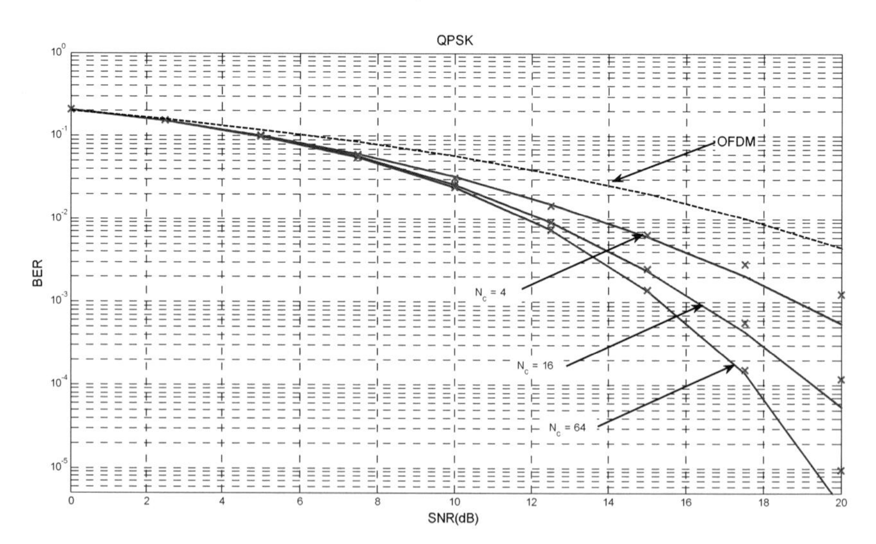

Figure 6.4 Analytical vs. simulation results for independent Rayleigh sub-carriers (4QAM) MMSE-FDE with N_C **=4, 16 and 64** sub-carriers.

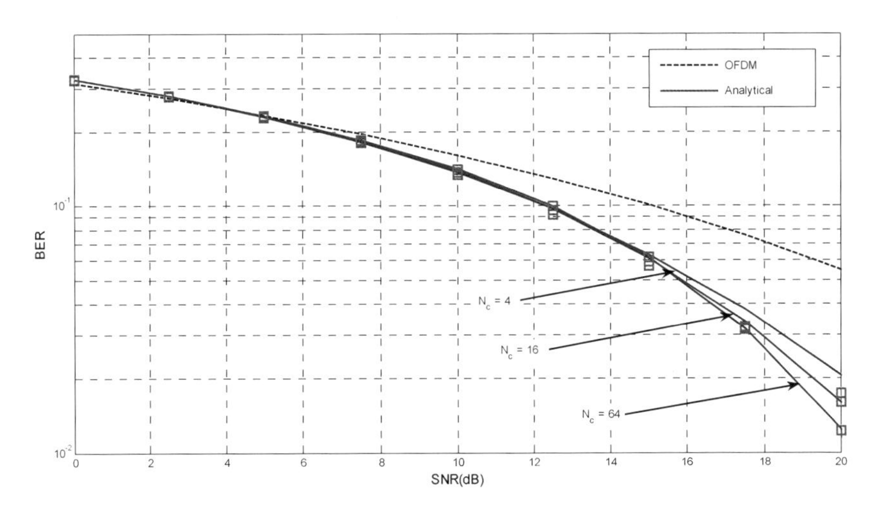

Figure 6.5 Analytical vs. simulation results for independent Rayleigh sub-carriers (16QAM) MMSE-FDE with $\boldsymbol{N_C}$ **=4, 16 and 64** sub-carriers.

reason is clear, as the approximation from the central limit theorem is only sensible for high number or terms.

Several insights can be obtained by inspection of the analytical and simulation results. It is known that ZF equalized OFDM has mostly the same BER performance than that of MMSE equalized OFDM, although the latter offers lower MSE (see [12] for a detailed justification). However, while OFDMA determines the upper bound for ZF equalized SC-FDMA BER values (see previous chapter), results for MMSE SC-FDMA are below those of OFDM. The behavior is kept for different constellation sizes. Moreover, using a high number of allocated sub-carriers improves MMSE SC-FDMA. We can find a reason in the fact that MMSE avoids frequency nulls by including a constant factor in the denominator. Thus, the effective SNR always increases with the number of allocated sub-carriers N_c, improving SC-FDMA performance. Results, consistent with those provided in [13], go against the extended idea of SC-FDMA as a degraded version of OFDM.

6.6 APPLICATION TO REALISTIC SCENARIOS

The analytical expressions for the BER presented in the previous section were derived under the assumption of independence among allocated sub-carriers. This assumption makes sense when coherence bandwidth of selective channels is

sufficiently small in comparison to the distance between assigned sub-carriers. Here we analyze the performance of SC-FDMA for actual selective fading channels and study under which circumstances the analytical expression may be used as an approximation for BER values in actual selective fading channels.
The scenario under study is an SC-FDMA transmission over some of the channel impulse response models proposed by the ITU in [14] and by the 3GPP [5] for several terrestrial test environments. Representative simulations parameters are shown in Table 6.1 for the channel models under consideration (see Table 6.2). For selective Rayleigh fading channels, the coefficients for the multipath fading model taps are zero centered mutually independent complex-valued Gaussian random variables. In the case of Rice fading, a constant part whose power depends on Rice factor K is added to each tap. Total power is normalized in all cases.
The delay spread and 50% coherence bandwidth for each test environment is presented in Table 6.3. The Coherence Ratio (CR), defined as the ratio between Coherence Bandwidth (CB) and frequency spacing between consecutive allocated sub-carriers (ΔF), is also provided in Table 6.3, first, for localized sub-carriers spaced 15 KHz apart and then, for 16 and 64 sub-carriers interleaved over a total of 2048. These ratios provide information about the similarity between channel responses for consecutive allocated sub-carriers: the higher its value, the more similar their channel responses are.

Figures 6.6 to 6.11 show the BER performance of the system for different constellation size, allocated subcarrier size and Rice factor K, for both MMSE and ZF equalization schemes. Results for those channel models given in Table 6.2 are shown in Figures 6.6 to 6.11. Moreover, BER results assuming independence between subcarriers are also shown in figures.

Table 6.1 Simulation Parameters

Parameter	*Value*
Carrier Frequency	2.5 GHz
System Bandwidth	20 MHz
Sampling Frequency	30.72 MHz
Sub-carrier Spacing	15 KHz
FFT Size	2048
Modulation Techniques	4 & 16QAM

Table 6.2 Tapped delay line parameters for considered channel models

PA		*PB*		*VA*		*VB*		*EPA*		*EVA*	
dB	ns	dB	ns	dB	ns	dB	ns	dB	Ns	dB	ns
0	0	0	0	0	0	-2.5	0	0	0	0	0
-9.7	110	-0.9	200	-1	310	0	300	-1	30	-1.5	30
-19.2	190	-4.9	800	-9	710	-12.8	8900	-1	70	-1.4	150
-22.8	4100	-8	1200	-10	1090	-10	12900	-3	80	-3.6	310
-	-	-7.8	2300	-15	1730	-25.2	17100	-8	110	-0.6	370
-	-	-23.9	3700	-20	2510	-16	20000	-17.2	190	-9.1	710
-	-	-	-	-	-	-	-	-20.8	410	-7	1090
-	-	-	-	-	-	-	-	-	-	-12	1730
-	-	-	-	-	-	-	-	-	-	-16.9	2510

Usually, interleaved transmission yields small coherence ratio values, that is, CR ≤ 1. In such cases, the provided analytical BER expressions serve as an approximation of actual values for interleaved SC-FDMA with MMSE-FDE. However, for greater values of CR (e.g., in the Pedestrian A (PA) channel) BER results obtained from simulations are far from the values obtained using the proposed analytical expressions as depicted in Figure 6.6. Nonetheless, expression (6.37) provide values of BER that come closer to the values obtained through simulation when CR decreases. This behavior can be observed in results depicted in Figures 6.6 to 6.11 with the remarkable exception of those corresponding to the Vehicular B (VB) channel with MMSE-FDE. In this case, the approximation is not as close to the simulation curve as its CB and CR values would suggest. We think that the reason behind this strange behavior lies in its peculiar channel impulse response (see Table 6.2).

Note that localized transmissions present greater values of coherence bandwidth as the sub-carrier spacing is minimum, thereby the assumption of independence no longer holds and BER values cannot be approximated using the proposed expressions.

Table 6.3 Delay Spread, Coherence Bandwidth and Coherence Ratio for considered channels

Channel Model	*Delay Spread (rms)*	*Coherence BW (50%)*	*CR Localized*	*CR Interleaved 16/2048*	*CR Interleaved 64/2048*
PA	46ns	4.35MHz	290	2.264	9.06
VB	4001ns	50KHz	3.33	0.026	0.104
ETU	990.93ns	160.6KHz	13.45	0.105	0.420
EVA	356.65ns	560.8KHz	37.8	0.292	1.168

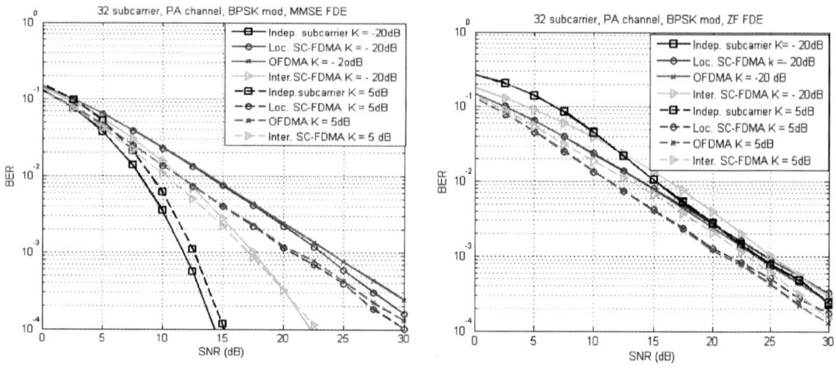

(a) Different K values with MMSE and ZF FDE

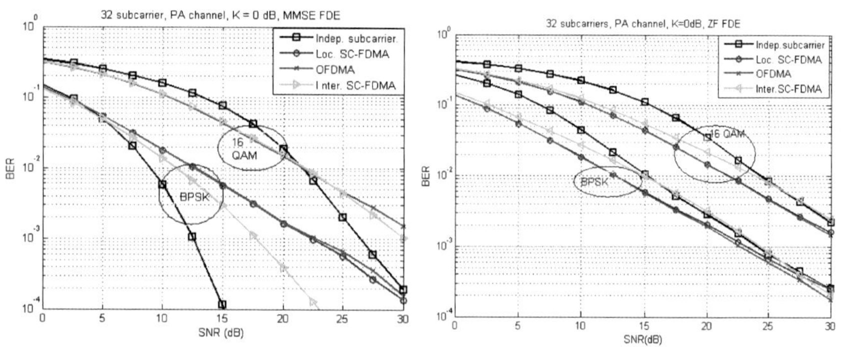

(b) Different constellation size with MMSE and ZF FDE

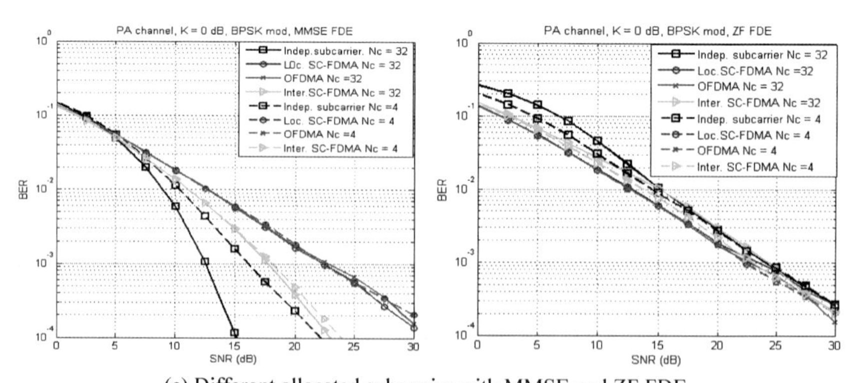

(c) Different allocated subcarrier with MMSE and ZF FDE

Figure 6.6 BER performance of SC-FDMA over PA channel.

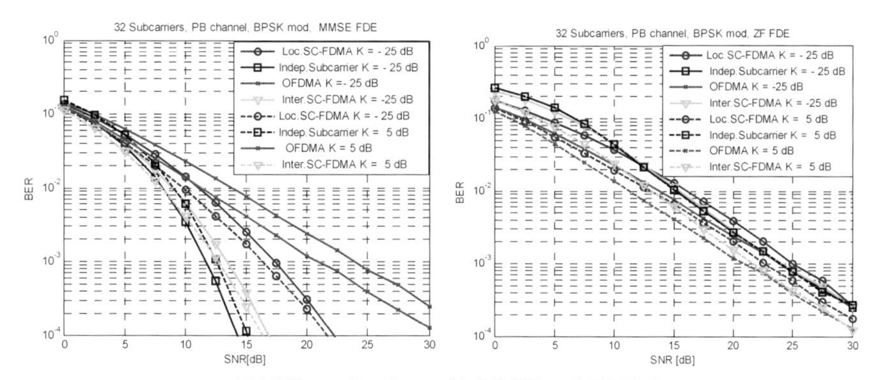

(a) Different K values with MMSE and ZF FDE

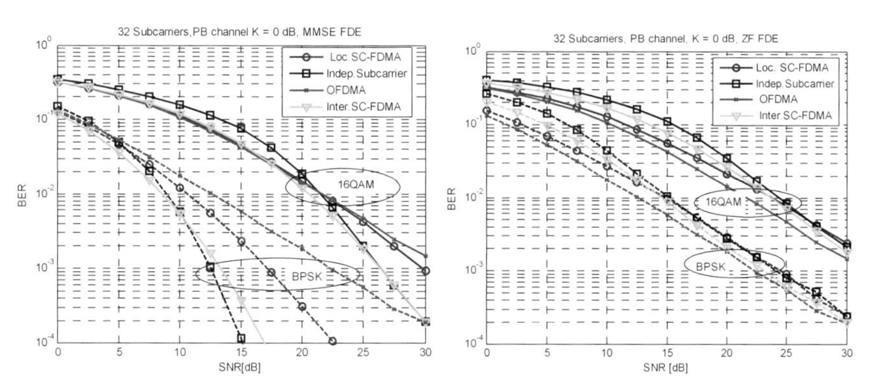

(b) Different constellation size with MMSE and ZF FDE

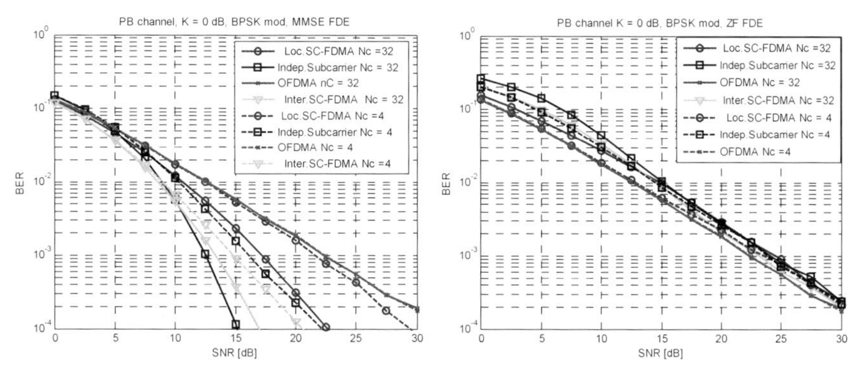

(c) Different allocated subcarriers with MMSE and ZF FDE

Figure 6.7 BER performance of SC-FDMA over PB channel.

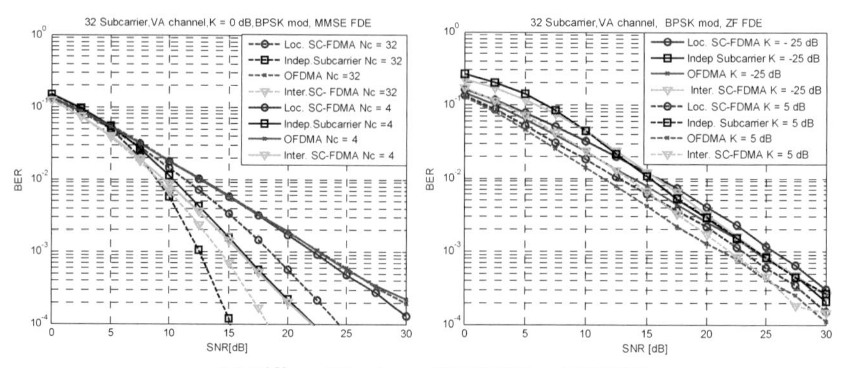

(a) Different K values with MMSE and ZF FDE

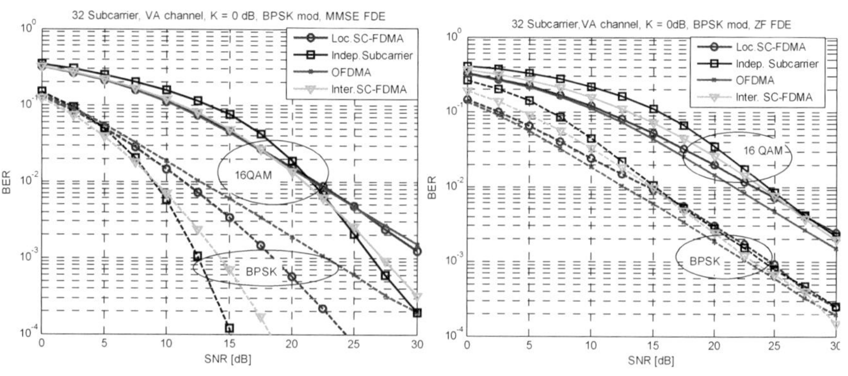

(b) Different constellation size with MMSE and ZF FDE

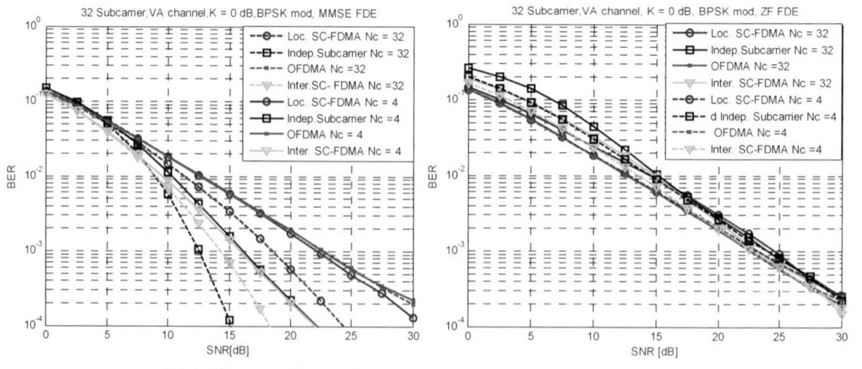

(c) Different allocated subcarriers with MMSE and ZF FDE

Figure 6.8 BER performance of SC-FDMA over VA channel.

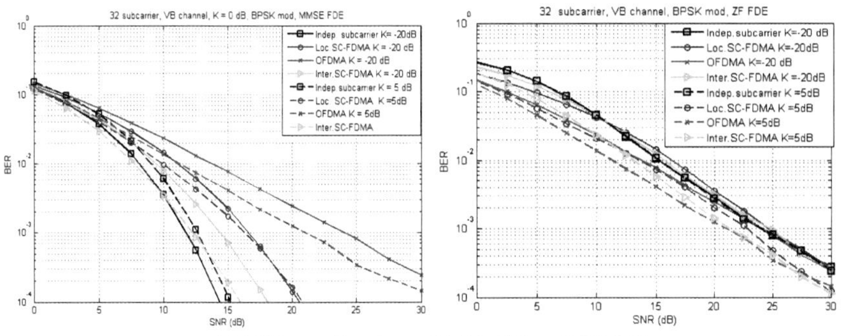

(a) Different K values with MMSE and ZF FDE

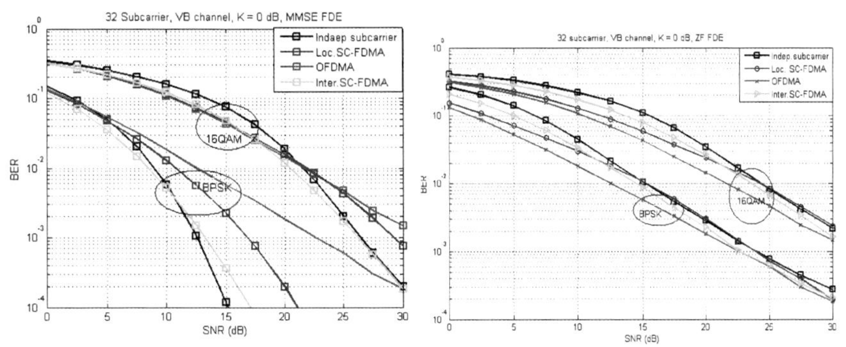

(b) Different constellation size with MMSE and ZF FDE

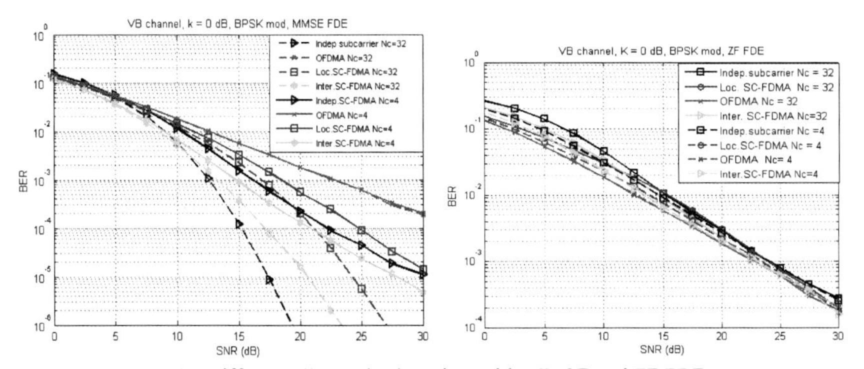

(c) Different allocated subcarriers with MMSE and ZF FDE

Figure 6.9 BER performance of SC-FDMA over VB channel.

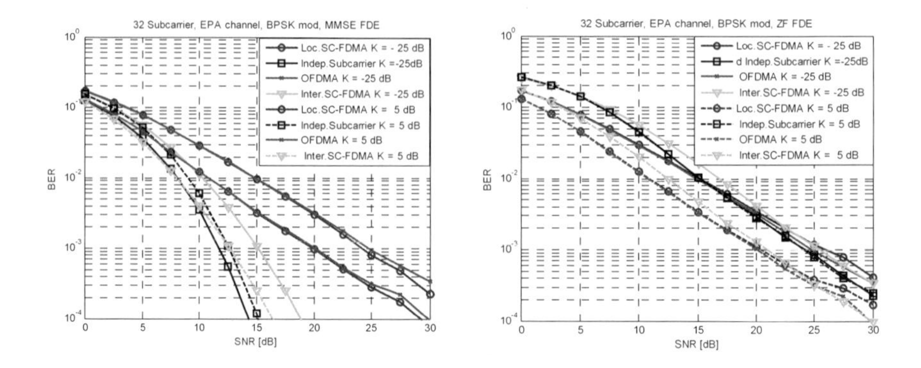

(a) Different K values with MMSE and ZF FDE

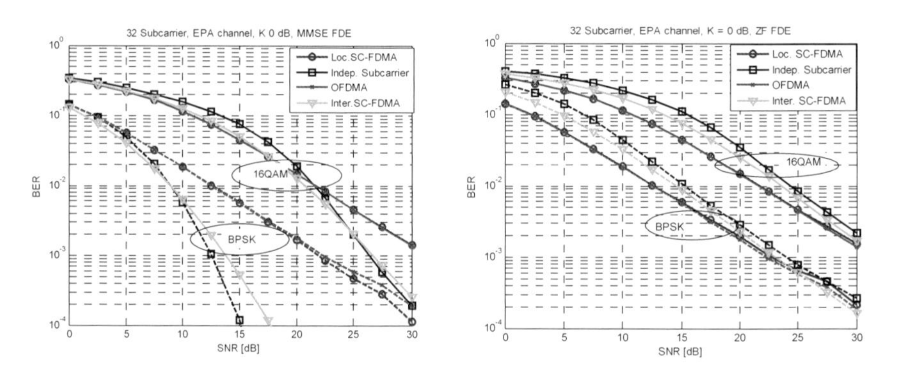

(b) Different constellation size with MMSE and ZF FDE

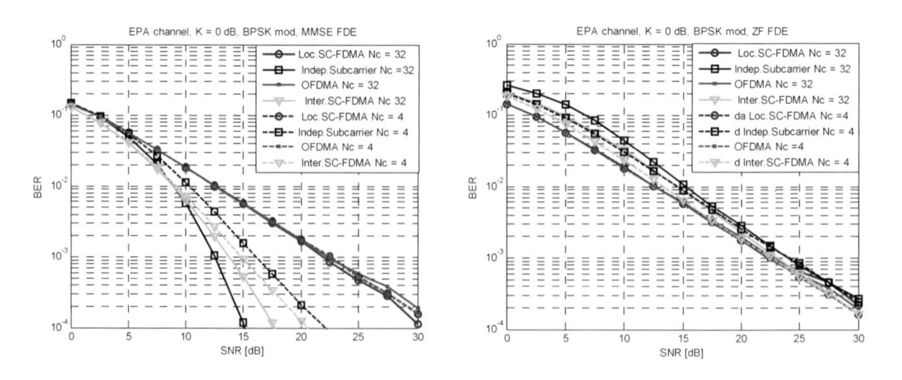

(c) Different allocated subcarriers with MMSE and ZF FDE

Figure 6.10 BER performance of SC-FDMA over EPA channel.

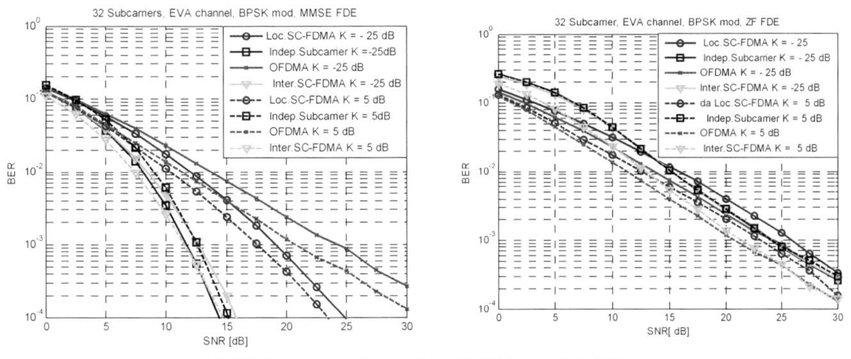

(a) Different K values with MMSE and ZF FDE

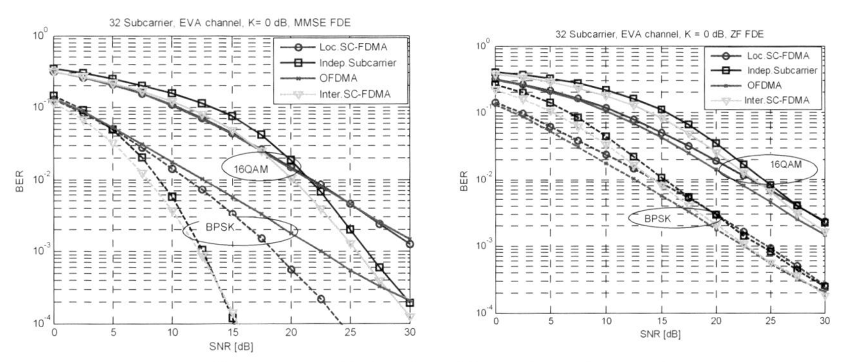

(b) Different constellation size with MMSE and ZF FDE

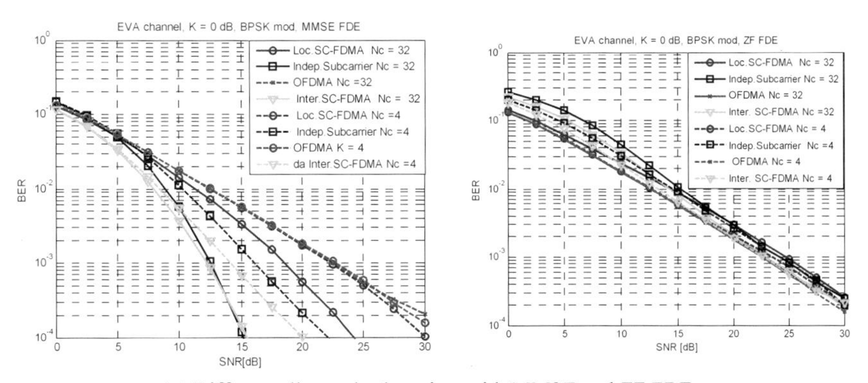

(c) Different allocated subcarriers with MMSE and ZF FDE

Figure 6.11 BER performance of SC-FDMA over EVA channel.

6.7 CONCLUSIONS

The study of MMSE equalized SC-FDMA performance has been addressed in this chapter. We have derived an approximation for the density of the effective noise applying the CLT under the assumption of frequency response independent among allocated subcarriers. We are then able to compute numerically values the BER for SC-FDMA over Nakagami-m channels. Contrary to what is often described, it has been shown that SC-FDMA performance is better than that of OFDM for the case of MMSE equalized systems. Moreover, MMSE SC-FDMA performance is improved by increasing the number of allocated subcarriers.

SC-FDMA performance over realistic Rayleigh and Rice channels has been also addressed. Although specific correlation function among frequency responses at allocated subcarriers influence SC-FDMA behavior, it was shown that provided BER expressions for independent subcarriers are suitable as approximations for realistic channel models if the ratio between coherence bandwidth and frequency spacing among consecutive allocated sub-carriers is small (close to or smaller than 1).

References

[1] J.D. Parsons, The Mobile Radio Propagation Channel, John Wiley & Sons, 1992.

[2] A. Doulas, G. Kalvias, Ricean K factor Estimation for Wireless Communication Systems, International Conference on Wireless and Mobile Communications, Budapest, July 2006.

[3] M.K. Simon, M.S. Alouini, Digital communication over fading channels, Wiley Interscience, 2005.

[4] Z. Kang, K. Yao, and F. Lorenzelli, Nakagami-m fading modeling in the frequency domain for OFDM system analysis, IEEE Communications Letters, vol. 7, pp. 484 - 486, Oct. 2003.

[5] 3GPP, 36series: Evolved Universal Terrestrial Radio Access (E-UTRA), 3rd Generation Partnership Project (3GPP), http://www.3gpp.org/ftp/Specs/html-info/36-series.htm, 2008.

[6] B. Picinbono, On circularity, IEEE Transactions on Signal Processing, vol. 42, pp. 3473 - 3482, Dec. 1994.

[7] D. Falconer, S. L. Ariyavisitakul, A. Benyamin-Seeyar, and B. Eidson, Frequency Domain Equalization for Single-Carrier Broadband Wireless Systems, IEEE Commun. Mag., vol. 40, pp. 58 - 66, 2002.

[8] M. Nisar, H. Nottensteiner, and T. Hindelang, On performance limits of DFT spread OFDM systems, Proceedings of 16th IST Mobile and Wireless Communications Summit, 2007, pp. 1 – 4.

[9] M. Abramowitz and I. A. Stegun, Handbook of Mathematical Functions with Formulas, Graphs, and Mathematical Tables, New York, Dover, 1964.

[10] F. López-Martínez, E. Martos-Naya, J. Paris, and U. F. Plazaola, Generalized BER Analysis of QAM and Its Application to MRC Under Imperfect CSI and Interference in Ricean Fading Channels, IEEE Transactions on Vehicular Technology, vol. 59, pp. 2598 - 2604, Jun 2010.
[11] M. Mori & M. Sugihara. The double-exponential transformation in numerical analysis, Journal of Computational and Applied Mathematics, vol. 127, pp. 287 - 296, 2001.
[12] Z. Wang, X. Ma & G. Giannakis, OFDM or single-carrier block transmissions, IEEE Transactions on Communications, vol. 52, no. 3, pp. 380 - 394, Mar. 2004.
[13] C. Ciochina, D. Castelain, D. Mottier & H. Sari, Single-Carrier Space-Frequency Block Coding: Performance Evaluation, IEEE 66th Vehicular Technology Conference, 2007, VTC-2007 Fall, pages 715 - 719, Sept. 30 2007 - Oct. 3 2007.
[14] ITU-R. M. 1225, Guidelines for Evaluation of Radio Transmission Technologies for IMT-2000, 1997.

Chapter 7.

Codes in Third and Fourth Generation Networks

Johnson I Agbinya

Department of Electronic Engineering, La Trobe University, Australia
J.Agbinya@latrobe.edu.au

7.1 INTRODUCTION

A significant number of codes are used for the implementation of CDMA, UMTS and LTE networks and for various purposes including multiple access (identification of users signals, base stations and nodes), for synchronisation, combating noise, interference and security. This chapter is a short summary of how the codes are generated and used in modern cellular communication systems.

7.2 MAXIMUM LENGTH SEQUENCES

Perhaps one of the most widely used code in 3G and 4G networks belong to the family of maximum length sequences, the so-called m-sequences. The m-sequences behave as pseudo-noise sequences and exhibit random noise like behaviour. They also possess several unique properties which make them suitable for CDMA operations. The sequences possess

1. Balanced property: they have equal number of 1's and 0's in the sequence
2. Run-length property
3. Shift-and-Add property

These three properties ensure that the sequences provide excellent auto-correlation behaviour suitable for spread spectrum CDMA operations. Specifically, such codes need to have

4G Wireless Communication Networks: Design, Planning and Applications, 141-160.

a) Half the occurrences of one '1s and 0's have length 1, a quarter have two 1's and 0's, one eight have three 1's and 0's and so on
b) Balanced number of zeros and ones (nearly equal numbers of zeros and ones). This property originates from property a)
c) Periodic auto-correlation functions that are two-valued, with major peak at zero shifts of the sequences (inner product of same code) and about zero everywhere else. This property allows the signal to be spread or encrypted into a larger bandwidth and to combat noise.

The auto-correlation function is between the sequence and its shifts is given as

$$\rho(k)=\sum_{n=1}^{N} a'_n a'_{n+k} = N\delta(k) \tag{7.1}$$

$\delta(k)$ is the Kronecker delta and the binary value a_n and sequence (+1 and -1) a'_n are related by the expression $a'_n = 1-2a_n$.

While there are many binary sequences which fail to meet these properties m-sequences have been found to possess most of them. Binary sequences which possess the properties are mostly pseudo-random in nature and thus noise-like.

The traditional method of generating m-sequences uses generator polynomials to govern the circuit used for generating the sequences. Primitive polynomials (irreducible polynomials) are used. A primitive polynomial always ensures that the resulting sequence is an m-sequence. The number of primitive polynomials was shown by Golomb [1] to be equal to

$$N_p(r)=\frac{2^r-1}{r}\prod_{i=1}^{k}\frac{P_i-1}{P_i} \tag{7.2}$$

Where $\{P_i, i=1, \ 2, \ \cdots, \ k\}$ is the prime decomposition of the number 2^r-1, so that

$$2^r-1=\prod_{i=1}^{k} P_i^{r_i} \tag{7.3}$$

r_i is an integer [2]. The practical approach used for generating the m-sequences or implementation is through the use of linear feedback shift registers (LFSR). The tap weights of the LFSR are specified by the terms of the generator polynomial normally derived from Galois field. The LFSR ensures that for each clock the contents of the shift registers are shifted one place to the right. The m-sequence is

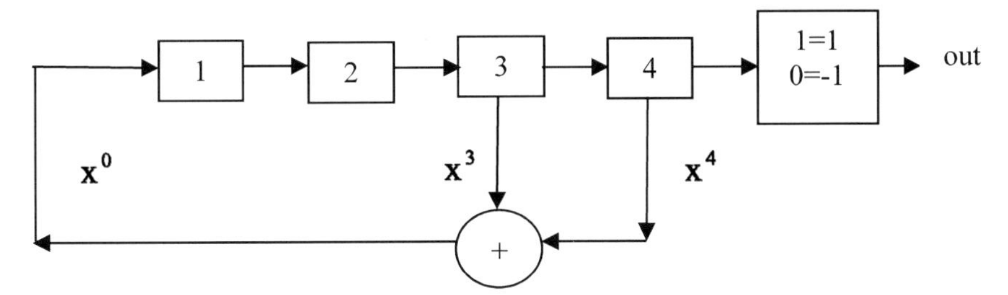

Figure 7.1 Four Bits Linear Feedback Shift Register.

generated from the expression

$$a_i = c_1 a_{i-1} + c_2 a_{i-2} + \cdots + c_n a_{i-n} = \sum_{k=1}^{n} c_k a_{i-k} \tag{7.4}$$

Where the arithmetic in equation (7.4) is in mod 2 and all the terms are either 0 or 1. The generating functions (polynomials) are also of the form

$$G(x) = a_0 + a_1 x + a_2 x^2 + \cdots = \sum_{i=0}^{\infty} a_i x^i \tag{7.5}$$

Here x is the delay variable. How to create the generator polynomials is given in [2]. Consider the generator polynomial $1 + x^3 + x^4$ and the resulting linear feedback shift register shown in Figure 7.1.

The first shift register in the left hand side contains the lowest significant bit and the last register on the right hand side contains the most significant bit of the number inside the LFSR. The maximum length of the sequence created from this register is 15 bits or $2^r - 1$ where r=4 is the degree of the polynomial used for creating the sequence. An example of how a 15 bit sequence is created from this LFSR is shown in Table 7.1. Two operations are performed to obtain this table. The exclusive OR is performed with the current values in the registers 3 and register 4. Table 7.1 shows the XOR operation used on register 3 and 4.

Table 7.1 XOR Operation

Register 3 content	Register 4 content	XOR
0	0	0
0	1	1
1	0	1
1	1	0

The result of the XOR is held momentarily. When a clock cycle arrives all the contents of registers 1, 2, 3 and 4 are shifted to the right, to issue an output bit. This shift creates a gap in register 1. The result of the XOR is now placed in register 1. Now the registers have all been updated and we repeat the operation in this case 15 times to generate the whole sequence. The clock signal is not shown in Figure 7.1.

The sequence created by this operation is given by the output column in Table 7.2 and is 11-1-1-11-1-111-11-111. We have thus created the pseudonoise sequence $PN(0)=\{a_1,\ a_2,\ \cdots,\ a_{15}\}$ and zeros are replaced with -1. This allows a waveform of ± 1 to be created. Observe that this sequence satisfies the properties listed in this section. It has 7 zeros and 8 ones, about the same number of pairs of zeros and ones. Its autocorrelation is 15. The cross-correlation between its one bit cyclic shifted (to the right) version is -1. This is much lower than +15. When the generator polynomial has many terms, the simple LFSR of Figure 7.1 is no more adequate. For such cases there will be more than one XOR operation required for calculating the contents of registers and the output bit sequence. From Table 7.1, several properties of the m-sequences are revealed.

i) They are periodic, with period $2^r - 1$. After this period, the sequence is repeated.

ii) The cross-correlation of the PN(n) with PN(n+k) has the following format

Table 7.2 Linear Feedback Shift Register (15 bits)

Register content	1	2	3	4	Bit shifted to the Output
Initial state	0	0	1	1	
1st shift	0	0	0	1	1 (=1)
2nd shift	1	0	0	0	1
3rd shift	0	1	0	0	0 (=-1)
4th shift	0	0	1	0	0(=-1)
5th shift	1	0	0	1	0(=-1)
6th shift	1	1	0	0	1
7th shift	0	1	1	0	0=-1
8th shift	1	0	1	1	0=-1
9th shift	0	1	0	1	1
10th shift	1	0	1	0	1
11th shift	1	1	0	1	0=-1
12th shift	1	1	1	0	1
13th shift	1	1	1	1	0=-1
14th shift	0	1	1	1	1
15th shift	0	0	1	1	1

$$\phi(k) = \langle PN(n), PN(n+k) \rangle = \frac{1}{N} \sum a(n+k) a^*(n) = \begin{cases} 1; & k = 0 \\ \dfrac{-1}{N}; & k \neq 0 \end{cases} \quad (7.6)$$

iii) The correlation function is periodic with period N chips and has the general shape as shown in Figure 7.2.

Equation (7.6) means that identical codes will lead to large auto-correlation and any other code pairs will result to nearly zero cross-correlation. This feature of m-sequences is used for multiple access and for identification of users and nodes in CDMA networks. In the multiple access situation only the user with the exact code (m-sequence) used to scramble the data meant for it is able to reproduce the data by multiplying the scrambled signal with the exact replica of the code. For this to happen, the code must be synchronised to the received signal.

iv) The power spectral density of m-sequences has a main peak with sidelobes as shown in Figure 7.3.

There are 2N spectral lines between the first two nulls $((-1/T_c, 1/T_c)$,where T_c is one chip duration.

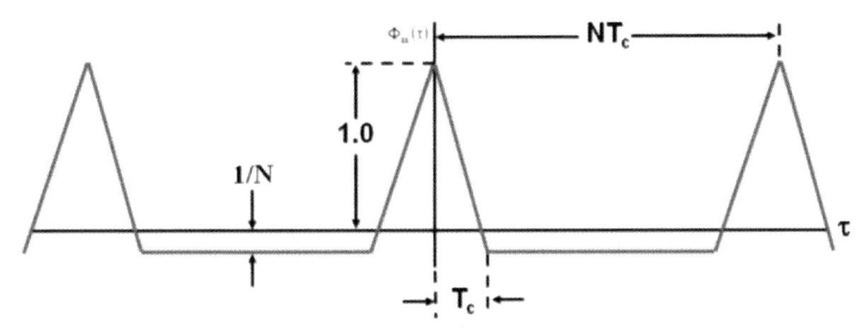

Figure 7.2 Correlation Functions of m-sequences.

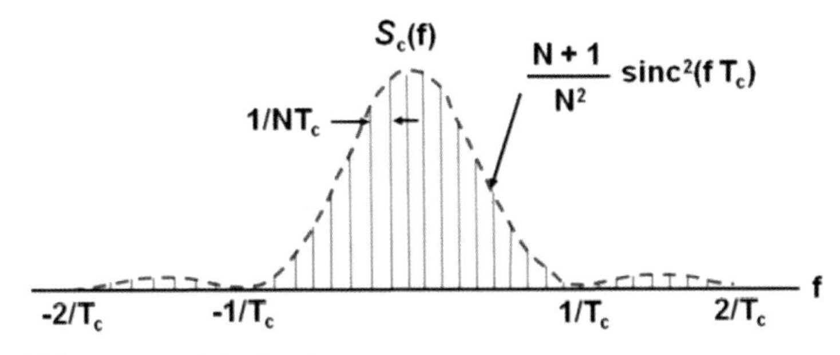

Figure 7.3 Power spectral density of m-sequences.

7.3 GOLD CODES

The auto-correlation properties of m-sequences are generally poorer compared to gold codes. We have seen from the example above that the cross-correlation between one m-sequence and its shifted versions is not really exactly zero but a small number. Gold codes permit the construction of longer sequences with better auto-correlation and cross-correlation properties. Gold codes are formed by linearly combining two m-sequences with different offsets. For example when two m-sequences of length $2^r - 1$ are combined the resulting code is of length $(2^r - 1)^2$. The outputs from two m-sequences are XORed to form the gold sequence. The resulting code is a gold code (Figure 7.4). Not all pairs of m-sequences can result to gold codes. The pairs of m-sequences which yield gold codes are termed 'preferred pairs'. Gold codes possess several desirable properties:

i) Gold codes have three-valued auto-correlation and cross-correlation values which are

$$\{-1, -\rho(m), \rho(m) - 2\}$$

Where

$$\rho(m) = \begin{cases} 2^{(m+1)/2} + 1; & m = odd \\ 2^{(m+2)/2} + 1; & m = even \end{cases} \tag{7.7}$$

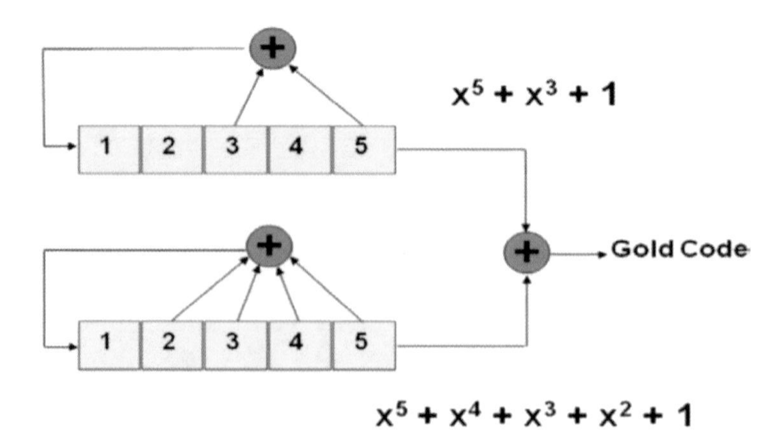

Figure 7.4 Generating Gold Codes.

ii) The number of gold codes obtained is $2^r + 1$

In Figure 7.4, two fifth order polynomials are used to generate the gold code.

7.4 GOLAY CODES

While m-sequences, pseudo random numbers and Gold codes are used for data transformation to match the channel, there are other codes which are used for combating noise and errors in the channel. Coding theory deals with such codes and they are of two types: block codes and convolution codes. Block codes are derived from sequences of n bits of which k are information bits and the remaining n-k are redundant bits, the so-called parity bits. The redundant bits are deployed to protect the information bits. The parity bits do not carry information, but are rather used for error correction encountered when the code words are transmitted. Block code encoders are memoryless. The n digits in the codewords do not depend on previous code words, but depend only on each other. In convolution codes, the n digits in the codewords depend on previous codewords encoded and thus the convolution encoder has memory. This section discusses only a type of block codes, specifically Golay codes.

The telecommunication channel is a largely dynamic environment in which a lot of signal degradation takes place including bit and byte error rates (BER). Most modern transmissions are binary in nature and hence the possibility of lost bits, pulse spreading as well as bit flipping is very high. This type of error is mostly random and hence there are no specific locations within a codeword transmitted at which we should expect errors. Errors may in fact occur anywhere within a codeword. When a channel is noisy and changes the bits that were transmitted, how do we identify the bits and also correct the errors? This type of error in transmission is mostly handled with error correcting codes. First we need some information about the nature of the channel. Next we also need information on the type of error correcting code to use. How do we create error correcting codes? Lastly we also need information on which error correcting codes are best and how many bit errors can they detect and how many of them can the error correcting code correct in practical terms?

Communication channels are mostly continuous which may carry both continuous and discrete data and also produce continuous or discrete outputs at the receiver. Error correcting codes mostly target discrete inputs to the channel and discrete received bits or codewords. The Gaussian channel is a popular choice for modelling a channel with real input and real output and its conditional probability distribution is Gaussian in nature. For a real input x and output y, the Gaussian distribution is:

$$P(y|x) = \frac{1}{\sqrt{2\pi\sigma^2}} \exp\left[-(y-x)^2/2\sigma^2\right] \tag{7.8}$$

This is the probability that y was received given that x was transmitted and σ^2 is the noise power. Although this channel has a continuous input and output, the channel has a discrete nature. The distribution function tells us that there is error when the received signal is compared with the transmitted signal. When the errors are bit errors, we need a practical code to tell us where the errors are and then to correct them. A practical error correcting code can be encoded and decoded within a reasonable amount of computing time. In practice binary codes are used. The role of the code is then to tell the receiver how close to the transmitted vector the received vector is. The measure of 'how close' is given by the so-called Hamming distance between the received and transmitted vectors. The Hamming distance between two binary codewords is the number of bit positions where the two code words differ.

For example, the Hamming distance between the code words
10001111 and 11011011
is 3. The code words differ in positions three, five and seven from the right.

Golay codes in general are error-control codes that can be used to detect bit errors and also to correct some of them. Golay codes belong to a class of codes called 'perfect' or quasi-perfect code. For perfect t-error correcting codes, every code word lies within a Hamming distance t to exactly one and only one codeword. This means that the minimum distance for perfect codes is $d_{\min} = 2t + 1$. Hence a code with distance d can correct $\lfloor (d-1)/2 \rfloor$ errors. The symbol $\lfloor x \rfloor$ is the integer smaller than x.

Golay code symbols are taken from an alphabet of q element where q result to two sub-classes of Golay codes. In binary Golay codes the value of q is two alphabets 0 and 1 while q is three alphabets for ternary Golay codes. We discuss first binary block codes and then progress to discussing binary and ternary Golay codes.

7.4.1 Binary block codes

Block codes provide an ordered rule for converting a sequence of bits c of length k bits created by a source to another sequence of length n bits for transmission. The length of the codewords is n bits implying that there are 2^n possible codewords. Each code word contains k information bits and the remaining n-k bits are parity or redundant bits. In essence this means that of the 2^n possible codewords, only 2^{n-k} are used. Formally the block code is described as an (n,k) block code. The 2^n codewords form a vector field the so-called finite field or the Galois field GF(2).

Definition 1: The code rate R is defined as $R = \frac{k}{n}$ where $R \leq 1$ for binary codes. Formal discussion of Galois field is not the subject of this chapter. We are interested in the subclass of linear binary block codes called cyclic codes that can be generated using polynomials.

Definition 2: Linear Cyclic code C = (n,k) is a code that has codeword c with the following property,

a) Given that

$$c = (c_0, \quad c_1, \quad \cdots, \quad c_{n-2}, \quad c_{n-1}) \\ where\ c_i \in \{0,1\}; 0 \leq i \leq n-1 \tag{7.9}$$

then the codeword $c^{(1)}$ created from the codeword c is a cyclic rotation of c given by the expression

$$c^{(1)} = (c_{n-1}, \quad c_0, \quad \cdots, \quad c_{n-3}, \quad c_{n-2})$$

b) there is a code polynomial c(X) of degree n-1 defined by the expression

$$c(X) = c_0 + c_1 X + c_2 X^2 + \cdots + c_{n-1} X^{n-1} \tag{7.10}$$

In other words, every digit of the codeword is a coefficient of the code polynomial.

c) The code polynomial that has minimum degree in a given (n,k) cyclic code is called the generator polynomial g(X). The generator polynomial has the form

$$g(X) = 1 + g_1 X + g_2 X^2 + \cdots + g_{n-k-1} X^{n-k-1} + X^{n-k} \tag{7.11}$$

The general form of the generator matrix forming the (n,k) code set is given by the matrix

$$\mathbf{G} = \begin{bmatrix} \mathbf{c}_0 \\ \mathbf{c}_1 \\ \vdots \\ \mathbf{c}_{k-2} \\ \mathbf{c}_{k-1} \end{bmatrix}$$

As shown there are k codewords in the rows of G that form the codewords. Thus if s are the bits created by the source, then the transmitted bits t would be given by the expression:

$$t = G^T s \tag{7.12}$$

Both s and t are column vectors.

7.4.2 Binary Golay Codes

In search of a perfect code, Golay noticed a code that can be derived from a sum of selected binomial terms given by the expression

$$\binom{23}{0} + \binom{23}{1} + \binom{23}{2} + \binom{23}{3} = 2^{11} = 2^{23-12} \tag{7.13}$$

The binomial sum suggested to him the possibility of an error correcting code of the form (23, 12) that could correct up to 3 errors. His discovery of the code in 1949 was the first time it was shown that indeed a perfect code does exist that could be used to correct any combination of up to three bits in a 23 bit sequence. Golay code is generated with the pair of polynomials

$$g_1(X) = 1 + X^2 + X^4 + X^5 + X^6 + X^{10} + X^{11} \tag{7.13a}$$

and

$$g_2(X) = 1 + X + X^5 + X^6 + X^7 + X^9 + X^{11} \tag{7.13b}$$

The two polynomials are factors of the polynomial $1 + X^{23}$ in GF(2) where

$$\frac{X^{23} + 1}{1 + X} = g_1(X) g_2(X) \tag{7.14}$$

Golay codes may be decoded using the systematic search decoder or the Kasami decoder. Golay codes may be extended by adding a parity bit to the end of each code. The parity bit is '1' if there are even numbers of bit '1' in the code. The parity bit is a '0' if the number of bit '1' is odd. Extension of the (23, 12) code this way leads to the (24, 12) Golay code. The resulting code is however not a perfect code. The resulting code has the generator matrix G = [I B] where I is the 12x12 identity matrix and B is the matrix shown below:

$$B = \begin{bmatrix} 1 & 1 & 0 & 1 & 1 & 1 & 0 & 0 & 0 & 1 & 0 & 1 \\ 1 & 0 & 1 & 1 & 1 & 0 & 0 & 0 & 1 & 0 & 1 & 1 \\ 0 & 1 & 1 & 1 & 0 & 0 & 0 & 1 & 0 & 1 & 1 & 1 \\ 1 & 1 & 1 & 0 & 0 & 0 & 1 & 0 & 1 & 1 & 0 & 1 \\ 1 & 1 & 0 & 0 & 0 & 1 & 0 & 1 & 1 & 0 & 1 & 1 \\ 1 & 0 & 0 & 0 & 1 & 0 & 1 & 1 & 0 & 1 & 1 & 1 \\ 0 & 0 & 0 & 1 & 0 & 1 & 1 & 0 & 1 & 1 & 1 & 1 \\ 0 & 0 & 1 & 0 & 1 & 1 & 0 & 1 & 1 & 1 & 0 & 1 \\ 0 & 1 & 0 & 1 & 1 & 0 & 1 & 1 & 1 & 0 & 0 & 1 \\ 1 & 0 & 1 & 1 & 0 & 1 & 1 & 1 & 0 & 0 & 0 & 1 \\ 0 & 1 & 1 & 0 & 1 & 1 & 1 & 0 & 0 & 0 & 1 & 1 \\ 1 & 1 & 1 & 1 & 1 & 1 & 1 & 1 & 1 & 1 & 1 & 0 \end{bmatrix}$$

The minimum distance for the extended code is 8 and the code rate is $R = \frac{1}{2}$ and it can be used to correct at most t=3 error bits. The extended Golay code is quasi-perfect and there are 4096 or 2^{12} of them.

7.4.3 Ternary Golay Codes

In addition to the binary Golay codes, he also discovered a ternary Golay code. The ternary Golay codes may also be defined although one of such codes is perfect. The (11, 6) Golay code is the only known ternary perfect non-binary code. The code is defined over the GF(3) field where

$$\binom{11}{0} + 2\binom{11}{1} + 4\binom{11}{2} = 3^{11-6} = 3^5 = 243 \tag{7.15}$$

The ternary (11, 6) Golay code has minimum Hamming distance 5 and can correct t=2 errors. The (11, 6) ternary Golay code is extendible to the (12, 6) code by adding parity bits. In general, of all the known codes (n, k) the Golay codes have the largest minimum distance.

7.5 SCRAMBLING CODES

7.5.1 Downlink Scrambling Codes

Scrambling codes [3-5] are central to the operation and performance of WCDMA networks. They are used to scramble the signals before transmission and help to protect data and also could be used for identification of users in a multiple access system.

The scrambling code sequences are constructed by combining two real sequences **x** and **y** into a complex sequence. Each of the two real sequences is constructed as the position wise modulo 2 sum of 38400 chip segments of two binary *m*-sequences generated by means of two generator polynomials of degree 18. The resulting sequences constitute segments of a set of Gold sequences. The scrambling codes are repeated for every 10 ms radio frame.

Let *x* and *y* be the two sequences respectively. The ***x*** sequence is constructed using the primitive (over GF(2)) polynomial $1+X^7+X^{18}$. The **y** sequence is constructed using the polynomial $1+X^5+X^7+X^{10}+X^{18}$. The sequence depending on the chosen scrambling code number n is denoted by z_n in the following procedure. This procedure originates from [3-5].

Let x(i), y(i) and $z_n(i)$ denote the ith symbol of the sequence x, y, and z_n, respectively. The m-sequences x and y are constructed as follows. We start from an initial condition.

Initial Conditions:

a) Start constructing sequence x with x (0)=1, x(1)= x(2)=...= x (16)= x (17)=0. In other words, we start from an all zero x sequence (000000000000000000).
b) Start constructing sequence y with y(0)=y(1)= ... =y(16)= y(17)=1. The y sequence on the other hand starts from an all '1' sequence (111111111111111111).
c) For full construction of sequences x and y we use their generator polynomials:

$$\begin{cases} 1+x^7+x^{18} \\ 1+y^5+y^7+y^{10}+y^{18} \end{cases}$$

d) The Recursion is:

$$x(i+18)=x(i+7)+x(i) \bmod ulo\, 2; \{i=0,1,\ldots,2^{18}-20$$

$$y(i+18)=y(i+10)+y(i+7)+y(i+5)+y(i) \bmod ulo\, 2; \{i=0,1,\ldots,2^{18}-20$$

The first recursion generates the x sequence and the second generates the y sequence.

e) The nth Gold code sequence z_n is then created as:

$$z_n(i)=\left(x(i+n)\,\mathrm{mod}\,ulo\left(2^{18}-1\right)\right)+y(i)\,\mathrm{mod}\,ulo\,2 \quad (7.16)$$
$$for\ (n,i)=0,1,\ldots,2^{18}-2$$

f) The binary sequence are converted to real-valued sequences z_n using the following procedure:

$$Z_n(i)=\begin{cases}+1 & if\ z_n(i)=0\\ -1 & if\ z_n(i)=1\end{cases} \quad for \quad i=0,1,\ldots,2^{18}-2 \quad (7.17)$$

The nth complex scrambling code $S_{dl,n}$ for DL is finally obtained with the equation [3-5]:

$$S_{dl,n}(i)=Z_n(i)+jZ_n\left((i+131072)\,\mathrm{mod}\,ulo\left(2^{18}-1\right)\right) \quad (7.18)$$
$$i=0,1,\ldots,38399$$

7.5.2 Uplink Scrambling Codes

All uplink physical channels on an activated uplink frequency are scrambled with a complex-valued scrambling code. Scrambling codes are either formed from Gold codes or maximal-length codes. Dedicated physical channels may be scrambled by either a long or a short scrambling code. The PRACH message part is scrambled with a long scrambling code. There are 2^{24} long and 2^{24} short uplink scrambling codes. Uplink scrambling codes are assigned by higher layers.

The long scrambling sequences $c_{long,1,n}$ and $c_{long,2,n}$ are constructed from position wise modulo 2 sum of 38400 chip segments of two binary *m*-sequences generated by means of two generator polynomials of degree 25.

7.5.2.1 Generation of Uplink Long Scrambling Codes

Let x, and y be the two *m*-sequences respectively. The x sequence is constructed using the primitive (over GF(2)) polynomial $X^{25}+X^3+1$.

The y sequence is constructed using the polynomial $X^{25}+X^3+X^2+X+1$. The resulting sequences thus constitute segments of a set of Gold sequences. The sequence $c_{long,2,n}$ is a 16777232 chip shifted version of the sequence $c_{long,1,n}$.

Let n_{23} ... n_0 be the 24 bit binary representation of the scrambling sequence number n with n_0 being the least significant bit. The x sequence depends on the chosen scrambling sequence number n and is denoted x_n, in the following section. Let $x_n(i)$ and $y(i)$ denote the ith symbol of the sequence x_n and y, respectively.

The m-sequences x_n and y are constructed as:

Initial conditions:

$x_n(0)=n_0$, $x_n(1)= n_1$, ... $=x_n(22)= n_{22}$,$x_n(23)= n_{23}$, $x_n(24)=1$.

$y(0)=y(1)=$ ... $=y(23)= y(24)=1$.

Recursive definition of symbols :

$x_n(i+25) =x_n(i+3) + x_n(i)$ modulo 2, $i=0,\ldots, 2^{25}-27$.

$y(i+25) = y(i+3)+y(i+2) +y(i+1) +y(i)$ modulo 2, $i=0,\ldots, 2^{25}-27$.

Define the binary Gold sequence z_n by:

$$z_n(i) = x_n(i) + y(i) \text{ modulo } 2, \; i = 0, 1, 2, \ldots, 2^{25}-2. \tag{7.19}$$

The real valued Gold sequence Z_n is defined by:

$$Z_n(i) = \begin{cases} +1 & \text{if } z_n(i) = 0 \\ -1 & \text{if } z_n(i) = 1 \end{cases} \quad \text{for } i = 0,1\cdots,2^{25}-2 \tag{7.20}$$

The real valued long scrambling sequences $c_{long,1,n}$ and $c_{long,2,n}$ are defined as follows:

$$C_{long,1,n}(i) = Z_n(i) \quad \text{for } i = 0,1\cdots,2^{25}-2 \tag{7.21a}$$

$$C_{long,2,n}(i) = Z_n\left((i+16777232) \bmod ulo\left(2^{25}-1\right)\right) \quad \text{for } i = 0,1\cdots,2^{25}-2 \tag{7.21b}$$

Finally, the complex-valued long scrambling sequence $C_{long,\,n}$, is defined as:

$$C_{long,n}(i) = c_{long,1,n}(i)\left(1 + j(-1)^i c_{long,2,n}\left(2\lfloor i/2 \rfloor\right)\right)$$

$$i = 0,1\cdots,2^{25}-2$$

$\lfloor i/2 \rfloor$ denotes rounding to nearest lower integer. The architecture for generating the uplink scrambling sequence is shown in Figure 7.5.

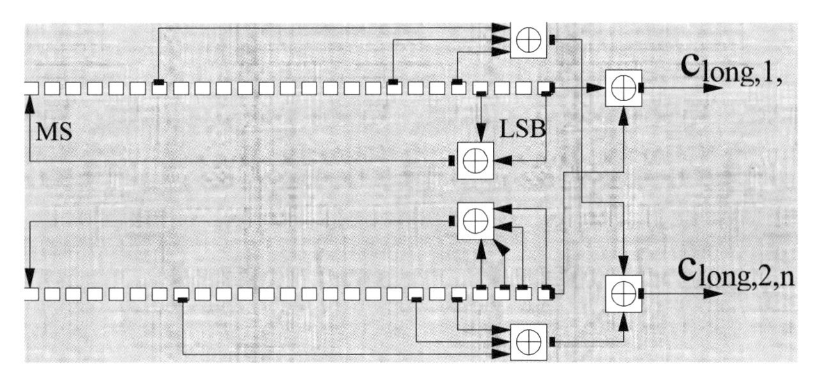

Figure 7.5 Architecture for Generating Uplink Scrambling Sequence.

7.5.2.2 Generation of Uplink Short Scrambling Codes

Apart from the long scrambling sequence, short scrambling codes are also used for uplink scrambling. Figure 7.6 shows how the uplink scrambling codes are generated.

The short scrambling sequences $c_{\text{short},1,n}(i)$ and $c_{\text{short},2,n}(i)$ are defined from a sequence in the family of periodically extended S(2) codes.

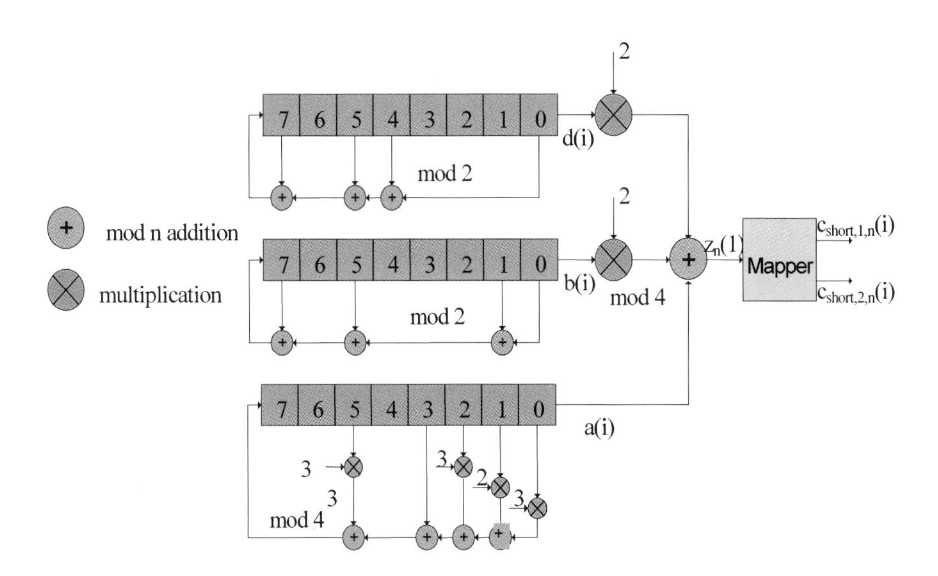

Figure 7.6 Generation of Uplink Short Scrambling Codes.

Let $n_{23}n_{22}\ldots n_0$ be the 24 bit binary representation of the code number n.
The nth quaternary S(2) sequence $z_n(i)$, $0 < n < 16777215$, is obtained by modulo 4 addition of three sequences: a quaternary sequence $a(i)$ and two binary sequences $b(i)$ and $d(i)$
The initial loading of the three sequences is determined from the code number n.
The sequence $z_n(i)$ of length 255 is generated according to the following relation:
$z_n(i) = a(i) + 2b(i) + 2d(i)$ modulo 4, $i = 0, 1, \ldots, 254$;

The quaternary sequence $a(i)$ is generated recursively by the polynomial
$g_0(x)= x^8+3x^5+x^3+3x^2+2x+3$
as:
$a(0) = 2n_0 + 1$ modulo 4;
$a(i) = 2n_i$ modulo 4, $i = 1, 2, \ldots, 7$;
$a(i) = 3a(i\text{-}3) + a(i\text{-}5) + 3a(i\text{-}6) + 2a(i\text{-}7) + 3a(i\text{-}8)$ modulo 4, $i = 8, 9, \ldots, 254$;

The binary sequence $b(i)$ is generated recursively by the polynomial $g_1(x)= x^8+x^7+x^5+x+1$ as
$b(i) = n_{8+i}$ modulo 2, $i = 0, 1, \ldots, 7$,
$b(i) = b(i\text{-}1) + b(i\text{-}3) + b(i\text{-}7) + b(i\text{-}8)$ modulo 2, $i = 8, 9,\ldots,254$,
The binary sequence $d(i)$ is generated recursively by the polynomial $g_2(x)= x^8+x^7+x^5+x^4+1$ as:
$d(i) = n_{16+i}$ modulo 2, $i = 0, 1, \ldots, 7$;
$d(i) = d(i\text{-}1) + d(i\text{-}3) + d(i\text{-}4) + d(i\text{-}8)$ modulo 2, $i = 8, 9,\ldots,254$
The sequence $z_n(i)$ is extended to length 256 chips by setting $z_n(255) = z_n(0)$.

The uplink short scrambling codes are mapped. The mapping from $z_n(i)$ to the real-valued binary sequences $c_{\text{short},1,n}(i)$ and $c_{\text{short},2,n}(i)$, , $i = 0, 1, \ldots, 255$ is defined in Table 7.3.

Finally, the complex-valued short scrambling sequence $C_{\text{short, n}}$, is defined as:

$$C_{short,n}(i) = c_{short,1,n}(i \bmod 256)\left(1 + j(-1)^i c_{short,2,n}\left(2\lfloor (i \bmod 256)/2 \rfloor\right)\right) \qquad (7.23)$$

where $i = 0, 1, 2, \ldots$

Table 7.3 Mapping Format

$z_n(i)$	$c_{short,1,n}(i)$	$c_{short,2,n}(i)$
0	+1	+1
1	-1	+1
2	-1	-1
3	+1	-1

7.5.2.3 DPCH Uplink Scrambling Codes

The uplink dedicated physical channels may be scrambled with either short or long scrambling codes. The nth long uplink scrambling code is defined by the expression:

$$S_{dpch,n}(i) = C_{long,n}(i); \quad i = 0,1,2,\ldots,38399 \tag{7.24a}$$

The nth short uplink scrambling code is defined by the expression:

$$S_{dpch,n}(i) = C_{short,n}(i); \quad i = 0,1,2,\ldots,38399 \tag{7.24b}$$

where the lowest index refers to the chip transmitted first in time.

7.1.1.1 PRACH Message Part Scrambling Code

The scrambling code used for the message part is 10ms long. There are 8192 different PRACH scrambling codes. The nth PRACH message part scrambling code is defined as

$$S_{r-msg,n}(i) = C_{long,n}(i+4096); \quad \begin{cases} i = 0,1,2,\ldots,38399 \\ n = 0,1,2,\ldots,8191 \end{cases} \tag{7.25}$$

The lowest index corresponds to the chip that is transmitted first in time.

7.5.2.4 PRACH Preamble Codes

A complex valued code sequence is used. It is built from a preamble scrambling code and a preamble signature code as follows:

$$C_{pre,n,s}(k) = S_{r-pre,n}(k) \times C_{sig,s}(k) \times e^{j\left(\frac{\pi}{4}+\frac{\pi}{2}k\right)}; \quad k = 0,1,2,\ldots,4095 \tag{7.26}$$

The preamble code is $S_{r-pre,n}(k)$

The signature code is $C_{sig,s}(k)$.

7.5.3 Preamble Scrambling Code

The scrambling code for the PRACH preamble part is constructed from the long scrambling sequences.

$$S_{r-pre,n}(i) = C_{long,1,n}(i); \quad \begin{cases} i = 0,1,2,\ldots,4095 \\ n = 0,1,2,\ldots,8192 \end{cases} \tag{7.27}$$

There are 8192 of them. They are divided into 512 groups (16 per group).

Each group is matched to the scrambling code used in the DL of a cell, so that for each, m and k.

The kth preamble scrambling code $S_{r-pre,n}(i)$ has index n as

$$k = 0,1,2,\ldots,15; \quad m = 0,1,2,\ldots,511$$
$$n = 16 \times m + k$$

7.5.4 Preamble Signature Code

The preamble signature corresponding to a signature s consists of 256 repetitions of a length 16 signature $P_s(n)$, n=0…15

$$C_{sig,s}(i) = P_s(i \text{ } modulo \text{ } 16); \quad \{i = 0,1,2,\ldots,4095 \tag{7.28}$$

The signature $P_s(n)$ is from the set of 16 Hadamard codes of length 16.

Preamble signatures are shown in Table 7.4.

7.6 ORTHOGONAL VARIABLE SPREADING FACTOR CODES

7.6.1 Generation of OVSF codes

The OVSF codes are fairly easy to generate using the Hadamard matrices. Given the Hadamard matrix for level n as H_n, the matrix H_{n+1} for level n+1 is given by the expression:

$$H_0 = (1); \; H_1 = \begin{pmatrix} 1 & 1 \\ 1 & -1 \end{pmatrix}$$
$$H_{n+1} = \begin{pmatrix} H_n & H_n \\ H_n & -H_n \end{pmatrix} \tag{7.29}$$

By using the recursive expression, it is a lot easy to generate the codes for each level of the OVSF tree.

Table 7.4 Preamble Signatures

Preamble signature	Value of *n*															
	0	1	2	3	4	5	6	7	8	9	10	11	12	13	14	15
$P_0(n)$	1	1	1	1	1	1	1	1	1	1	1	1	1	1	1	1
$P_1(n)$	1	-1	1	-1	1	-1	1	-1	1	-1	1	-1	1	-1	1	-1
$P_2(n)$	1	1	-1	-1	1	1	-1	-1	1	1	-1	-1	1	1	-1	-1
$P_3(n)$	1	-1	-1	1	1	-1	-1	1	1	-1	-1	1	1	-1	-1	1
$P_4(n)$	1	1	1	1	-1	-1	-1	-1	1	1	1	1	-1	-1	-1	-1
$P_5(n)$	1	-1	1	-1	-1	1	-1	1	1	-1	1	-1	-1	1	-1	1
$P_6(n)$	1	1	-1	-1	-1	-1	1	1	1	1	-1	-1	-1	-1	1	1
$P_7(n)$	1	-1	-1	1	-1	1	1	-1	1	-1	-1	1	-1	1	1	-1
$P_8(n)$	1	1	1	1	1	1	1	1	-1	-1	-1	-1	-1	-1	-1	-1
$P_9(n)$	1	-1	1	-1	1	-1	1	-1	-1	1	-1	1	-1	1	-1	1
$P_{10}(n)$	1	1	-1	-1	1	1	-1	-1	-1	-1	1	1	-1	-1	1	1
$P_{11}(n)$	1	-1	-1	1	1	-1	-1	1	-1	1	1	-1	-1	1	1	-1
$P_{12}(n)$	1	1	1	1	-1	-1	-1	-1	-1	-1	-1	-1	1	1	1	1
$P_{13}(n)$	1	-1	1	-1	-1	1	-1	1	-1	1	-1	1	1	-1	1	-1
$P_{14}(n)$	1	1	-1	-1	-1	-1	1	1	-1	-1	1	1	1	1	-1	-1
$P_{15}(n)$	1	-1	-1	1	-1	1	1	-1	-1	1	1	-1	1	-1	-1	1

References

[1] S.W. Golomb, Shift Register Sequences, Aegean Park Press, 1992

[2] Esmael H Dinan and Bijan Jabari, "Spreading Codes for Direct Sequence CDMA and Cellular Networks", IEEE Communications Magazine, September 1998, pp. 48-54.

[3] 3GPP TS 25.301 V4.2.0 (2001-12) ;Universal Mobile Telecommunications System (UMTS); Radio Interface Protocol Architecture (Release 4)
[4] 3GPP TS 25.211 V11.0.0 (2011-12); Physical channels and mapping of transport channels onto physical channels (FDD) (Release 11)
[5] 3GPP TS 25.213 V11.1.0 (2012-03); Spreading and modulation (FDD) (Release 11)

Chapter 8.

Code Planning in WCDMA (UMTS)

Johnson I Agbinya

Department of Electronic Engineering, La Trobe University, Australia
J.Agbinya@latrobe.edu.au

8.1 INTRODUCTION

A great deal of re-engineering was required to enhance the performance of existing GSM networks. Initially it was thought that the evolution from GSM to GPRS and then to EDGE was enough to position GSM networks to deliver large data rates suitable for carrying both voice and data (video, images and graphics). At that time text messaging and MMS were becoming ubiquitous as data services on GSM networks. Apart from this the need to reduce or eliminate co-channel interference as well as adjacent channel interference and yet efficiently identify users and base stations properly proved too much to be achieved by mere evolution of GSM. Also lessons learned from IS-95 deployment in the US showed that CDMA networks were more secure and also devoid of adjacent channel interference. It was also well known that CDMA as a technology is robust in noisy environments and able to withstand jamming and interception. It was also considered that packet switching is desirable rather than existing circuit-switched networks. Furthermore it was considered that a fully operational IP cellular network would lead to new services and also help to consolidate the Internet. The Internet was allowing file download, music and providing gaming, services which were unavailable in the existing 2G networks. This made it necessary to consider IP core and radio IP networks because doing so would lead to network convergence. This motivated early proposals to design a third generation network which uses CDMA as the basic air interface.

Adopting code division multiple access system instead of TDMA/FDMA as used in GSM networks implies that the means for identifying and separating users and base stations needs to be changed as well. CDMA systems traditionally use

4G Wireless Communication Networks: Design, Planning and Applications, 161-168.

one carrier frequency for the downlink and another carrier frequency for transmissions in the uplink and all users share this frequency. CDMA networks provide a suite of orthogonal codes which could be used to separate and identify users and base stations. Using them requires code allocation and planning, a replacement for frequency planning in FDMA/TDMA systems. This is in addition to planning of the radio and core networks. This chapter discusses exclusively WCDMA codes, how they are allocated and their characteristics and usage in UMTS networks.

8.2 CODES IN WCDMA NETWORKS

Three types of codes are used in CDMA networks. They are scrambling codes, channelization codes and synchronisation codes. Orthogonal scrambling codes are used as the means for identifying sectors with which users (multiple access technique) are able to communicate or access. This is then used by the network to decide the sectors that will communicate with the handset and also how to manage handover.

In WCDMA systems a set of 16 scrambling codes are assigned to one sector. One of these codes is a primary code and the remaining 15 are secondary codes. The set is taken from a total of 512 different code sets. A network however usually contains thousands of sectors. Hence the planning of codes to ensure no code-collision and to remove scrambling code confusion is essential. Later in this chapter we describe several situations in faulty code planning which could result to scrambling code confusion between sectors. Two types of codes are employed, orthogonal variable spreading codes and pseudo-random codes. (See Chapter 7)

The roles the orthogonal variable spreading codes play vary in the downlink and the uplink. In the downlink scrambling codes are used for mitigating interference and the channelization codes are used for rate matching and separation of users (Figure 8.1a). In the uplink (Figure 8.1b) scrambling codes are used for user identification and interference mitigation and the channelization codes are used for rate matching. Channelization codes are the so-called orthogonal variable spreading codes (OVSF). Walsh codes are traditionally used for this. Scrambling codes are traditionally pseudo noise (PN) codes. Complex-valued gold codes are used for scrambling.

The uplink scrambling codes are either long/short of length 38400/256 chips of total duration of 10ms/66.67micro seconds. There are 16,777,216 of these codes. They are used for separation of terminals (See Figures 8.1a and b).

The scrambling codes in the downlink have length 38400 chips or 10ms duration. There are 512 code sets. One code set contains one primary code and 15 secondary codes. Therefore, there are actually (512x16=8192) codes used for downlink scrambling. Neither the downlink nor uplink scrambling changes the bandwidth of the signal. They are used for separation of sectors.

Channelization codes are of length 4-512 chips (1.04 – 133.34 microseconds). There are 4 to 256 channelization codes in the uplink and 4 to 512 channelization codes in the downlink. The spreading factors for the channelization codes are also

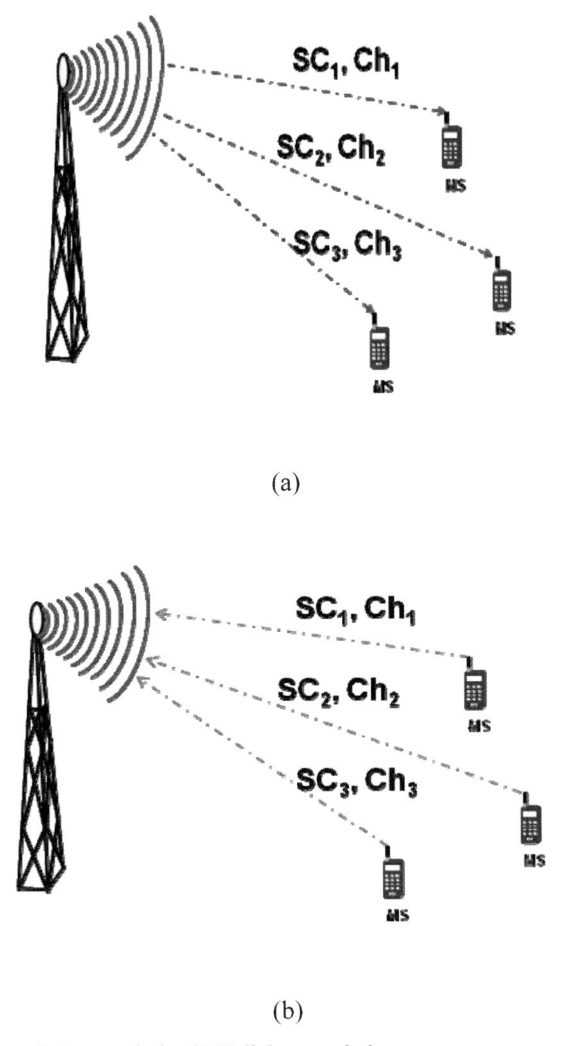

Figure 8.1 a) Downlink transmission b) Uplink transmission.

of these values. Channelization codes increase bandwidth. In the uplink they are used to separate the physical data and control data from the same terminal. In the downlink however, they are used to separate the connection to different terminals in the same cell.

The third type of codes used in UMTS is called synchronisation codes. They are used to enable terminals to locate and synchronise to the main control channels of cells and do change bandwidth. Gold codes are used for this purpose consisting of primary synchronisation codes (PSC) and secondary synchronisation codes (SSC). They are typically of length 256 chips and duration

66.67 microseconds. There is one primary and 16 secondary synchronisation codes.

Table 8.1 is a summary of codes used in UMTS FDD (frequency domain duplex) systems as given by 3GPP.

8.3 SPREADING

As shown in Figure 8.2, UMTS systems use two coding operations, spreading and

Table 8.1 UMTS Codes (3GPP) [1]

	Synchronisation Codes	**Channelization Codes**	**Scrambling Codes, UL**	**Scrambling Codes, DL**
Type	Gold Codes Primary Synchronization Codes (PSC) and Secondary Synchronization Codes (SSC)	Orthogonal Variable Spreading Factor (OVSF) codes sometimes called Walsh Codes	Complex-Valued Gold Code Segments (long) or Complex-Valued S(2) Codes (short) Pseudo Noise (PN) codes	Complex-Valued Gold Code Segments Pseudo Noise (PN) codes
Length	256 chips	4-512 chips	38400 chips / 256 chips	38400 chips
Duration	66.67 μs	1.04 μs - 133.34 μs	10 ms / 66.67 μs	10 ms
Number of codes	1 primary code / 16 secondary codes	= spreading factor 4 ... 256 UL, 4 ... 512 DL	16,777,216	512 primary / 15 secondary for each primary code
Spreading	No, does not change bandwidth	Yes, increases bandwidth	No, does not change bandwidth	No, does not change bandwidth
Usage	To enable terminals to locate and synchronise to the cells' main control channels	UL: to separate physical data and control data from same terminal DL: to separate connection to different terminals in the same cell	Separation of terminal	Separation of sector

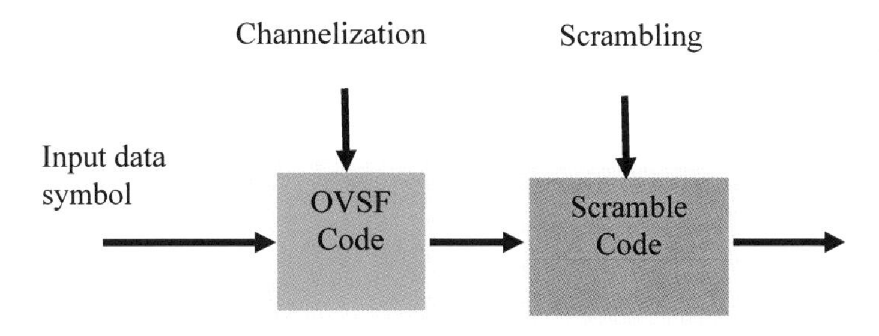

Figure 8.2 Spreading Operations in WCDMA.

scrambling. The first, a spreading operation is called channelization in which the input symbols are spread with the selected OVSF. The resulting signal is further scrambled using a scrambling code.

The channelization spreading transforms each symbol into a number of chips. The number of chips per symbol is called the spreading factor. This process therefore provides different data rates according to the spreading factors. The second operation is scrambling using pseudo-noise codes. This operation does not lead to bandwidth change.

8.4 ORTHOGONAL VARIABLE SPREADING CODES

The spreading (channelization) operation uses the so-called orthogonal variable spreading codes. The orthogonal codes are assigned and also planned as discussed in this section.

By assigning a code to any user equipment (UE), all its ancestors and descendants cannot be allocated to other UEs. This is to prevent code confusion and interference. UMTS systems generally support various data rates and these are supported through the allocation of codes to base stations and user equipment. The general structure of OVSF is a binary tree of codes as shown in Figure 8.3.

By using orthogonal codes (OVSF), orthogonal transmission is preserved between different channels. The spreading factors for data transmission for the downlink are shown in light shadow and for the uplink in dark shadow in Table 8.2. The code at the lowest levels provide the largest data rates. When a user requires more data rate, more codes (parallel code channels) may be allocated to the user upon request. In general the system capacity is equal to the maximum spreading factor (SF_{max}).

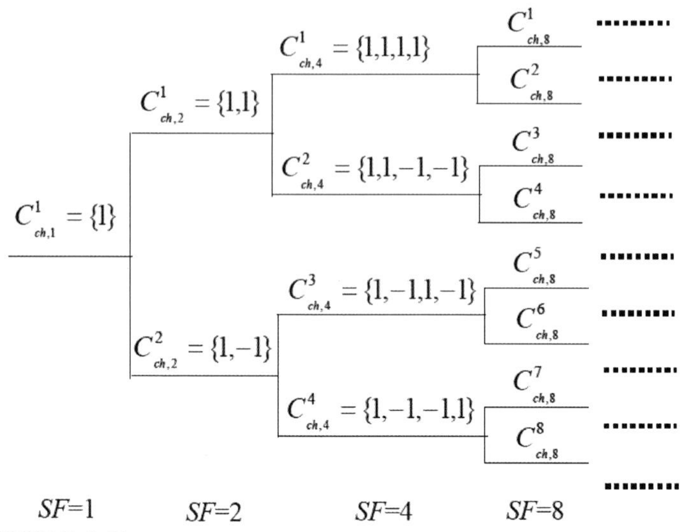

Figure 8.3 OVSF Code Tree.

Table 8.2 Layers of OVSF Code-Tree and Their Spreading Factors

Level	1	2	3	4	5	6	7	8	9	10
Number of Codes	1	2	4	8	16	32	64	128	256	512
Spreading Factor (SF)	1	2	4	8	16	32	64	128	256	512
	1	2	4	8	16	32	64	128	256	512
Data Rate	$2^9 R$	$2^8 R$	$2^7 R$	$2^6 R$	$2^5 R$	$2^4 R$	$2^3 R$	$2^2 R$	$2^1 R$	$2^0 R$

8.5 GENERATION OF OVSF CODES

The OVSF are sequentially labelled and numbered as in Figure 8.3. This code plan enables the determination of code parents and children and which codes may or not be concurrently allocated for use.

Figure 8.3 shows that codes with the same SF are located in the same level of the tree, where level is defined as $\log_2(SF)$.

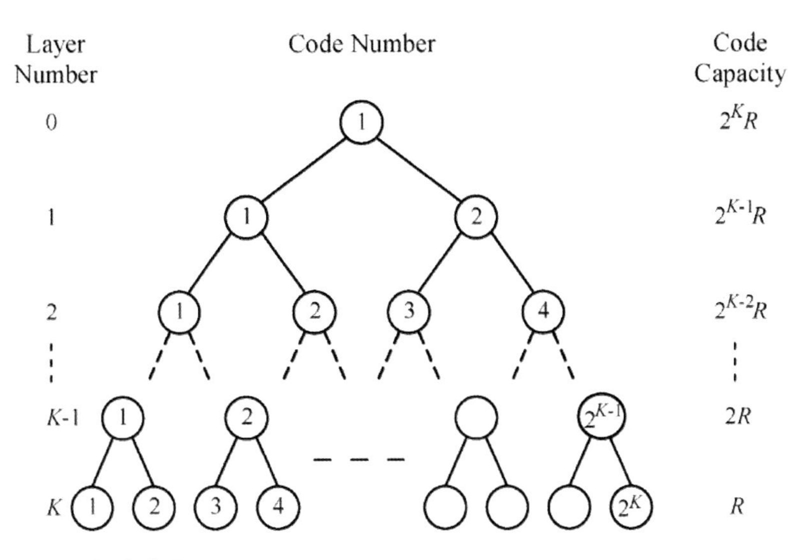

Figure 8.4 OVSF Code Rates.

Figure 8.4 shows how the system data rates vary with the OVSF code rates. Also the data rate for codes at this level is given by the expression: $\frac{SF_{\max}}{SF} R_b$ where R_b is the basic data rate. In Figure 8.3, each OVSF code is given by the variable $C^k_{ch,SF}$, with k being the code number $\left(1 \le k \le SF = 2^n\right)$and SF the its spreading factor.

In a K layer tree, the total capacity of codes in each layer is equal to $2^K R_b$.

References

[1] 3GPP, Spreading and modulation (FDD) (Release 7), TR25.213, v7.4.0; www.3gpp.org

[2] ETSI,http://www.etsi.org/deliver/etsi_ts/125200_125299/125213/04.04.00_60/ts_125213v040400p.pdf

[3] Howard H. Xia, IMT-2000 and G3G CDMA technologies, IEEE International Conference on 3G Wireless Communications, Silicon Valley, USA, Vodafone AirTouch Plc, 2999 Oak Road, MS 900, Walnut Creek, CA 94596

[4] Randy Roberts, RF/Spread Spectrum Consulting, http://www.sss-mag.com/ss.html

[5] The principles of Spread Spectrum communication, http://cas.et.tudelft.nl/

[6] Jacobus Petrus Franciscus Glas, "The principles of spread spectrum communications" ,http://cas.et.tudelft.nl/~glas/

Chapter 9.

Spread Spectrum

Johnson I Agbinya

Department of Electronic Engineering, La Trobe University, Australia
J.Agbinya@latrobe.edu.au

9.1 SPREAD SPECTRUM

Spread spectrum (SS) was first used by the military to prevent jamming. It now covers secure digital communications, commercial and industrial purposes as well. Applications for commercial spread spectrum range from WLAN, integrated bar code scanner/palm top, computer/radio modem devices for warehousing, digital dispatch, digital cellular telephone communications 2G/3G, wide area networks for passing faxes, computer data, e-mail and multimedia.

Spread-spectrum radio communications resists jamming and is hard to intercept the transmission. Spread-spectrum signals are intentionally distributed over a wide range of frequencies and then collected onto their original frequency at the receiver. This makes them inconspicuous. They are not likely to interfere with other signals intended for business and consumer users just as they are unlikely to be intercepted by a military opponent. This applies to spread-spectrum signals transmitted on the same frequencies.

Spread Spectrum uses wide band, noise-like signals which makes them hard to detect. Spread Spectrum signals are harder to jam (interfere with) than narrow band signals. These low probability of intercept (LPI) and anti-jam (AJ) features are why the military has used spread spectrum for so many years. Spread-spectrum signals are intentionally made to have much wider band than the information they carry to make them more noise-like.

Spread spectrum uses fast codes that run many times the information bandwidth or data rate. These special "spreading" codes are called "pseudo random" or "pseudo noise" codes as described in chapters seven and eight. They are called "pseudo" because they are not real Gaussian noise.

4G Wireless Communication Networks: Design, Planning and Applications, 169-188.

Spread spectrum transmitters use similar power levels to narrow band transmitters. Spread spectrum signals are so wide, they transmit at a much lower spectral power density, measured in watts per hertz, than narrow band transmitters. Spread and narrow band signals can occupy the same band, with little or no interference. This capability is the main reason for all the interest in spread spectrum.

Spread-spectrum modulation techniques are characterised by their wide frequency spectra. The modulated output signals occupy a much greater bandwidth than the signal's baseband information bandwidth. Two criteria must be satisfied for a signal to qualify as a spread spectrum:

1. The transmitted signal bandwidth must be much greater than the information bandwidth.
2. Some function other than the information being transmitted is employed to determine the resultant transmitted bandwidth.

Commercial spread spectrum systems with bandwidth many times the information rates are common. Military systems use spectrum widths from 1000 to 1 million times the information bandwidth. There are two very common spread spectrum modulations techniques techniques: frequency hopping and direct sequence. Two other types of spreading modulations have also been used: time and chirp hopping.

A SS receiver uses a locally generated replica pseudo noise code and a correlator receiver to separate only the desired coded information from all possible signals. A SS correlator can be thought of as a very special matched filter. It responds only to signals that are encoded with a pseudo noise code that matches its own code. Thus, an SS correlator can be "tuned" to different codes simply by changing its local code. The correlator does not respond to man-made, natural or artificial noise or interference.

The spread of energy over a wide band, or lower spectral power density, makes SS signals less likely to interfere with narrow band communications. Narrow band communications, conversely, cause little to no interference to SS systems because the correlation receiver effectively integrates over a very wide bandwidth to recover an SS signal. The correlator then "spreads" out a narrow band interferer over the receiver's total detection bandwidth. The total integrated signal density or SNR at the correlator's input determines whether there will be interference or not. All SS systems have a threshold or tolerance level of interference beyond which useful communication ceases. This tolerance or threshold is related to the SS processing gain. Processing gain is essentially the ratio of the RF bandwidth to the information bandwidth expressed as a logarithm to base ten. A typical commercial direct sequence radio might have a processing gain of from 11 to 16 dB, depending on data rate. It can tolerate total jammer power levels of from 11 to 16 dB stronger than the desired signal.

Besides being hard to intercept and jam, spread spectrum signals are hard to exploit or spoof. Signal exploitation is the ability of an interceptor to listen in to a transmission and use information from the network without being a valid network member or participant. Spoofing is the act of falsely or maliciously introducing misleading or false traffic or messages to a network. SS signals can be made to have any degree of message privacy that is desired. Messages can be cryptographically encoded to any level of secrecy desired. The very nature of SS allows military or intelligence levels of privacy and security to be had with minimal complexity.

9.2 DIRECT-SEQUENCE SPREAD SPECTRUM PROCESSING

Consider a binary discrete time communication system with the received time sequence given by the expression

$$r_m = \varepsilon\, b_m + w_m; \quad \{r_m\}_{m=0}^{\infty} \tag{9.1}$$

where, b_m is the sequence of information symbols of antipodal binary nature, $b_m \varepsilon \{\pm 1\}$, and w_m is an additive white Gaussian noise (AWGN) of zero mean. This means that the noise average and power are given by the expressions

$$E[w_m] = 0, \quad E[w_m w_{m+l}] = \sigma^2 \delta(l). \tag{9.2}$$

We assume the transmitter is connected to an information source which produces regular outputs of +1 and -1 with equal probability. To determine if a +1 or -1 was transmitted for the time instant m, a correlator receiver is used. A level detector can be used as the receiver for this system, such that when the received signal is more than zero, it outputs a +1 and when negative a -1. The decision variables $r_m = y_m$ are:

$$r_m \geq 0; \quad \textit{decision is that} \quad +1 \textit{ was sent}$$
$$r_m < 0; \quad \textit{decision is that} \quad -1 \textit{ was sent}$$

The mean of r_m a normal random variable is $\varepsilon\, b_m$ and its variance σ^2. Having transformed the symbol sequence to ± 1, we need to modulate the sequence (signal). The modulation procedure that results to spreading of the symbol energy into a wider spectrum is called *spread spectrum.* To achieve that we use a spreading sequence consisting of ± 1 to modulate each symbol. Let the representative spreading sequence be $\{c_n\}_{n=0}^{n=N-1}$ such that each symbol b_m results to the transmission of either

$$c_0, c_1, \cdots, c_{N-1} \quad or \quad -c_0, -c_1, \cdots, -c_{N-1}$$

The sign of the sequence transmitted depends on the value of b_m. Certain niceties of spread spectrum need more explanations. Each bit of duration T must be coded into a sequence of N chips (duration of one bit or symbol being transmitted a code) of duration $T_c = T / N$. The chip is a rectangular pulse that occupies the whole bandwidth (T = 1/W, where W is the CDMA bandwidth). This essentially means an increase in bandwidth or signalling rate and spreads the spectrum of the transmitted signal by a factor of N (hence 'spread spectrum'). Compare this with analog modulation schemes whereby a narrow band signal is modulated with a high frequency carrier to lift its frequencies to a higher band. Ideally, the comparison falls short in one aspect. In the traditional modulation schemes, the signal energy is not spread into a wide band, but rather lifted from one low spectral band to a higher spectral band. In spread spectrum, the symbol energy is actually spread out. Figure 9.1 illustrates the process. The data signal of low bandwidth is multiplied with the PN-code having a wide bandwidth and the result is the coded signal having the bandwidth of the PN-code. This is typically a binary multiplication.

This signal may be further angular modulated for transmission so that each chip is pre-multiplied by a cosine function and transmitted. In the channel the transmitted bit signal encounters channel degradation effects such as fading and addition of thermal noise. In general, the received signal will contain thermal noise, multipath effects, interference from other users and the environment and can be described with the expression:

$$r_n = \varepsilon_c \, bc_n + w_n; \quad n = 0,1,\cdots,N-1 \tag{9.3}$$

where, ε_c the chip energy is $\varepsilon_c = \frac{\varepsilon}{N}$ *and* $E\left[w_n^2\right] = \frac{\sigma^2}{N}$. There are two essential properties of spreading codes. The spreading sequence must have a mean of zero and signals modulated with different codes must cancel at the

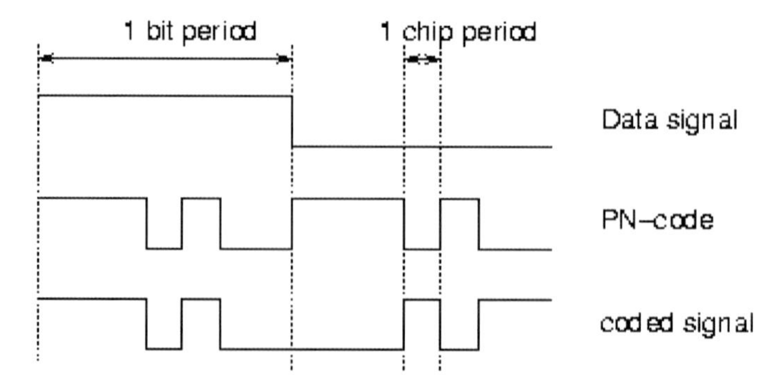

Figure 9.1: Illustration of Direct Sequence Spreading [3].

receiver. These conditions give the sequences a noise-like appearance (pseudo noise). The second condition is necessary for interference cancellation. The two conditions are stated as:

Zero mean: $\sum_{n=0}^{N-1} c_n \approx 0.$

Autocorrelation: $\sum_{n=0}^{N-1} c_n c_{n+i} \approx \begin{cases} N, & i=0 \\ 0, & otherwise \end{cases}.$

In practice, these ideal conditions may not hold but need to be followed to a close approximation. To see how these conditions are applied in the spread spectrum receiver, we will go through the process of spread spectrum demodulation.

9.2.1 Demodulation of Spread Spectrum

Demodulation of spread spectrum is called *despreading*. This is basically a restoration of the base band signal bandwidth and energy to what they were before SS operation. To obtain the decision variable, the correlator in the receiver must carry out the following discrete operation

$$y = \sum_{n=0}^{N-1} r_n c_n = \sum_{n=0}^{N-1} (\varepsilon_c b c_n + w_n) c_n = \varepsilon_c b \sum_{n=0}^{N-1} c_n c_n + \sum_{n=0}^{N-1} w_n c_n \tag{9.4}$$

or

$$y = \varepsilon_c b N + \sum_{n=0}^{N-1} w_n c_n \tag{9.5}$$

The decision variable y has mean $N\varepsilon_c b = \varepsilon b$ and variance σ^2. We observe from the despreading that in real terms, spreading has not improved upon the ideal AWGN channel. Although the signalling rate has increased by a factor of N, the signal bandwidth has also increased, so also the noise power has increased by the factor N. The major advantage of spreading is in correlated signals. Signals with codes other than the one the receiver is tuned to, have 'no correlation' with the desired signal. This includes other mobile terminals, multipath signals, thermal noise and any other interferers.

In a normal CDMA based system, there will be interference from other users. Let us consider the case when the channel contains interference, a constant additional sequence to the desired signal as:

$$r_n = \varepsilon_c\, bc_n + i_n + w_n; \quad n = 0,1,\cdots,N-1 \tag{9.6}$$

Let this interference have a constant amplitude I. What will the receiver produce as an output? The output from the receiver is:

$$y = \sum_{n=0}^{N-1} r_n c_n = \sum_{n=0}^{N-1} (\varepsilon_c bc_n + i_n + w_n) c_n = (N\varepsilon_c b) + I\sum_{n=0}^{N-1} c_n + \sum_{n=0}^{N-1} w_n c_n \tag{9.7}$$

Since the interferer is not correlated with the spread sequence, the receiver must output the variable:

$$y \approx (N\varepsilon_c b) + 0 + \sum_{n=0}^{N-1} w_n c_n \tag{9.8}$$

The interference from other users is suppressed by the despreading operation. The decision variable has the same statistics as in the ideal case with mean $N\varepsilon_c b = \varepsilon b$ and variance σ^2. If spread spectrum were not used, the decision variable would have a mean $\varepsilon b + I$ which could make the signal unusable if the magnitude $|I|$ of the interferer is large. We have assumed in the previous analysis that the received signal is synchronised to the despreading sequence. In practice, this may not be the case because there are no ideal oscillators and system effects do lead to de-synchronisation in the channel. Synchronisation is a key requirement in spread spectrum applications. IS-95 systems use GPS for synchronisation and other systems use pilot sequences (added to the transmitted signal) for the same purpose. For a discussion of synchronization codes, see Chapter 7 of this book.

9.3 INTERFERENCE CANCELLATION USING SPREAD SPECTRUM

A further analysis of interference cancellation by spread spectrum is necessary. After all, it is the corner stone of SS applications. Let us consider a normal channel with multipath reflection (specular component) in addition to the line of sight signal. The delayed copy of the signal at the receiver experiences delay l seconds and unknown attenuation β. We assume that the delay in the channel is less than one symbol duration (ie., $l < N$). Assuming that the chip energy ε_c is normalised to unity, the received signal is:

$$r_n = \begin{cases} b_m c_n + \beta b_{m-1} c_{N-l+n}, & n = 0,1,\cdots,l-1 \\ b_m c_n + \beta b_m c_{n-l}, & n = l, l+1,\cdots,N-1 \end{cases} \tag{9.9}$$

This expression shows that the specular path causes interference from both a delayed version of the desired symbol and the previously transmitted symbol at time m-1 (b_{m-1}). The decision variable is given by the expression:

$$y_m = Nb_m + \beta b_{m-1} \sum_{n=0}^{l-1} c_{N-l+n} c_n + \beta b_m \sum_{n=l}^{N-1} c_{n-l} c_n + \sum_{n=0}^{N-1} w_n c_n \tag{9.10}$$

with solution

$$y_m \approx Nb_m + 0 + 0 + \sum_{n=0}^{N-1} w_n c_n \tag{9.11}$$

This shows that the multipath signal has been suppressed by the despreading operation. If this signal were not spread, the specular part could cause serious inter-symbol interference. To illustrate these concepts, consider Figure 9.2 originally from [3].

One of the major attractions of spread spectrum is its ability to cancel multi-user interference. Assume there are K users or transmitters. The data from the kth user is modulated with the spreading sequence $\left\{c_n^{(k)}\right\}$. The correlation of these signals is given by the expression

$$\sum c_n^{(k)} c_{n+i}^{(j)} \approx \begin{cases} N, & k = j, i = 0 \\ 0, & otherwise \end{cases} \tag{9.12}$$

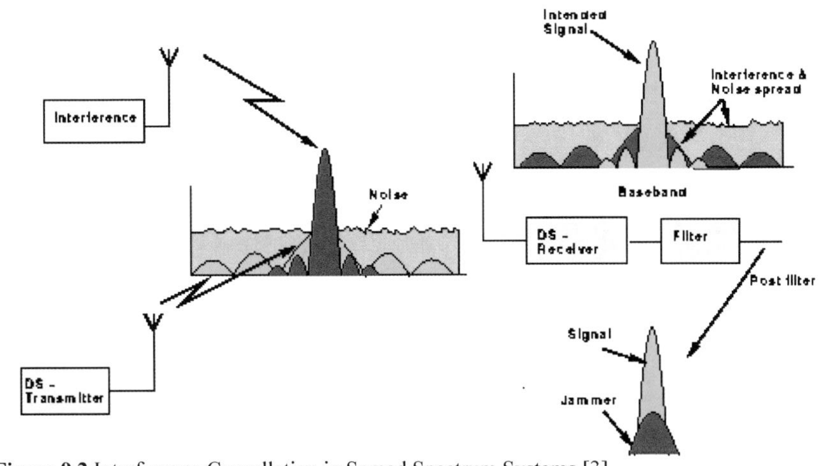

Figure 9.2 Interference Cancellation in Spread Spectrum Systems [3].

9.3.1 Processing Gain

Processing gain is the logarithmic value of the spreading factor. In spread spectrum, a narrowband signal is spread to a wideband signal. The main parameter in spread spectrum systems is the processing gain: the logarithm to base 10 of ratio of transmitted to baseband information bandwidth:

$$G_P = 10 \times Log\left(\frac{Transmission\ bandwidth}{bit\ rate}\right) \tag{9.13}$$

The processing gain determines the number of users that can be allowed in a system, the amount of reduction in multi-path effect and difficulty to jam or detect a signal. For spread spectrum systems it is advantageous to have a processing gain as high as possible. The higher the bit rate, the less the processing gain and higher transmit power or smaller coverage.

9.4 PSEUDO NOISE CODES

A pseudo noise code is generated usually by using shift registers. Chapter 7 provides a good coverage on how such codes may be created. With a shift register of length n, the period of the pseudo noise code obtained is:

$$N_{DS} = 2^n - 1$$

For the direct sequence case, the length of the code is the same as the spreading factor so that the processing gain is: $G_P(DS) = N_{DS}$. The bandwidth of the signal is therefore multiplied by a factor N_{DS}. The power content is not changed, but the power spectral density is lowered.

9.5 IMPLEMENTATION OF DIRECT SEQUENCE SPREAD SPECTRUM

High-speed code sequence is used directly to modulate the carrier, thereby directly setting the transmitted RF bandwidth. Binary code sequences as short as 11 bits or as long as [2^{89} - 1] have been employed for this purpose, at code rates from under a bit per second to several hundred megabits per second.

The result of modulating an RF carrier with such a code sequence is to produce a signal centred at the carrier frequency, direct sequence modulated spread spectrum with a $\left(\frac{\sin x}{x}\right)^2$ frequency spectrum. The main lobe of this spectrum has a bandwidth twice the clock rate of the modulating code, from null to null. The side lobes have a null to null bandwidth equal to the code's clock rate. Direct sequence spectra vary somewhat in spectral shape depending upon the actual carrier and data modulation used. The signal illustrated is that for a binary phase shift keyed (BPSK) signal, which is the most common modulation signal type used in direct sequence systems.

Direct sequence spectrum spreading (DSSS) alters the detection statistics of a communication system. The major result is the interference averaging property that is normally assumed when taking estimates of system capacity. Commercial CDMA spreading is quadrature in both forward and reverse channels. As a prelude to analysis of that system we first consider BPSK spreading. Extension to QPSK case is easy once we have the BPSK results.

9.6 PSEUDO-RANDOM SPREADING SEQUENCES

In this section we approximate the actual binary PN spreading sequences by an ideal Bernoulli (coin-tossing) sequence with equi-probable outcomes. That is Pr[0] = Pr[1] = 1/2, with all trials independent. Mapping the (0, 1) sequence to a (+1, -1) discrete modulation sequence $\{a_n\}$, the autocorrelation of the sequence is a Kronecker delta function, that is

$$R_a(m,n) = E[a_m a_n] = \delta_{m,n} = \begin{cases} 1 & if \quad m = n \\ 0 & if \quad m \neq n \end{cases} \tag{9.14}$$

The actual sequences have off-time correlations of the order of 1/N, where N is the length of the sequence. This approximation is well justified in practice because the random relative RF phases of the interferers tend to remove the small bias that the approximation might otherwise introduce.

9.7 ALTERNATIVE REPRESENTATIONS

In transmitters the most convenient way to impose spreading on data is usually modulo-two addition (exclusive OR) in conventional binary-valued logic. This is illustrated in the first part of Figure 9.3. In the analog world those binary values are represented by bipolar signals. The modulo-two binary addition is equivalent to analog multiplication by ±1 (shown in the second part of Figure 9.3), provided binary 1 maps to bipolar -1 and binary 0 maps to +1.

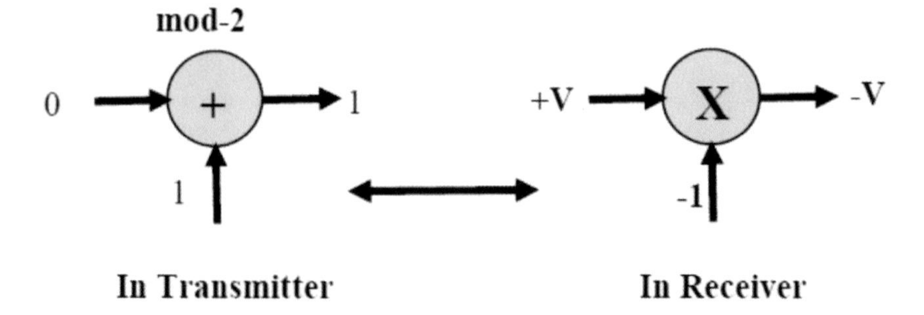

Figure 9.3 Modulo-two binary addition.

We also consider **impulse modulators.** The impulse modulator accepts a binary-valued input, but produces a bipolar impulse output. These alternative representations should not cause any confusion. They come up often enough that we will not specifically call attention to them.

9.8 SYNCHRONISATION

For the purposes of this analysis we assume that time and frequency synchronization has been achieved.

The QPSK model that we consider here is similar to the Forward CDMA Channel except we neglect the orthogonal channelization. We assume here that the spreading sequences are completely uncorrelated between users. There are two consequences of this assumption. First, it means that the users active in the various channels of one base station interfere with one another just as though they would if they came from different stations. Second, it means that the expectations of the chip (time it takes to transmit one bit or symbol of a pseudo noise code) detection amplitudes depend only on the user being addressed, and have no contributions from the other users.

Neither of these is true in the real Forward CDMA Channel. First, not only are the spreading sequences correlated, they are specifically designed to be rigorously orthogonal over the span of 64 chips, which is the span of an FEC code symbol. This means that the mutual interference terms are correlated in such a way that when the amplitudes are summed to make a soft code symbol, the other-channel interference terms rigorously cancel. Second, there is a contribution to the mean detection amplitude from all the code channels. It is precisely that property that lets us separate the code channels in the receiver by selectively de-covering with the desired code channel.

The effect of the orthogonal channelization is to reduce the mutual interference between users. This means for two codes c_1 and c_2 the following relationship holds.

$$\int_a^b c_1(t)c_2(t)dt = 0 \tag{9.15}$$

While the cancellation is not perfect in a real system due to unavoidable multipath, it does help, and contributes somewhat to the forward capacity.

9.8.1 Reverse CDMA Channel

The QPSK model that we consider here is very similar to the Reverse CDMA Channel except for the offset modulation. This does not affect the primary conclusion about interference averaging in any significant way.

9.8.1.1 Statistics

The calculations here pertain only to the second-order (i.e., mean and variance) of a single *chip* from the DSSS demodulator. The overall system performance depends on the coding that takes place outside the domain of the spreading and despreading operations. The forward link, for purposes of calculating symbol energy-to-noise ratios, is essentially repetition coded. That is, each FEC code symbol is repeated 64 times. The SNR of its detection statistic is approximately 64 times (18 dB greater than) the per-chip SNR because the 64 chips sum coherently, while the variances sum rms fashion. Subsequent to the soft decision detection statistic, one must consider the performance of the Viterbi decoder to ascertain the overall SNR (E_b/N_0 performance).

The reverse link is somewhat more difficult because the receiver must form 64 decision metrics from 256 chips, and the noise contributions to those 64 metrics are correlated.

Our purpose here is to show how the effects of multiple access interference and possible jamming and interference can be accounted for in our capacity calculations and link budgets by a simple calculation of an effective noise power spectral density. The unwanted signals can be modelled, as far as the communication link is concerned, including thermal noise within the spreading bandwidth.

The primary result of all the mathematical rigour is that the effect of mutual interference and jamming is, for most purposes, the same as an effective total noise level of

$$
\begin{aligned}
N_{eff} &= N_0 + \sum_{other\ users} \varepsilon_m + \int_{-\infty}^{\infty} J(f)|H(f)|^2 df \\
&\approx N_0 + \sum_{other\ users} \varepsilon_m + P_{interf}/W
\end{aligned}
\tag{9.16}
$$

where N_0 is the thermal noise level of the receiver, and P_{interf} is the total in-band (within the spreading band of width W = 5 (or 1.25) MHz) interference power. This interference averaging property is the primary direct benefit of the use of CDMA.

9.8.2 BPSK Spreading - Receiver Statistics

Consider a modulator as shown in Figure 9.4. The data input to this process is a sequence $\{x_n\}$ and the spreading sequence $\{a_n\}$. The indices denote periods of the spreading sequence.

The input sequence $\{x_n(k)\}$ might be, as in the forward link, repetitions of the code symbols from a convolutional coder after application of the orthogonal cover, or it might be some other encoding of the data. The important thing is that, in all cases, the input sequence $\{x_n\}$ is comprised of many chips per information bit. Alternatively, the spreading (chip) rate is many times the data rate. If the modulo-two addition of the spreading sequence can be reversed in the receiver, then a summation of these "despread" chips will recover an approximation of the original code symbol.

For this analysis we assume a channel model that consists only of attenuation and phase shift, different for each user. Additive white Gaussian noise of one-sided spectral density N_0 Watts per Hertz accounts for the receiver front end

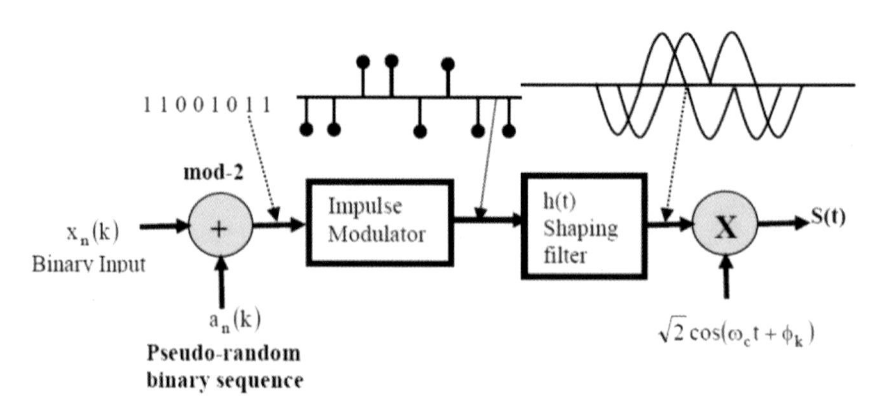

Figure 9.4 Direct sequence BPSK modulator.

thermal noise. Possible jamming is represented by q(t), which we assume to be wide-sense stationary.

Multipath with a delay spread that can be resolved by the receiver contributes some diversity gain. Within each correlator the distinct delay components behave like other-user interference. Multipath with a delay spread when unresolved by the receiver introduces inter-chip interference, which complicates the analysis considerably. Its effect is similar to the other-user noise, but with a much smaller magnitude.

A possible demodulator that is suitable for removing the spreading is shown in Figure 9.6. This structure is optimal if the noise is Gaussian and uniform over the signal bandwidth. By analysing the detection statistics of a single chip in this demodulator, analysis of the actual receiver can proceed according to the uses of the demodulator output $\{y_n\}$.

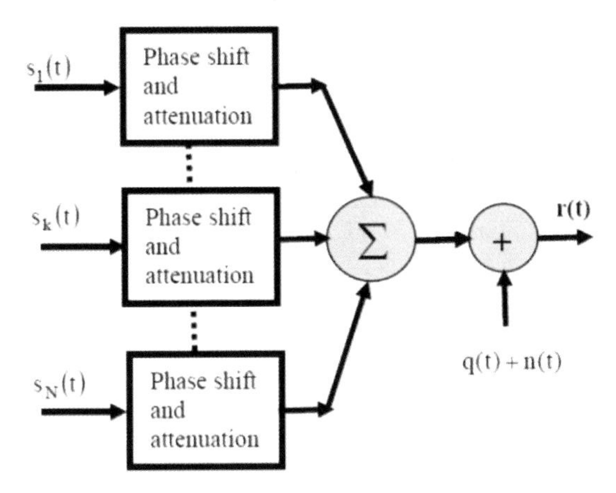

Figure 9.5 Channel model.

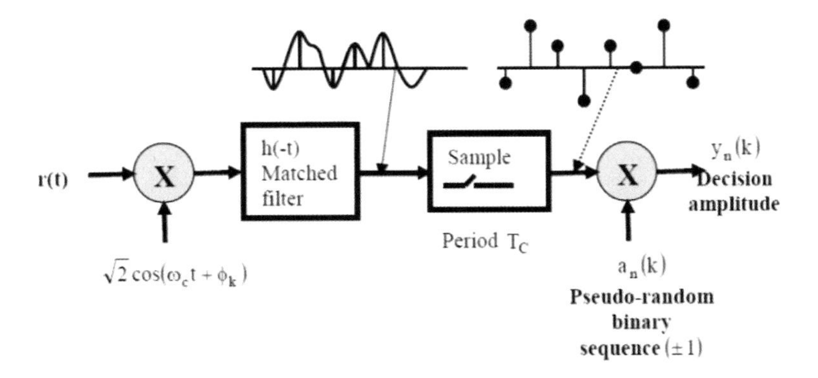

Figure 9.6 Direct sequence demodulator.

The decision amplitudes are further processed in accordance with the nature of the coding that preceded the spreading. Normally this will include the creation of one or more detection statistics that are linear combinations of y_n. Forward error correcting codes are used by a decoder for processing detection statistics. The received signal is

$$r(t)=\sum_{k}\sqrt{\varepsilon_k}\,h_k(t)+n(t)+q(t) \tag{9.17}$$

where the h_k are randomly phase-shifted replicas of *h(t)*, *n(t)* is the thermal noise of the receiver front end, and *q(t)* is jamming, if any.

All spreading sequences are assumed white and uncorrelated with one another. The jamming q(t) is assumed wide-sense stationary, but otherwise arbitrary. Normalisation of the band limiting filter characteristic is assumed to be

$$\int_{-\infty}^{\infty}\left|H(f)\right|^2 df=1 \tag{9.18}$$

This is a little unusual. It makes the impulse response of the filter have dimensions of the square root of bandwidth.

The expectations of y_n for user k, with this scaling and normalisation is

$$E\left[y_n(k)\right]=\sqrt{\varepsilon_k} \tag{9.19}$$

The other users contribute nothing to the expectation because their spreading sequences are uncorrelated with that for user k. All interference, noise, and jamming terms contribute nothing because none are correlated to the spreading sequence.

Let the noise and interference part of y_n (k) be

$$v_k=a_k\sum_{m\neq k}\sqrt{\varepsilon_m}\,a_m\cos(\phi_m-\phi_k)+\int_0^{\tau_0}\left[n(t)+q(t)\right]h_k(T_c-t)dt \tag{9.20}$$

The variance of any $y_n(k)$ is

$$\sigma_v^2=E\left[v_k^2\right]=\sum_{m\neq k}\frac{\varepsilon_m}{2}+\int_0^{\tau_0}\int_0^{\tau_0}\phi_{n'n'}(\tau-\eta)h(\tau)h(\eta)d\tau d\eta \tag{9.21}$$

where

$$\phi_{n'n'}(t-s)=E\{[n(t)+q(t)][n(s)+q(s)]\} \tag{9.22}$$

The variance of the other-user interference contributes a noise factor because of the assumption of random phase. Using the well-known relationship between the auto correlation functions and their spectra, (9.21) is equivalent to

$$\sigma_v^2=\sum_{m\neq k}\frac{\varepsilon_m}{2}+\frac{1}{2}\int_{-\infty}^{\infty}[N_0+J(f)]|H(f)|^2\,df \tag{9.23}$$

where J(f) is the one-sided power-spectral density of the interference. If we compare the AWGN term to the interference term, we see that the effect of the interference on the variance is the same as a white noise level having the same total power in the passband:

$$J_{eff}=\int_{-\infty}^{\infty}J(f)|H(f)|^2\,df \tag{9.24}$$

Equations (923) and (9.24) represent the central conclusion from this calculation. It underlies all the calculations relating to capacity and loading that treat interference only in a total power sense. In calculations of capacity we can replace N_0 by N_{eff} = N_0 + Total other user energy per chip + J_{eff}.

9.8.3 QPSK Spreading - Receiver Statistics

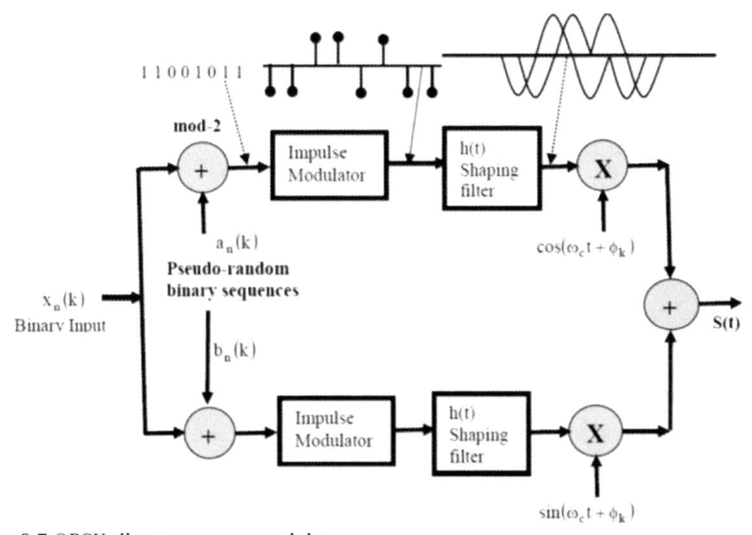

Figure 9.7 QPSK direct sequence modulator.

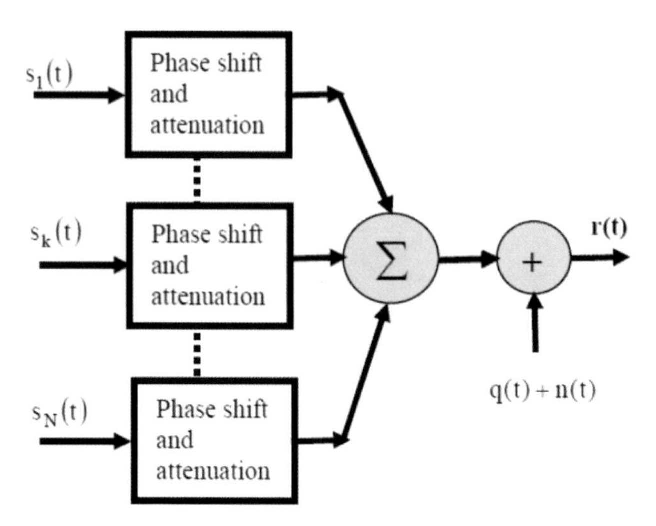

Figure 9.8 Channel model.

The QPSK version of the modulator duplicates the BPSK structure, but with a applied quadrature RF carrier s in (Figure 9.5).

The channel model is the same as the BPSK case (Figure 9.6).

The demodulator also duplicates the BPSK demodulator with a quadrature local oscillator (Figure 9.7). The sign changes in the quadrature channel can be thought of as conjugating the complex carrier and spreading sequence amplitudes. Figure 9.8 shows the channel model and affects the signal in three ways. First due to multipath, the channel changes the phase of the signal. Second, it also attenuates it. Furthermore, noise is added to the signal by the channel as shown in Figure 9.8.

The demodulation process is shown in Figure 9.9. We can shortcut most of the mathematics here by noting that each of the QPSK channels, I and Q, is exactly like the BPSK model, except that half the signal energy appears in each channel, so we replace with /2. The noise and interference, on the other hand are the same in each channel. Therefore those expressions are unchanged. The detection amplitude means and variances are now given by the expressions

$$
\begin{aligned}
E[y'_n(k)] &= 2\sqrt{\varepsilon_k / 2} = \sqrt{2\varepsilon_k} = \sqrt{2}E[y_n(k)] \\
\sigma_{v'}^2 &= \sum_{m \neq k} \varepsilon_m + \int_{-\infty}^{\infty} [N_0 + J(f)]|H(f)|^2 \, df = 2\sigma_v^2
\end{aligned}
\tag{9.25}
$$

The point being made here is that the signal-to-noise ratio

$$E[y'_n(k)]/\sigma_{v'} = E[y_n(k)]/\sigma_v \tag{9.26}$$

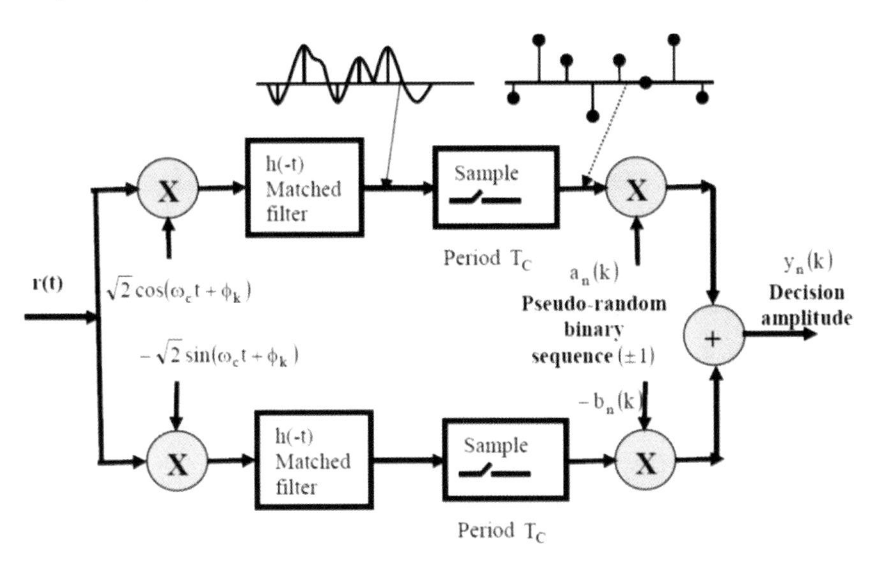

Figure 9.9 QPSK direct sequence demodulator.

is unchanged. The reason for the quadrature spreading is really to make sure that the mutual interference between users and between stations is uniformly distributed in phase. It otherwise contributes nothing to performance.

9.8.4 Frequency Hopping

The main problem with applying Direct Sequence spreading is the so-called Near-Far effect which is illustrated in the Figure 9,10. This effect is present when an interfering transmitter is much closer to the receiver than the intended transmitter [4]. Although the cross-correlation between codes A and B is low, the

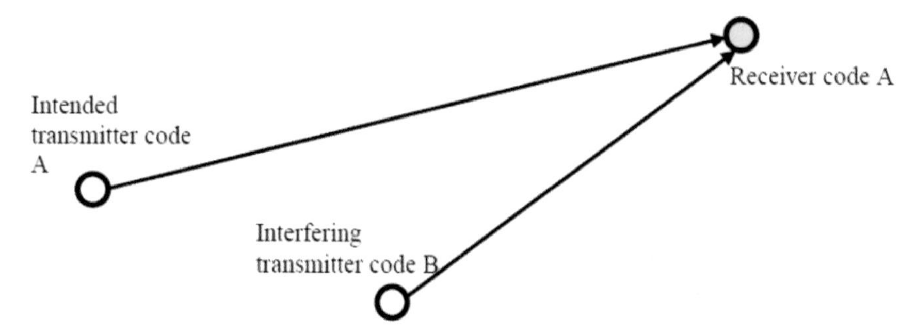

Figure 9.10 Interfering Transmitters.

correlation between the received signal from the interfering transmitter and code A can be higher than the correlation between the received signal from the intended transmitter and code A. The result is that proper data detection is not possible.

Another spread spectrum technique: frequency-hopping is less affected by Near-Far effect. When applying frequency hopping, the carrier frequency is 'hopping' according to a unique sequence (an FH-sequence of length). In this way the bandwidth is increased by a factor (if the channels are non-overlapping). The process of frequency hopping is illustrated in Figure 9.11. A disadvantage of frequency-hopping as opposed to direct-sequence is that obtaining a high processing-gain is hard. There is need for a frequency-synthesiser able to perform fast-hopping over the carrier-frequencies. The faster the ``hopping-rate" is, the higher the processing gain which is given by the expression.

$$G_P(FH) = N_{FH}$$

Frequency-hopping sequences have only a limited number of “hits”" with each other.

This means that if a near-interferer is present, only a number of “frequency-hops”" will be blocked instead of the whole signal. From the ``hops" that are not blocked it should be possible to recover the original data-message.

The wideband frequency spectrum desired is generated in a different manner in a frequency hopping system. It does just what the name implies. That is, it

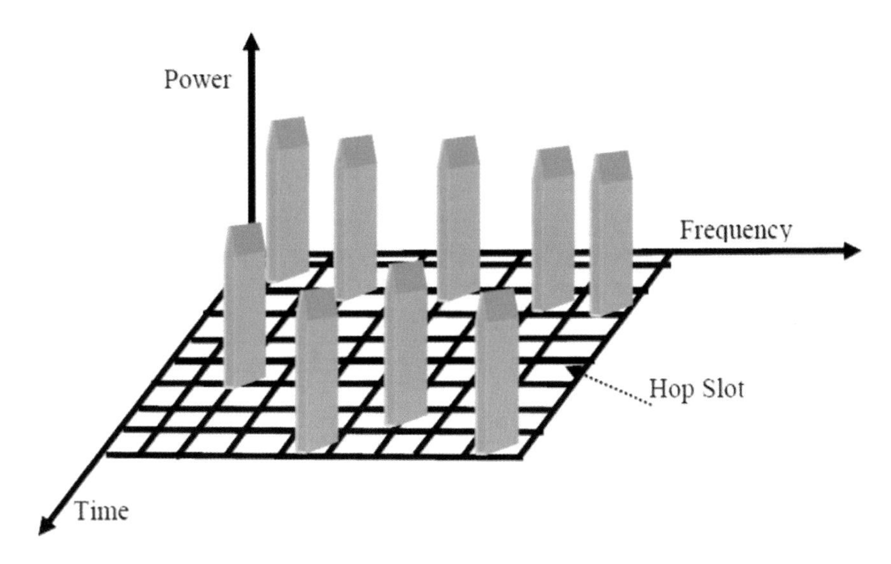

Figure 9.11 Frequency Hopping.

"hops" from frequency to frequency over a wide band. The specific order in which frequencies are occupied is a function of a code sequence, and the rate of hopping from one frequency to another is a function of the information rate. The transmitted spectrum of a frequency-hopping signal is quite different from that of a direct sequence system. Instead of a $\left(\frac{\sin x}{x}\right)^2$-shaped envelope, the frequency hopper's output is flat over the band of frequencies used. The bandwidth of a frequency-hopping signal is simply w times the number of frequency slots available, where w is the bandwidth of each hop channel. Frequency hopping is the easiest spread spectrum modulation to use. Any radio with a digitally controlled frequency synthesiser can, theoretically, be converted to a frequency hopping radio. This conversion requires the addition of a pseudo noise (PN) code generator to select the frequencies for transmission or reception. Most hopping systems use uniform frequency hopping over a band of frequencies. This is not absolutely necessary, if both the transmitter and receiver of the system know in advance what frequencies are to be skipped. Thus a frequency hopper in two meters could be made that skipped over commonly used repeater frequency pairs. A frequency-hopped system can use analog or digital carrier modulation and can be designed using conventional narrow band radio techniques. A synchronised pseudo-noise code generator is used for de-hopping in the receiver. This drives the receiver's local oscillator frequency synthesiser.

The most practical, all digital version of SS is direct sequence. A direct sequence system uses a locally generated pseudo noise code to encode digital data to be transmitted. The local code runs at much higher rate than the data rate. Data for transmission is simply logically modulo-2 added (an EXOR operation) with the faster pseudo noise code. The composite pseudo noise and data can be passed through a data scrambler to randomise the output spectrum (and thereby remove discrete spectral lines). A direct sequence modulator is then used to double sideband suppressed carrier modulate the carrier frequency to be transmitted. The resultant DSB suppressed carrier AM modulation can also be thought of as binary phase shift keying (BPSK). Carrier modulation other than BPSK is possible with direct sequence. However, binary phase shift keying is the simplest and most often used SS modulation technique.

9.9 HYBRID SYSTEM (DS/FAST FREQUENCY HOPING)

The DS/FFH spread spectrum technique is a combination of direct-sequence and frequency hopping. The data is divided over frequency-hop channels (carrier frequencies). In each frequency-hop channel one complete PN-code of length is multiplied with the data signal (Figure 9.12).

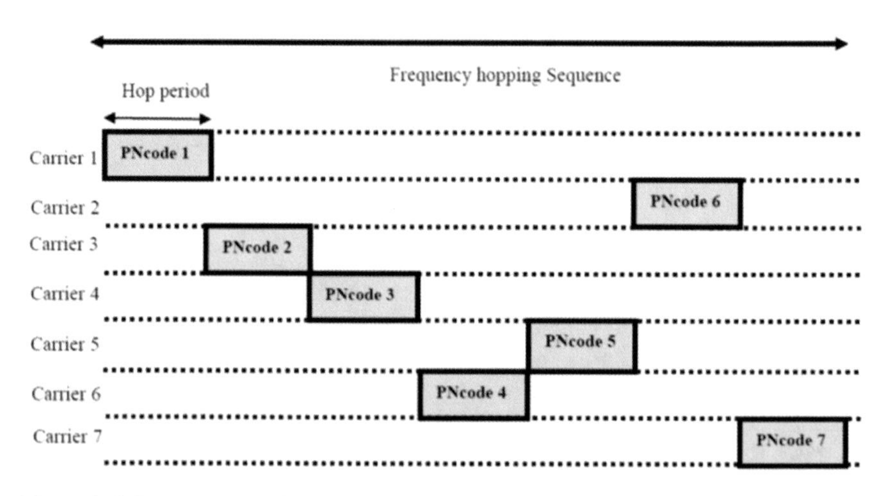

Figure 9.12 Frequency Hopping Sequence.

As the FH-sequence and the PN-codes are coupled, an address is a combination of an FH-sequence and PN-codes. To bind the hit-chance (the chance that two users share the same frequency channel in the same time) the frequency-hop sequences are chosen in such a way that two transmitters with different FH-sequences share at most two frequencies at the same time (time-shift is random).

References

[1] Howard H. Xia, IMT-2000 and G3G CDMA technologies, IEEE International Conference on 3G Wireless Communications, Silicon Valley, USA, Vodafone AirTouch Plc, 2999 Oak Road, MS 900, Walnut Creek, CA 94596

[2] Randy Roberts, RF/Spread Spectrum Consulting, http://www.sss-mag.com/ss.html

[3] The principles of Spread Spectrum communication, http://cas.et.tudelft.nl/

[4] Jacobus Petrus Franciscus Glas, "The principles of spread spectrum communications", http://ens.ewi.tudelft.nl/pubs/glas_phdthesis.pdf

Chapter 10.

Planning of WCDMA Networks

Johnson I Agbinya

Department of Electronic Engineering, La Trobe University, Australia
J.Agbinya@latrobe.edu.au

10.1 INTRODUCTION

The previous chapters provide details of requirements for designing and planning mobile networks and in particular of WCDMA networks. In this chapter the information provided in those chapters will be used. The planning of WCDMA networks may be undertaken from two points of view, the coverage limited design plan or the capacity limited design plan point of view. For the capacity limited plans, the pole capacity as a function of system parameters is used. The system parameters of interest include the chip rate for the WCDMA system, voice and data rates, energy per bit per noise, voice activity factor and also influence of interference on other sectors and nodes around the base station of interest. The pole capacity is therefore developed as a function of these parameters. For this approach historical statistical information on the geographical area in which the network is to be deployed is essential. Prior input data such population of the area, the voice and data usage of customers or potential customers in such areas are useful for planning. Examples are given in sections 10.1.

For the coverage limited plans, estimation of the cell range is undertaken as a function of the system parameters and the environment. Hence system link budgets are used. This means the antenna properties such as gain, directivity and polarisation are important. Furthermore, consideration for fading in the channel and transmitted power leads to a value for the received power through the link budget. This approach allows for estimation of losses in building, human body, vehicles in which communications are undertaken and also the topography and morphology of the area. Software planning tools traditionally use coverage

4G Wireless Communication Networks: Design, Planning and Applications, 189-208.

limited as the design approach for the characteristics and performance of base stations.

10.2 CAPACITY LIMITED PLANNING

A reasonable starting point for planning a WCDMA network is through the use of the so-called pole capacity of the individual base station. A pole capacity gives an estimate of the number of users that a base station can serve based on defined data rate. The uplink load factor of a WCDMA base station can be estimated by the expression:

$$\eta_{uplink} = \frac{E_b/N_0}{W/R}.N.\alpha(1+\beta) \tag{10.1}$$

Where
α is the activity factor for a user for the uplink ($\alpha = 1$ for data and varies for voice with values ranging from around 0.4 to 0.7). At certain times during voice communication there is silence and then words are spoken. The measure of how long the non-silence period is to the overall time for voice communication is defined as the voice activity factor. This also means that when data is transmitted, the line is busy during the overall data transmission and the activity factor then is unity (for data).

β is other cell to own cell interference and represents the influence of other cells. This factor is impacted by the characteristics of the sector and base station. Suggested values for a 3-sector base station is 0.93; for omni-directional antennas the value is around 0.67 and for micro cells a value of 0.4 is advised. For the rest of the discussions in this chapter a value of 0.5 will be used. By defining noise rise as the ratio of the total interference power to thermal noise power, then noise rise is given by the expression:

$$\text{Noise rise} = \frac{I_{total}}{N_{thermal}} = \frac{1}{1-\eta_{uplink}}$$

At the point when the uplink load factor is close to unity, the noise rise is infinite and at that point, the maximum pole capacity is defined as the approximation:

$$N_{pole} = \left(\frac{W/R}{(E_b/N_0)\alpha}+1\right)\frac{1}{(1+\beta)} \approx \frac{W/R}{(E_b/N_0)(1+\beta)\alpha}+1 \tag{10.2}$$

W is the WCDMA chip rate (3.84 Mcps) for spreading and R is the required data rate per user. The remaining two variables are energy per bit per noise, E_b and N_0 respectively.

10.1.1 Estimation of Energy per bit per Noise Ratio

Energy per bit per noise is traditionally used to estimate the required voltage level of a bit of information in a noisy environment that will allow accurate reproduction of the bit in the receiver. It is estimated with the following expression:

$$E_b / N_0 = \frac{b}{\sigma_{thermal} + \sigma_{interference}} \tag{10.3}$$

Where b is energy per bit

$\sigma_{thermal}$ is thermal power spectral density

$\sigma_{interference}$ is the power spectral density of interference

E_b/N_0 is a dimensionless quantity. It is the required signal to noise/interference ratio for transmitting bits. E_b/N_0 for the uplink is influenced by the sum of all interferers from the same cell or sector and from other cells or sectors. As shown in equation (10.3), when the power spectral density for the interferers is high, the energy per bit per noise is very low and affects efficient reproduction of the transmitted bits at the receiver.

Energy per bit can be accurately estimated. It is expressed in terms of the power in the signal P_S divided by the transmission data rate R and given by the expression

$$E_b = \frac{P_s}{R} \tag{10.4a}$$

The noise component N_0 in equation (10.3) is the sum of the total noise spectral densities originating from thermal noise N_{th} plus noise inside the cell or sector I_{int} and the noise from all the interferers outside the sector or cell I_{ext}. It is given by the expression:

$$N_0 = N_{th} + I_{int} + I_{ext} \tag{10.4b}$$

The internal and external interference powers can also be estimated. Assume that all the users in the cell (sector) internal to the cell and external users transmit at the same power. This is possible because of power control. Due to spreading, the power in the signal from each interferer is spread into the system chip rate W (3.84Mcps for UMTS) so that the power spectral density for an interferer is P_S/W. Assume also that there are N_f frequency carriers in the sector or cell and there are N_u users in the cell or sector. If they are uniformly distributed over the carriers then the number of interferers inside the sector or cell is

$$N_{interferers} = \left(\frac{N_u}{N_f} - 1 \right) \tag{10.4c}$$

Therefore the internal noise power in the cell or sector is

$$I_{in} = \left(\frac{N_u}{N_f} - 1 \right) \frac{P_s}{W} . \alpha \tag{10.4d}$$

Where α is the voice activity factor for the cell or sector. The external interference noise power is often taken as a fraction of the internal noise power. In general this can be written as

$$I_{ext} = \beta I_{\text{int}} ; 0 \leq \beta \leq 1 \tag{10.4e}$$

A typical estimate for $\beta = 0.6$. If the separation between the serving cell and its neighbours is large, meaning that the path loss between interferers and the cell is large, the proportion of external interference could be less than 0.6. This happens traditionally in rural cells of large cell radius and with sparse population of users. By substituting equations (4b) to (4e) into equation (4a) we can show that

$$\frac{E_b}{N_0} = \frac{P_s / R}{N_{th} + I_{\text{int}} \left(1 + \frac{I_{ext}}{I_{\text{int}}} \right)} = \frac{P_s / R}{N_{th} + \alpha \left(\frac{N_u}{N_f} - 1 \right) \frac{P_s}{W} (1 + \beta)} \tag{10.5a}$$

Or

$$\frac{E_b}{N_0} = \frac{W / R}{\frac{WN_{th}}{P_s} + \alpha (1 + \beta) \left(\frac{N_u}{N_f} - 1 \right)} \tag{10.5b}$$

By using equation (10.5b) to solve for the number of users N_u we have

$$N_u = N_f\left[\frac{1}{(1+\beta)\alpha}\left(\frac{W/R}{E_b/N_0} - \frac{WN_{th}}{P_s}\right) + 1\right] \tag{10.6}$$

The ratio $1/(1+\beta)$ is called the frequency reuse efficiency. Also when the ratio $W.N_{th}/P_s$ in equation (10.6) tends to zero an upper limit on the number of users per cell or sector is obtained. The upper limit of the number of users is a theoretical upper bound that can be supported by the cell and is called the pole number of the cell given by the expression:

$$N_{pole} = N_f\left[\frac{W/R}{\left(E_b/N_0\right)(1+\beta)\alpha} + 1\right] \tag{10.7}$$

Equation (10.7) is identical to equation (10.2) provided that we take the number of frequency carriers $N_f = 1$ per cell or sector. This is done in practice. Formally, the pole capacity of a cell is given by the expression:

$$N_{pole} = \left[\frac{W/R}{\left(E_b/N_0\right)(1+\beta)\alpha} + 1\right] \tag{10.8}$$

The ratio W/R in equation (10.8) is called the spreading factor. The noise rise which we defined earlier is also given by the expression

$$Noise\ rise = \frac{N_{th} + I_{\text{int}} + I_{ext}}{N_{th}} \tag{10.9}$$

Equation (10.9) shows that noise in a cell rises above the thermal noise level and that the noise rise is due to the influences of other users within the cell and also users external to the cell.

10.3 WCDMA POLE CAPACITY FOR VOICE CALLS

WCDMA defines a pole capacity for a base station. This is the expected raw capacity (number of subscribers that a site can accommodate concurrently) of a site. It will be used in the discussions as the foundation for determining the capacity of the sectors and sector efficiency.

This chapter is presented as a series of exercises. Each exercise provides an incremental idea on previous exercises or further illustration on a recent solution. The exercises are given with the aim of helping a reader to know how to estimate a WCDMA cell capacity. The goal is to determine how the range of a cell is limited by its capacity and vice versa. First we will overview

a) How to determine pole capacity of WCDMA
b) How to determine the capacity of a sector
c) How to determine the number of channels required per sector
d) How to use Erlang B Table to estimate the number of channels required

We will use equation (10.2) to first calculate the pole capacity for voice transmission for a WCDMA cell. Observe that equation (10.2) does not involve frequency because frequency is a common commodity for all subscribers on a cell. The choices of parameters for the estimate are therefore:

$\alpha = 0.45$

$W = 3.84 Mcps$ (UMTS chip rate)

$R = 12 kbps$ (voice rate)

$E_b/N_0 = 5.7 dB$ or $10^{\frac{5.7}{10}} \approx 3.72$ (energy per bit per noise)

$\beta = 0.5$

With the above parameters, the pole capacity is given as:

$$N_{pole} = \frac{3840/12}{3.72(1+0.5)0.45} + 1 \approx 128.44$$

This result applies to omni-directional cells and presents the pole load as the number of subscribers that it can serve concurrently. In practice it is unwise to load the base station to capacity. Hence assume that the site is de-rated to 0.75 (75%) of the pole capacity. Assume also that this cell is sectored with sector efficiency $\eta = 0.85$. We will use 3-sectors for our network design so that all of the capacity that would be available in an omni-directional cell is available in each of the sectors. So the number of simultaneous users in a sector reduces to 0.75 of the pole load which is:

$N_{sector} = N_{pole} \times \eta \times 0.75 = 128.44 \times 0.85 \times 0.75 \approx 81.88$

We take $N_{sector} = 81$

We have selected the lowest integer value $\lfloor N_{sector} \rfloor$ for the number of subscribers per sector. The small redundancy introduced by choosing 81 instead of 82 provides some buffer for times when the cell is congested. Suppose one subscriber completely occupies a channel, then we need 81 channels. In practice this would not be the case because channels are trunked. Trunking is used to share a channel between several subscribers. This means that we are in some sense willing to live with some dropped calls during the busy hour in the channel. The percentage of dropped calls is represented as the system grade of service. If the maximum grade of service (GoS) is 5%, we are then prepared to live with five calls in a hundred being dropped during the busiest hour. We could therefore use the Erlang loss function (Erlang B Table) to determine the offered traffic as follows:

$A_{\sec tor} = 75.84E$

Also the voice capacity per cell becomes $A_{cell} = 3 \times 75.84 = 227.52E/cell$. This cell should be dimensioned to a total of about 228Erlangs. The choice of 228Erlangs (over-dimension) provides a small extra capacity which could be useful during congestion times in the cell.

10.4 CELL CAPACITY (DATA) LIMITED RANGE

The geographical area covered by a network is naturally segmented by its morphology and also by its topography. Also, data other than voice are carried by WCDMA networks and indeed all modern mobile communication networks. Since data requires more capacity than voice, determining the pole capacity and hence cell capacity for voice only would not work well in practice. In the next exercise we take into account the variations of the area covered by the network in terms of the population density and also with the demand for both voice and data calls.

Assume we are given the following system parameters:
Voice demand per user is 0.01Erlang
The maximum average packet delay is 0.5 seconds.
Total land area of network is 15,000square km
Voice capacity per cell is $A_{cell} = 3 \times 75.84 = 227.52E/cell$ (from the analysis in section (10.3), we know this pole capacity)

$$N_{pole} = \frac{W/R}{E_b/N_0(1+\beta)\alpha} + 1$$

Suppose then that we have the following traffic and population distributions in the area covered by our network as shown in Table 10.1. In this case the region is divided into four sub-regions with different populations, different traffic demands and also different land areas.

Table 10.1 System Parameters

Region	Population	Proportion of traffic demand	% Land area of region
I	1, 200, 000	0.6	25%
II	1, 200, 000	0.25	40%
III	1, 200, 000	0.10	25%
IV	1, 000, 000	0.05	10%

The task in this section is to compute:

a) The number of cells for each region
b) Radius of each cell

As in section 10.3 we first compute the voice traffic for each region. A useful starting point is to consider the demand for voice services. Voice calls are basic requirements of all cellular networks and getting that right from the start is essential. The voice traffic is estimated using the expression:
Voice Traffic="Voice Demand per User × Total Number of Users × Fraction of Total Demand in Region 1 $=0.01\times 1200\times 10^3\times 0.6=7200E$

Compute The Number of Cells Per Region

a) Number of cells required in Region I:

$$(N_{cell})_1 = \frac{7200E}{227.52E} \approx 31.65 = 32 \text{ cells}$$

We require 32 cells in region I.

b) Compute the land area of Region I
Network area in Region I is

$$S_1 = 25\%\times 15000km^2 = 3750km^2.$$

c) Compute the Cell Range (uniform) in Region I
Cell range in region I is:

$$(S_{cell})_1 = \frac{S_1}{(N_{cell})_1} = \frac{3750}{32} \approx 117.19km^2 / cell$$

d) Compute The Radius of Each Cell
Radius of the cell in region I is

$$(R_{cell})_1 = \sqrt{\frac{(S_{cell})_1}{\pi}} = \sqrt{\frac{117.19}{\pi}} \approx 6.11km$$

Observe that we have assumed that the area of each cell is circular! A better template to use is the hexagonal template. For hexagonal cells we will replace pi in the divisor of these equations by 2.598 (see section 1.1 of chapter 1). We have now computed the raw variables for our network in Region I. Naturally if these variables are optimum, the network will perform as expected. Unfortunately, the above design is not optimum because there are other factors which affect the transmissions. The factors include the morphology, topography, fading, interference sources, heights of base stations to be used and their power ratings. Hence, this network needs to be optimized using iterative design. One of such first iteration would be to modify the range or radius of each cell to ensure coverage at the required reception threshold. Other choices related to fading and mobility of terminals and interferences will follow this first optimization procedure. The good part is that network planning software tools such as CelPlanner can be used to optimize this network for this region. Let us complete this design procedure for the remaining three regions in the following sections.

10.4.1 Cell Capacity Limited Range in Region Ⅱ

The traffic demand of region II is: $0.01\times1200\times10^3\times0.25=3000E$

So the number of cells required in region II is $(N_{cell})_2=\dfrac{3000E}{227.52E}\approx 13.19$ cells

We take 14 cells in region Ⅱ.

The network area in region II is $S_2=40\%\times15000km^2=6000km^2$.

So the cell range in region II is: $(S_{cell})_2=\dfrac{S_2}{(N_{cell})_2}=\dfrac{6000}{14}\approx 428.57km^2/cell$

The radius of the cell in region II is $(R_{cell})_2=\sqrt{\dfrac{(S_{cell})_2}{\pi}}=\sqrt{\dfrac{428.57}{\pi}}\approx 11.68km$

10.4.2 Cell Capacity Limited Range in Region Ⅲ

According to the network specifications given in Table 10.1 the traffic demand of region 3 is: $0.01\times1200\times10^3\times0.1=1200E$

The number of cells required in region III is $(N_{cell})_3=\dfrac{1200E}{227.52E}\approx 5.27=6$ cells

We take 6 cells in region III.

Network area in region Ⅲ is $S_3=25\%\times15000\,km^2=3750\,km^2$.

Cell range in region Ⅲ is: $(S_{cell})_3=\dfrac{S_3}{(N_{cell})_3}=\dfrac{3750}{6}=625km^2/cell$

Radius of the cell in region 3 is $(R_{cell})_3=\sqrt{\dfrac{(S_{cell})_3}{\pi}}=\sqrt{\dfrac{625}{\pi}}\approx 14.10km$

This cell radius is an optimistic estimate as it does not account for losses due to fading and the effects of the area where the cell will be deployed.

10.4.3 Cell Capacity Limited Range in Region IV

According to Table 10.1, the traffic demand of region IV is: $0.01\times1200\times10^3\times0.05=600E$

Number of cells required in region IV is $(N_{cell})_4=\frac{600E}{227.52E}\approx2.64$ cells

We take 3 cells in region IV.

Network area in region IV is $S_4=10\%\times15000km^2=1500km^2$.

Cell range in region IV is: $(S_{cell})_4=\frac{S_4}{(N_{cell})_4}=\frac{1500}{3}=500km^2/cell$

And the radius of the cell in region IV is $(R_{cell})_4=\sqrt{\frac{(S_{cell})_4}{\pi}}=\sqrt{\frac{500}{\pi}}\approx12.62km$

By accounting for the effects of the terrain and fading, a more reliable cell radius can be obtained and will be a lot less than the above result.

10.5 COVERAGE LIMITED CELL RANGE CALCULATION

In the previous sections we were only interested in the cell capacity from the traffic point of view and paid no attention to losses in the channel and also not to interference in the system. Such a network design will not perform optimally. In a good design, consideration for signal power loses in the channel must be undertaken through the use of a link budget. A link budget is a summary of how the transmitted signal power is spent in the link. The link is formed by three distinct system entities: the transmitter (TX), the channel and the receiver (RX).

In this section we will account for losses in the channel. These are path loss due to the separation between the TX and RX, feeder losses L_f, fading margins (L_{FM}), losses if the device is being used inside a building (in-building losses), and shadowing losses. Figure 10.1 contains most of the important losses to address in the design. Note that body loss, diffraction and in-vehicle losses are not shown in Figure 10.1. They should be considered when they are likely to

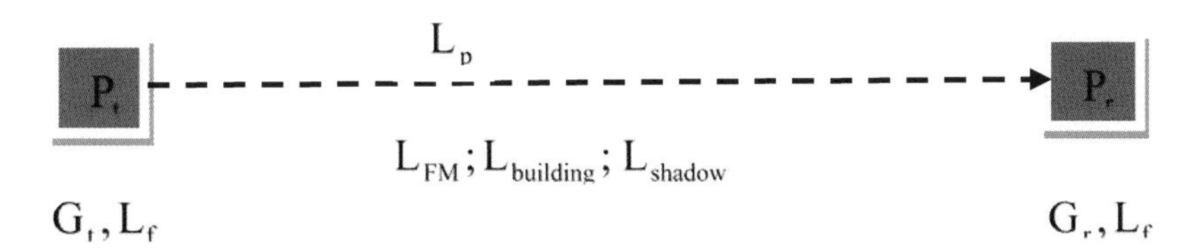

Figure 10.1 Uplink Budget.

come into play to affect the system performance. A good cell planning tool will give you estimates for most of these losses shown in Figure 10.1.

Radio links can be either interference limited or noise limited. This is a reference to the performance of such networks. When the link is interference limited, the network is referred to as capacity limited. For capacity limited networks interference sets the limit on the achievable spectral efficiency. When a network is noise limited the network will be referred to as coverage limited. Hence instead of focusing on spectrum efficiency in the network performance, noise limitation and cell range are the focus.

In the current design of WCDMA macro cell, the uplink coverage is more limited, because mobile devices are traditionally low powered devices. For example for voice terminals, the power levels are around 250mW limited. In practice therefore handsets face more channel limitations than NodeBs. Therefore, we will focus on the uplink budget which will affect the maximum cell radius. This is referred to coverage limited cell range consideration. Just as in the previous sections, we will estimate the uplink budget and the coverage limited cell range for our regions.

10.5.1 Calculation of Link Budget

Consider the uplink budget model shown in Figure 10.1. Since we have regions with different topography and morphology, we expect them to also have different fading margins and shadowing. If the regions contain buildings which create losses (concrete buildings) then in-building margins should be included. If the mobile device is used inside a vehicle then margins due to losses in the vehicle must be considered. The same consideration should given to losses in the human body.

Firstly we only consider the shadow fading margin ($M_{In-Building} = 0dB$) to calculate the link budget, and then we consider the in-building margin to compute the coverage limited cell range.

From the above diagram we can determine the link budget where

$$P_r = P_t + G_t - L_f - L_p - L_{FM} - L_{shadowing} + G_r - L_f \tag{10.10}$$

This equation helps us to determine the maximum mean path loss for the uplink

$$L_p = P_t + G_t - L_f - P_r - L_{FM} - L_{shadowing} + G_r - L_f \tag{10.11}$$

In the uplink, the mobile equipment is the transmitter and the base station is the receiver. Hence we expect the gain of the receiver antenna in this case to be a lot bigger than the gain of the ME antenna.

The power transmitted by a UMTS (WCDMA) handset can be estimated as: $P_t = 250mW$

When converted to decibels, this becomes $P_t(dB) = 10\log_{10}\left(\frac{250mW}{1mW}\right) \approx 23.98dBm$,

The gain of the transmitting antenna is estimated to be $G_t = -3dBd = -3 + 2.15 = -0.85dBi$,

$G_{Div} = 4dB$, (diversity gain)

Because we use 3-sectored cell for our network design, the gain of the sector antennas with respect to a dipole antenna is first converted to an equivalent isotropic gain as follows: $G_r = 17dBd = 17 + 2.15dBi = 19.15dBi$ (Sector antenna gain)

$L_f = 3dB$ - This is our feeder loss. The feeder cable connects the system electronics to the antenna.

In order to get the link budget, we need the minimum receive power P_r. This may be estimated with the following expression:

$$P_r = \frac{(E_b/N_0)N_{th}}{W/R + (E_b/N_0)(1+\beta)\alpha} n \tag{10.12}$$

Here $E_b/N_0 = 5.7dB$ or $10^{\frac{5.7}{10}} \approx 3.72$, $W = 3.84Mcps$, $R = 12kbps$, $\alpha = 0.45$, $\beta = 0.5$, $n = 2dB$ or $10^{\frac{2}{10}} = 1.58$.

Thermal noise power is $N_{th} = KTB$

Here K = Boltzmann's Constant = $1.38\times10^{-23}\,Joule/^{\circ}K$

T = Temperature in degrees Kelvin: $^{\circ}K = 300^{\circ}K$

B = 5MHz (the nominal bandwidth for all third-generation proposals in our project is 5 MHz). Therefore the thermal noise power becomes

$$N_{th} = KTB = 1.38\times10^{-23}\times300\times5\times10^{6} = 2.07\times10^{-14} \text{ Watt}$$

The received power at the base station can be estimated with the expression:

$$P_r = \frac{(E_b/N_0)N_{th}}{W/R + (E_b/N_0)(1+\beta)\alpha} n$$

$$= \frac{3.72\times2.07\times10^{-14}}{3840/12 + 3.72\times0.45\times(1+0.5)}\times1.58 \approx 3.77\times10^{-16}W$$

Or $P_r = 10\log\left(\frac{3.77\times10^{-16}}{1\times10^{-3}}\right) \approx -124.23dBm$

This is an extremely small power level. Bearing in mind that the caller could be standing between two very tall buildings or in a canyon, we will further try to estimate shadowing losses due to the obstruction of the direct line of sight path by the tall buildings (canyon).

10.5.1.1 Estimation of Shadowing Loss

We could calculate the fading margin for shadowing with the objective of achieving a probability of 90% outdoor coverage along the cell border. The signal level for this coverage is required to be greater than -124.23dBm when log-normal shadowing is assumed. Let us assume that the standard deviation for log-normal shadowing is σ=8dB. This value needs to be used to estimate any amount of macro diversity gain when equal power is received from two base stations.

The probability that shadowing increases the median path loss by more than kσ is given by the expression:

$$p\left(L_{shadowing} > y\right) = \frac{1}{\sqrt{2\pi}} \int_{x=y/\sigma}^{\infty} \exp\left(\frac{-x^2}{2}\right) dx = Q\left(\frac{y}{\sigma}\right) \qquad (10.13)$$

Where Q(.) is the complementary cumulative normal distribution and Q is given by the expression:

$$Q\left(\frac{y}{\sigma}\right) = \frac{1}{\sqrt{2\pi}} \int_{x=y/\sigma}^{\infty} \exp\left(\frac{-x^2}{2}\right) dx = \frac{1}{2} erfc\left(\frac{y}{\sigma\sqrt{2}}\right) \qquad (10.14)$$

Equation (10.14) is often called the *error function* in telecommunication text books and listed in tables. It provides a measure of the probability of error that the signal power in our 90% coverage area is going to be less than -124.23 dBm. In our case in this exercise, we need to compute the probability that the received signal is smaller than -124.23dB. We need to calculate a value for $x = y/\sigma$ at which the path loss is less than the maximum value or at least 90% of the coverage is achieved. This is obtained from the Q(.) function as

$Q(k) = p(X > m + k\sigma) = p(X < m - k\sigma) = p(P_r < -124.23dBm) = \sqrt{1-0.9} \approx 0.3162$

Or $Q(k) \approx 0.3162$

When Q(k)=0.3162, k = 0.48. This therefore implies that the shadowing loss is $L_{shadowing} = k\sigma = 8x0.48 = 3.84dB$

So using the above calculated values of P_t, G_t, $L_{shadowing}$, G_r, L_f, and P_r, we can get the maximum mean path loss as:

$$\begin{aligned} L_p &= P_t + G_t - L_{FM} + G_r - L_{shadowing} - 2L_f - P_r \\ &= 23.98dBm + (-0.85dBi) - 3.84dB + 4dB + 19.15dBi - 3dB - (-124.23dBm) \\ &= 163.67dB \end{aligned} \tag{10.15}$$

Knowing how to compute the Q(.) value is essential for understanding the performance of most communication systems. The error function is also used for estimating bit error rates in normal communication systems including mobile networks. In a nutshell the steps required for estimating shadowing loss in our case are:

a) Find the sensitivity of the receiver (here it is -124.23dBm)
b) Estimate the Q(k) value based on the required coverage ($Q(k) \approx \sqrt{1 - coverage}$. In our case coverage is 90% = 0.9.
c) Determine the value of k from a Q-table
d) Find shadowing loss with the expression: $L_{shadowing} = k\sigma$; we have assumed in our case that $\sigma = 8$ standard deviations.
e) Insert the shadowing loss in your path loss equation and determine the equivalent path loss (see equation (10.15))
f) Using the value of L_p, go ahead to estimate the cell range based on a propagation model. In section 10.5.2, we do exactly this by using the COST 231 model over four different regions.

10.5.2 Cell Range Estimation: Coverage Limited

We will base our design on the COST 231/Hata model. Assume the height of the mobile station antenna is 1.5m and the base station antenna height is 30m. Previously we have arrived at an estimate for the total path loss in the system. In this section, we will use the values obtained.

10.5.2.1 Region I

Consider Region I to be in a CDB area with a lot of high rise offices, residential buildings and shopping centres. The COST 231 path loss model for a CBD area is given by the expression:

$$L_p = L_c + 3 = 46.3 + 33.9\log(f_{MHz}) - 13.82\log(h_{BS}) + [44.9 - 6.55\log(h_{BS})]\log(d_{km}) - a(h_{MS}) + 3 \tag{10.16}$$

Where $f = 2GHz$, $h_{BS} = 30m$ and $h_{MS} = 1.5m$.

$$a(h_{MS}) = [1.1\log(f_{MHz}) - 0.7]h_{MS} - [1.56\log(f_{MHz}) - 0.8]$$
$$= [1.1\log(2\times10^3) - 0.7]\times1.5 - [1.56\log(2\times10^3) - 0.8] \approx 0.05dB \quad (10.17)$$

Therefore

$$L_p = 163.67dB = 46.3 + 33.9\log(2\times10^3) - 13.82\log(30) + [44.9 - 6.55\log(30)]\log(d_{km}) - 0.05 + 3$$
$$= 140.74 + 35.22\log(d_{km})$$

The coverage limited cell radius is obtained from the above results as

$$d = 10^{\frac{163.67-140.74}{35.22}} \approx 4.48km$$

If the in-building margin is 15dB, then

$$L_p = P_t + G_t - L_{FM} + G_r - L_{shadowing} - 2L_f - P_r - L_{in-building} = 163.67\text{-}15 = 148.67\text{dB}$$

The coverage limited cell radius is $d = 10^{\frac{148.67-140.74}{35.22}} \approx 1.7km$

Observe how the cell radius shrinks when various loss margins are considered. Indeed these margins should not be ignored during design because in some practical situations we encounter severe shadowing losses, fading, body loss, and in-building losses.

10.5.2.2 Region II

Since region II is a region comprising a mix of medium density multi-storey residential buildings with some office and low rise shopping areas, we use COST-231 Urban path loss model:

$$L_p = L_{cost231} = 46.3 + 33.9\log(f_{MHz}) - 13.82\log(h_{BS}) + [44.9 - 6.55\log(h_{BS})]\log(d_{km}) - a(h_{MS}) \quad (10.18)$$

Where $f = 2GHz$, $h_{BS} = 30m$, $h_{MS} = 1.5m$, and $a(h_{MS}) = 0.05dB$

So

$$L_p = 163.67dB = 46.3 + 33.9\log(2\times10^3) - 13.82\log(30) + [44.9 - 6.55\log(30)]\log(d_{km}) - 0.05$$
$$= 137.74 + 35.22\log(d_{km}) \quad (10.19)$$

The coverage limited cell range (cell radius) can now be calculated as

$$d = 10^{\frac{163.67-137.74}{35.22}} \approx 5.45km$$

With 15dB in-building margin

$$L_p = P_t + G_t - L_{shadowing} + G_r - 2L_f - P_r - L_{in-building} = 163.67\text{-}15 = 148.67\text{dB}$$

The coverage limited cell radius is $d = 10^{\frac{148.67-137.74}{35.22}} \approx 2.1km$. Note once again how the cell range has reduced in length.

10.5.2.3 Region III

Because region III is a region comprising low density residential area with shops and factories we could use COST-231 Suburban path loss model. This path loss model is

$$\begin{aligned} L_p &= L_c - 2[\log(f_{MHz}/28)]^2 - 5.4 \\ &= 46.3 + 33.9\log(f_{MHz}) - 13.82\log(h_{BS}) + [44.9 - 6.55\log(h_{BS})]\log(d_{km}) \\ &\quad - a(h_{MS}) - 2[\log(f_{MHz}/28)]^2 - 5.4 \end{aligned} \tag{10.20}$$

Where $f = 2GHz$, $h_{BS} = 30m$, $h_{MS} = 1.5m$, and $a(h_{MS}) = 0.05dB$

So $L_p = 163.67dB = 137.74 + 35.22\log(d_{km}) - 2[\log(2\times10^3/28)]^2 - 5.4$

$= 125.47 + 35.22\log(d_{km})$

The coverage limited cell range (cell radius) can now be calculated as

$d = 10^{\frac{163.67-125.47}{35.22}} \approx 12.15km$

The in-building fade margin for this area is in practice smaller than the fade margin used for a CBD area. We will estimate this to be 10dB in-building margin $L_p = P_t + G_t - L_{shadowing} + G_r - 2L_f - P_r - L_{in-building}$ = 163.67 - 10 = 153.67dB

The coverage limited cell radius therefore is $d = 10^{\frac{153.67-125.47}{35.22}} \approx 6.4km$

10.5.2.4 Region IV

In Region IV the area is assumed to be rural consisting primarily of farm lands with low density (scattered) housing. The area may also contain highways. For this case we may use COST 231 Open Rural path loss model:

$$\begin{aligned} L_p &= L_c - 4.78[\log(f_{MHz})]^2 + 18.33\log(f) - 40.94 \\ &= 46.3 + 33.9\log(f_{MHz}) - 13.82\log(h_{BS}) + [44.9 - 6.55\log(h_{BS})]\log(d_{km}) \\ &\quad - a(h_{MS}) - 4.78[\log(f_{MHz})]^2 + 18.33\log(f) - 40.94 \end{aligned} \tag{10.21}$$

Where $f = 2GHz$, $h_{BS} = 30m$, $h_{MS} = 1.5m$, and $a(h_{MS}) = 0.05dB$

Therefore

$$L_p = 163.67dB = 137.74 + 35.22\log(d_{km}) - 4.78[\log(2\times10^3)]^2 + 18.33\log(2\times10^3) - 40.94$$

The coverage limited cell radius is also given by the expression

$$d = 10^{\frac{163.67-105.22}{35.22}} \approx 45.66km$$

In-building margins may be ignored in this region. This improves the coverage limited cell radius in this region and is d=45.7km.

10.5.3 Effects of Increased System Heights

Apart from the range of the cells, the COST 231 link budget provides three other variables which could be used to optimise the system coverage. Assuming that the frequency is fixed as in WCDMA networks, we could optimise the network by raising the height of the base station structure, the height of the mobile station and the transmitting power. In this section the transmitting power and the frequency of operation are assumed fixed and we vary either h_{MS} or h_{BS}. We have observed from the previous design that the cell ranges are functions of the heights of the base station and that of the mobile equipment. In this section the impact of the height of the base station is investigated. For the sake of analysis, we will increase the height of the base station to 50m ($h_{BS} = 50$). We will keep the height of the mobile station constant at 1.5m. Hence by recomputing the new cell radii using equations (10.16) to (10.21) we have:

Table 10.2 Cell Range as Function of Height of NodeB

Region	R_{cell} (km) $h_{BS} = 30m$	R_{cell} (km); (In Building margin) $h_{BS} = 30m$	R_{cell} (km) $h_{BS} = 50m$	R_{cell} (In Building margin) $h_{BS} = 50m$	R_{cell} (km); $h_{MS} = 5m$ $h_{BS} = 30m$;bldg margin;
I	4.48	1.7	5.88	2.12	0.85
II	5.45	2.1	7.22	2.6	1.04
III	12.15	6.4	17.07	8.43	3.21
IV	45.66	45.66	66.29	66.29	23.35

When the base station heights are set to 50m the range of the cells is increased dramatically. However, with in-building margin used in the link budgets, the increase is rather modest but significant enough to affect the number of base stations used to service an area of coverage.

Column 6 of Table 10.2 shows when the height of the mobile equipment is 5m about the height of an average first floor of a storey building. Its net effect is reduction of the cell range by a significant amount. We have now accounted for most of the margins in the channel and in-building. With the cell ranges in Table 10.1, we can now estimate the number of cells required for the four regions of interest to us. For this we use the areas of the regions given in the earlier sections.

We will calculate the range of each cell range by assuming hexagonal cell shapes. The number of cells in each region is given as Figure 10.2.

Given a hexagonal shaped cell, its total area is given by the expression

$$A_{cell} = 6x\frac{R}{2}\sqrt{3}x\frac{R}{2} = \frac{3\sqrt{3}R^2}{2} = 2.598R^2$$

The total land area for the network is 15, 000 square km. This is divided in proportions of 25%, 40%, 25% and 10% into the four regions. Therefore:

Region I: Since $R_{Icell} = 1.7\,km$, hence the area of one cell in Region I is

$$A_{Icell} = 2.598\,R_{Icell}^2 \approx 2.598\,x1.7^2 = 7.50822\,km^2$$

The number of cells in region I is

$$N_{Icell} = 0.25x15000\,km^2/7.50822\,km^2 \approx 500$$

Region II: $R_{IIcell} = 2.1\,km$, hence the area of one cell in Region II is

$$A_{IIcell} = 2.598\,R_{IIcell}^2 \approx 2.598\,x2.1^2 = 11.45718\,km^2$$

The number of cells in region II is

$$N_{IIcell} = 0.4x15000\,km^2/11.45718\,km^2 \approx 524$$

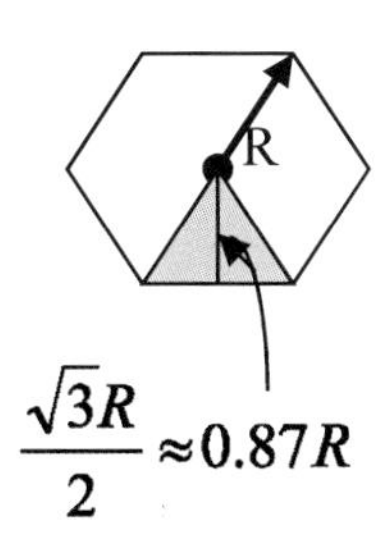

Figure 10.2 Hexagonal Cell Template.

Region III: $R_{IIIcell} = 6.4\,km$, hence the area of one cell in Region III is

$$A_{IIIcell} = 2.598\,R^2_{IIIcell} \approx 2.598\,x6.4^2 = 106.41408\;km^2$$

The number of cells in region III is

$$N_{IIIcell} = 0.25x15000\,km^2/106.41408\,km^2 \approx 36$$

Region IV: $R_{IVcell} = 45.66\,km$, hence the area of one cell in Region IV is

$$A_{IVcell} = 2.598\,R^2_{IVcell} \approx 2.598\,x45.66^2 = 5416.4028888\;km^2$$

The number of cells in region IV is

$$N_{IVcell} = 0.1x15000\,km^2/5416.4\,km^2 \approx 1$$

Although it appears that region IV will need only one node B, a single large base station in a large area without some form of redundancy is technically not advisable. This is because if it breaks down, then the whole area is left without access. Therefore we will deploy smaller Node Bs by dividing the radius of this large cell into 4 and use a cell with radius 11.415km. We then need to compute afresh the number of cells required. The new number of cells becomes:

$$A_{IVcell} = 2.598\,R^2_{IVcell} \approx 2.598\,x11.415^2 = 338.525\,km^2$$

$$N_{IVcell} = 0.1x16000\,km^2/338.525\,km^2 \approx 5$$

Five small cells will be used to cover this region.

10.6 SUMMARY

We have presented two approaches for estimating the coverage of WCDMA base stations or sectors based on considerations in the first case for population and data requirements for different kinds of regions. In the second case the impact of the terrain, the heights of base stations and losses in the link are considered. This approach normally provides better estimates for the cell range and most mobile network software planning tools tend to adopt some forms of this approach. In this approach fading and losses in the vehicles, human body losses, in building losses, diffraction and losses due to the morphology of the area under consideration should be taken into account. This is in addition to the traditional path loss expected of normal radio wave transmissions.

Chapter 11.

WiMAX Technology

Johnson I Agbinya

Department of Electronic Engineering, La Trobe University, Australia
J.Agbinya@latrobe.edu.au

11.1 OVERVIEW OF WIMAX

11.1.1 Introduction to WiMAX Technology

The mobile communication industry is a rapidly evolving commercial landscape. As a result it hardly takes up to a decade before new technologies emerge forcing re-design of mobile communication networks. The third generation (3G) technologies depended heavily on the use of orthogonal codes and direct sequence spread spectrum or DS-CDMA systems. Compared to GSM, CDMA is useful for better separation of users and base stations. It is however not that much efficient in offering high data rates suitable for data communications and Internet access. For this reason the need to develop fourth generation (4G) technologies arose. Two contending technologies emerged called Long Term Evolution (LTE) and WiMAX. LTE is the brain child of mobile phone operators and equipment manufacturers and WiMAX emerge from research laboratories spearheaded by universities and dreamed into being through IEEE support. This chapter deals only with WiMAX technology. According to the World wide interoperability for Microwave Access (WiMAX) is a potential solution to the "Digital Divides" [1].

WiMAx is a new technology which provides broadband communication and supports user's mobility. By being cost effective and interoperable with other existing technologies, it provides support for different user's mode such as fixed, nomadic, portable and mobile, flexible architecture and easy deployment while

4G Wireless Communication Networks: Design, Planning and Applications, 209-240,

providing a high data rate communication in a wide coverage area; WiMAX is supposed to be the first truly global wireless broadband network. WiMAX can provide services such as Voice over IP (VOIP), IPTV, video conferencing, multiplayer interactive gaming and web browsing [2]. In addition, it is a functional alterative for the fixed networks in the case of natural disaster when fixed networks are demolished and unable to perform.

Based on the standard IEEE802.16 known as WiMAX, it can support data rate up to 100Mbps by using OFDM (Orthogonal Frequency Division Multiplexing) modulation scheme, which will be discussed in detail in this chapter. WiMAX base stations also can support a coverage range of up to 70Km, by using smart antennas such as Multi Input Multi Output (MIMO) technique, which also results in higher throughput and performance enhancement [2]. This chapter will briefly overview the WiMAX technology first and then progress onto how to plan WiMAX networks.

11.1.2 WiMAX Physical Layer

According to the IEEE802.16, WiMAX can support four physical specifications at the physical layer as follows [3], [4]:

- Wireless MAN-SC (Signal Carrier, 10-66 GHz)
- Wireless MAN-Sca (below 11GHz), which uses single carrier modulation
- Wireless MAN-OFDM (below 11GHz), which uses orthogonal frequency division multiplexing and a 256-carrier OFDM scheme.

Wireless MAN-OFDMA (below 11GHz), stands for Orthogonal Frequency Division Multiple Access. This air interface supports multiple user accesses, as it allocates a subset of the carriers to one receiver. However, a 2048-carrier OFDM is the scheme that is used by MAN-OFDMA [4].

Three air interfaces were defined, wireless MAN-SCa, wireless MAN-OFDM and MAN-OFDMA, which operates in the frequency band below 11GHz, can support Non-line-of-sight communication [3].

11.1.3 WiMAX MAC Layer

The IEEE802.16 standard suggests, the MAC layer of WiMAX can support:

- Reliability, as it is a connection-oriented protocol.
- Quality of service for subscriber, by adaptive allocation of uplink and downlink traffic [4].

- Compatibility with different transport protocols such as IPv4, IPv6, Ethernet, and Asynchronous Transfer Mode (ATM) [4].
- Two different connection modes, which are known as point-to-multipoint (PMP) mode and Mesh mode for multihop ad hoc networks.
- Frequency Division Duplex (FDD) and Time Division Duplex (TDD) transmission modes.

11.1.4 Quality of Service in WiMAX

WiMAX provides three different services with different required quality of services [2]. The first service is Unsolicited Grant Service (UGS), which needs guaranteed quality of service (QoS), therefore WiMAX assigns a certain amount of bandwidth to each connection statically to reduce delay and jitter. Commercial IPTV is an example which requires UGS. The second is Polling Service (PS), which requires less QoS guarantee compared to UGS; hence WiMAX allocates the bandwidth dynamically. Finally, Best Effort service (BE) requires no QoS guarantees; as a result unused bandwidth by UGS and PS traffic is assigned to BE service. For instance, web browsing and email traffic use this type of service [2].

11.2 WIMAX NETWORK ARCHITECTURE

Network architecture refers to a framework design in which the physical components of the network are connected and their functions are specified in the manner that all the components operate properly and work with each other to provide different services to the users. WiMAX architecture is based on the Internet Protocol (IP) which means the data transmission is all based on the packet switched technology, and therefore provides flexibility and modularity for the network deployment [1]. With support of IP, WiMAX can provide a range of coverage options for small, large scale, urban, suburban and also rural areas. However, it also supports fixed, nomadic, portable and mobile usage models [1]. Through its core network it can provide internetworking and interoperability with other networks.

In this section a general WiMAX network architecture and its components and the functions associated with each component is introduced.

11.2.1 Different Sub-networks

From Figure 11.1, the WiMAX network is divided into two main subnets, which work seamlessly with each other to provide network access for the different types of users. The network includes Access Service Network (ASN) and Core Service Network (CSN). The networks have their own components and functions, which are explored in the following sections.

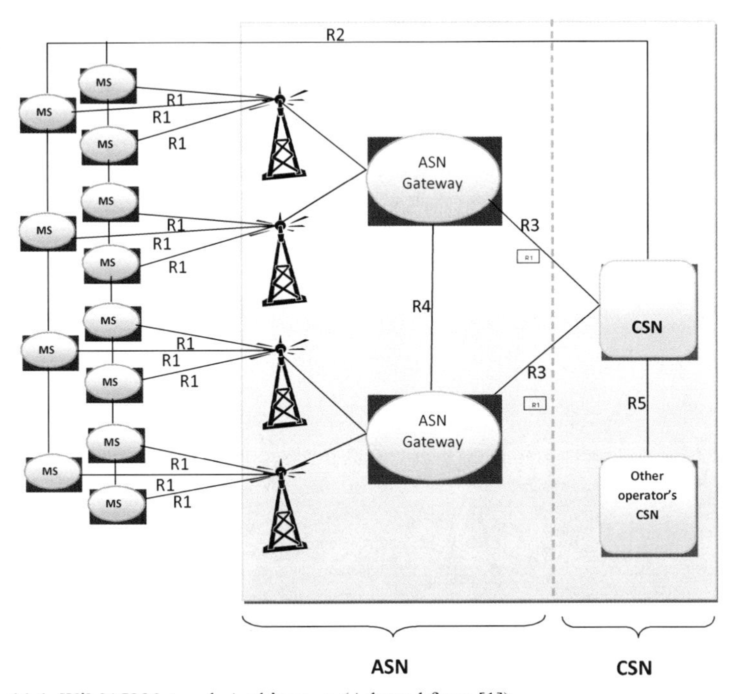

Figure 11.1 WiMAX Network Architecture (Adapted from [1]).

To plan a logical WiMAX network, the architecture needs to address some important requirements. The architecture provides support for interoperability of different vendors by dividing the network in terms of functionality. Moreover it should support divergent deployment schemes such as centralized, fully and semi distributed implementations and variety of usage models such as fixed, nomadic, portable and mobile. Internetworking with the other communication technologies such as Wi-Fi, 3GPP/3GPP2 and wired networks is essential. However, WiMAX architecture needs to underpin different business modes such as NAP (Network Access Provider), NSP (Network Service Provider) and ASP (Application Service Provider) [1].

11.2.1.1 Access Service Network

The ASN consists of one or a number of WiMAX Base Stations (BS) and one or more WiMAX Gateways (GW) [1]. It provides the interface between the user and CSN. Base station is responsible for provision of an interface between the user and the WiMAX network. A base station is typically located in the center or close to the borders of a WiMAX cell and have different range based on the elements such as the coverage area, environmental effects and the number of the users, also mobility of the subscribers. As well as the base station, there are radio equipment and the base station link to the backbone network in each cell.

Gateway is another component of ASN, which provides the connectivity between the ASN and CSN. It is responsible for mobility management and connection management. By processing the user's control and bearer data traffic it performs as a provider of the inter-service network boundaries.

As WiMAX forum suggests, ASN performs a number of functions which are outlined as follows [1], [5]:

- *Radio resource management* [5]: ASN enhance the efficiency of the radio resources employment by radio resource management. This function is divided into two functional entities, which are radio resource agent and radio resource controller. Radio resource agent gathers and measures radio resource indicators to communicate with the radio resource controller if necessary also to control the local radio resource. Radio resource controller collects the radio resource indicators from the radio resource agents associated with it also to communicate with other radio resource controllers. A radio resource agent is embedded in the base station while a radio resource controller is located in each base station and gateway.
- *Mobility management* [5]*:* In case of deployment of a mobile WiMAX network, it is crucial to have mobility management to minimize undesirable effects of mobility such as handoff delay or packet loss.
- *IP address assignment [5]:* Based on the usage model, ASN uses static or dynamic IP assignment. To perform that, ASN uses DHCP (Dynamic Host Configuration Protocol) for dynamic IP allocation or either via DHCP or manually in mobile Station in static IP allocation.
- *Authentication through proxy Authentication, Authorization and Accounting (AAA) Service [1]:* ASN provides authenticating and security service by an AAA server. It is also responsible for collecting the information about the usage of the resources by the subscribers and facilitates the charging process.

Handoff, interoperability with the other ASNs and relay of functionality between CSN and mobile station are other functions of the ASN [1].

NAP (Network Access Provider) is a service provider that provides access to the network service provider (NSP) and benefits through the marketing of applications and services that WiMAX can offer to users [1], [5]. Some of the services that WiMAX provides to subscribers are VoIP, video conferencing, streaming video, interactive gaming, mobile instant message, IPTV and basic broadband wireless Internet.

11.2.1.2 Core Service Network

Core Service Network (CSN) is the other part of the WiMAX network, which consists of the Home Agent (HA), which provides roaming through the network, AAA system to ensure the unique identification of the customers, IP server and gateway to other networks such as 3G for internetworking. CSN is mostly responsible for authentication, switching and transport. Some of the main functions of the WiMAX CSN are briefly mentioned below [1], [5]:

- Connectivity with other networks also the Internet and ASPs
- Management of the IP addressing
- Providing AAA services to the users
- Tunneling support
- Roaming between NSPs also between ASNs
- Mobility and location management
- Quality and policy control according to the contracts with users

A network service provider (NSP), owns the CSN, and provides core network services to the WiMAX network [5].

11.2.2 WiMAX Air Interfaces

Figure 11.1 presents the main open interfaces used by WiMAX. As can be seen, interface R1 provides a connection between the subscribers and the base station within the ASN. A logical interface between the mobile station and the core service network, which corresponds to authentication, IP configuration, mobility and service management, is defined as R2. R3 is the air interface between the ASN and CSN, which is associated with the AAA, policy enforcement and mobility management. Different ASNs are connected together through the R4 interface and mobility of the mobile stations between two ASNs is the concepts that R4 is dealing with. Finally R5, which deals with internetworking between different CSNs, such as roaming, is the interface between the CSNs.

11.3MOBILE WIMAX

11.3.1 Mobile WiMAX PHY

The mobile WiMAX PHY uses a combination of TDD and OFDMA for downlink and uplink signaling and multiple user access. The unique features within the TDD/OFDMA frame provide frequency diversity, frequency reuse, and cell segmentation which improves the performance against fading and inter-cell interference. WiMAX networks use a combination of TDMA and FDMA. WiMAX symbols are assigned in the time domain using QPSK or some form of digital modulation. This process creates complex signals which are processed in the frequency domain by using orthogonal sub-carriers through the Inverse FFT. This process is equivalent to orthogonal frequency division multiplexing access (OFDMA).

- ✓ *TDD*
  The WiMAX OFDMA frame is configured to support a point-to-multipoint network. The 802.16e PHY supports TDD, FDD, and half-duplex FDD operation. The initial release of the Mobile WiMAX profilel only includes TDD as shown in Table 11.2. Later releases includes FDD variants to match spectrum regulatory requirements in specific countries. For interference mitigation, system-wide synchronization is required when using TDD. Synchronization is typically achieved using a global positioning system (GPS) reference at the BS. In the event that network synchronization is lost, the BS will continue to operate until synchronization is recovered, using a local frequency reference. TDD, as specified in the WiMAX profile, enables asymmetric DL and UL traffic. Asymmetric traffic using TDD may improve the spectrum utilization and system efficiency as compared to FDD operation which typically requires equal UL and DL bandwidths. TDD uses a common channel for both UL and DL transmission allowing for a lower cost and less complex transceiver design. TDD also assures channel reciprocity which may benefit applications such as MIMO and other advanced antenna technologies. The TDD form of WiMAX uses a frame that is divided into a DL and an UL section.
- ✓ *Time and frequency parameters*
  The IEEE 802.16e air interface as adopted by the WiMAX Forum specifies channel bandwidths ranging from 1.25 to 20MHz. The first release of the mobile WiMAX system profile incorporated 5, 7, 8.75, and 10 MHz bandwidths as shown in Table 11.1.
- ✓ The bandwidth scalability in Mobile WiMAX OFDMA is achieved by adjusting the FFT size and the subcarrier spacing. For a given channel bandwidth, the subcarrier spacing is inversely proportional to the number of subcarriers and, therefore, the FFT size. The time duration of the OFDMA symbol is set by the inverse of the subcarrier spacing. Therefore by fixing the subcarrier spacing, the symbol time is automatically specified. The inverse relationship between subcarrier spacing and symbol duration is a necessary and sufficient condition to ensure that the subcarriers are orthogonal. Table 11.2 shows the subcarrier spacing and symbol time for the Mobile WiMAX 10 and 8.75 MHz (WiBRO) profiles using nominal bandwidths of 10 and 8.75 MHz respectively.

Table11.1 WiMAX and WiBRO time and frequency parameters, using a1024-point FFT

Parameter	Mobile WiMAX		WiBRO
Nominal bandwidth	10 MHz	7 MHz	8.75 MHz
Subcarrier spacing	10.9375 kHz	7.8125 kHz	9.7656 kHz
Useful symbol time (Ts=1/ Subcarrier spacing)	91.4 μs	128 μs	102.4 μs
Guard time (Tg=Ts/8)	11.4 μs	16 μs	12.8 μs
OFDMA symbol duration (Ts+Tg)	102.9 μs	144 μs	115.2 μs
Number of symbols in frame	47	33	42
TTG+RTG	464 PS	496 PS	404 PS
Frame length	5 ms	5 ms	5 ms
Sampling frequency (Fs=FFT points × subcarriers spacing)	11.2MHz	8 MHz	10 MHz
Physical slot (PS) ($/Fs)	357.14 ns	500 ns	400 ns

The Mobile WiMAX frame contains 48 symbols. The symbol contains the actual user data and a small extension called the guard time. The guard time is a small copy of data from the end of the symbol that is inserted before the start of the symbol. This guard time is also called the cyclic prefix (CP) and its length is chosen based on assumptions about the wireless channel. As long as the CP interval is longer than the channel delay spread, inter-symbol interference (ISI) introduced by the multi-path components can be eliminated.

The 802.16 standard specifies a set of CP values but the initial profile specifies a CP value of 1/8, meaning that the guard time is 1/8 the length of the symbol time. Table 11.1 shows the guard time and symbol duration for the Mobile WiMAX and WiBRO using the nominal bandwidth of 10 and 8.75 MHz respectively.

There are a number of significant differences in the UL signal compared to the DL. They reflect the different tasks performed by the BS and MS, along with the power consumption constraints at the MS. Differences include:

- No preamble, but there are an increased number of pilots. Pilots in the UL are never transmitted without data subcarriers
- The use of special CDMA ranging bursts during the network entry process

- Data is transmitted in bursts that are as long as the uplink sub-frame zone allows, and wrapped to further sub-channels as required

✓ *Preamble*
The DL sub-frame always begins with one symbol used for BS identification, timing synchronization, and channel estimation at the MS. This symbol is generated using a set of 114 binary pseudo random number (PN) sequences, called the preamble ID, of 568 length. The data in the preamble is mapped to every third subcarrier, using BPSK, giving a modest peak-to-average power ratio (compared to the data sub-channels). The preamble subcarriers are boosted by a factor of eight over the nominal data subcarrier level. There are no preambles in the UL except for systems using adaptive antenna systems (AAS). For the case when there is no UL preamble, the BS will derive the required channel information based on numerous pilot subcarriers embedded in the UL sub-channels.

✓ *FCH*
The FCH follows the DL preamble with a fixed location and duration. The FCH contains the downlink frame prefix (DLFP). The DLFP specifies the sub-channelization, and the length and coding of the DL-MAP. The DLFP also holds updates to the ranging allocations that may occur in subsequent UL sub-frames. In order that the MS can accurately demodulate the FCH under various channel conditions, a robust QPSK rate 1/2 modulation with four data repetitions is used.

✓ *DL-MAP and UL-MAP*
The DL-MAP and UL-MAP provide sub-channel allocations and control information for the DL and UL sub-frames. The MAP will contain the frame number, number of zones, and the location and content of all bursts. Each burst is allocated by its symbol offset, sub-channel offset, number of sub-channels, number of symbols, power level, and repetition coding.

✓ *Channel coding*
There are various combinations of modulations and code rates available in the OFDMA burst. Channel coding includes the randomization of data, forward error correction (FEC) encoding, interleaving, and modulation. In some cases, transmitted data may also be repeated on an adjacent subcarrier.

✓ *Randomization*
Randomization of the data sequence is typically implemented to avoid the peak-to average power ratio (PAPR) increasing beyond that of Gaussian noise, thus putting a boundary on the nonlinear distortion created in the transmitter's power amplifiers. It can also help minimize peaks in the spectral response.

- ✓ *FEC*
  The mobile WiMAX OFDMA PHY specifies convolutional coding (CC), convolutional turbo coding (CTC), and repetition coding schemes. When repetition coding is used, additional blocks of data are transmitted on an adjacent sub-channel. CTC can give about a 1 dB improvement in the link performance over CC.
- ✓ *Interleaving*
  Interleaving is a well-known technique for increasing the reliability of a channel that exhibits burst error characteristics. Interleaving involves reordering the coded data, which spreads any errors from burst of interference over time, increasing the probability of successful data recovery.
- ✓ *Modulation*
  There are three modulation types available for modulating the data onto the subcarriers: QPSK, 16QAM, and 64QAM. In the UL, the transmit power is automatically adjusted when the modulation coding sequence (MCS) changes to maintain the required nominal carrier-to-noise ratio at the BS receiver. 64QAM is not mandatory for the UL. Binary phase shift keying (BPSK) modulation is used during the preamble, on the pilots, and when modulating subcarriers in the ranging channel.

Table11.2 Mobile WiMAX PHY data rates

Modulation	Code rate	5MHz Channel		10MHz Channel	
		DL Data rate (Mbps)	UL Data rate (Mbps)	DL Data rate (Mbps)	DL Data rate (Mbps)
QPSK	1/2CTC, 6x	0.53	0.38	1.06	0.78
	1/2 CTC, 4x	0.79	0.57	1.58	1.18
	1/2CTC, 2x	1.58	1.14	3.17	2.35
	1/2 CTC, 1x	3.17	2.28	6.34	4.70
	1/2CTC	4.75	3.43	9.50	7.06
16QAM	1/2CTC	6.34	4.57	12.07	9.41
	3/4CTC	9.50	6.85	19.01	14.11
64QAM	1/2CTC	9.50	6.85	19.01	14.11
	2/3CTC	12.67	9.14	26.34	18.82
	3/4CTC	14.26	10.28	28.51	21.17
	5/6CTC	15.84	11.42	31.68	23.52

The BS scheduler determines the appropriate data rates and channel coding for each burst based on the channel conditions and required carrier-to-interference plus noise ratio (CINR) at the receiver. Table 11.2 shows the achievable data rates using a 5 and 10 MHz channel for both DL and UL transmissions.

Matrix A space time coding, Matrix B spatial division multiplexing (also known as multi input multi output (MIMO)) Zones can be configured to make use of multi-antenna technology, including phased array beamforming, STC, and MIMO techniques. Matrix A is an Alamouti-based transmit diversity technique, which involves taking pairs of symbols and time-reversing each pair for transmission on a second antenna. Matrix B uses MIMO spatial division multiplexing to increase the channel capacity. For downlink MIMO, user data entering the BS is split into parallel streams before being modulated onto the OFDMA subcarriers. As with the single channel case, channel estimation pilots are interleaved with the data subcarriers. For MIMO operation, the pilots are made unique to each transmit antenna to allow a dual receiver to recover four sets of channel coefficients. This is what is needed to remove the effect of the signal coupling that inevitably occurs between transmission and reception.

The MS is initially only required to have one transmit antenna and support open loop MIMO. More advanced, closed loop MIMO operates by the MS transmitting regular encoded messages back to the BS, which provide the closest approximation to the channel seen by the MS. The BS then pre-codes the MIMO signal before transmission, according to the channel state information (CSI) provided by the MS. The BS may also have the facility to control the single transmitters from two MSs to act together to create a collaborative MIMO signal in the UL.

Matrix A and Matrix B techniques can be applied to PUSC and AMC zones, to be described next.

11.3.2 IP-Based WiMAX Network Architecture

The IEEE 802.16e-2005 standard provides the air interface for WiMAX but does not define the full end-to-end WiMAX network. The WiMAX Forum's Network Working Group (NWG), is responsible for developing the end-to-end network requirements, architecture, and protocols for WiMAX, using IEEE 802.16e-2005 as the air interface.

The WiMAX NWG has developed a network reference model to serve as an architecture framework for WiMAX deployments and to ensure interoperability among various WiMAX equipment and operators.

The network reference model envisions unified network architecture for supporting fixed, nomadic, and mobile deployments and is based on an IP service model. Below is simplified illustration of an IP-based WiMAX network architecture. The overall network may be logically divided into three parts:

1. Mobile Stations (MS) used by the end user to access the network.

2. The access service network (ASN), which comprises one or more base stations and one or more ASN gateways that form the radio access network at the edge.
3. Connectivity service network (CSN), which provides IP connectivity and all the IP core network functions.

The network reference model developed by the WiMAX Forum NWG defines a number of functional entities and interfaces between those entities. Figure 11.2 shows some of the more important functional entities.

- **Base station (BS):** The BS is responsible for providing the air interface to the MS. Additional functions that may be part of the BS are micromobility management functions, such as handoff triggering and tunnel establishment, radio resource management, QoS policy enforcement, traffic classification, DHCP (Dynamic Host Control Protocol) proxy, key management, session management, and multicast group management.
- **Access service network gateway (ASN-GW):** The ASN gateway typically acts as a layer 2 traffic aggregation point within an ASN. Additional functions that may be part of the ASN gateway include intra-ASN location management and paging, radio resource management and admission control, caching of subscriber profiles and encryption keys, AAA client functionality, establishment and management of mobility tunnel with base stations, QoS and policy enforcement, foreign agent functionality for mobile IP, and routing to the selected CSN.
- **Connectivity service network (CSN) or Core Service Network:** The CSN provides connectivity to the Internet, ASP, other public networks, and corporate networks. The CSN is owned by the NSP and includes AAA servers that support authentication for the devices, users, and specific services. The CSN also provides per user policy management of QoS and security. The CSN is also responsible for IP address management, support for roaming between different NSPs, location management between ASNs, and mobility and roaming between ASNs.

The WiMAX architecture framework allows for the flexible decomposition and/or combination of functional entities when building the physical entities. For example, the ASN may be decomposed into base station transceivers (BST), base station controllers (BSC), and an ASNGW analogous to the GSM model of BTS, BSC, and Serving GPRS Support Node (SGSN).

The WiMAX physical layer is based on orthogonal frequency division multiplexing. OFDM is the transmission scheme of choice to enable high-speed data, video, and multimedia communications and is used by a variety of commercial broadband systems, including DSL, Wi-Fi, Digital Video Broadcast-Handheld (DVB-H), and MediaFLO, besides WiMAX.

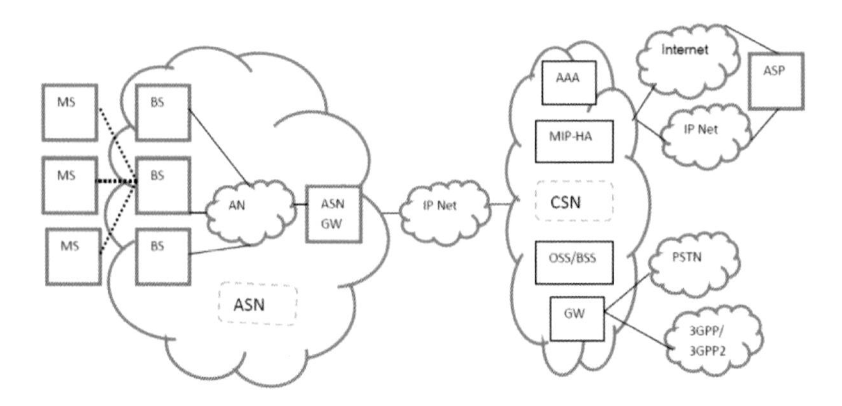

Figure 11.2 IP-based WiMAX Network Architecture

OFDM is an elegant and efficient scheme for high data rate transmission in a non-line-of-sight or multipath radio environment

11.3.3 Sub-carrier Grouping and Sub-channels

Grouping of multiple subcarriers into sub-channels is used to improve system performance. There are two types of subcarrier allocations to form sub-channels. These are distributed and adjacent subcarrier allocations.

In *distributed allocations* the subcarriers are pseudo-randomly distributed over the available bandwidth. Distributed allocation provides inter-cell interference averaging and frequency diversity in frequency-selective fading channels. Distributed allocation of subcarriers is preferred for mobile applications.

In *adjacent allocations* subcarriers adjacent to each other in the frequency domain are grouped to form sub-channels. Adjacent allocation has advantages in slowly fading channels, frequency nonselective channels and for implementing adaptive modulation and coding (AMC). Adjacent allocation is used in low mobility and fixed applications. For this the subscriber may be assigned the sub-channel with the best frequency response.

11.3.4 Slot Allocation and Data Regions

The WiMAX PHY layer allocates slots and framing over the air. A slot is defined as the minimum time-frequency resource that can be allocated by a WiMAX system to a given link. Each slot consists of one sub-channel over one, two, or three OFDM symbols. This depends on the sub-channelization scheme used. A user data region is a contiguous series of slots assigned to the given user. This is done by the scheduling algorithms. Data regions are allocated to users based on quality of service requirements, demand and the condition of the channel. Data is mapped to physical subcarriers in two steps.

a) The first step is controlled by the scheduler. Data is mapped to one or more logical sub-channels called slots. Slots may be grouped and assigned to segments based on applications. This can be used by a BS for different sectors in a cellular network.
b) In the second step, logical sub-channels are mapped to physical subcarriers. During this process, pilot subcarriers are also assigned. Data and pilot subcarriers are uniquely assigned based on the type of sub-channelization.

To understand the sub-channel planning, consider Table 11.3 with the parameters shown. The size of the FFT varies with the throughput sought and the available frequency bandwidth.

In Mobile WiMAX the size of the FFT varies from 128, 256, 512, 1024 to 2048 subcarriers (tones). These sizes of FFT correspond to the bandwidths of 1.25 MHz, 3.5, 5, 10 and 20MHz respectively. These choices allow the subcarrier spacing to be constant at 15.625kHz and 10.94 kHz for fixed and Mobile WiMAX respectively.

The basic resource unit in WiMAX is the symbol duration. The subcarrier value defines the symbol duration. In the fixed case the useful symbol duration is

$$T_{sf} = \frac{1}{f_{subcarrier}} = \frac{1}{15625Hz} = 64\mu s \tag{11.1}$$

This duration does not account yet for the pilot and null subcarrier times. For the Mobile WiMAX case, the useful symbol duration is

$$T_{sm} = \frac{1}{f_{subcarrier}} = \frac{1}{10940Hz} = 91.4\mu s \tag{11.2}$$

Mobility has been considered in the definition of the carrier spacing for Mobile WiMAX. The value of 10.94kHz was chosen as a balance between fulfilling the requirements for Doppler spread and delay spread. In the Table 11.3, the mobile WiMAX sub-carriers listed are for downlink PUSC.

When a terminal moves its carrier frequency changes in proportion to its speed. The chosen subcarrier spacing can support delay spreads of 20 microseconds for vehicular speeds of up to 125km/hr when operating around the 3.5GHz range. In addition to the bandwidths specified for Mobile WiMAX additional bandwidth profiles are permitted (Table 11.3). As an example, WiBro uses 8.75MHz with a 1024 FFT size.

Table 11.3 OFDM Parameters Used in Fixed and Mobile WiMAX

Parameter	Fixed WiMAX OFDM-PHY	Mobile WiMAX Scalable OFDMA-PHY			
Channel bandwidth (MHz)	3.5	1.25	5	10	20
FFT size	256	128	512	1,024	2,048
Number of used data sub-carriers	192	72	360	720	1,440
Number of pilot sub-carriers	8	12	60	120	240
Number of null/guard band sub-carriers	56	44	92	184	368
Cyclic prefix or guard time (Tg/Tb)	1/32, 1/16, 1/8, 1/4	None	None	None	None
Oversampling rate (Fs/BW)	Depends on bandwidth: 7/6 for 256 OFDM, 8/7 for multiples of 1.75MHz, and 28/25 for multiples of 1.25MHz, 1.5MHz, 2MHz, or 2.75MHz.				
Sub-carrier frequency spacing (kHz)	15.625	10.94			
Useful symbol time (μs)	64	91.4			
Guard time assuming 12.5% (μs)	8	11.4			
Duration of OFDM symbol (μs)	72	102.86			
Number of OFDM symbols in a 5 ms frame	69	48.0			

In a WiMAX transmission therefore there is a DC subcarrier at the transmission frequency and it is not used for carrying data, the pilot subcarriers that are used for synchronization, the guard subcarriers and the data carrying subcarriers. The guard subcarriers are the outer subcarriers. Thus the total number of subcarriers used for a WiMAX system is:

$$N_{FFT} = N_{Guard\ left} + N_{used(\max)} + N_{DC} + N_{Guard\ right}.$$

11.3.5 Permutation Zones

Permutation zones, or zones, are groups of contiguous symbols that use a specific type of sub-channel assignment. The OFDMA PHY specifies seven subcarrier permutation zone types: FUSC, OFUSC, PUSC, OPUSC, AMC, TUSC1, and TUSC2 [6]. PUSC, FUSC, and AMC are widely used in practical systems. "Except for AMC, the other zones use distributed allocation of subcarriers for sub-channelization. A frame may contain one or more zones. The DL sub-frame requires at least a zone and always starts with PUSC" [7]. Sub-channel allocation need not be contiguous. Subcarriers may be common to all sub-channels (in FUSC) or each sub-channel may allocate its own pilot carriers (PUSC).

DL FUSC: In the DL Fully Used Sub-channelization (FUSC) the pilot subcarriers are common to all sub-channels and are first allocated. After this the remaining subcarriers are divided into sub-channels. In the DL and UL Partially used sub-channelization (PUSC), the set of subcarriers that are available for use for data transmission and pilot are first divided into sub-channels. Then within each sub-channel, pilot subcarriers are allocated. It is reasonable for the DL FUSC to clearly specify the pilot tones because in a DL, sub-channels may be intended by a base station for different or groups of receivers while in UL, Subscriber Stations (SS) may be assigned one or more unique sub-channels by a base station and several transmitters may transmit simultaneously with the base station. This requires the specification of also known pilot subcarriers for each SS within its set of subcarriers.

DL PUSC [7]: Clusters of 14 contiguous subcarriers per symbol are grouped together and two clusters form a sub-channel. The slot in this case is one sub-channel over two OFDM symbols. This default zone is required at the start of all DL sub-frames following the preamble. In this zone, on alternate symbols, pairs of pilots swap positions, averaging one in seven of the subcarriers. For dedicated pilots they are only transmitted for corresponding data. In a DL PUSC zone the sub-channels can also be mapped into larger groups called segments. The first PUSC zone is always a single input single output (SISO). Further PUSC zones can be specified for other forms of multi-antenna MIMO systems.

UL PUSC [7]: "For this zone type, four contiguous subcarriers are grouped over three symbols. This grouping is called *a tile*. Six tiles make a sub-channel. For the UL PUSC, the slot is defined as one sub-channel that occurs over the three symbols. Pilots are incorporated within the slot, their position changing with each symbol. Over the course of one tile, one in three subcarriers is a pilot"[7].

AMC: The AMC zone occupies a wider bandwidth compared with PUSC and FUSC [719. The sub-channel is in this case a contiguous block of subcarriers. The zone structure is the same for both the DL and the UL. The slot is defined as a collection of bins using the N x M formula, where M is the number of OFDM symbols and N is the number of bins. A bin or symbol consists of nine contiguous subcarriers. A slot is one sub-channel wide and the length changes according to the zone. The pilots in the DL change positions periodically in a rotating pattern every fourth symbol.

DL FUSC: The DL FUSC zone uses all subcarriers to provide a high degree of frequency diversity. The subcarriers are divided into 48 groups of 16 subcarriers and a sub-channel is formed by taking one subcarrier from each group. For this zone a slot is defined as one sub-channel over one OFDMA symbol. "The pseudo-random distribution of data changes with each OFDMA symbol over the length of the zone, which can be useful when attempting to mitigate interference through the use of what is effectively a type of frequency hopping. The pilots are regularly distributed. Their position alternates with each symbol" [7].

In the *DL optional FUSC (DL OFUSC)* pilot subcarriers are evenly spaced by eight data subcarriers from each other. In the UL OPUSC the zone is identical with the UL PUSC "except that it uses a tile size that is three subcarriers wide by three symbols long" [7].

***TUSC1 and TUSC2*:** The total usage of sub-channels (*TUSC1 and TUSC2)* zones is only available in the DL using AAS. "They are both optional and similar to DL PUSC and OPUSC but use a different equation for assigning the subcarriers within the sub-channel" [7].

With the exception of the DL PUSC, which is assigned after the DL preamble, all of the zones can be assigned in any order within the frame. The switching points between zone types are listed in the DL MAP. Figure 11.3 shows an example of an OFDMA frame with several different types of zones [7]. From Figure 11.3 the mandatory DL PUSC zone follows the preamble in the

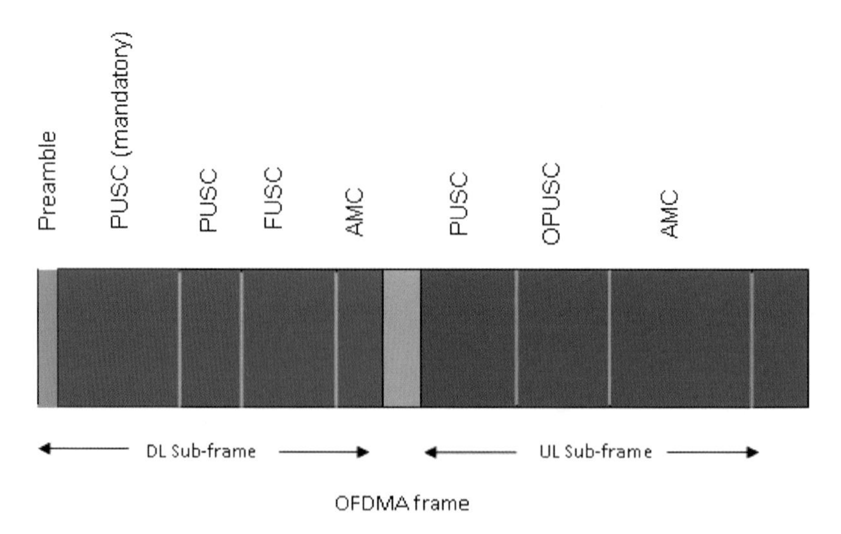

Figure 11.3 Example of an OFDMA frame with multiple zones.

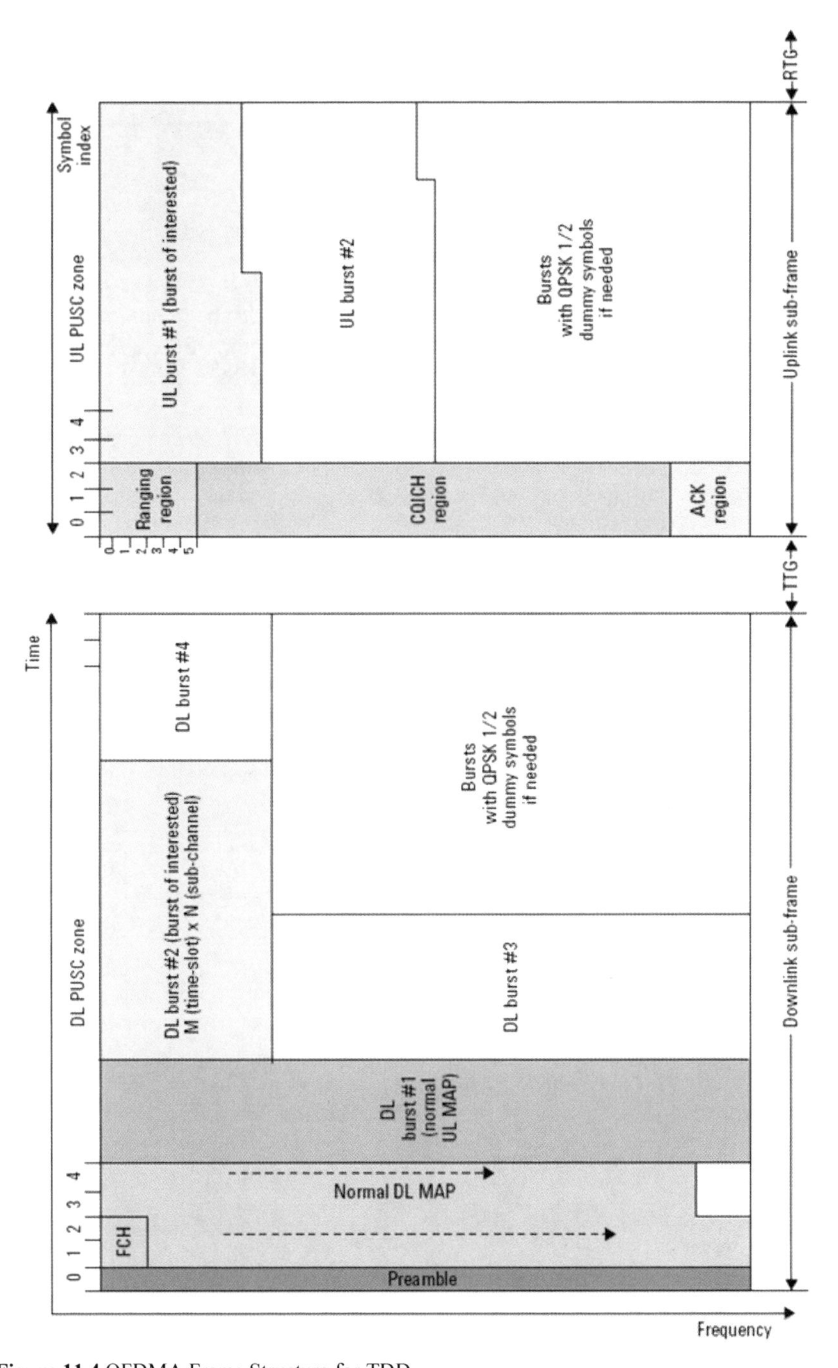

Figure 11.4 OFDMA Frame Structure for TDD.

frame. "The DL sub-frame also shows a second PUSC zone, a FUSC zone, and an AMC zone. The UL sub-frame follows the TTG and, in this example, contains a PUSC, OPUSC and AMC" [7].

11.3.6 Frame Structure

The OFDMA frame consists of two main parts, a DL sub-frame and an UL sub-frame plus two sub-parts the TTG and RTG. The flexible frame structure of the TDD signal consists of a movable boundary between the DL and UL sub-frames. A short transition gap is placed between the DL and UL sub-frames and is called transmit-receive transition gap (TTG). After the completion of the UL sub-frame, another short gap is added between this sub-frame and the next DL sub-frame. This gap is called the receiver-transmit transition gap (RTG). The minimum time durations for these transition gaps are called out in the 802.16 standard and are a function of the channel bandwidth and the OFDM symbol time. It is typical to define these transition gaps in terms of physical slot (PS) units. A PS is a unit of time defined as 4/(sampling frequency).
An example of a mobile WiMAX frame is shown in Figure 11.4. This figure shows the time-frequency relationship where the symbol time is shown along the x-axis and the logical sub-channels along the y-axis. Logical sub-channels are groupings of frequency subcarriers assigned to individual users. The concept of sub-channels and zones will be covered later in this chapter. Figure 11.4 shows the DL and UL sub-frames separated by the TTG and ending with the RTG. The figure also shows the relative position of the preamble, frame control header (FCH), downlink media access protocol (DL-MAP), and uplink media access protocol (UL-MAP) whose functions will be discussed in the next section.

11.3.7 How OFDM Operates

To allocate a number of sub-channels to the sub-streams, without the need for independent radio frequencies, OFDM uses the DFT (Discrete Fourier Transform) technique and specifically an efficient form of DFT known as FFT (Fast Fourier Transform) and the inverse of the technique, IFFT. Using these two techniques OFDM can divide a single radio channel into L sub-channels.

In the physical layer the WiMAX signal to be transmitted can be represented by the expression:

$$y(t) = \mathrm{Re}\left\{ e^{j2\pi f_c t} \times \sum_{\substack{k=-N_{used}/2 \\ k\neq 0}}^{N_{used}/2} C_k e^{j2\pi k\Delta f\left(t-T_g\right)} \right\} \tag{11.3}$$

Where in this expression the variables are

Δf is the subcarrier frequency spacing

f_c is the central carrier frequency and

T_g is the guard time

N_{used} is the number of used subcarriers

C_k is a complex number representing the data to be transmitted

k is the frequency offset or subcarrier index

An OFDM symbol is formed from L data symbols and lasts for T seconds, which is equal to LT_S. To prevent the interference between the OFDM symbols transmitting through a channel, a guard time (Figure 11.5) should be considered between each two OFDM symbols. The guard time (T_g), needs to be larger than the delay spread of the channel and also large enough to overcome the interference between the subsequent OFDM symbols.

However, the intra OFDM-symbols interference still remains. OFDM technique solves this problem by using circular convolution. Assuming x[n] as the channel input, y[n], the channel output and h[n] as the Finite Impulse Response of the channel and that the channel is linear time-invariant, then:

$$y[n] = x[n] * h[n] \tag{11.4}$$

Re-writing equation (11.15) in terms of circular convolution, we obtain

$$y[n] = x[n] \otimes h[n] = h[n] \otimes x[n] \tag{11.5}$$

and

$$y[n] = \sum_{k=0}^{L-1} h[k]x[n-k]_L \tag{11.6}$$

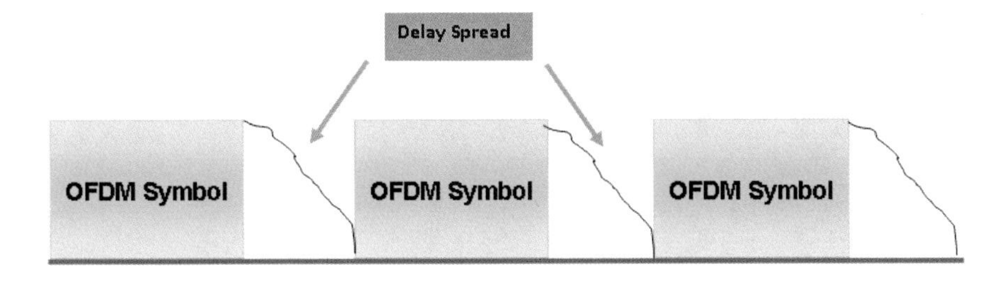

Figure 11.5 Guard time should be larger than delay spread (Adapted from[8]).

Where,

$$x[n]_L = x[n \bmod L] \tag{11.7}$$

Thus *x[n]* is periodic with the period *L samples.*

Taking the DFT of the channel output results in the output in the frequency domain:

$$DFT\{x[n]\} = X[m] = \frac{1}{\sqrt{L}} \sum_{n=0}^{L-1} x[n] e^{-j\frac{2\pi n}{L}} \tag{11.8}$$

The inverse discrete Fourier transform is given by the expression:

$$IDFT\{X[m]\} = x[n] = \frac{1}{\sqrt{L}} \sum_{n=0}^{L-1} X[m] e^{j\frac{2\pi n}{L}} \tag{11.9}$$

Equation (11.9) shows that each input *X[m]* in the frequency domain is extended by *H[m].* It can be demonstrated that the channel is ISI free. Although a circular convolution is deployed in this technique, the natural linear channel produces a linear convolution. However the circular convolution is forged by adding a cyclic prefix to the transmitted data. A cyclic prefix technique is when bits at the end of the symbol are included at the beginning of an OFDM symbol where ISI is expected. The technique is used to form a circular convolution from a linear convolution to achieve more robustness in the system. The length of the cyclic prefix is usually equal to the guard interval or the length of the anticipated ISI. The guard interval should have at least v samples, when the delay spread of the channel has the duration of *v+1* samples. After adding a cyclic prefix to the original samples with the length of *L*, the input of the channel is $x_{cp} = [x_{L-V}\, x_{L-V+1\ldots}\, x_{L-1} x_0 x_1 \ldots x_{L-1}]$, where x_0 to x_{L-1} stands for the original data while x_{L-v} to x_{L-1} at the beginning of the vector shows the cyclic prefix symbols.

Based on the input of the channel the output would be $y_{cp} = h * x_{cp}$, which has the length $(V+1)+(V+L)-1 = L+2V$. However, *2v* samples or bits will be discarded at the receiver in order to eliminate interference with the preceding and subsequent OFDM symbols. Therefore L samples remain at the end. In [8], it is proved that the L sample corresponds to $y = h \otimes x$, therefore the output y is disintegrated into a channel frequency response of $H = DFT\{h\}$ and channel frequency domain input of $X = DFT\{x\}$. Although the cyclic prefix provides a circular convolution and assists to provide an ISI free channel, it is associated with some drawbacks. To implement that, more bandwidth and transmission power are needed, as the v additive samples burdens the bandwidth. The required bandwidth is extended by $(L+V/L)$, in other words the bandwidth will be $(L+V/L)$ times more. Similarly the transmission power required for transmission of the cyclic prefix is $10\log_{10}(L+V/L)dB$. In conclusion, using the cyclic prefix is equivalent to a loss of power and data rate of $L/L+v$.

In this section, a step by step operation of the OFDM is described [8], based on the assumption that the synchronization between the receiver and the transmitter is perfect and receiver faultlessly knows the channel. The steps are

- Dividing the wideband channel into L narrowband channels and L subcarriers for each channel, to achieve the ISI free transmission for each sub stream, with using a proper cyclic prefix.
- Modulation of L subcarriers using IFFT technique for overcoming the need for L independent radio frequencies
- After IFFT operation, a cyclic prefix with length of v has to be attached to the OFDM symbols and sent to the wideband channel serially to provide orthogonality.
- At the receiver, using FFT demodulation technique, the original data is identified by dispensing of the cyclic prefix.
- Each subcarrier is equalized via a frequency domain equalizer to estimate the data symbols.

11.3.8 Synchronization in OFDM Systems

Two types of synchronization must be considered at the receiver for successfully demodulating the OFDM data symbols which are timing and frequency synchronizations [8]. Determination of the OFDM symbol's offset and optimal timing instant is known as the timing synchronization and frequency synchronization is the arrangement of the subcarriers at the receiver in the manner that is as close as possible to the transmitter's subcarriers. Thanks to the cyclic prefix the timing synchronization is relatively acquitted, however, the frequency synchronization is very crucial as the individual orthogonal data symbols need to be implemented at the frequency domain. Development of the frequency synchronization algorithm is one of the most important issues in this technique.

11.3.8.1 Timing Synchronization

Although timing errors can occur due to improper timing synchronization, the system can tolerate some imperfection in timing synchronization, as long as the timing offset of τ is larger than zero and smaller than the difference in time between the maximum channel delay spread (T_m) and the guard time (T_g). If τ >0, sampling occurs at the later time compared to the ideal instant but if sampling occurred at the earlier time, it is assumed that $\tau<0$. However, ISI happens if $\tau<0$ or $\tau>T_m-T_g$. According to the amount of τ, the receiver will lose some energy and that energy degradation results in $\Delta SNR(\tau)\approx -2\left(\tau / LT_s\right)^2$, which is the signal to noise ratio loss.

It is derived from the equation above that to minimize the timing error, τ should be as close as possible to zero or as small as possible in comparison with the guard time.

11.3.8.2 Frequency Synchronization

Frequency synchronization is a very serious issue in OFDM systems and lack of an efficient algorithm for the synchronization may result in undesirable overlapping of the subcarriers instead of being isolated from each other. The frequency offset *(δ)* should be equal to zero to provide an interference free channel, however, the frequency offset is unequal to zero in practical situations, which result from inconsistencies in the oscillators at the transmitter and receiver and also Doppler frequency shift due to mobility. Using crystal oscillator is very costly, therefore systems must tolerate to some extent frequency offsets. The formula presented below expresses that as the frequency offset increases, the SNR decreases:

$$\Delta SNR=\frac{\varepsilon_x / N_0}{\varepsilon_x / \left(N_0+C_0\left(L\,T_S\delta\right)^2\right)}=1+C_0\left(L\,T_S\delta\right)^2 SNR \tag{11.10}$$

- ε_x : Average synchronization energy
- C_0 : Constant depends on the assumption

11.4 OFDMA (ORTHOGONAL FREQUENCY DIVISION MULTIPLE ACCESS)

Whereas OFDM is a modulation technique, OFDMA is a multiple access technique used primarily in mobile WiMAX and allows different users to have access to the available channel at the same time [8]. In OFDM based system, entire subcarriers are allocated to only one user, while in OFDMA technique subcarriers are divided into different groups of subcarriers and each group has a number of subcarriers, known as sub-channels [9]. To understand the OFDMA technique, a brief overview of different multiple access techniques are provided in this section.

11.4.1 Multiple Access Techniques

To provide multiple access to the channel for the users, three different strategies [8] have been used known as FDMA (Frequency Multiple Access), TDMA (Time Division Multiple Access), and CDMA (Code Division Multiple Access). A subset of carrier frequency is allocated to each user in FDMA technique, which can be performed statically and dynamically. To provide the sub-channels to the users statically, channel assignment is conducted by a multiplexer in the digital environment and before the IFFT accomplishment. However, in dynamic channel allocation, each user is assigned a specific channel, which is more appropriate for the particular user and it can use the channel more efficient. The second form of multiple access is TDMA. In this technique, each user can access entire bandwidth for a specific amount of time called time slot. Although TDMA suits circuit switched-based data transmission the most, it can perform well in packet switched networks using more complex and improved algorithms. CDMA is another method for multiple access which provides each user a unique code. Therefore each user can use the entire bandwidth as long as the data transmission is occurring. However it is not suitable for high data rate transmission because it is sensitive to interference.

OFDMA is a hybrid form of the FDMA and TDMA, which allocates a user a subset of subcarriers in a proportion of time dynamically and provides robustness, scalability, spectrum efficiency and better frequency reuse [8]. It assigns a channel to the user according to the conditions and suitability of the channels for particular users; hence it provides more efficient resource allocation. Based on the important features of OFDMA, such as down link and uplink sub-channelization, better frequency reuse and also scalability, it is more suitable for mobile broadband wireless networks such as mobile WiMAX [9] and LTE-A. However, OFDM performs better in fixed networks. Moreover, OFDMA offers two important principles known as multiuser diversity and adaptive modulation, which result in high performance of OFDMA.

11.4.2 Multiuser Diversity

Multiple user diversity defines the availability of the gain by choosing a user of a group of users with a 'good' channel condition [8]. It enhances the capacity also in some cases, link reliability and coverage area. Assuming k number of users, each one has the channel gain (h_k) independent from others; the probability density function (PDF) of *the* k_{th} user's channel gain is defined as:

$$p(h_k) = \begin{cases} 2h_k\ e^{-h_k^2} & \text{if } h_k \geq 0 \\ 0 & \text{if } h_k < 0 \end{cases} \tag{11.11}$$

Based on the assumption that the base station transmits to the user with the highest channel gain *(h_{max})*:

$$p(h_{\max}) = 2kh_{\max}\left(1 - e^{-h_{\max}^2}\right)^{k-1} e^{-h_{\max}^2} \tag{11.12}$$

The equation exhibits that increasing the number of users results in more probability of getting a large channel and consequently enhancing the capacity and also BER (Bit Error Rate).

11.4.3 Adaptive Modulation

Adaptive modulation refers to the selection of different modulation and coding schemes by the transmitter and allows for the scheme to change on a burst-by-burst basis per link, based on the state of the channel [8]. In other words, a transmitter has to transmit the data in as high as possible data rate if the channel has a good and proper condition. However, if the condition of the channel is weak, the data rate must be as low as possible. The condition of the channel is mostly determined through the SINR measurement. To perform that, the transmitter has to know the channel information through the receiver feedback. The SINR at the receiver is attained by multiplying the transmitter power and SINR at the transmitter. According to the SINR, the best configuration is selected. For example if the SINR is high, larger constellation and higher error correcting rate should be deployed for achieving the higher data rate; Whereas, if the SINR is low, to maintain a lower data rate, lower constellation and error correcting rate has to be chosen.

In OFDMA systems, based on the value of the SINR and its variety among users, different blocks of subcarriers are allocated to users. Each block provides the best configuration for a specific user or group of users. Different configurations are called *burst profiles* [8]. Table 11.4 displays different burst profiles in WiMAX.

The different modulation schemes employed in WiMAX are shown in Table 11.5.

11.4.4 Scalable OFDMA (SOFDMA)

SOFDMA is the scalable form of the OFDMA, which performs well in mobile WiMAX networks. In this scheme, the number of subcarriers which is equal to the size of the FFT, scales with the bandwidth, while maintaining each sub-channel's bandwidth constant [11]; which means the subcarrier spacing and the number of the subcarriers are independent of the bandwidth. This leads to less complexity for smaller channels while enhancing the performance of wider channels. SOFDMA can provide bandwidths of 1.25Mbps to 20 Mbps for each sub-channel [11].

Table 11.4 Different burst profiles in WiMAX [10].

Channel Bandwidth	3.5 MHz		1.25 MHz		5 MHz		10 MHz	
PHY Mode	256 OFDM		128 OFDM		512 OFDM		1024 OFDM	
Oversampling	8/7		28/25		28/25		28/25	
Modulation &Code Rate	PHY –Layer Data Rate (Kbps)							
	DL	UL	DL	UL	DL	UL	DL	UL
BPSK,1/2	946	326	Not applicable					
QPSK,1/2	1,882	653	504	154	2,520	653	5,040	1,344
QPSK,3/4	2,822	979	756	230	3,780	979	7,560	2,016
16 QAM, 1/2	3,763	1,306	1,008	307	5,040	1,306	10,080	2,688
16 QAM,3/4	5,645	1,958	1,512	461	7,560	1,958	15,120	4,032
64 QAM,1/2	5,645	1,958	1,512	461	7,560	1,958	15,120	4,032
64 QAM,2/3	7,526	2,611	2,016	614	10,080	2,611	20,160	5,376
64 QAM,3/4	8,467	2,938	2,268	691	11,340	2,938	22,680	6,048
64 QAM,5/6	9,408	3,264	2,520	768	12,600	3,264	25,200	6,720

Table 11.5 Different modulation and coding schemes supported by WiMAX

	Downlink	Uplink
Modulation	BPSK, QPSK, 16 QAM, 64 QAM; BPSK optional for OFDMA-PHY	BPSK, QPSK, 16 QAM; 64 QAM optional
Coding	Mandatory: convolutional codes at rate 1/2, 2/3, 3/4, 5/6 Optional: convolutional turbo codes at rate 1/2, 2/3, 3/4, 5/6; repetition codes at rate 1/2, 1/3, 1/6, LDPC, RS-Codes for OFDM-PHY	Mandatory: convolutional codes at rate 1/2, 2/3, 3/4, 5/6 Optional: convolutional turbo codes at rate 1/2, 2/3, 3/4, 5/6; repetition codes at rate 1/2, 1/3, 1/6, LDPC

It is mentioned in [12] that *11 kHz* is the optimum sub-carrier spacing for wireless mobile networks, which is the tradeoff between some parameters such as the level of protection to multipath effects, Doppler shift, design cost or complexity. SOFDMA assists to maintain optimum subcarrier spacing, by changing the FFT size based on the bandwidth, which can underpin the NLOS operations. OSFDMA also supports MIMO techniques, Advanced Modulation and Coding and some other important features.

11.5 ADAPTIVE ANTENNA

In practice, a transmitted signal may be received at the receiver from different paths due to phenomena such as reflection, refraction, shadowing and scattering, which may result in two scenarios. A stronger signal may be received at the receiver, if different signal components enhance each other in phase. If they change each other's phase the received signal will be a weak signal which leads to reduction of the link quality by degradation of the SNR and consequently decreasing the total capacity of the channel, according to the Shannon's capacity formula $C = B\log_2\left(1+{}^{S}\!/_{N}\right)$. To overcome the problem of signal degradation arising from the multipath effects and utilize this effect, a technology known as Smart Antenna or Adaptive Antenna is used in the transmitter, receiver or both.

Using arrays of antennas as well as sophisticated smart signal processing algorithm at the transmitter or receiver is referred to as adaptive antenna. Based on the use of the antenna array at the receiver, transmitter or both, this technology is divided into three different types [13]:

- SIMO (Single Input, Multiple Output): using the array of antennas at the receiver
- MISO (Multiple Input, Single Output):using the array of antennas at the transmitter
- MIMO (Multiple Input, Multiple Output): using arrays of antennas at receiver and transmitter (Figure 11.6). In nature, many insects communicate using some form of MIMO technique (for example the two butterflies in Figure 11.6).

In this technology, the signal processor unit integrates all the signal components received at different antennas, resulting from the multipath effects and makes use of the effect and derive benefit from it to produce a strong signal which increases the channel capacity and link quality. Employing adaptive antenna leads to less multipath fading and co-channel interference [13].

Adaptive antenna gain is n times more than the gain of individual element of the antenna array, if each array consists of n individual antennas, with the equal antenna gains [14].

Figure 1.6 Butterfly Communication System (MIMO in Nature).

$$G_{(AdaptiveAn\ tenna)} = nG_{(AntennaEle\ ments)} \quad (11.13)$$

According to the equation, the signal at the receiver increases through the combination of the signals from different antennas. This also results in extension in the coverage range and strengthening of the SNIR (Signal per Noise Interference Ration) result in more capacity. Adaptive antenna also magnifies the signal strength and the antenna gain at a specific desired direction to promote the link quality and reliability at that particular side and decrease the interference by minimizing the signal propagation at the directions which the interference is more likely to occur [13].

Spectral efficiency is another important advantage of adaptive antenna technology, which is accomplished through reducing the number of the required cells per area unit also increasing the amount of billable services per spectrum that profits the operators economically. Less handoff is also needed, since cell splitting is not necessary thanks to the smart signal propagation toward the desired directions [13].

Although adaptive antenna technology offers many advantages, it has some disadvantages such as complexity of the transceiver, resource management and enlarging the physical size due to using a number of antennas at the transceiver.

As described in [13] a number of techniques are applied to the technology such as conventional beamformer, null Steering beamformer, minimum variance distortionless response beamformer , minimum mean square error beamformer and least square despread multitarget array.

11.6 MIMO (MULTIPLE INPUT-MULTIPLE OUTPUT) TECHNIQUE

MIMO, used in WiMAX technology, is a form of adaptive antenna which employs antenna arrays in both transmitter and the receiver to combine the signal components resulting from multipath effect and provide a strong signal at the receiver. To achieve that, the MIMO system needs to be implemented in a rich-scattering environment, where a signal reaches the destination from different paths [15]. MIMO technique increases the capacity, in turn the link quality, for a given capacity and transmission power compared to SISO system (Single Input-Single Output). By using a number of antennas, MIMO technique provides three different diversities [16]. Through antenna diversity, the link quality increases by exploiting the multipath effect [16].

The first type of diversity is known as spatial diversity. Antennas located in different positions far enough from each other so that they experience different signal strengths to promote C/I (Carrier to Interference Ratio) by integrating different signal components. The optimum distance between the antennas is suggested to be about 38% of the transmission wavelength [16]. Beam diversity is another type of diversity, where the transceiver uses multiple directional antennas. Each antenna experiences different signal strength level of the original signal based on the angle of the received signal, then a combination of all received signal components is used to provide more C/I. However the distance between the antennas is not as serious as spatial diversity. Finally, polarization diversity uses an orthogonal pattern for locating the antenna elements to provide different channel gains in the incoming signal direction and increases the C/I. Figure 11.7 illustrates different diversity patterns.

Typically, combinations of different diversity techniques are used to achieve the best implementation of the antenna depending on the propagation path between the transmitter and receiver [15], [16]. However, achieving a perfect diversity is impossible in practical environments. To achieve a desirable diversity, isolation

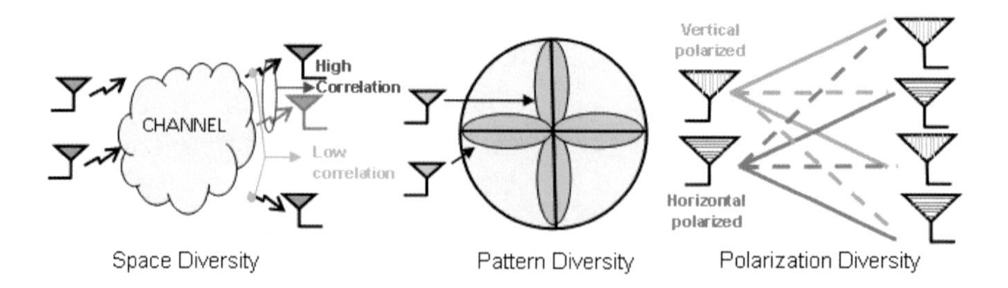

Figure 11.7 different diversities [15]

of the antenna elements is crucial. Since the mutual coupling intensifies the correlation between antenna ports, the efficiency of the diversity technique drops because for achieving a high diversity gain, correlation between the ports needs to be minimized. Maintaining the orthogonality of the antenna elements as much as possible assists in the achievement of more isolation. However, according to Chu-Harrington [16], to maintain the orthogonality, the size of the antenna needs to be enlarged. Therefore, in the concept of the MIMO antenna technology, one of the most important challenges is to implement the multiple antennas in an ever-decreasing space [16].

References

[1] Yan Zhang. WiMAX Network Planning and Optimization. USA: CRC Press, 2009

[2] Johnson I Agbinya. IP Communications and Services for NGN. New York: Taylor and Francis, 2009

[3] Kejie Lu, Yi Qian, Hsiao-Hwa Chen, Shengli Fu. " WiMAx Networks: From Access to Service Platform". IEEE Computer Society,vol. 22.,pp.38-45, May/June 2008

[4] Zakhia Abichar, Yanlin Peng, J. Morris Chang. "WiMAX: The Emergence of Wireless Broadband". IEEE Computer Society,vol. 8., pp.44-48, July/August 2006

[5] Kosta Tsagkaris, Panagiotis Demestichas. " WiMAX Network". Vehicular Technology Magazine, IEEE. Vol.4, pp:24-35, June 2009

[6] Eklund, Carl, et.al. WirelessMAN, Inside the IEEE802.16 Standard for Wireless Metropolitan Networks, IEEE Press, 2006.

[7] Peter Cain, "Mobile WiMAX PHY (RF) operation and measurement, Part 1", Agilent Technologies, http://www.eetimes.com/design/microwave-rf-design/4018993/Mobile-WiMAX-PHY-RF-operation-and-measurement-Part-1

[8] Jeffrey G. Andrews, Arunabha Ghosh, Rias Muhamed. Fundamentals of WiMAX, USA: Prentice Hall, 2007

[9] Hujun Yin, Siavash Alamouti. "OFDMA: A Broadband Wireless Accss Technology." Sarnoff Symposim. 2006, pp: 1-4

[10] "WiMAX-Physical Layer." Internet: http://digitaxis.com/web/sb-files/1201343773phy.JPG , Jan.26, 1008 [Sep. 18, 2009]

[11] "Introduction to FDM, OFDM, OFDMA, SOFDMA". Internet: http://WWW.conniq.com/wimax/fdm-fdma-ofdma-sofdma-01.htm, 2009 [Sep.16,2009]

[12] Vlodimir Bykovnikov, "the advantages of SOFDMA for WiMAX". Internet: http://my.com.nthu.edu.tw/~jmwu/LAB/sofdma-for-wimax.pdf, 2005 [Sep.16, 2009]

[13] Hafeth Hourani. "An Overview of Adaptive Antenna System" . Postgraduate course in Radio Communication, 2004/2005

[14] M.Viberg, T.Boman, U.Carlberg, L.Pettersson, S.Ali, E.Arabi, M.Bilal, O.Moussa . "Simulation of MIMO Antenna System in Sinulink and Embeded Matlab", Internet: http://www.mathworks.se/company/events/conferences/mlnordic_conf08/proceedings/papers/mimo_antenna_systems.pdf , [Sep. 17, 2009]

[15] Aakanksha Pandey, " Researching FPGA Implementations of Baseband MIMO Algorithms Using Acceldsp" Internet: http://www.rimtengg.com/coit2008/proceedings/MS32.pdf [Sep. 18, 2009]

[16] Crown, "Antenna Design for MIMO System" Internet: http://info.awmn.net/users/images/stories/Library/Antenna%20Theory/Feeder/antenna_designs_mimo.pdf , [Sep. 18. 2009]

[17] Rhode and Schwarz, WiMAX General Information about the standard 802.16 (application note), 2006, pp. 1 – 34

[18] WiMAX Capacity White Paper, SR Telecim, 2006

Chapter 12.

Planning and Dimensioning of WiMAX Networks

Johnson I Agbinya

Department of Electronic Engineering, La Trobe University, Australia
J.Agbinya@latrobe.edu.au

12.1 WIMAX NETWORK PLANNING PARAMETERS

OFDM has been covered extensively in chapters 5, 6 and 11. In this chapter it is used. The chapter demonstrates how to plan an OFDM-based network. The key air interface technology used in WiMAX is OFDM. OFDM is an efficient form of multi-carrier modulation schemes, which is used in most of the modern communication technologies, such as 3G-LTE, DSL, 4G Cellular Systems and WiMAX.

To understand OFDM we introduce a few pertinent terminologies that are specific to it. OFDM system bandwidths are normally purchased by telecommunication operators through spectrum bids. The main system bandwidths range from 1.25 MHz, 3.5 MHz, 5 MHz, 8.75 MHz (WiBRO), 10 MHz to 20MHz. Except for a small region of the spectrum (guard band), the rest is used for carrying data signals and for synchronization of the receiver to the transmitter. The FFT therefore occupies the signal spectrum in the frequency domain. The bandwidth is given by the expression

$$B = N_{c(\max)} \cdot \Delta f \tag{12.1}$$

Where B is the OFDM signal bandwidth and Δf is the subcarrier spacing in Hz. This bandwidth does not include the guard band area. The analog to digital

4G Wireless Communication Networks: Design, Planning and Applications, 241-252,

converters in OFDM systems sample the incoming signal at a rate which permits for adding guard bands (cyclic prefix).

OFDM is sampled using a so-called sampling factor. The sampling factor is the ratio of the sampling frequency to the OFDM bandwidth. Usually OFDM is sampled at a frequency a bit larger than the critical bandwidth to make provision of guard bands. The sampling factor is

$$n = \frac{F_s}{OFDM\ bandwidth} \tag{12.2}$$

Where Fs is the sampling frequency. Some of the common sampling factors are $8/7$, $28/25$, $78/75$, ...

The available spectrum is sampled using the equation

$$F_S = Floor\left(\frac{n.BW}{8000}\right).8000 \tag{12.3}$$

Where Floor implies rounding down the result of the equation and n is the sampling factor and depends on the available spectrum (bandwidth). For WiMAX the channel spacing is

$\Delta f = F_S / N_{FFT}$, where N_{FFT} is the size of the FFT used to implement WiMAX. For example in fixed WiMAX $N_{FFT} = 256$. The symbol time is given by the relationship

$$T_S = \frac{(1+G)}{\Delta f} \ and\ G = \frac{1}{2^m}; \quad m = \{2,3,4,5\} \tag{12.4}$$

12.2 OFDM SYMBOLS

To understand how OFDM operates, in some of the important terms are defined and their relationships.

- *Channel delay spread (τ):* the time difference between the arrival of the first multipath component and the last one.
- Useful Symbol Time (T_b): The information bearing signal samples occupy the useful symbol time given by the relationship $T_b = \frac{1}{\Delta f}$. This time does not include the guard band time.

- Guard Period Interval / Ratio: To provide for multipath effect, a guard time proportional to the anticipated multipath is added to the useful symbol time. The guard time is related to the useful symbol time by the relation: $T_g = G \times T_b$, where $G = \frac{1}{2^m}$; $m = \{2,3,4,5\}$. The cyclic prefix samples occupy this time period.
- *Overall OFDM Symbol time (T_s)*: This is the time over which an OFDM symbol is valid. The symbol time is given by two times, the guard band time and the useful symbol time over which the actual information bearing signal samples are present. Therefore the total time over which the OFDM FFT is performed is given by the sum of the guard band time and the useful symbol time

$$T_S = T_b + T_g = (1+G)T_b$$

Or (12.5)

$$T_S = \frac{(1+G)}{\Delta f}$$

- Coherence bandwidth (B_c): the bandwidth in which the channel is assumed to experience a flat fading propagation and it is approximately equal to $2\pi/\tau$

If $\tau \gg T_s$, number of symbols per second is high, which means high data rate transmission. In this case, the system is very fragile to ISI and it is desirable to minimize the ISI. Therefore, to address this problem, multicarrier modulation schemes reduce the T_s for the individual sub-bit streams to have $T_s \gg \tau$.

The raw capacity of the channel per symbol can therefore be estimated and is

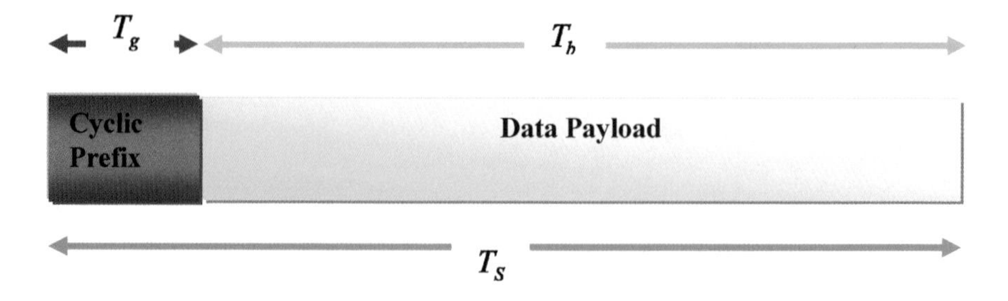

Figure 12.1 OFDM Symbol Time.

$$C_{raw} = \frac{k.N_{subcarriers}}{T_S} \quad (12.6)$$

Where k is the modulation order and different modulation schemes are used to cover different areas of a cell as explained in Figure 12.1 of chapter 3. For a 3.5MHz bandwidth, the symbol time neglecting the guard band is $64\mu s$ and if G=1/32, the total symbol time is $66\mu s$. With 192 subcarriers and using 64QAM modulation with 6 bits per symbol, the capacity is

$$C_{raw} = \frac{6x192}{66\mu s} = 17.45Mbps \quad (12.7)$$

This is a theoretical result which assumes that the transmission channel does not introduce errors. In practice transmission channel errors are introduced and forward error correction (FEC) is required to safeguard against errors. FEC is implemented by using redundant bits in each symbol. This is an increase in the bits used to represent the symbol. The coding rate for the FEC is the ratio of the number of information bearing bits to the total number of bits including the redundant bits. The mandatory FEC rates for different modulation schemes are given in Table 12.1.

The effect of FEC is to reduce the raw capacity by the coding rate. Thus we have

$$C = C_{raw} xCR = \frac{CR \times k \times N_{subcarriers}}{T_S} \quad (12.8)$$

Table 12.1 Mandatory FEC Coding Rates Per Modulation

Modulation	Uncoded block size(byetes)	Coded block size (byetes)	Overall coding rate	RS code	CC code rate
BPSK	12	24	1/2	(12,12,0)	1/2
QPSK	24	48	1/2	(32,24,4)	2/3
QPSK	36	48	3/4	(40,36,2)	5/6
16-QAM	48	96	1/2	(64,48,8)	2/3
16-QAM	72	96	3/4	(80,72,4)	5/6
64-QAM	96	144	2/3	(108,96,6)	3/4
64-QAM	108	144	3/4	(120,108,6)	5/6

Thus in the previous example the useful capacity when $CR = 3/4$ becomes $C = (17.45 \times 3)/4 = 13.1Mbps$. As a measure of how effective the 3.5MHz bandwidth has been used, the spectral efficiency is given

$$\eta_S = \frac{C}{B} = \frac{13.1Mbps}{3.5MHz} = 3.74b/\sec/Hz \tag{12.9}$$

The spectral efficiency improves with higher coding rates. In other words, the less the number of redundant bits used for forward error correction, the better the spectral efficiency.

The theoretic WiMAX capacity was estimated based on this understanding and with a 20MHz bandwidth, symbol time of $11.3\mu s$, 192 used subcarriers and (k) 6 bits per symbol at $CR = 3/4$ as

$$C = \frac{CR \times k \times N_{subcarriers}}{T_S} = \frac{0.75 \times 192 \times 6}{11.3\mu s} = 76.46Mbps \tag{12.10}$$

This theoretical WiMAX capacity estimate overlooks the overheads from the MAC layer and PHY. It also fails to highlight the fact that the range of a base station with such a capacity will be a lot shorter than often advertised for WiMAX. The range 70km or more is normally for BPSK modulation and the large capacity is for 64QAM modulation and not distinguishing the two often leads to wrong conclusions about WiMAX capacity versus range. This confusion is created not by the standards body but by various writers on WiMAX.

12.3 WIMAX FRAME

WiMAX symbols are normally sent in groups of symbols called a frame. Rather than sending the symbols as a stream, they are formatted into a time domain multiple access (TDM) frame. In practice the symbol time varies with the width of the channel. Therefore whole numbers of symbols do not necessarily fit into a frame snugly. There are often small gaps at the ends of frames that occur and not used. This overhead is often less than a symbol period in a frame and wastes capacity. The impact of this unused gap is more prominent in small frames.

As specified in the standard frames may be of lengths $T_F = [2.5,\ 4,\ 5,\ 8,\ 10,\ 12.5,\ 20ms]$. The number of symbols per frame N is

$$N = FLOOR\left(\frac{T_F}{T_S}\right) \tag{12.11}$$

For example in a 3.5MHz channel with a cyclic prefix (CP) of 1/8, the symbol time is $72\mu s$ and the number of symbols in 8ms frame is

$$N = FLOOR\left(\frac{8\times10^{-3}}{72\times10^{-6}}\right) = FLOOR\left(\frac{1000}{9}\right) = 111 symbols \tag{12.12}$$

The unused gap at the end of the symbol is $T_F - (111\times 72\mu s) = 8ms - 7.992ms = 8\mu s$, in this case a small reduction in the capacity (0.1% reduction).

12.4 MULTICARRIER MODULATION SCHEMES

In modern wideband high data rate transmission systems such as WiMAX, the effect of the ISI is very crucial, because in this technologies the delay spread $\tau >> T_s$. In WiMAX this can easily happen when the range of the cell is large enough to cause significant multipath delay spread. The result is more ISI and consequently more errors occur. Therefore modern communication technologies tend to use multicarrier modulation schemes in the physical layer to reduce the ISI. In the modulation schemes, a high data rate bit stream is divided into L lower rate sub-streams and transmitted on L parallel sub-channels or sub-carriers, which are orthogonal to each other in an ideal channel. The symbol time for each sub-stream is $T_s/L >> \tau$, thus the ISI will be decreased. Although the data rate for each sub-channel is equal to R_{Total}/L, the total required data rate will be retained as all sub-channels are parallel to each other. Similarly the bandwidth of each sub-channel is derived by B_{total}/L.

To overcome the ISI in each sub-channel with a flat fading propagation, the bandwidth of each must be far less than the coherence bandwidth ($B_{total}/L<<B_c$). It can be derived from this relationship that the more the number of sub-carriers, the less bandwidth and data rate for each sub-stream but the more the sub-streams symbol time and resulting in less ISI.

Although, there are some limitations associated with the multicarrier modulation such as high cost of low pass filters to achieve orthogonality and the requiring of a number of independent frequency channels for each sub-streams, OFDM technique copes with those problems and efficiently uses the available spectrum.

12.4.1 Guarding Against Inter-Symbol Interference

Figure 12.2 a) and b) show the orthogonality, inserted guard intervals and the arrangement of symbols in the frame. OFDM is well known for creating Inter Symbol Interference (ISI) free channels. The interference between a symbol and the subsequent symbols is referred to as ISI, which is an undesirable effect of the

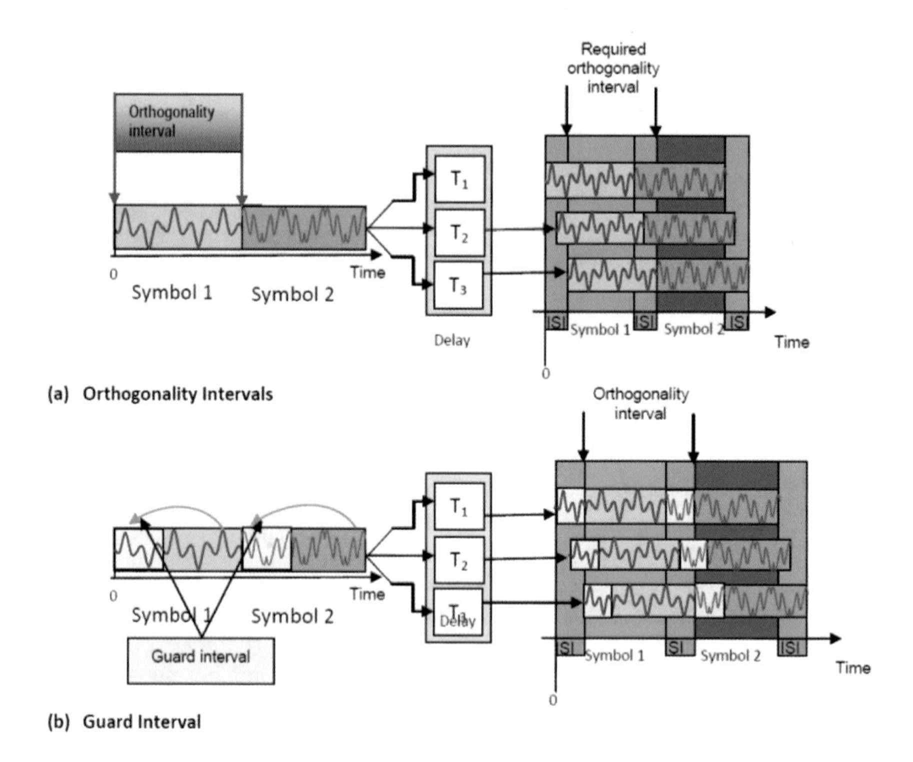

Figure 12.2: Explanation of OFDM Guard Intervals and Prefix.

environment, caused by either multipath propagation or Doppler effect. When a signal reaches the destination from different paths due to reflection, refraction and scattering, the propagation process is called multipath propagation.

In multicarrier modulation schemes a high data rate stream of bits is divided into several lower data rate streams and transmitting each bit stream via parallel individual subcarriers.

Radio signals often propagate from the transmitter to the receiver through many paths due to obstructions from objects in the terrain (the space separating the transmitter and receiver). Hence several copies of the same transmitted signal arrive at the receiver time shifted from each other. The time shifts are proportional to the paths taken by the signal and the effects of the channel (see Figure 12.3) on the signal. Assume that the direct path signal arrives at the receiver first. Thus delays encountered by signals arriving later can be estimated relative to the time of arrival of the direct path signal. The effect of adding the different copies of the received signal at the receiver causes signal power degradation (known as fading) due to intersymbol interference as shown in Figure

12.3. In this Figure three components of the same signal are received with the channel delaying them by $T_{,1}$ T_2, *and* T_3 seconds respectively.

Two separate regions can be identified in the received signals. Firstly, one due to time shifted version of the same symbol, called self-symbol-interference (SSI) and secondly interference from neighbouring symbols called inter-symbol interference (ISI). OFDM resists SSI because it uses orthogonal frequencies in the transmitter leading to zero correlation for two different frequencies. Hence SSI can be constructive in OFDM systems as the signal in such regions produce higher signal to noise ratio. This advantage is more prominent with the use of cyclic prefixes, which ensures SSI is removed.

ISI is provided for in OFDM systems through the use of the guard period (orthogonality interval, Figure 12.3). The guard band is proportional to the maximum delay expected for all the paths the signal could take to arrive at the receiver. The guard period is implemented by adding the last part of an OFDM symbol to its front as a guard against the delay expected on arrival. Thus the desired signal could arrive at the receiver without error and at a good time distance away from copies of the same signal that have been received.

What should the length of the guard period be? The length of the guard period is variable and can only be assumed and it is proportional to the largest delay path the signal could take from transmitter to receiver. A good estimate of this value requires simulation of the channel (path) the signal should take from transmitter to receiver. Thus information about the nature of the terrain separating them is required. If the reflection paths are long, the guard interval should also be long and vice versa. There is penalty in using large guard times as it causes a

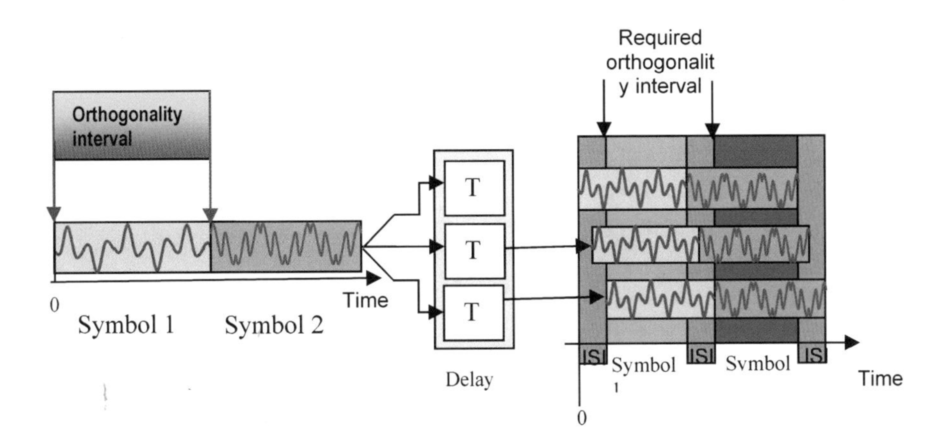

Figure 12.3 Intersymbol Interference at Receiver.

Table 12.2 Example of System Parameters

Parameter	Variable and Calculations	Value
Bandwidth	B	7 MHz
Sampling factor	N	$8/7$
Sampling frequency	$F_S = n.B$	8 MHz
Size of FFT	N_{FFT}	256
Subcarrier spacing	$\Delta f = F_S / N_{FFT}$	31.25 kHz
Useful symbol time	$T_b = 1/\Delta f$	$32 \mu s$
The guard intervals	G	$1/4$, or $1/32$
Cyclic prefix time	$T_g = GxT_b = 32x\frac{1}{4}$	$8 \mu s$ or $1.0 \mu s$
delay path (given by the speed of light)	$d_{delay} = cxT_g = 3x10^8 x8\mu s$ $d_{delay} = cxT_g = 3x10^8 x1\mu s$	2.4 km or 0.3 km
Overall symbol time (G=1/4)	$T_S = T_b + T_g = 32\mu s + 8\mu s$	$40 \mu s$
Number of Symbols	$N_{Symbols}$	25
Length of subframe	$T_{sym.frame} = N_{Symbols} xT_S = 25x40\mu$	$1000 \mu s$
Length of Frame	T_F	10ms
Number of symbols per frame	$N = FLOOR\left(\frac{10ms}{40\mu s}\right)$	250
User subcarriers	N_{user}	200
Pilot subcarriers	N_{Pilot}	8
Used data subcarriers	$N_{Used} = N_{User} - N_{Pilot}$	192
Modulation (QPSK), bits in subframe	$N_{bit} = N_{Used} x25x2$	9600 bits
Raw bit rate (no coding)	$R_{b(raw)} = \frac{9600}{1000\mu s}$	9.6 Mbps

reduction in throughput due to wasted bandwidth.

Long range base stations in suburban areas should use larger guard interval. Table 12.2 is a summary of typical WiMAX system parameters.

12.4.2 Overcoming Doppler Shift

A moving radio transmitter relative to a receiver or vice versa causes the transmission frequency to change. This change in frequency and its effects on the received signal power is termed Doppler effect. Reducing the distance between the transmitter and receiver causes the carrier frequency to increase and increasing the distance between them causes the frequency to decrease. In an OFDM system when the transmitter and receiver move closer the carrier frequency increases and when they move apart it reduces. The change in carrier frequency is proportional to the relative velocity between the transmitter and receiver. This change is given by the expression

$$\delta f = \frac{v}{c} \cdot f_C \qquad (12.13)$$

Where, f_C is the carrier frequency and c is the speed of light. Consider a fast moving car at 120 km/hr carrying a terminal transmitting at 6GHz. The Doppler shift is as high as 670 Hz. This change is substantial for a sub-carrier frequency of 31.25kHz (a 2.144% change in carrier position). Thus the demodulator at the receiver must be able to track this change and make corrections for it.

12.5 GENERAL CONSIDERATIONS FOR WIMAX NETWORK PLANNING

Before the establishment of a communication network, a careful planning and dimensioning of the network is necessary. Through this process, network designers gather the analytical data and using practical design methodologies to obtain an optimum network design to increase the network performance, while reducing the time and expenses required for network implementation [1]. Network dimensioning and planning is the most significant part of the network design, which should be considered from three different perspectives of dimensioning for service, coverage and capacity. To achieve a proper planning, required information and Key Performance Indicators (KPI) need to be well defined. Some of the most important information that has to be identified is as follows:

- Geographical area, which is due for service delivery, needs to be identified in terms of the size in km^2; also the profile that defines the area as either urban, suburban or rural environment. Furthermore the coverage area should be distinguished in terms of the network being fixed, nomadic, portable, and mobile or any combination of them.

- Subscriber's profile such as residential, small businesses and corporation customers needs to be identified.
- How subscribers are distributed through the network, which can be based on different parameters such as the number of the users per a service area, per a particular profile or per a period of time.
- Service profile is also very important, particularly for the service dimensioning. VOIP, broadband Internet access and IPTV are some of the services that WiMAX can offer.
- Available spectrum, in which WiMAX network is allowed to communicate within.
- Cartographic data to illustrate the service area through a digital map.
- KPIs as a measure of performance
- Customer requirements such as bandwidth, number of base stations and type of site.

Using the information outlined above, a WiMAX network planning and dimensioning for service, coverage and capacity can be done following the methods in Chapter 10, 11 and 12.

Chapter 13.

Introduction to LTE-Advanced

Johnson I Agbinya

La Trobe University, Melbourne, Australia
J.Agbinya@latrobe.edu.au

13.1 LONG TERM EVOLUTION

The mobile communication technology has undergone rapid technical innovations over the last two decades. Starting from modest analog networks broadcasting and using time division and frequency division multiple access, new technologies have been developed which take advantage of new telecommunication network technologies and signal processing. Traditionally, the architectures of GSM, UMTS and WiMAX networks were very complex and require many system components to deploy. This complexity is mostly in the radio network architecture. Designers of LTE made a concerted decision to simplify the LTE architecture to enable simpler integration with the Internet.

To address the ever increasing appetite for larger data rates and throughput it is essential for mobile communication networks to operate at larger system bandwidths. This appetite has gradually been satisfied from the modest 200 kHz for GSM, to 5MHz bandwidth for UMTS, higher system bandwidths up to 20MHz were specified for WiMAX. The increasing bandwidth not only allows operators to offer higher data rates in line with Shannon's capacity expression but also to best use their scarse resource – the available spectrum to them. At its peak WiMAX boasts scalable bandwidth up to 20MHz and also scalable data rates up to 100Mbps data rate. This is a lot higher than the 2Mbps first offered by WCDMA. The application of MIMO in WCDMA networks increased the data rates to a lot larger than the 2Mbps. LTE also supports scalable bandwidths up to 100MHz and promises 300Mbps data rates in the downlink and 75Mbps in the uplink. The ultimate goal for the future is to achieve up to 1Gbps for downlink and 500 Mbps for the uplink. In addition LTE promises reduced service latency

4G Wireless Communication Networks: Design, Planning and Applications, 253-264,

and optimized packet delivery. In other words, LTE services should be a lot faster than previously offered in earlier networks at lower bit error rates. Although as large as 100MHz may be available to an operator, a contiguous bandwidth of such magnitude is rarely available to the operators. Hence the promised large data rates are offered by aggregating smaller bandwidths or sub-carriers from different parts of the spectrum. This has the advantages of backward compatibility because existing spectrum for older networks can be reused.

13.2 CARRIER AGGREGATION

LTE supports multi-carrier aggregation [1]. Carrier aggregation is the technique whereby chunks of available carrier frequencies from various spectrum bands are aggregated for transmission. The aggregated carriers may be from the same band or from different bands. For the sake of supporting higher data rates, each user's equipment (UE) may be allocated aggregated component carriers. This allows operators to deliver higher throughput that could support applications which require larger bandwidth. Through a UE specific configuration, the component carriers that it could support are allocated. Carrier component allocation and de-allocated is undertaken dynamically. Normally each UE is configured with a primary carrier component known as Primary Cell (PCell). All other carrier components are allocated as secondary component carriers and also referred to as Secondary Cells (SCells). Each UE maintains SCell de-activation timer for de-allocating the secondary component carriers. Allocation and de-allocation may also be undertaken by the activation/deactivation medium access control. In the next paragraphs brief overviews of component carrier aggregation technique is discussed.

Intra-band Contiguous components carrier aggregation:
In Figure 13.1 we find two carriers from band A of the electromagnetic spectrum are aggregated. The two bands intersect each other. They do not have to intersect in practice.

Intra-band non-contiguous components carrier aggregation:
In the non-contiguous case, the carrier components although belonging to the same band are decoupled from each other and do not mix as in the previous case.

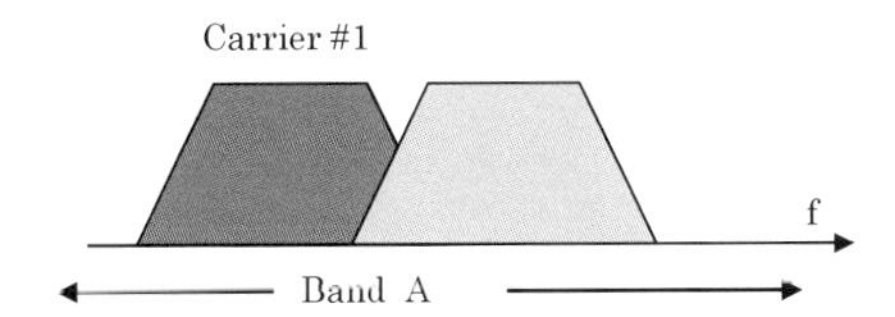

Figure 13.1 Intra-band Contiguous component Carrier Aggregation.

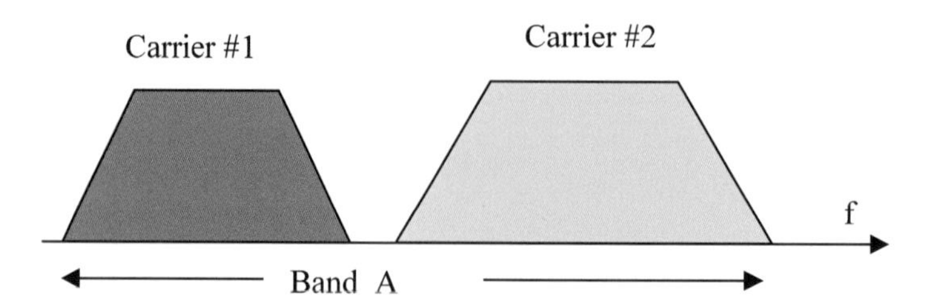

Figure 13.2 Intra-band Non-contiguous component Carrier Aggregation.

Figure 13.2 shows this case. As shown in Figure 13.2, the components may be of the same size or different. In Figure 13.2 the carrier #2 component has a larger bandwidth than the carrier #1 component.

Inter-band component carrier aggregation:
In this case the aggregated component carriers are derived from different frequency bands. Figure 13.3 shows the case when the two bands are derived from band A and band B. The components may also have the same or different bandwidths.

Configured component carriers:
In the configured component carriers, aggregation may consist of component carriers from different bands. Some secondary component carriers are activated to offer larger bandwidth while other component carriers are de-activated. This is illustrated in Figure 13.4.

The fact that part of the carrier spectrum used can be activated or deactivated allows user terminals to turn on and off parts of the chain of transceivers such as the RF front end and FFT. The process of activating and deactivating component carriers changes the state of SCells frequently. This makes monitoring of the radio links rather difficult when SCells are used. As a result radio link monitoring is supported only for PCells. This reduces system complexity in the UEs.

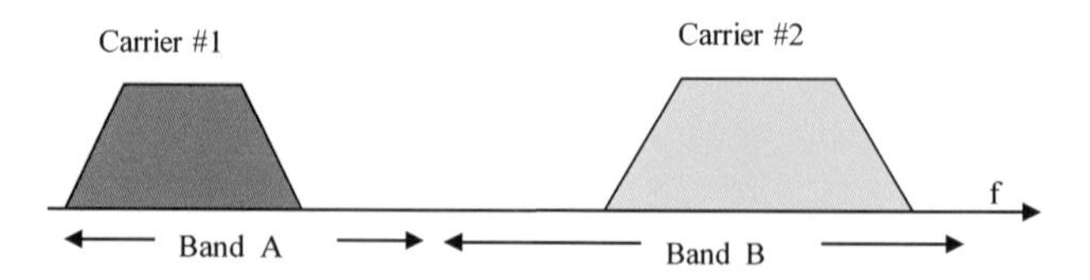

Figure 13.3 Inter-band component carrier aggregation.

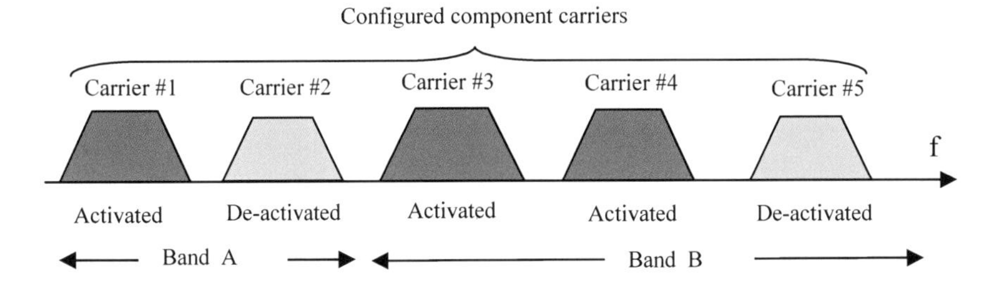

Figure 13.4 Configured Component Carriers.

In all the cases of component carrier aggregation discussed the RF filter design depends on the type of component carrier aggregation used. Table 13.1 gives a summary of some of the available frequency bands in the US from which component carrier aggregation can be derived as in Table 13.2 (US).

Table 13.1 LTE Component Carrier Bands

Band Number	Frequency Range (UL)	Frequency Range (DL)
2	1850 – 1910 MHz	1930 – 1990 MHz
3	1710 – 1785 MHz	1805 – 1880 MHz
4	1710 – 1785 MHz	2110 – 2155 MHz
5	824 – 849 MHz	869 – 894 MHz
7	2500 – 2570 MHz	2620 – 2690 MHz
12	699 – 716 MHz	729 – 746 MHz
13	777 – 787 MHz	746 – 756 MHz
17	704 – 716 MHz	734 – 746 MHz
20	2500 – 2570 MHz	2620 – 2690 MHz
38	2570 – 2620 MHz	2570 – 2620 MHz
41	2496 – 2690 MHz	2496 – 2690 MHz

Table 13.2 Aggregated Component Carrier Bands (US)

Carrier Aggregation Band	Operator	E-UTRAN Bands	Duplex Mode
Band 2 + band 17	AT&T	Band 2: 1850 – 1910 MHz (UL) 1930 – 1990 MHz (DL) Band 17: 704 – 716 MHz (UL) 734 – 746 MHz (DL)	FDD
Band 3 + band 7	TeliaSonera	Band 3: 1710 – 1785 MHz (UL) 1805 – 1880 MHz (DL) Band 7: 2500 – 2570 MHz (UL) 2620 – 2690 MHz (DL)	FDD
Band 4 + band 5	AT&T	Band 4: 1710 – 1785 MHz (UL) 2110 – 2155 MHz (DL) Band 5: 824 – 849 MHz (UL) 869 – 894 MHz (DL)	FDD
Band 4 + band 12	US Cellular	Band 4: 1710 – 1785 MHz (UL) 2110 – 2155 MHz (DL) Band 12: 699 – 716 MHz (UL) 729 – 746 MHz (DL)	FDD
Band 4 + band 13	Verizon	Band 4: 1710 – 1785 MHz (UL) 2110 – 2155 MHz (DL) Band 13: 777 – 787 MHz (UL) 746 – 756 MHz (DL)	FDD
Band 4 + band 17	AT&T	Band 4: 1710 – 1785 MHz (UL) 2110 – 2155 MHz (DL) Band 17: 704 – 716 MHz (UL) 734 – 746 MHz (DL)	FDD
Band 5 + band 17	AT&T	Band 5: 824 – 849 MHz (UL) 869 – 894 MHz (DL) Band 17: 704 – 716 MHz (UL) 734 – 746 MHz (DL)	FDD
Band 5 + band 12	US Cellular	Band 5: 824 – 849 MHz (UL) 869 – 894 MHz (DL) Band 12: 699 – 716 MHz (UL) 729 – 746 MHz (DL)	FDD
Band 20 + band 7	Orange	Band 20: 832 – 862 MHz (UL) 791 – 821 MHz (DL) Band 7: 2500 – 2570 MHz (UL) 2620 – 2690 MHz (DL)	FDD
Band 38	CMCC	Band 38: 2570 – 2620 MHz	TDD
Band 41	Clearwire	Band 41: 2496 – 2690 MHz	TDD

13.3 NETWORK ARCHITECTURE OF LTE-A

The LTE-A network architecture is highly simplified. The radio network (UTRAN) is simplified to consist of mainly base stations called eNodeB. The interface between two eNodeB is called X2 and the interface linking an eNodeB to the gateways are called S1.

The MME/S-GW is the mobility management entity in the control plane and the serving gateway in the user plane. One of its roles is to enable LTE to offer smooth handover and integration between to and from existing 3GPP and 3GPP2 networks. Handover between existing networks is therefore facilitated through the MME/S-GW. Furthermore, growing the network is a lot easier. Thus to expand network capacity or coverage a new MME/S-GW may be deployed to serve added eNodeBs. Therefore a dramatic evolution is all-but eliminated.

LTE offers IP core network (CN), the so-called evolved packet core (EPC). The core network is formed from the combination of the S1 and X2 interfaces with the MME/S-GW and the radio network (E-UTRAN provides access network (AN). It is also an evolved packet system (EPS). Thus LTE is completely a packet network. Therefore users are connected to the packet network through IP enabled devices, thereby offering Internet access.

13.4 TECHNICAL FEATURES OF LTE-A

In this section the roles of the system components in Figure 13.5 are discussed.

The MME serves as the control plane in the enhanced packet system. Its major functions include inter CN node signalling for mobility between 3GPP access networks, roaming, selection of S-GW, authentication, bearer management functions and Non Access Stratum (NAS) signalling.

The S-GW terminates the CN towards the E-UTRAN. Each user is associated with an EPS, at a given point in time. The Serving GW for a user is responsible for transferring user IP packets, lawful interception and mobility anchor for inter-eNodeB handover and for inter-3GPP mobility.

13.4.1 Spectrum and Throughput

LTE is an evolution of the 3GPP 3G/HSDPA with the following features. The transmission bandwidth ranges from 1.4MHz to 20MHz and scalable as (1.4, 3, 5, 10, 15, 20MHz) in both FDD and TDD modes. The Downlink data rates from 100 Mbps to 326.4 Mbps and the uplink rate up to 75Mbps. These rates depend on the modulation scheme used. LTE provides mobility support up to 300km/hr to 500km/hr depending on the frequency band. LTE promises more than 200 active users for every 5MHz spectrum.

LTE physical access is based on OFDMA in the downlink and single carrier

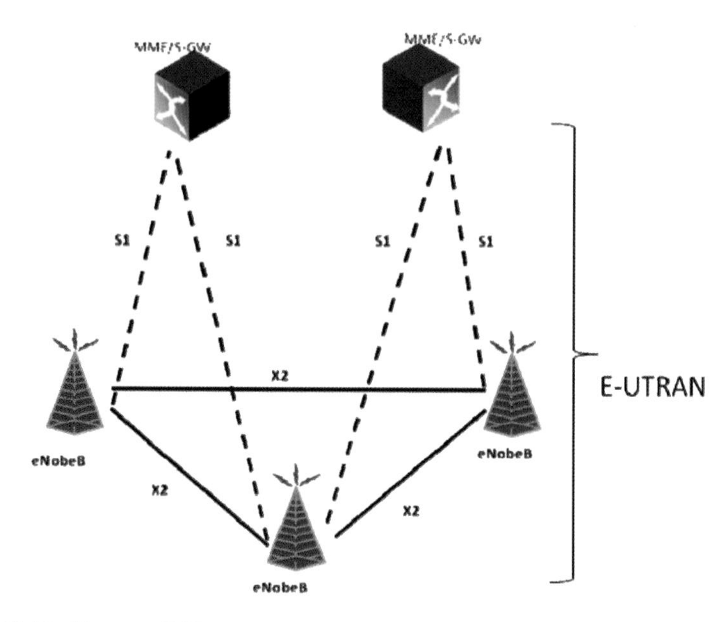

Figure13.5 Architecture of LTE-A.

FDMA (SC-FDMA) in the uplink and the number of OFDMA subcarriers range from 72, 180, 300, 600, 900, to 1200, with 15 kHz subcarrier spacing. Because of component carrier aggregation to use a single FFT to process the contiguous component carriers, the carrier spacing should be integer multiples of 300 kHz. LTE also supports various digital modulation schemes including phase shift keying (QPSK), 16-quadrature amplitude modulation, and 64-QAM signal constellations. Currently it also supports 2×2 and 4×4 multiple-input multiple-output (MIMO) modes and space frequency block coding (SFBC).

Table 13.3: LTE-A Spectrum Ranges

Region	Downlink (MHz)	Uplink (MHz)
Japan	830-840; 1749.9-1784.9; 1427.9 – 1452.9	875-885; 1844.9-1879.9; 1475.9-1500.9
Europe, Asia	880-915; 1710-1785; 1920-1980; 2500-2570	925-960; 1805-1880; 2110-2170; 2620-2690
America	698-716; 777-787; 788-798;824-849; 1710-1755;1710-1770; 1850-1910	728-746;746-756;758-768;869-894;1930-1990;2110-2155;2110-2170

Like in any other existing mobile network, the range of spectrum used differs from country to country. Table 13.3 shows the spectrum ranges used in various countries.

13.5 HETNETS

A heterogeneous network (HetNet) consists of a mix of low-power nodes and macrocells. Some of the nodes may have restricted access and some may also have no backhaul. Heterogeneous networks are discussed in [2]. New deployments strategies for LTE-A will be based on a heterogeneous network (HetNet) topology. HetNets blends macrocells with picocells, “metrocells”, relays and femtocells. In a HetNet a macro coverage area is overlaid with multiple smaller cells such as picocells and femtocells. Picocells and femtocells transmit at much lower power than macro cells. These are deployed in unplanned manner and bring the network closer to the user with improved signal-to-interference-plus-noise ratio. This improves system data rates and is more robust to noise and better QoS. The smaller cells are used to eliminate coverage holes in the network and help to improve capacity in hotspots (shops, mall, stadiums and indoor).

13.6 INTERCELL INTERFERENCE COORDINATION (ICIC)

In homogeneous networks, UEs at the edge of macro cells suffer from near-end-far-end effects and thus operate at reduced capacity. To alleviate the problem users may deploy femtocells. Operators on the other hand alleviate this problem by deploying picocells. These smaller cells that are used for alleviating the problem create intercell interference. With a frequency reuse of one, LTE eNodeBs transmit on all available time-frequency resource blocks simultaneously and hence the reuse ratio of one pose interference limitations. The UEs are however uplink limited due to their transmit power limitations. Therefore smaller cell sizes a lot smaller than the 2G and 3G cells need to be deployed. This is because at shorter range, higher data rates can be supported for both near-end and far-end terminals because the signal-to-noise ratios are higher. In a nutshell if LTE networks are to achieve their full data rate potential interference mitigation and cancellation techniques are required. The LTE standards therefore includes ICIC to help solve this problem.

A static reuse ratio of one (N=1) for a fully loaded network in an interference prone environment creates zones where the signal to interference noise ratio (SINR) become negative or the signal power levels are a few milliwatts. For a fully loaded network, the severity of the SINR degradation is a function of the average path loss exponent [3]. Areas with higher propagation path loss exponents experience less severe overall interference when compared with areas with lower path loss exponents. This is because potential interfering signals from neighbouring cells are less attenuated in the low path loss exponent areas. For LTE the ICIC techniques include power control, static fractional frequency reuse (FFR), adaptive fractional frequency reuse (AFFR) and AFFR with tiering [3]. Other ICIC mitigation and cancellation techniques include intelligent SINR based scheduling, MIMO, space division multiple access (SDMA) antenna techniques, opportunistic scheduling and access techniques and beamforming [3].

13.6.1 Power Control

The use of power control is well known in 2G and 3G technologies. Although not optimal, it however ensures that user terminals at the edge of cells are not over-shouted by terminals close to the eNodeBs. Power control is discussed in chapter 1.

13.6.2 Adaptive Frequency Reuse

A static frequency reuse is defined as a constant frequency without adaptation. The frequency reuse is an integer number (eg. N=1). The same frequency band is used across all cells. Figure 13.6 a) illustrates static frequency reuse. Static fractional frequency reuse (FFR) is defined as the partitioning of the usable spectrum "into a number of static sub-bands and assigning a given sub band to a cell in a coordinated manner that minimizes inter- cell interference" [3]. Two types of FFR are shown in Figure 13.6 b) and c). In Figures 13.6 b) and c) the FFR are 1/3 and 2/3 respectively. FFR improves SINR by reducing the interference at reduced bandwidth. Consequently the throughput is also reduced due to reduced bandwidth. The aggregate throughput is about 75 – 80% of the throughput obtainable when the reuse factor N=1.

13.6.3 Adaptive Fractional Frequency Reuse

In adaptive fractional frequency reuse (AFFR), measurements of channel quality and interference levels are used to adapt the reuse factor. Information on the channel quality is fed back to the UE. An implementation of AFFR may be based on power bandwidth profiles (PBP) as in Figure 13.6 c) [3]. As shown in Figure 13.6 c), power splits are manifested. The difference in power spectral density (PSD) within a PBP is defined as power split (PS). In Figure 13.6 c) the power split is 3 dB, meaning that the lower power level is half of the upper power level. Comparatively AFFR with PBP out performs AFFR without PBP [3] in terms of throughput. When compared with the throughput at N=1, the adaptive fractional frequency reuse for N=1/3 and for N=2/3 with PBP outperform the case for N=1 at the cell edge where interference is expected to have the most negative effect on received signal power.

Adaptation of the FFR is undertaken by each base station by measuring the throughput in going through the reuse modes starting from N=1 and then N=1/3

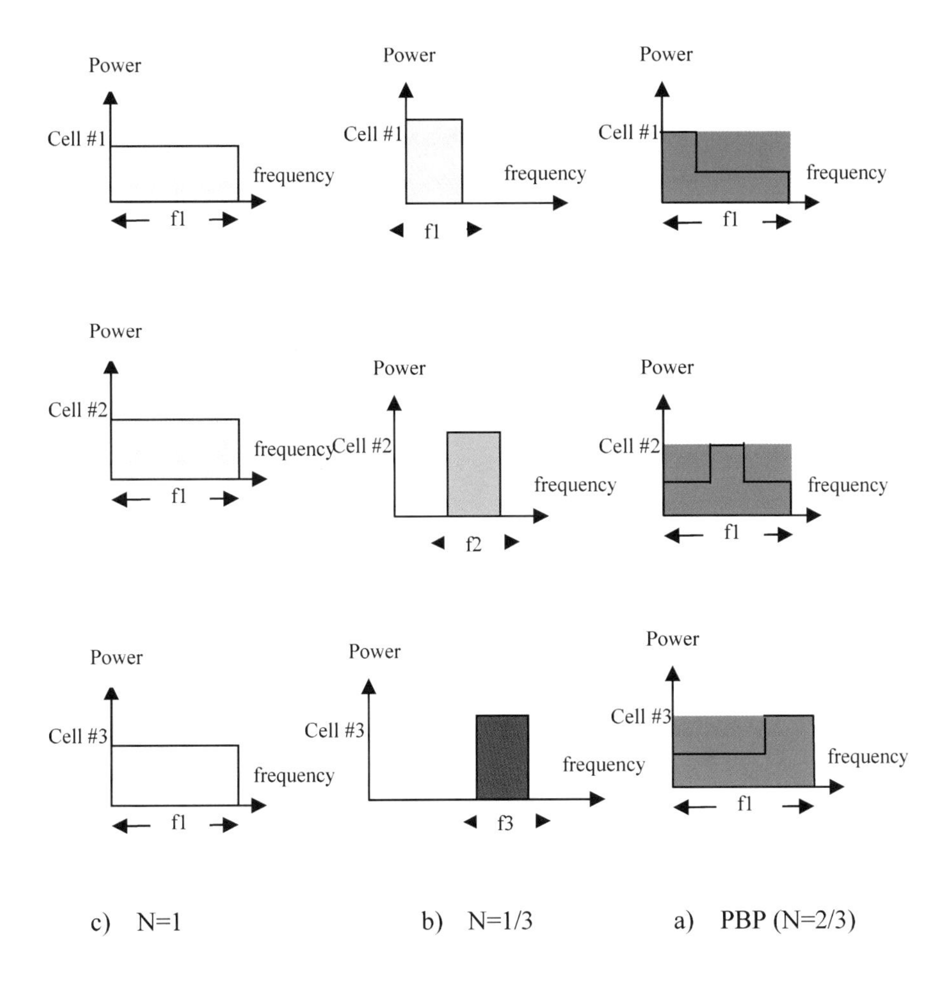

Figure 13.6 Static Frequency Reuse.

and finally N=2/3 including the use PBP. Each base station maintains a record of the achieved throughput and coverage across the FFR modes by using the channel quality indicator (CQI) and handover feedback from the UE it is serving. If in mode 1 (N=1) a base station detects a problem such as the number of users at the cell edge with data to transmit or receive exceeding a set threshold, then the FFR is adapted to mode 2. The base station reports this information to neighbouring base stations to adapt to mode 2. This adaptation is over the X2 interface linking the eNodeBs. For neighbouring base stations already using mode 2 or mode 3, mode changes are not necessary.

13.6.4 Network MIMO

Network MIMO is transmission on multiple paths to and from a mobile device from multiple base stations. In the uplink normally transmissions from a mobile go through different paths to several nearby base stations. In the down link, multiple base stations may also intentionally transmit through several paths to target mobiles. Associated with network MIMO is the problem of latency in the exchange of information between base stations. The minimum acceptable X2 latency for exchange of information between base stations set by the LTE standard is 20msec. Resource blocks are however allocated on every 1msec subframe. Hence real time processing of interference cancellation data between base stations is difficult.

13.6.5 Interference Regeneration and Cancellation Methods

In this method, interfering signal is generated and used for interference cancellation (IC) in the desired signal. Interference cancellation is an UL signal processing task and is implemented in the base station receiver. None real time IC through the X2 interface is supported in the present version of LTE standard. Figure 13.7 illustrates IC at eNodeB receiver.

In Figure 13.7, UE1 is the target receiver. UE2 to UEn are sources of interference. These interfering sources could be within the cell (inter-cell

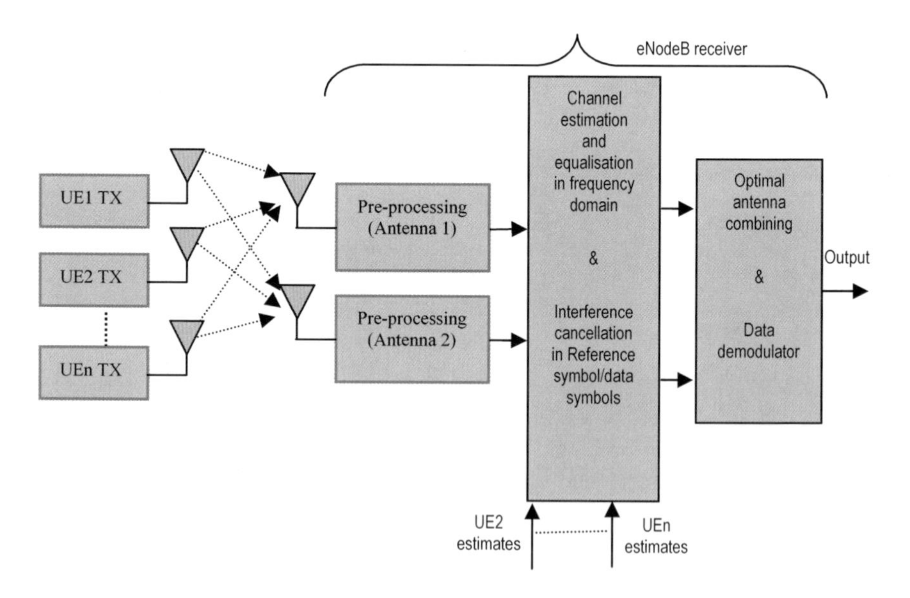

Figure 13.7 Interference Cancellation at eNodeB Receiver.

interferors) or outside the cell (intra-cell interferors). Figure 13.7 shows that the interfering signals from all sources (UE2 to UEn) are regenerated and from them the channel information related to the interfering sources can be estimated. To achieve this the base station acquires the information on the reference symbol sequences used by the interfering sources UE2 to UEn. From the RS for each source the base station estimates the channel gain, phase and time delays. This means that the channels (multi-paths) are now known fairly well. With the information the base station estimates the interfering signals for the individual interfering UE which is subsequently subtracted from the received signal. Assuming that these estimates are accurate, the regenerated signals are accurate approximations of the interfering signals. Hence subtracting them from the received signal will cancel the interference in the signal and the useful signal for user UE1 is therefore correctly obtained. It is possible to use either maximum likelihood sequence estimation or minimum mean squared estimation methods.

By cancelling the interference in the desired signal, the signal to noise ratio for UE1 is improved and hence the obtainable throughput by UE1 is also greatly enhanced. Up to 9.5dB and 13.5dB is obtainable with a 2-branch and 4-branch MMSE-IC respectively at pedestrian speeds of about 3km/hr [3].

References

[1] Dongwoon Bai et al., "LTE-Advanced Modern Design: Challenges and Perspectives", IEEE Communications Magazine, February 2012, pp. 178 – 186.

[2] Alexander Damnjanovic, Yongbin Wei, Tingfang Ji, Tao Luo, Madhavan Vajapeyam, Taesang yoo, Osok Song and Durga Malladi, "A Survey on 3GPP Heterogeneous Networks," IEEE Wireless Commun., vol. 18, no. 3, June 2011, pp. 10–21.

[3] Gary Boudreau, John Panicker, Ning Guo, Rui Chang, Neng Wang, and Sophie Vrzic, "Interference Coordination and Cancellation for 4G Networks", IEEE Commun. Magazine, April 2009, pp. 74 – 81.

Chapter 14.

Developing a Computationally-Efficient Dynamic System-Level LTE Simulator

P. Muñoz, I. de la Bandera, F. Ruiz, S. Luna-Ramírez, J. Rodríguez, R. Barco, M. Toril, P. Lázaro and M. Fernández

University of Málaga, Communications Engineering Dept., Málaga, Spain - Campus de Teatinos. 29071. Málaga
{pabloml, ibanderac, ferv, sluna, rbm, mtoril, plazaro, mariano}@ic.uma.es, jaime.rodriguez.membrive@ericsson.com

14.1 INTRODUCTION

Long-Term Evolution (LTE) is the evolution of the current UMTS mobile communication network. This 3GPP standard is the combination of the all-IP core network known as the evolved packet core (EPC) and the evolved UMTS terrestrial radio access network (E-UTRAN). The key benefits of LTE can be summarized in improved system performance, higher data rates and spectral efficiency, reduced latency and power consumption, enhanced flexibility of spectral usage and simplified network architecture.

In LTE, the multiple access scheme is Orthogonal Frequency-Division Multiplexing (OFDM) in the downlink and Single Carrier Frequency Division Multiple Access (SC-FDMA) in the uplink [1]. These techniques achieve a reduction in the interference, thus increasing network capacity. A Physical Resource Block (PRB), which has a bandwidth resolution of 180 kHz, is the minimum amount of frequency resources that can be scheduled for transmission. LTE performs channel-dependent scheduling in both time and frequency, with a resolution of one subframe (1ms) and one PRB, respectively. The scheduler is a fundamental part of the base station due to its influence over system performance.

The access network E-UTRAN [2] is composed basically of just one type of node: the base station called evolved NodeB (eNB). To reduce network elements,

4G Wireless Communication Networks: Design, Planning and Applications, 265-296,

all functions that were included in the Radio Network Controller (RNC) in UMTS are located in the eNB in LTE (e.g., radio protocols, mobility management, header compression and security algorithms). The eNBs are connected by standardized interfaces, called X2, which allow multivendor interoperability. In addition, information such as traffic load can be exchanged between eNBs over the X2 interface.

LTE defines a set of advanced functions for Radio Resource Management (RRM) in order to achieve an efficient use of the available resources. These functions include radio bearer control, radio admission control, radio mobility control, scheduling and dynamic allocation of resources. At layer 2, Link Adaptation and Dynamic Packet Scheduling are key features to ensure high spectral efficiency [3] based on user connection quality. On the one hand, Link Adaptation dynamically adjusts the data rate (modulation scheme and channel coding rate) to match the radio channel capacity for each user. The Channel Quality Indicator (CQI) transmitted by the UE, which is an indication of the data rate that can be supported by the channel, is an important input for link adaptation algorithms. On the other hand, Dynamic Scheduling distributes the PRBs among the UEs and the radio bearers of each UE every Transmission Time Interval (TTI) of 1 ms.

At RRM layer 3 [4], Admission Control and Mobility Management are crucial to ensure seamless service as the user moves. Admission Control decides whether the requests for new bearers are granted or rejected, taking into account the available resources in the cell, the QoS requirements for the bearer, the priority levels and the provided QoS to the active sessions in the cell. Mobility management includes procedures for idle and connected UEs. For both types of procedures, the UE periodically performs not only serving cell quality measurements, but also neighboring cell measurements. In idle mode, cell selection selects a suitable cell to camp based on radio measurements. In connected mode, handover decides whether the UE should move to another serving cell. The main difference between UMTS and LTE is that in LTE only hard handovers are defined.

This chapter describes the design of a dynamic system-level LTE simulator conceived for large simulated network time. A physical layer abstraction is performed to predict link-layer performance with a low computational cost. Thus, realistic OFDM channel realizations with multi-path fading propagation conditions have been generated to obtain an accurate value of Signal-to-Interference Ratio (SIR) for each subcarrier. Then, a method is used to aggregate SIR measurements of several OFDM subcarriers into a single scalar value. Subsequently, Block Error Rate (BLER) is estimated from those SIR values, which is used in the Link Adaptation and Dynamic Scheduling functions. Additionally, functions for admission control and mobility management are included in the simulator. For computational efficiency, the tool is focused on the downlink of E-UTRAN.

The design of efficient simulation tools for LTE networks has been addressed in the literature. In [5], a simple physical layer model is proposed for LTE in order to reduce the complexity of system level analysis. In [6], a link-level simulator for LTE downlink is presented as an appropriate interface to a system level simulator. In [7], an LTE downlink system-level simulator is proposed for free under an academic, noncommercial use license. The physical layer model is described in [8]. The main purpose of the MATLAB-based simulation tool presented in [7] is to assess network performance of new scheduling algorithms.

This chapter proposes the design of a system-level LTE simulator that can be used to evaluate optimization algorithms for the main network-level functionalities, namely handover, admission control and cell reselection. For this purpose, simulations are composed of epochs or optimization loops, where the modification of network parameters can be evaluated. Each epoch is composed of a configurable number of iterations, whose duration is determined by the simulator time resolution. In addition, the size of the simulation scenario must be larger than that in most of the existing LTE simulators, which usually consider only a few cells in the network layout.

This chapter is organized as follows. Section 14.2 presents the general simulator structure. Section 14.3 describes the Physical Layer, focusing on the calculation of the OFDM channel realizations needed for resource planning. Section 14.4 is devoted to the Link Layer, where link adaptation and resource planning are performed. Section 14.5 outlines the Network Layer, including admission control, congestion control and mobility management. In Section 14.6, simulation results are presented. Finally, in Section 14.7, the main conclusions are highlighted.

14.2 SIMULATOR GENERAL STRUCTURE

This section presents the general structure of the LTE simulator developed in MATLAB. Figure 14.1 shows the main functional blocks of the simulator.

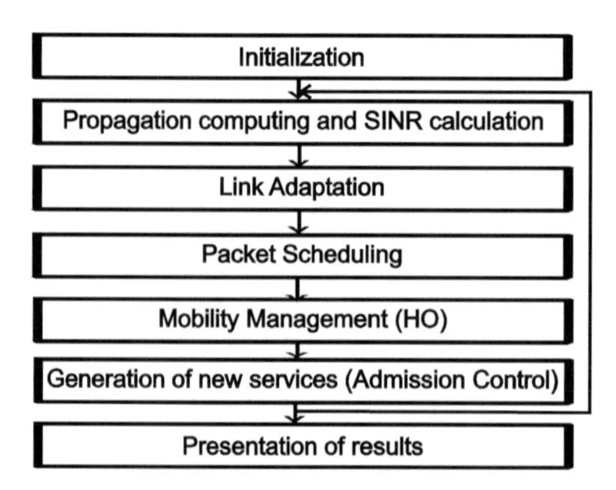

Figure 14.1 Block diagram of the simulator.

The first stage of a simulation is the initialization of the main simulation parameters, defining the behavior of the main functions in the simulator. The scenario to be simulated is generated here. A warm-up distribution of users is also created by this function, which allows to obtain meaningful network statistics from the first iterations of the simulation.

The next function calculates the propagation losses. This function calculates the power received by each user from the base stations of the scenario. The simulator includes a propagation loss model, a slow fading model and a fast fading model. During this phase, the interference suffered by each user is also calculated. From this information, the value of the SINR experienced by each user for different frequency subbands is obtained. Once the main parameters of interest have been obtained, the functions of radio resource management are executed. At link level, the simulator includes link adaptation and resource scheduling functions.

The link adaptation function selects the most appropriate modulation and coding scheme for each user to transmit the information, maximizing spectral efficiency. This decision is based on the propagation conditions experienced by the user. The CQI indicator is used to represent the environment conditions.

The radio resource scheduling function assigns available radio resources to users based on channel conditions experienced by each user for different frequency subbands. This function is also based on the CQI indicator.

At network level, the simulator includes several functions. The main ones are handovers and admission control. Lastly, the main results and statistics are shown.

14.2.1 Simulation scenarios

In the simulator, two different scenarios have been developed: a macrocell scenario and a Manhattan scenario. The first one consists of a configurable number of hexagonal cells. In a Manhattan, different types of elements such as buildings and roads are additionally defined and distributed along the cells of the scenario, shaping a rectangular grid.

<u>Macrocell scenario</u>

This simulation scenario models a macro-cellular environment. Figure 14.2 illustrates the layout for a scenario with 19 tri-sectorized sites evenly distributed.

To avoid border effects in the simulation, the simulator incorporates the wrap-around technique described in [9]. Wrap-around consists in creating replicas of the scenario surrounding the original one. Only the original scenario is considered when collecting results and statistics. Figure 14.3 shows the simulation scenario with the wrap-around technique.

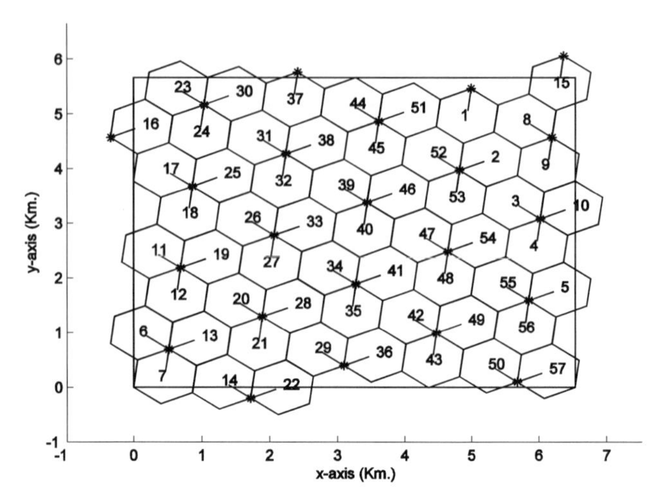

Figure 14.2 Simulation scenario.

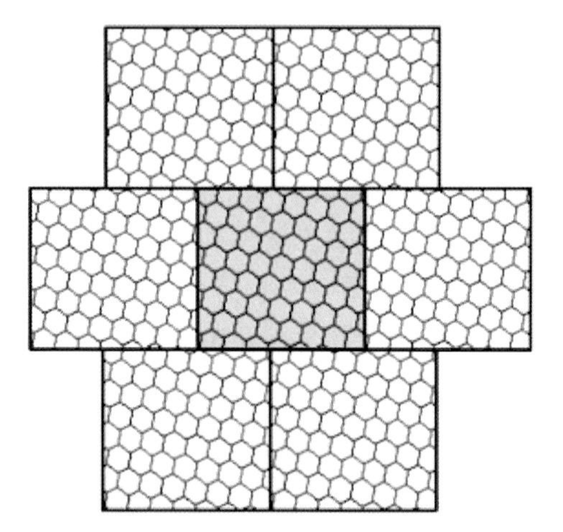

Figure 14.3 Simulation scenario with wrap-around.

Finally, it is necessary to define the set of interfering cells for each cell of the scenario. For each cell, an ordered set of interfering cells are constructed in terms of power received from each interferer by static system-level simulations.

Manhattan scenario

In this scenario, a set of buildings and roads are additionally defined. This scenario also includes the wrap-around technique. When defining the roads and the buildings, it is necessary to ensure that a user that leaves the original scenario, continues in a permitted area (e.g., a vehicle is not allowed to enter in the replica

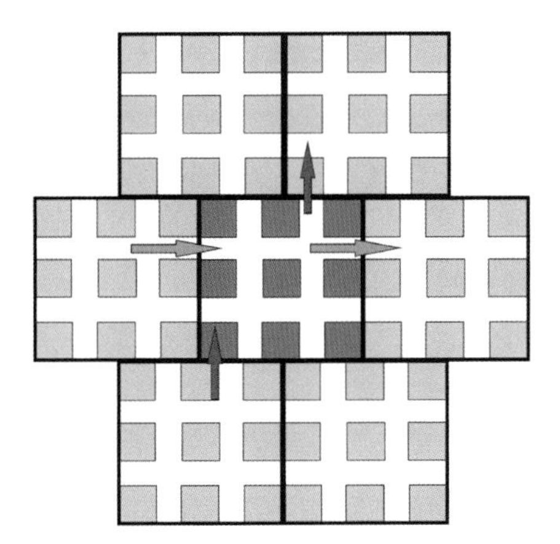

Figure 14.4 Wrong design of the Manhattan scenario with wrap-around.

on a sidewalk or a block) in the replica scenario. Figure 14.4 shows the case in which the wrap-around technique does not consider the difference between blocks and sidewalks in the definition of the scenario.

14.2.2 Spatial Traffic Distribution

Users can be spatially distributed in both an uniform or non-uniform way over the scenario. In the case of uniform spatial distribution, users are located in whatever point of the scenario with the same probability. However, to reproduce a realistic

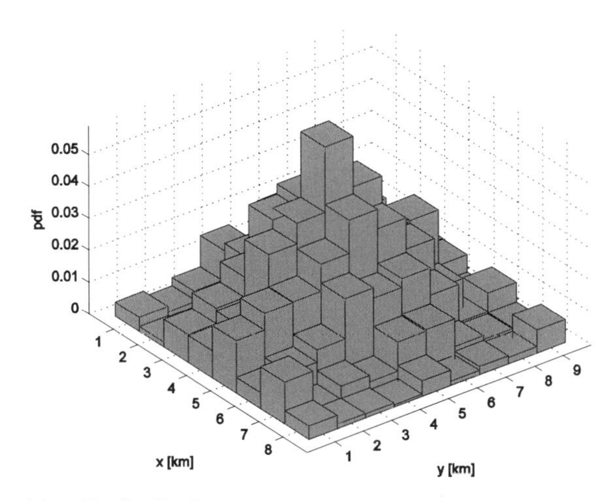

Figure 14.5 Spatial traffic distribution

situation, it is recommended to use a non-uniform distribution. The typical spatial distribution in urban areas can be described by a lognormal distribution at a cell level. A function of traffic estimation is created adding to the log-normal distribution a Gaussian random variable [10]. Figure 14.5 shows the probability of starting a call in any location of the scenario. It is observed that the spatial traffic distribution has a central peak, creating a congested area with higher traffic density. It is noted that the spatial traffic distribution will be slightly affected by the mobility model, explained in the next section.

14.2.3 Mobility model

The simulator includes two user mobility models. About the user move directions, a first model does not define any constraint about user directions. User can freely move over the scenario. The second model distinguishes between buildings and roads at the time of moving users along the scenario.

The first mobility model considers random constant paths for the users in the simulation scenario. Users move at constant speed, set to 3, 50 or 120 km/h. This model also includes the effect of the wrap-around technique, which means that when a user reaches the limit of the original scenario, appears in the correct position of this scenario, as shown in Figure 14.6.

The second mobility model developed in the simulator emulates the realistic behavior of users in an urban environment. This model implements the behavior of different types of users in the urban environment: vehicles, buses, pedestrians and indoor users. Manhattan Mobility Model is used for outdoor users and Random Waypoint Model for indoor users [11]. The dimensions of buildings, streets and sidewalks are taken from a real urban environment (the 'Ensanche' area in Barcelona, Spain). Such values of elements are defined in Table 14.1.

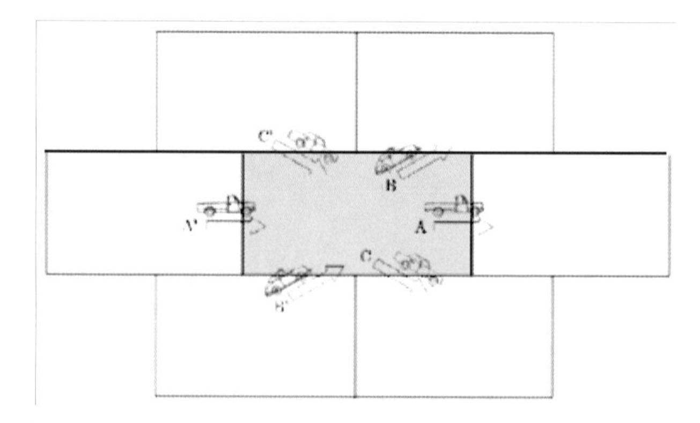

Figure 14.6 First mobility model: random constant paths with wrap-around.

Table 14.1 Dimension of elements in a real scenario.

Block size	*Street width*	*Road width*	*Sidewalks width*
120 x 120 m	20 m	10 m	10 m

Manhattan Mobility model is implemented to describe the outdoor users (vehicles, buses, pedestrians) movement. This model is widely used to describe the movement of mobiles in urban areas by means of a grid road topology, as shown in Figure 14.7. Users move along the streets and can turn at crosses with a given probability, that is recommended to be 25% probability to turn right and the same for a left turn [12]. In vehicles turn it is ensured that they are at the correct lane of the road for their new direction. The scenario is composed of vertical and horizontal streets perpendicular to each other. Every street has two lanes (each for one movement direction): North/South for vertical streets and East/West for horizontal streets. Vehicles move along a particular area of the street (road) and pedestrians move along another area (sidewalks). Buses behavior in the model is the same as those of vehicles, but providing the possibility to allocate multiple users sharing the same position, movement direction and speed. Studies on "moving hotspot" can be made with this type of users [13].

Additional features are included in the above-described Manhattan model to give more realism to the model. One of the features is the implementation of traffic lights at the street corners. Every street at an intersection has a traffic light [14]. Thus, the behavior of the network under conditions of user agglomeration in the corner where vehicles are stopped can be studied. To complete the Manhattan model for outdoor users, the wrap-around technique is implemented to avoid the border effects [9].

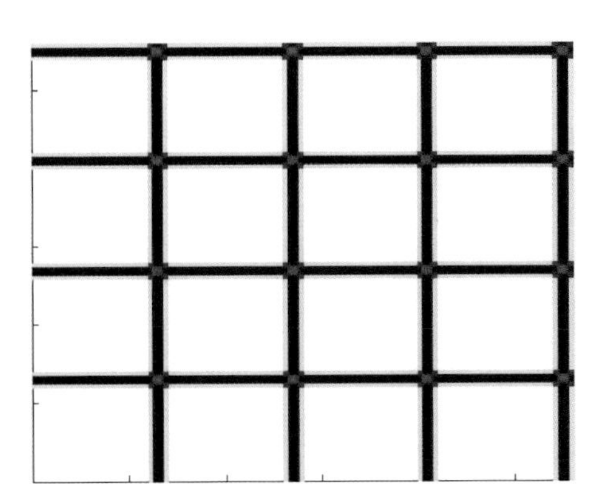

Figure 14.7 Movement of outdoor users in a Manhattan scenario.

Indoor users follow a Random Waypoint mobility model [15]. Random Waypoint is a simple but very useful model to simulate users in buildings. The goal of implementing these users is to create some traffic load in specific cells. These users are modeled as static load, i.e. they do not leave the building in which they were created.

14.2.4 Traffic model

Two types of service are considered: Voice over IP (VoIP) service and Best Effort (BE) service (similar to Full Buffer). The VoIP service is defined as a source generating packets of 40 bytes every 20 ms [16], reaching a bit rate of 16 kbps. As it will be seen later, the radio resource allocation in the simulator is performed for time intervals of 10ms. For this reason, the voice service has been implemented as users that transmit packets of 20 bytes every 10 ms. For this service, it is necessary to determine when a call is dropped, that is, when the service is interrupted. Such an event occurs when a user does not receive packets during a specific time interval. In particular, user packets are not scheduled when the connection quality is below a certain threshold or there are not enough resources, so the call may be dropped.

The BE service is similar to the Full Buffer service. The Full Buffer service is defined as a user that has infinite data to transmit. For this reason, a Full Buffer user will always transmit with the maximum available bit rate if radio resources are assigned to him. This BE service allows to assess network performance in terms of throughput. The service implemented in the simulator maintains these features with the only difference that the user is active only during a period of time, that is, the service starts at a certain time of simulation and ends some time later. During that activity time, the user has unlimited information to transmit. Once the service is finished, both the time that the user has been active and the experienced bit rate are known so that it is possible to calculate the size of the packet received by the user.

14.3 PHYSICAL LAYER

14.3.1 Channel model

The mobile radio channel can be described as a time-varying linear filter [17]. Therefore, it can be represented in the time domain by its impulse response, $h(\tau,t)$, where τ stands for delay of each path in h, and the amplitude of each path varies with time t.

Also, the channel can be characterized by the time-variant transfer function, $H(f,t)$, which is related with impulse response through the Fourier transform with respect to the delay variable, τ.

When the behavior of the channel is randomly time variant, the above-mentioned channel functions become stochastic processes. A realistic approach to the statistical characterization of such a channel may be accomplished in terms of correlation of channel functions since it enables channel output autocorrelation to be determined. Channel autocorrelation functions are related through Fourier transform as well.

For typical physical channels, time fading statistics can be assumed stationary over short periods of time and channel correlation function is invariant under a translation in time t, thus being categorised as wide-sense stationary (WSS). In addition, frequency-selective behaviour is stationary in frequency f being the autocorrelation function invariant under frequency translations. This condition is termed uncorrelated scattering (US), and most practical channels satisfy it fairly well.

Autocorrelation functions of wide-sense stationary uncorrelated scattering (WSSUS) channels exhibit the property that the time-variant transfer function autocorrelation is stationary both in time t and frequency f variables, i.e. its value does not depend on the absolute time or frequency considered but only on the time or frequency shift between time or frequency points of observation.

As a consequence, a WSSUS channel can be simulated generating the impulse response, $h(\tau,t)$, with stationary variation in time t for each path and no cross-correlation between different values of delay τ (i.e. generating independent stochastic processes for different paths). Stationarity is achieved by applying Doppler filters to the amplitude time t variation on each path. These filters perform spectrum shaping according to Doppler effect experimented by any radio signal propagating from a transmitter to a moving receiver (or vice versa). Afterwards, the frequency transfer function, $H(f,t)$, can be computed easily by applying the Fourier transform to the impulse response with respect to delay variable.

To simulate non-constant speed mobiles, fading realizations cannot be performed over time as an independent variable. Alternatively, space variables have to be used so that channel varies according to the current position of the mobile at each iteration of simulation.

Therefore, a fading channel spatial grid has been generated. This grid provides channel responses for every physical position in the simulated scenario, regardless of mobiles speed. In fact, mobiles can stay at a static position for a time interval, and then can start moving at any speed. This allows simulation of urban mobility pattern, where vehicle mobiles stop and afterwards go on because of traffic lights, or pedestrian users wandering inside a shopping centre.

Narrow band fading grid is generated to get a Lord Rayleigh universe [18]. In other words, following Clarke's model [17], a spatial bidimensional complex Gaussian variable is filtered by a bidimensional Doppler filter. The bandwidth of 2-D Doppler filter can be obtained as a function of spatial grid resolution and wavelength size.

Once narrowband channel behavior for each spatial position is obtained, extension to wideband is possible performing the same procedure for every path in power delay profiles described in the specification for Extended Typical Urban (ETU), Extended Pedestrian A (EPA) and Extended Vehicular A (EVA) channels

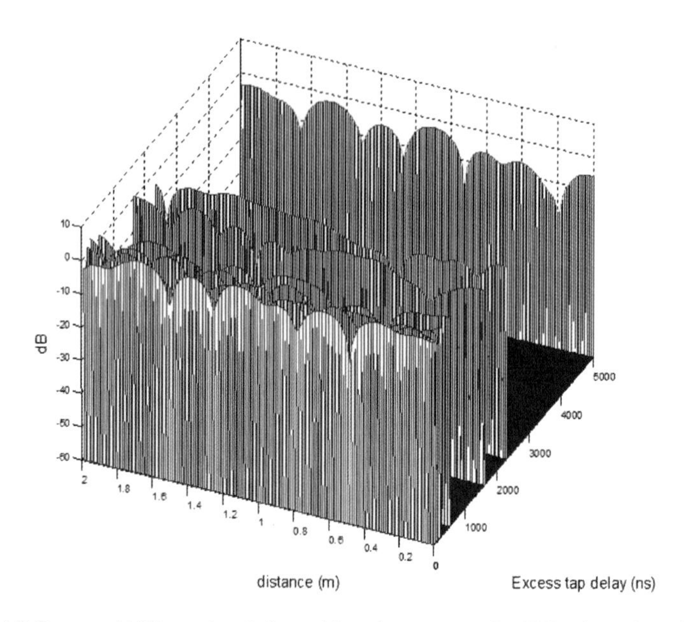

Figure 14.8 Generated bidimensional channel impulse response for ETU channel model in [19].

in [19]. Thus, different (uncorrelated) Rayleigh universes are generated for each delay in wideband channel scenario. This results in a distance-variant impulse response $h(\tau,d)$ (autocorrelation) of the channel instead of a time-variant impulse response $h(\tau,t)$ described in [17] as one of the four system functions for complete WSSUS (Wide Sense Stationary Uncorrelated Scattering) channel characterization. The only difference is the time to distance (t to d) variable change made. A realization of the function is shown in Figure 14.8.

Since the simulator requires channel realizations for different frequency bands (corresponding to OFDM subcarriers), the distance-variant impulse response has to be transformed into a distance-variant transfer function $H(f,d)$ at each position, by applying Fourier transform respect to delay variable τ. An example of this function can be seen in Figure 14.9.

The only remaining step is to extend the space variable d of the generated function $H(f,d)$ to a bidimensional (x,y) space variable, obtaining $H(f,x,y)$, a tridimensional function that provides frequency response for each spatial position given by coordinates, x and y.

14.3.2 Radio propagation channel

The simulator includes two alternatives for obtaining propagation calculations. As a first option, the calculations are performed at each iteration and whenever necessary (e.g., in the function that evaluates the channel conditions of each link or in the admission control function). Alternatively, propagation calculations are made from a set of pre-computed matrices. In this case, it is not necessary to

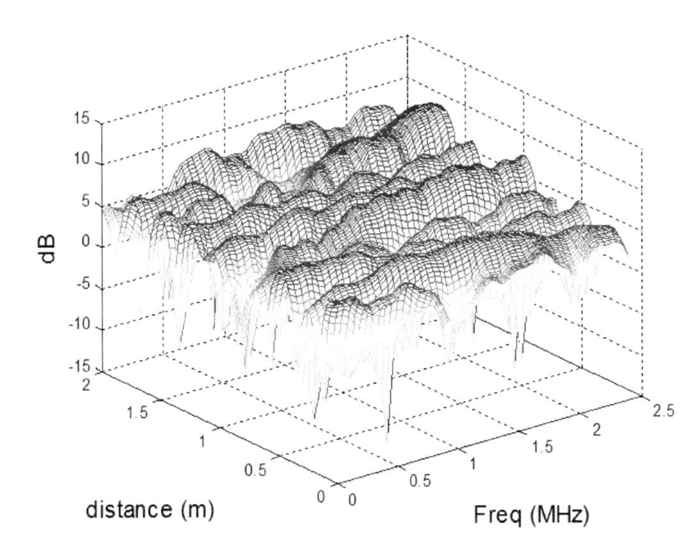

Figure 14.9 Generated distance-variant transfer function for ETU channel model in [19].

perform the calculations during the simulation.

For the definition of the pre-computed propagation matrix, the scenario is divided into a grid, whose resolution is given by the correlation distance of the slow fading (20 m). To know the values of the propagation loss a user is experiencing, it is only necessary to read the position of the matrix corresponding to the position occupied by the user in the scenario relative to every base station and then interpolate it with other values of the matrix depending on the relative position in the grid. The propagation matrices include the path loss calculations and the slow fading.

In both options, the radio propagation model is the COST 231 extension of Okumura-Hata model [20]. This model is applicable for frequencies in the range from 1500 to 2000 MHz. The effective height of the base station or eNB antenna has been set to 30 meters, while the effective height of the UE antenna has been set to 1.5 meters. With these assumptions and setting the operating frequency to 2 GHz, the expression for the propagation loss as a function of the distance is given by:

$$L = 134.79 + 35.22 \log d \tag{14.1}$$

where d represents the distance in km between the UE and the eNB which the user is connected to.

In addition to the propagation loss, the simulator includes a slow fading model based on the fact that the local average of the radio signal envelope can be modeled by a lognormal distribution, i.e., the local average, in dB, is a Gaussian random variable. The standard deviation of the distribution depends on the considered environment. A typical value for the macrocell urban area analyzed is 8 dB [21].

For the choice of the propagation matrices, the value of shadowing is included in these matrices. The other alternative requires some additional calculations. The dynamic nature of the simulator leads to the implementation of a correlation model between the successive samples which represent the slow fading. An ARMA(1,1) model [22] has been selected for the simulator in this work

$$z_t = \theta z_{t-1} + (1-\theta)a_t \tag{14.2}$$

where z_t represents the slow fading sample at the current simulation step, z_{t-1} is the slow fading sample at the previous simulation step, a_t is a Gaussian random variable uncorrelated with z_t and θ and $(1-\theta)$ are the coefficients of the ARMA(1,1) model.

The coefficients of this model are determined from the probability that a user terminal suffers fading caused by the same obstacle at the time interval Δ/v. That probability can be modeled as an exponential distribution:

$$\theta = P(\tau < \Delta / v) = \exp(-\Delta \cdot \lambda) \tag{14.3}$$

where Δ is the distance moved by the user terminal at a time interval, v is the UE velocity and λ is the interruption rate of the line of sight. The interruption rate of the line of sight, λ, is the inverse of the correlation distance. A typical value of the correlation distance for the macrocell urban area simulated is 20 m (or 50m) [23].

Finally, the Gaussian random variable, a_t, must be defined based on its mean and standard deviation. This variable provides a statistical distribution of zero mean and a standard deviation, σ_a, that relates to the standard deviation of the slow fading, σ_z, as follows:

$$\sigma_z^2 = \frac{\sinh(\Delta \cdot \lambda / 2)}{\cosh(\Delta \cdot \lambda / 2)} \cdot \sigma_a^2 \tag{14.4}$$

Once the propagation calculations have been carried out, it is possible to study the link quality experienced by each user in terms of Signal to Interference Ratio (SIR). The next section describes the process to calculate the value of SIR for each user.

14.4 LINK LAYER

14.4.1 SIR calculation

The SIR is a representative measurement of the link quality that the user is experiencing. To calculate the SIR in the simulator, it is first necessary to calculate the interference experienced by each user. It is assumed that intracell interference is negligible in LTE because the scheduler assigns different frequencies and time slots to each user. Thus, only co-channel intercell interference due to the interfering cells using the same subcarriers is considered. This requires knowing the signal arriving to each user from all interfering cells. To calculate the interference from each base station to the terminal, the channel response is not taken into account, but only the path loss and slow fading are considered here.

The SIR calculation for a given subcarrier k, γ_k, is computed using the expression proposed in [24],

$$\gamma_k = P(k) \times \overline{G} \times \left(\frac{N}{N + N_p} \right) \times \frac{R_D}{N_{SD} / N_{ST}} \tag{14.5}$$

where $P(k)$ represents the frequency-selective fading power profile value for the kth subcarrier, G includes the propagation loss, the slow fading, the thermal noise and the experienced interference, N is the FFT size used in the OFDM signal generation, N_p is the length of the cyclic prefix, R_D indicates the percentage of maximum total available transmission power allocated to the data subcarriers, N_{SD} is the number of data subcarriers per Transmission Time Interval (TTI) and N_{ST} is the number of total useful subcarriers per TTI.

If it is assumed that the multipath fading magnitudes and phases are constant over the observation interval, the frequency selective fading power profile value for the kth subcarrier can be calculated using the expression

$$P(k) = \left| \sum_{p=1}^{paths} M_p A_p \exp(j[\theta_p - 2\pi f_k T_p]) \right|^2 \tag{14.6}$$

where p is the multipath index, M_p and θ_p represent the amplitude and the phase values of the multipath fading respectively, A_p is the amplitude value corresponding to the long-term average power for the pth path, f_k is the relative frequency offset of the k_{th} subcarrier within the spectrum, and T_p is the relative time delay of the pth path. In addition, the fading profile is assumed to be normalized such that $E[P(k)]=1$.

The value of $\overline{G}$ is calculated from the expression:

$$\overline{G} = \frac{P_{\max} \dfrac{g_n(UE) \times g_{UE}}{PL_{UE,n} \times SH_{UE,n}}}{P_{noise} + \sum\limits_{k=1,k \neq n}^{N} P_{\max} \times \dfrac{g_k(UE) \times g_{UE}}{PL_{UE,k} \times SH_{UE,k}}} \quad (14.7)$$

where $g_n(UE)$ is the antenna gain of the serving base station in the direction of the user UE, g_{UE} is the antenna gain of the user terminal, P_{noise} is the thermal noise power, $PL_{UE,k}$ is the propagation loss between the user and the eNB k, $SH_{UE,k}$ is the loss due to slow fading between the user and the eNB k and N is the number of interfering eNBs considered (set to 43 in the simulator).

A PRB is the minimum amount of resources that can be scheduled for transmission in LTE. As a PRB comprises 12 subcarriers, it is necessary to translate those SIR values previously calculated for each subcarrier into a single scalar value. This can be made using the Exponential Effective SIR Mapping, which is based on computing the effective SIR by the equation

$$SIR_{eff} = -\beta \ln\left(\frac{1}{N_u} \sum_{k=1}^{N_u} \exp\left(-\frac{\gamma_k}{\beta}\right)\right) \quad (14.8)$$

where β is a parameter that depends on the Modulation and Coding Scheme (MCS) used in the PRB [25] assuming that all subcarriers of the PRB have the same modulation and N_u indicates the number of subcarriers used to evaluate the effective SIR. The values of β have been chosen so that the block error probability for all the subcarriers are similar to those obtained for the effective SIR in a AWGN channel [26]. The value of β for a particular MCS is shown in Table 14.2.

Table 14.2 Values of β depending on the modulation and coding scheme.

Modulation	*Coding*	*β factor*	*Modulation*	*Coding*	*β factor*
QPSK	1/3	1.49	16QAM	3/4	7.33
QPSK	2/5	1.53	16QAM	4/5	7.68
QPSK	1/2	1.57	64QAM	1/3	9.21
QPSK	3/5	1.61	64QAM	2/5	10.81
QPSK	2/3	1.69	64QAM	1/2	13.76
QPSK	3/4	1.69	64QAM	3/5	17.52
QPSK	4/5	1.65	64QAM	2/3	20.57
16QAM	1/3	3.36	64QAM	17/24	22.75
16QAM	1/2	4.56	64QAM	3/4	25.16
16QAM	2/3	6.42	64QAM	4/5	28.38

Once the effective SIR has been calculated, the BLER showing the connection quality can be derived. There exist curves that establish the relationship between the values of SIR and BLER defined for an AWGN channel for every modulation and coding rate combination. These curves can also be used to calculate the BLER because intercell interference is equivalent to AWGN as the value of β has been selected for this purpose.

14.4.2 Hybrid automatic repeat request scheme

The Hybrid Automatic Repeat reQuest (H-ARQ) is a function at link level that allows to perform retransmissions directly at physical or MAC layer in LTE. A low-complexity model capable of accurately predicting the H-ARQ gains on the physical layer is derived in [27]. When an H-ARQ retransmission occurs, an improvement of the BLER is expected. The result is that the BLER curves based on AWGN channel model are shifted providing a Signal to Noise Ratio (SNR) gain due to H-ARQ. Hence, the new SIR can be calculated as follows:

$$SIR(i) = SIR + SNR_{gain}(i) \tag{14.9}$$

where i represents the ith retransmission. The value of SNR_{gain}, which depends on the redundancy version index i and the given MCS can be derived from a specific table given in [27].

Once the value of BLER has been obtained and taking into account the MCS used in the transmission, it is possible to calculate the value of throughput, T_i, for each user as follows:

$$T_i = \left(1 - BLER(SIR_i)\right) \times \frac{D_i}{TTI} \tag{14.10}$$

where D_i is the data block payload in bits [28], which depends on the MCS selected for the user in that time interval, TTI is the transmission time interval and $BLER(SIR_i)$ is the value of BLER obtained from the effective SIR.

14.4.3 Link Adaptation

Before explaining the link adaptation function, the 3GPP standardized parameter known as CQI is described. Such an indicator represents the connection quality in a subband of the spectrum. The resolution of the CQI is 4 bits, although a differential CQI value can be transmitted to reduce the CQI signaling overhead. Thus, there is only a subset of possible MCS corresponding to a CQI value [29]. QPSK, 16QAM and 64QAM modulations may be used in the transmission scheme. In the simulator, the CQI is reported by the user to the base station each iteration (100 ms).

Based on CQI values, the link adaptation module selects the most appropriate modulation and coding scheme to transmit the information on the physical downlink shared channel (PDSCH) depending on the propagation conditions of the environment. To quantify the link quality for each user and for each subband of the spectrum, the CQI index is used to provide this information. If the experienced BLER value is required to be smaller than a specific value given by the service, it is possible to establish a SIR-to-CQI mapping that allows to select the most appropriate MCS from a given value of SIR [8]. The standard 3GPP defines a 5-bit modulation and coding scheme field of the downlink control information to identify a particular MCS. This leads to a greater variety of possible modulation and coding schemes. For simplicity, the developed LTE simulator includes only the same set of MCS given by the CQI index. From the effective SIR value, the index CQI is calculated and the MCS can be determined for the next time interval.

14.4.4 Resource Scheduling

The resource scheduling can be decomposed into a time-domain and frequency-domain scheduling. On the one hand, it is necessary to determine which user transmits at the following time interval. On the other hand, the frequency-domain scheduler selects those subcarriers within the system bandwidth whose channel response is more suitable for the user transmission. For this purpose, the channel response for each user and for each subcarrier of the system bandwidth has to be estimated. Such a piece of information is given by the channel realizations generated in the initialization phase of the simulation, assuming a perfect estimation of the channel response. To select the most appropriate frequency subband for the user, the CQI index is used.

The developed simulator includes different strategies for radio resource scheduling. In all of them, the CQI parameter gives the information of the channel quality experienced by each user. Likewise, scheduling is done for each cell at each iteration following the configured strategy [30]. The following paragraphs describe the scheduling algorithms implemented in the simulator.

Best Channel Scheduler (BC)

In this scheduler, both time-domain and frequency-domain scheduling are done for a more efficient use of resources. At each iteration, all users are sorted based on the quality experienced for each PRB, which is obtained from CQI values. Once the users are sorted, the allocation will proceed until there are not available radio resources or no more users to transmit.

The resource allocation is made following the expression:

$$\hat{\imath}[n] = \arg\max_{i} \{r_{ik}[n]\} \tag{14.11}$$

where $\hat{\imath}$ is the selected user i, and r_{ik} is the estimated achievable throughput for PRB k and user i obtained from the CQI.

This scheduling algorithm maximizes the overall system efficiency because the resource allocation is done looking for the combinations PRB-user with better channel conditions. The disadvantage of this algorithm is that harms users with bad channel conditions . Thus, if a user is far from the serving eNB or has a deep fading for prolonged periods of time, it cannot be scheduled and can suffer significant delays.

Round Robin to Best Channel Scheduler (RR-BC)

This scheduler uses different strategies for time-domain and frequency-domain scheduling. For time-domain scheduling, the Round Robin method is applied. Thus, users are selected cyclically without taking into account the channel conditions experienced by each of them. Then, each PRB is assigned to the user with a higher potential transmission rate for that PRB (transmission rate is estimated based on the user's CQI value for each PRB).

At each iteration and for each base station, the expressions to be evaluated are:

$$\hat{\imath}[n+1] = (\hat{\imath}[n]+1) \bmod N_u \tag{14.12}$$

$$\hat{k}[n] = \arg\max_{k}\{r_{ik}[n]\} \tag{14.13}$$

where $\hat{\imath}$ is the selected user, N_u is the number of users and $\hat{k}$ represents the PRB selected. In this case, the goal is to maximize system efficiency, but trying not to harm users with unfavorable channel conditions.

Large Delay First to Best Channel Scheduler (LDF-BC)

This scheduler is similar to the previous one only differing in time-domain scheduling. In this case, instead of cyclically selecting the users, they are sorted by the time they have spent without transmitting. Thus, if for some reason, such as a fading prolonged in time, the user has not been allocated in previous iterations, he will get a higher priority in the current iteration.

In the same way as in the previous case, at each iteration and for each base station the allocation is carried out based on the following terms:

$$\hat{i}[n] = \arg\max_{i}\{W_i[n]\} \tag{14.14}$$

$$\hat{k}[n] = \arg\max_{k} \{r_{ik}[n]\} \tag{14.15}$$

where $W_i[n]$ is the number of iterations without transmitting for user i. At the end of each iteration, the value of $W_i[n]$ is updated for all the users based on whether they have been allocated or not.

Proportional Fair Scheduler (PF)

The Proportional Fair scheduler is an algorithm similar to Best Channel, but it tries not to harm users with worse channel conditions. The objective of this algorithm is to find a balance between getting the maximum possible efficiency of the channel and keeping fairness between users. To this end, scheduling is not only based on the potential transmission rate but also takes into account the average transmission rate of the user in previous iterations. The algorithm follows the expression:

$$\hat{i}[n] = \arg\max_{i} \left\{ \frac{r_{ik}[n]}{\bar{r}_i} \right\} \tag{14.16}$$

Scheduler for different types of service

As mentioned before, the simulator includes two different types of service, i.e., voice service and best effort service. To schedule all users, a division between the two types of service is made. The scheduler strategy can be different for each service. Also, the radio resources are divided into two groups. One of them is reserved for voice users who need to meet a constant bit rate and the other group can be used for best effort users that can maximize their throughput based on channel conditions. The division of the PRB in these two groups is configurable.

14.5 NETWORK LAYER

Network level functionality comprises those techniques and algorithms giving a global entity to the software tool. Cellular networks are so successful as they care about user mobility giving a global support (coverage, access…). While physical and link level define the propagation and transmission characteristics along the UE-eNB link, network level manages all base stations, terminals and their resources as a whole.

Main network level functionalities rely on Radio Resource Management (RRM) processes. This section describes the Admission Control (AC), Congestion Control (CC) and HandOver (HO) techniques implemented in the simulation tool. It should be pointed out, although scheduling is also usually labeled as an RRM technique, it has been already described in previous sections since it is located at link level in the simulation tool.

14.5.1 Admission control

Once an UE decides to start a connection, a first decision is which cell will serve that connection. Such a decision is taken through two main steps:

- *Minimum Reference Signal Received Power* (RSRP). UE collects and sends to the network reference signal received levels from the camping cell and its neighbors. Cells are ordered from higher to lower levels and candidate cells are those fulfiling:

$$RSRP(i) \geq MinThreshddLEV(i) \quad (14.17)$$

  where *RSRP(i)* is a wideband measurement meaning the received level for the reference signals in cell *i*, and *MinThresholdLEV* is the minimum required signal level to be accepted. Minimum level is defined on a cell basis. Finally, the best '*i*' cell in the list is initially selected.

- Enough free resources. The availability of free PRBs in best cell is then checked. Note that the mobile network does not know how many PRBs will require the user data connection once it is admitted. Signal-level measurements are taken from the reference signals, but radio channel conditions could be quite different for the finally assigned data radio channel (e.g., fast fading, interference). That is the reason why a 'worst-case' criterion has been taken to accept UEs. Thus, the UE is finally accepted if:

$$freePRB(i) \geq MaxPRB(serv) \quad (14.18)$$

  where *freePRB(i)* is the number of PRBs available in cell *i*, and *MaxPRB(serv)* is the worst-case PRB requirement (i.e., the highest number of PRBs needed to maintain a connection) that a specific type of service, '*serv*', would demand along the entire connection.

If there are not enough free PRBs, the next candidate cell in the list is checked. A user connection is blocked when no cell fulfils (1.17) and (1.18).

14.5.2 Congestion Control

Congestion control avoids congestion situations in the mobile network. This technique usually defines a pool of resources which will be assigned differently than by admission control.

Operators give priority to ongoing connections over fresh calls [31]. If both fresh and ongoing users are in conflict for the same radio resources (e.g., a handover and a fresh connection occur simultaneously), existing users should be first scheduled. With that aim, fresh users will not be accepted in a cell if:

$$LR(i) \geq LRthreshold(i) \tag{14.19}$$

where *LR(i)* is the Load Ratio in cell *i*, and *LRthreshold* is the congestion threshold. There is a trade-off when selecting the *LRthreshold* value. A too low value might cause call dropping from rejected incoming handovers, but a very high level could lead to unnecessary call blocking while protected resources are idle.

14.5.3 Handover

The HO algorithm is the main functionality to manage the connected user mobility. HO algorithms are vendor specific. The following paragraphs describe classical handover algorithms proposed for LTE and implemented in the simulator.

Quality Handover (QualHO)

A QualHO is triggered when:

$$RSRQ(i) \leq RSRQ_{threshold}(i) \text{ for } \mathrm{TTT}^{\mathrm{Qual}} \text{ seconds} \tag{14.20}$$

and

$$RSRP(j) - RSRP(i) \geq \mathrm{Margin}_{Qual}(\mathrm{i}, \mathrm{j}) \tag{14.21}$$

where RSRQ is the Reference Signal Received Quality, usually measured by the SINR for the references signals, RSRP is the Reference Signal Received Power, TTT^{Qual} is Time-To-Trigger value, and $Margin_{Qual}$ is the level hysteresis between server and adjacent cells (*i* and *j*, respectively).

This QualHO aims to re-allocate connections which are experiencing a bad quality connection to other cells. $Margin_{Qual}$ is defined on an adjacency basis.

For monitoring purposes, a QualHO is classified as an Interference HO (IntHO) if

$$RSRP(i) \geq RxLEV_{threshold}^{Interf}(i) \quad \text{for } \mathrm{TTT}^{\mathrm{Qual}} \text{ seconds} \tag{14.22}$$

i.e., the UE have a high signal level but low SINR figures.

Minimum Level Handover (LevHO)
A LevHO is triggered when:

$$RSRP(i) \leq MinRxLEV_{LevHO}(i) \tag{14.23}$$

and

$$RSRP(j) - RSRP(i) \geq \text{Margin}_{Lev}(\text{i, j}) \tag{14.24}$$

where $MinRxLEV_{LevHO}$ is a minimum signal level threshold. LevHO aims to re-allocate connections experiencing a very low signal level (e.g., when the UE is getting out of coverage area).

LevHO is considered an 'urgent' HO and must be triggered as soon as possible. Thus, no Time-to-Trigger parameter has been considered.

Power Budget Handover (PBGT_HO)
A PBGT_HO is triggered when:

$$RSRP(j) - RSRP(i) \geq \text{Margin}_{PBGT}(\text{i, j}) \tag{14.25}$$

In this case, there is no first condition to be fulfiled. Equation (14.25) is only evaluated every N^{PBGT} seconds. PBGT_HO is not considered an urgent HO, but an optimization algorithm. At the end of a PBGT_HO process, the UE should be connected to the best cell in terms of signal level (provided that $Margin_{PBGT}$ is positive).

14.6 EVALUATION OF SYSTEM PERFORMANCE

In this section, several reference scenarios are simulated to evaluate system performance. These are termed 'reference' because network parameters (e.g., HO margin or load ratio threshold in the Admission Control) are set to a moderate default value. For voice service, system performance is evaluated by testing different levels of traffic demand and different strategies of scheduling. For best effort service, system performance is quantified in terms of cell throughput.

14.6.1 Key Performance Indicators

For voice service, a figure of merit widely used by network operators is the Call Dropping Ratio (CDR), defined as:

$$CDR = \frac{N_{dropped}}{N_{finished}} = \frac{N_{dropped}}{N_{dropped} + N_{succ}} \tag{14.26}$$

where $N_{dropped}$ is the number of dropped calls, N_{succ} is the number of successfully finished calls and $N_{finished}$ is the total number of finished calls. The simulation tool assumes that a call is dropped when a percentage of data packets are dropped during a specific time interval. Packet dropping may occur not only because there is no enough connection quality to be scheduled, but also because there are no available resources to be scheduled.

To quantify how efficiently resources are used, another performance indicator is the Call Blocking Ratio (CBR), which can be determined by the following expression:

$$CBR = \frac{N_{blocked}}{N_{offered}} = \frac{N_{blocked}}{N_{blocked} + N_{accepted}} \tag{14.27}$$

where $N_{blocked}$ and $N_{accepted}$ are the number of blocked and accepted calls by the admission control respectively, and $N_{offered}$ is the total number of offered calls.

Regarding BE traffic, cell throughput is a useful measure of spectral efficiency. The process to calculate this performance indicator was described in previous sections. Another indicator representing the link quality experienced by users is the CQI, which was also defined in previous sections.

To check the impact of traffic demand on network performance, several performance indicators are evaluated. To estimate the overall load level during the simulation, the average percentage of occupied PRBs in the system is monitored. This can be calculated as:

$$\rho = \frac{\sum_{k=1}^{N_{users}} N_k}{N_{total}} \tag{14.28}$$

where N_k is the current number of PRBs occupied by user k, N_{users} is the number of users and N_{total} is the total number of PRBs in the cell.

14.6.2 Simulation parameters

The simulated scenario includes a macro-cellular environment whose layout consists of 19 tri-sectorized sites evenly distributed in the scenario. The main simulation parameters for the simulations are summarized in Table 14.3.

Table 14.3 Simulation parameters.

Parameter	*Configuration*
Cellular layout	Hexagonal grid, 57 cells (3x19 sites), cell radius 0.5 km
Transmission direction	Downlink
Carrier frequency	2.0 GHz
System bandwidth	5 MHz
Frequency reuse	1
Propagation model	Okumura-Hata with wrap-around Log-normal slow fading, σ=8dB, correlation dist=20m Multipath fading, EPA model
Mobility model	Random direction, constant speed 3 km/h
Service model	VoIP: Poisson traffic arrival, mean call duration 120s, 16 kbps Best effort: full buffer, Transport Block size exact fit to PRB allocation
Base station model	Tri-sectorized antenna, SISO, $EIRP_{max}$=43dBm
Scheduler	Round Robin - Best Channel Large Delay First - Best Channel Resolution: 1 PRB
Power control	Equal transmit power per PRB
Link Adaptation	Fast, CQI based, Perfect estimation
RRM features	Directed Retry HO: QualHO, LevHO, IntHO, PBGT_HO
HO parameter settings	Time-To-Trigger = 100 ms HO margin = 3 dB
Traffic distribution	Log-normal distribution Unevenly distributed in space
Time resolution	Iteration time = 100 TTI (100 ms) Epoch time configurable

14.6.3 Performance results

The first simulation is carried out to evaluate performance when only the voice service is offered by the network. In this case, measurements of CDR, CBR and CQI are collected for different levels of traffic load and different types of schedulers. Figure 14.10 represents the dependency of the average CQI value on the distance for several load levels. As expected, the CQI value decreases as the distance from the base station is higher due to the path loss. Also it is noted that the range of variation of the CQI is much lower when the load level is higher. This is because the interference term in the expression of SIR becomes more important as the load level is increased, leading to a more restrictive set of CQI values in the cell coverage area.

Regarding call dropping, Figure 14.11 shows the CDR for several levels of traffic load and two different scheduling schemes. The solid blue line represents the measured CDR when the scheduling strategy is RR-BC, while the dotted red line depicts the measured CDR when the scheduling strategy is LDF-BC. The difference between these two approaches is that LDF-BC sorts the users by the time they have spent without transmitting instead of cyclically sorting them according to the Round-Robin strategy. The LDF-BC scheme leads to a lower CDR for the same load level due to those users that get more priority in scheduling as they experience higher delay, avoiding call dropping.

The CBR for the two previous scheduling strategies as a function of the traffic load is illustrated in Figure 14.12. As it is expected, the CBR increases as the traffic load level is higher. However, it is noted that the CBR is higher for the LDF-BC scheduling. This is because a lower CDR due to the scheduling strategy leads to a higher traffic load in the system, increasing the CBR. Logically, it is not appropriate to increase the network capacity by dropping ongoing calls. Thus, it can be concluded that the LDF-BC scheduler provides better performance than the RR-BC scheduler.

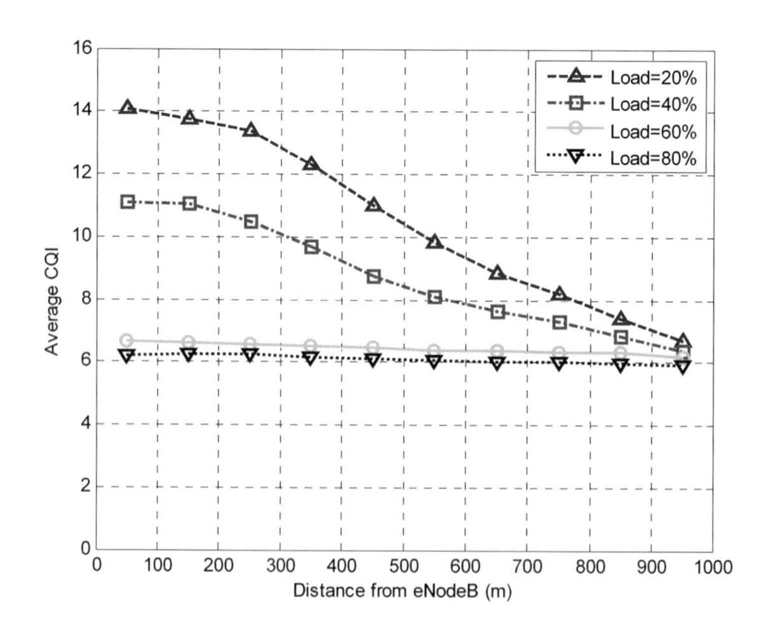

Figure 14.10 Average CQI as a function of the distance for different traffic load levels.

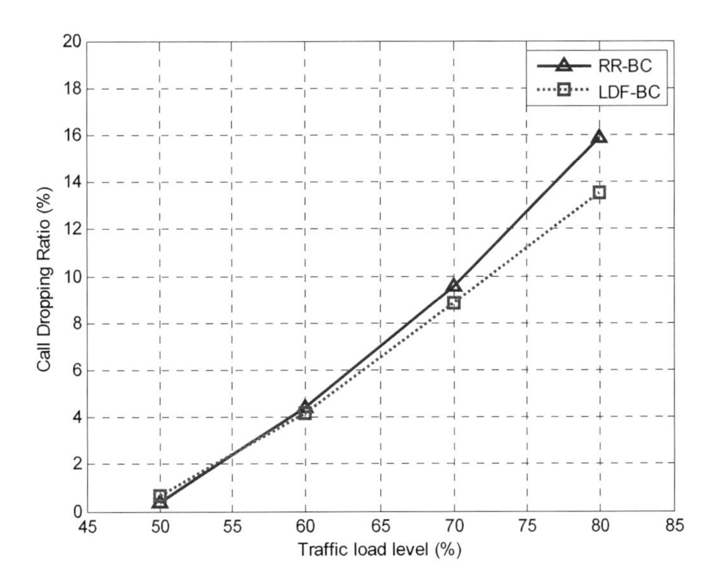

Figure 14.11 Call Dropping Ratio as a function of the traffic load for two scheduling schemes.

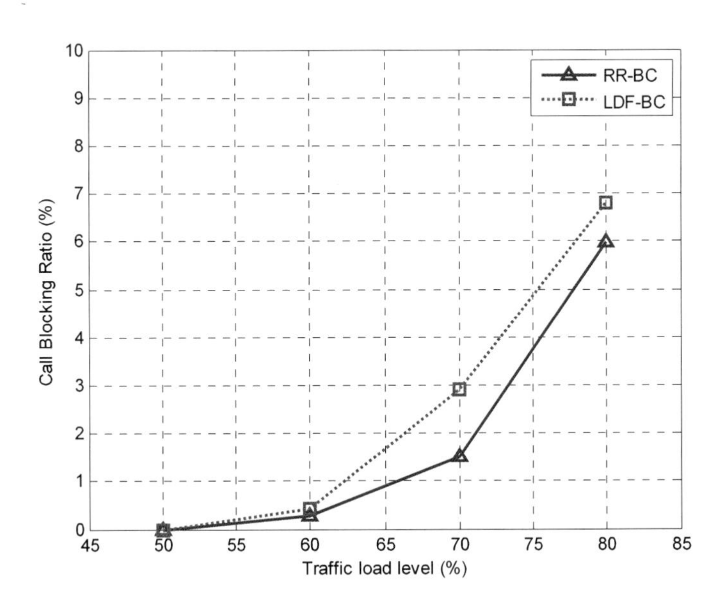

Figure 14.12 Call Blocking Ratio as a function of the traffic load for two scheduling schemes.

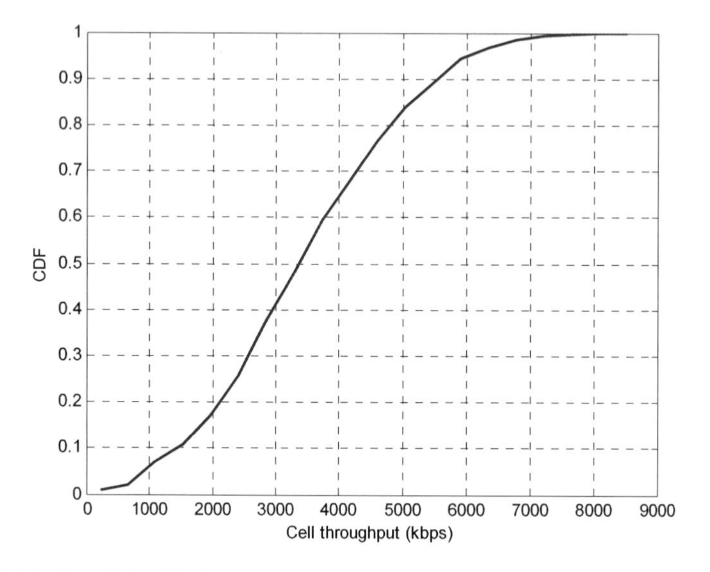

Figure 14.13 Cumulative distribution function of downlink cell throughput.

Another simulation has been carried out to quantify the network capacity. For this purpose, only the best effort service based on the full buffer traffic model is activated in the simulation tool. Thus, all PRBs are fully exploited and the cell throughput is a useful measurement to estimate the network capacity. The results are presented in Figure 14.13 as the overall cumulative distribution function of cell throughput.

14.7 CONCLUSIONS

In this chapter, a computationally-efficient dynamic system-level LTE simulator has been described. This simulator includes the main characteristics of the radio access technology as well as the radio resource management algorithms which provide notable improvements in the efficient use of the available radio resources. For this purpose, the simulator has been implemented so that simulations require a low computational cost. A simulation is composed of epochs to evaluate the modification of network parameters performed by optimization algorithms.

At physical and link layers, this work focuses on the calculation of several indicators with the purpose of evaluating the connection quality in a mobile communication. Those indicators are required in the execution of radio resource management functions. Hence, it is essential that these indicators reflect accurately the behavior of a real network. To achieve this goal, an OFDM channel model has been performed to characterize the temporary and frequency variation of the radio transmission environment for each user during the

Table 14.4 PRB assignment across the time

ITERATION	*USER 1*			*USER 2*			*USER 3*		
	PRB	*CQI*	*TX RATE*	*PRB*	*CQI*	*TX RATE*	*PRB*	*CQI*	*TX RATE*
1	1	8	322	1	10	459	1	6	198
	2	6	198	2	9	405	2	3	64
	3	7	249	3	10	459	3	5	148
	4	8	322	4	11	559	4	4	102
	5	5	148	5	8	322	5	3	64
	6	6	198	6	7	249	6	5	148
2	1	5	148	1	9	405	1	3	64
	2	4	102	2	10	459	2	2	40
	3	5	148	3	10	459	3	3	64
	4	3	64	4	8	322	4	2	40
	5	2	40	5	9	405	5	2	40
	6	5	148	6	10	459	6	1	26
3	1	4	102	1	10	459	1	4	102
	2	3	64	2	10	459	2	5	148
	3	4	102	3	11	559	3	3	64
	4	2	40	4	11	559	4	6	198
	5	3	64	5	8	322	5	2	40
	6	4	102	6	9	405	6	3	64

simulation.

The main functions of radio resource management have been also described in this chapter. At link level, the previous calculated indicators are inputs of the Link Adaptation and Dynamic Scheduling functions. At network level, the main functions are admission control and mobility management, whose parameters can be modified to evaluate optimization algorithms.

Finally, several simulations have been carried out. Results show network performance in terms of several indicators for different traffic load levels and scheduling schemes.

PROBLEMS AND EXERCISES

1. In the initialization stage of a simulation, a warm-up distribution of users is created. What is the purpose of this initial distribution?
2. Manhattan model aims to create a more realistic urban scenario, emulating a reasonable behavior of users in an urban environment. What use case in wireless network optimization can be studied when modeling the behavior of buses in the urban scenario? And how does the presence of traffic lights in the scenario affects?

3. The simulator implements the wrap-around technique to avoid border effects in the simulation. What condition must satisfy the elements of the Manhattan scenario (street and buildings) to apply the wrap-around technique in this scenario? Suppose that we use a square scenario of size M x M, square buildings with dimensions B_w x B_w, and the street width is S_w.
4. The mobile radio channel can be modeled as a Wide-Sense Stationary Uncorrelated Scattering (WSSUS) channel. How is time stationarity achieved in the simulation of the fading propagation channel?
5. Why is time variable converted into distance variable in the simulation of wideband propagation channel?
6. A fast-fading grid is implemented in the simulator to avoid excessive memory resources consumption. Why cannot the size of the multipath fading grid be smaller than the correlation distance of the shadowing?
7. Suppose there is a cell with 3 users served by an e-Node B using 6 PRB. Use the information in Table 14.4 to schedule the users for the next 3 iterations following the packet scheduling strategies Best Channel, Round Robin to Best Channel and Proportional Fair.
   In Table 14.4, it is shown the CQI value experienced by the users for each PRB during the next 3 iterations and the number of bits the users can transmit on each PRB (bits/PRB/ms). At the beginning, all the users have a cumulative number of bits transmitted equal to 100.
8. A cell service area is defined as the area where a cell is dominant, that is, *i*-cell service area covers every point where the highest signal received comes from cell *i*. Assume a very basic propagation equation is working in our scenario:

$$PL = \max(30\log(d[m]) + 30, 80)$$

PBGT handover is activated in the network, where $Margin_{PBGT}(i,j)$ is initially set to 3 dB for every *(i,j)* adjacency, and equation

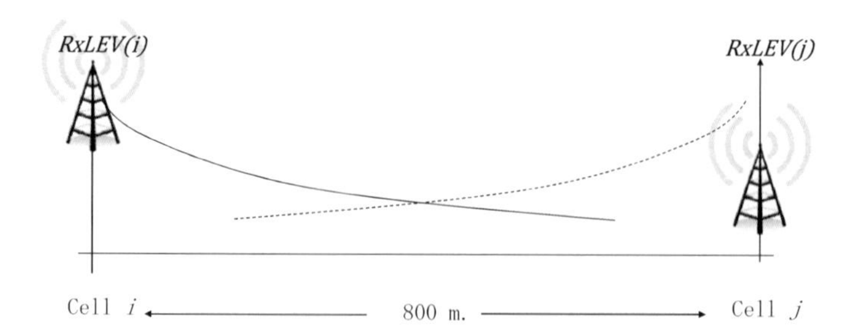

Figure 14.14 Scenario with two eNBs deployed at a distance of 800 m.

$$Margin_{PBGT}(i,j) + Margin_{PBGT}(j,i) = 6dB$$

must be always fulfiled (hysteresis area).
Depict i and j-cell service areas in Figure 14.14. Note that there is a hysteresis region. Calculate distances when possible. Assume BTS transmission power is the same for both cells.

9. $Margin_{PBGT}$ is frequently used to modify cell service areas.
   a. Use previous figure to assess changes in $Margin_{PBGT}(i,j)$, as well as cell power transmissions and hysteresis values (6 dB in the previous exercise).
   b. There exists 10 Erlangs uniformly distributed between both i and j eNBs, but cell i can only manage 2 Erlangs and cell j has enough capacity. Determine the new values for $Margin_{PBGT}(i,j)$ and $Margin_{PBGT}(j,i)$. Draw conclusions.
10. Adjusting coverage areas, through $MinThresholdLEV(i)$ modifications, also shares traffic during the admission control process. Imagine $MinThresholdLEV(i) >> MinThresholdLEV(j)$ and $Margin_{PBGT}(i,j)=7dB$. 10 Erlangs are uniformly distributed between both cells. Describes how the traffic is distributed during the admission process, and, later, after PBGT HOs take place. Draw conclusions.z

References

[1] 3GPP TS 36.201, "Evolved Universal Terrestrial Radio Access (E-UTRA); LTE physical layer; General description".

[2] 3GPP TS 36.300, "Evolved Universal Terrestrial Radio Access (E-UTRA) and Evolved Universal Terrestrial Radio Access Network (EUTRAN); Overall description".

[3] H. Holma & A. Toskala, *LTE for UMTS – OFDMA and SC-FDMA Based Radio Access*, Wiley, 2009.

[4] 3GPP TS36.133, "Evolved Universal Terrestrial Radio Access (E-UTRA); Requirements for Support of Radio Resource Management".

[5] J. Wu, Z. Yin, J. Zhang & W. Heng, "Physical layer abstraction algorithms research for 802.11n and LTE Downlink", *International Symposium on Signals Systems and Electronics (ISSSE)*, 2010.

[6] J. Olmos, A. Serra, S. Ruiz, M. García-Lozano & D. Gonzalez, "Link Level Simulator for LTE Downlink", *COST2100*, TD(09)779, 2009.

[7] J.C. Ikuno, M. Wrulich, & M. Rupp, "System Level Simulation of LTE Networks", *2010 IEEE 71st Vehicular Technology Conference (VTC 2010-Spring)*, 2010.

[8] C. Mehlführer, M. Wrulich, J. C. Ikuno, D. Bosanska & M. Rupp, "Simulating the Long Term Evolution Physical Layer", *17th European Signal Processing*

Conference (EUSIPCO 2009), 2009.

[9] T. Hytönen, "Optimal Wrap-around Network Simulation", Helsinki University of Technology Institute of Mathematics: Research Reports 2001.

[10] B. Ahn, H. Yoon & J.W. Cho, "A design of macro-micro CDMA cellular overlays in the existing big urban areas". *IEEE Proc. Vehicular Technology Conference (VTC 2001)*, pp.2094-2104.

[11] I. Khider, A.Saad & Wang Furong, "Study on Indoor and Outdoor Environment for Mobile Ad hoc Network Supported with Base Stations", *Wireless Communications, Networking and Mobile Computing (WiCom)*, 2007.

[12] ETSI TR 101 112 v.3.2.0, Selection procedures for the choice of radio transmission technologies of the UMTS (UMTS 30.03 version 3.1.0), ETSI April 1998.

[13] A. Lobinger, S. Stefanski & T. Jansen, "Load Balancing in Downlink LTE Self-Optimizing Networks", *IEEE 71st Vehicular Technology Conference, 2010 (VTC 2010-Spring)*, 2010.

[14] A. Mahajan, et al., "Urban mobility models for VANETs", *2nd Workshop on Next Generation Wireless Networks*, 2006.

[15] T. Camp, J. Boleng, & V. Davies, "A survey of mobility model for ad hoc network research", *Wireless Communication and Mobile Computing (WCMC): Special Issue on Mobile AdHoc Networking: Research, Trends and Applications*, Vol. 2, No. 5, pp.483-502, 2002.

[16] NGMN, "NGMN Radio Access Performance Evaluation Methodology", Version 1.0, Enero 2008, www.ngmn.org.

[17] J.D. Parsons, *The Mobile Radio Propagation Channel*, Pentech, 1992.

[18] W.C. Jakes, *Microwave Mobile Communications*, Wiley, 1974.

[19] 3GPP, "Evolved Universal Terrestrial Radio Access (E-UTRA); User Equipment (UE) Radio Transmission and Reception (Release 9)", 3GPP TS 36.101, Dec. 2009.

[20] E. Bonek, "Tunnels, corridors, and other special environments", *COST Action 231: Digital mobile radio towards future generation systems*, L. C. E. Damosso, Ed. Brüssel: European Union Publications, 1999, pp. 190–207.

[21] F. Khan, *LTE for 4G Mobile Broadband: Air Interface Technologies and Performance*, New York, NY, USA: Cambridge University Press, 2009.

[22] D. Huo, "Simulating slow fading by means of one dimensional stochastical process", *IEEE 46th Vehicular Technology Conference, 1996. 'Mobile Technology for the Human Race'*, vol. 2, apr-1 may 1996, pp. 620 –622 vol.2.

[23] M. Gudmundson, "Correlation model for shadow fading in mobile radio systems," *Electronics Letters*, vol. 27, no. 23, pp. 2145 –2146, nov. 1991.

[24] 3GPP, "Feasibility study for Orthogonal Frequency Division Multiplexing (OFDM) for UTRAN enhancement", 3GPP TR 25.892, 2004.

[25] 3GPP, "System Analysis of the Impact of CQI Reporting Period in DL SIMO OFDMA (R1-061506)", Shanghai, China, 3GPP TSG-RAN WG1 45, May 2006.

[26] E. Tuomaala, "Effective SINR approach of link to system mapping in OFDM/multi-carrier mobile network", *IEEE Mobility Conference, The Second International Conference on Mobile Technology Applications and Systems*, 2005.

[27] J. C. Ikuno, M. Wrulich, & M. Rupp, "Performance and Modeling of LTE H-ARQ", *International ITG Workshop on Smart Antennas WSA*, 2009.

[28] 3GPP, "OFDM-HSDPA System level simulator calibration (R1-040500)", Montreal, Canada, 3GPP TSG-RAN WG1 37, May 2004.

[29] 3GPP, "E-UTRA; UE conformance specification; Radio transmission and reception; Part 1: Conformance testing", 3GPP TS 36.521, 2009.

[30] J.T. Entrambasaguas, M.C. Aguayo-Torres, G. Gomez & J.F. Paris, "Multiuser Capacity and Fairness Evaluation of Channel/QoS-Aware Multiplexing Algorithms", *IEEE Network*, vol.21, no.3, pp.24-30, May-June 2007.

[31] D. Hong & S. S. Rappaport, "Traffic model and performance analysis for cellular mobile radio telephone systems with prioritized and nonprioritized handoff procedures", *IEEE Transactions on Vehicular Technology*, 35(3), 77–92, 1986.

Chapter 15.

Multi-Carrier Cooperative Wireless Communication

Muhammad Abrar, Xiang Gui, Amal Punchihewa

School of Engineering and Advanced Technology, Massey University, New Zealand
M.Abrar@massey.ac.nz

This chapter provides a brief introduction about multi-carrier cooperative wireless communication systems using relays. We will describe some basic protocols that are being considered in relay operation. We will focus on the discussion of performance analysis and resource allocation in these systems. The performance analysis will be based on the system capacity.

15.1 INTRODUCTION

During the last couple of decades the importance of wireless communication has been increasing remarkably in almost every field of life. The evolution of wireless cellular networks from the First Generation (1G) to the current third Generation (3G) has enabled more reliable, faster and secure communication services. We are now approaching the fourth generation (4G) which will provide even higher data rates and quality of service [1].
Signal fading in wireless communication systems due to multipath propagation is a major limitation in the performance of wireless communication systems. Various diversity techniques such as time, frequency and space diversities have been proposed and adopted in practical applications to mitigate this fading effect. Multi-input multi-output (MIMO) is a space diversity technique that plays an important role in the improvement of wireless systems. Multiple wireless paths are used in MIMO systems to transmit and receive signals via multiple antennas

4G Wireless Communication Networks: Design, Planning and Applications, 297-318,

at transmitters and receivers, respectively. Figure 15.1 illustrates the different classes of multi-antenna systems including single-input-multiple-output (SIMO), multiple-input-single-output (MISO), multiple-input-multiple-output (MIMO) and conventional single-input-single-output (SISO). Exploiting multiple wireless paths, MIMO systems improve the quality of the received signal and increase the data transmission rate by using digital signal processing methods [2].

It is widely accepted that the proposed high data rates for next generation wireless networks can only be achieved by MIMO users who have mobile terminals with multiple antennas [2]. MIMO systems also require sufficient separation between multiple antennas to achieve maximum benefits from this diversity. While antenna separation at the base station is easily satisfied, it may not be feasible at the mobile terminal due to its small physical size. In fact, the small form factor of portable mobile terminals may make it impossible to accommodate multiple antennas. Furthermore, when a mobile terminal is located far away from the base station, even the use of multiple antennas may not guarantee the communication link quality. To overcome these limitations, wireless communication systems are being developed departing from the conventional way of point to point communication. Stemming from the broadcast nature of wireless channels and the idea of allowing a network node to forward the information of other node or nodes in multi-hop networks such as sensor and adhoc networks, a new way of wireless communication known as cooperative communication has been developed.

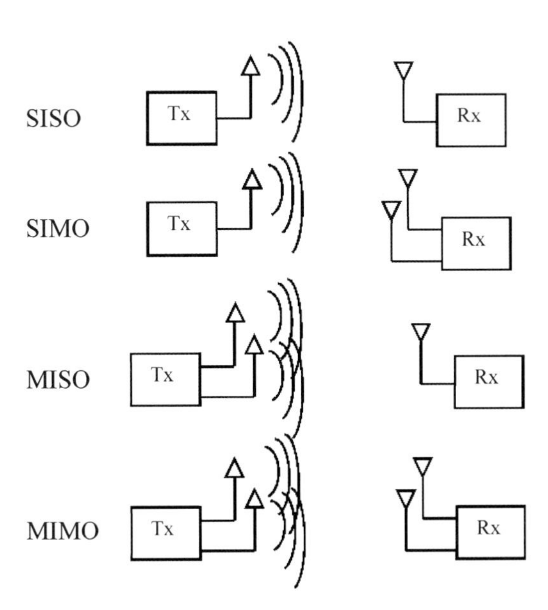

Figure 15.1 Multi-Antenna System [3].

15.1.1 What is Cooperative Network?

Cooperative communication generally refers to a system where the source transmits its signal to the destination with the cooperation of one or more relays to improve system performance. This idea of cooperative communication in wireless environments is more attractive due to the diverse quality of wireless channel and the limited number of usable resources. With this cooperation, users that experience a deep fade in their channel links can utilize better channels provided by relays to improve their quality of service.

As shown in Figure 15.2, cooperative communication technology can establish a reliable wireless communication link between a mobile terminal (MT) and a base station (BS) by creating independent wireless propagation paths with the help of other intermediate nodes known as relay terminals (RT). Cooperative communication can achieve almost the same advantages as MIMO systems.

15.1.2 Historical Background

Although the idea of cooperative communication originated from the concept of relaying that was introduced by Van der Meulen [3] in 1971 and Cover and El Gamal [4] in 1979, the early concept of using relays was different from the recent concept of relays being employed in cooperative communication. The authors in [3] and [4] considered AWGN channel only and there was no concept of diversity at that time. In [4], the authors analyzed the capacity of a three-node network consisting of a source, a relay and a destination. They elaborated on different

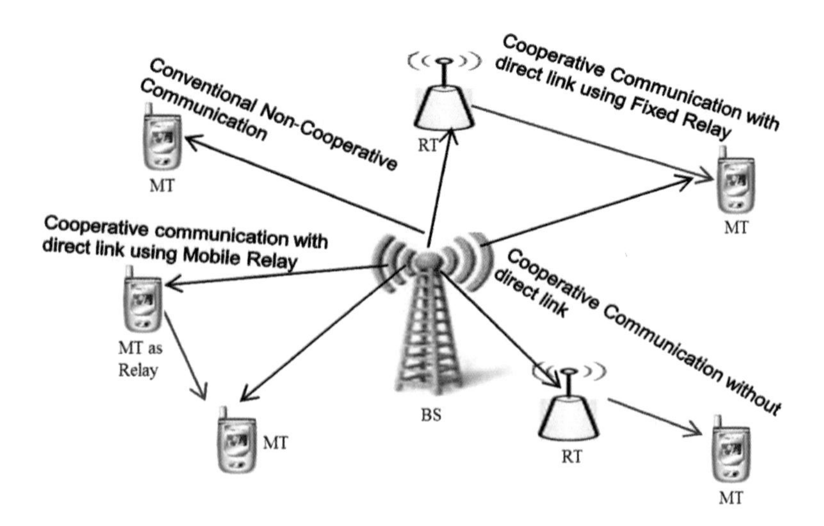

Figure 15.2 Overview of Cooperative Communication.

ways that a relay can assist a remote source in extending coverage. In cooperative communication, the relays are used to achieve diversity in fading wireless channels in addition to extending the coverage area. Despite these differences, as far as signal processing at the relay is concerned, they are almost the same. Therefore cooperative communication is interchangeably known as relay communication [5]. Sendonaris et al. were the first to introduce the concept of cooperative diversity in cellular networks [6], while Laneman et al. introduced cooperative communication in ad hoc networks [7]. Since the publication of [6] and [7] , in 2003 and 2004 respectively, cooperative communication has become a hot research topic and has received a lot of attention in the research community.

15.1.3 Multi-Carrier Transmission

OFDM is a multi-carrier scheme and is a special case of frequency division multiplexing (FDM). In a conventional FDM system, guard bands are introduced to separate adjacent sub-carriers and these result in reduced spectrum efficiency. Unlike FDM systems where guard bands are used to ensure that there is no spectral overlap between sub-carriers, in OFDM systems there is no inter-carrier interference (ICI) even though sub-carriers are spectrally overlapped as shown in Figure 15.3. This is due to the orthogonality between OFDM sub-carriers [8].

OFDM has inherent resistance to multipath propagation. In OFDM systems, data is demultiplexed into N parallel streams and each symbol duration is made N times longer, which helps to reduce the effect of Inter-Symbol Interference (ISI) due to multi-path. To eliminate ISI, a guard interval known as the cyclic prefix

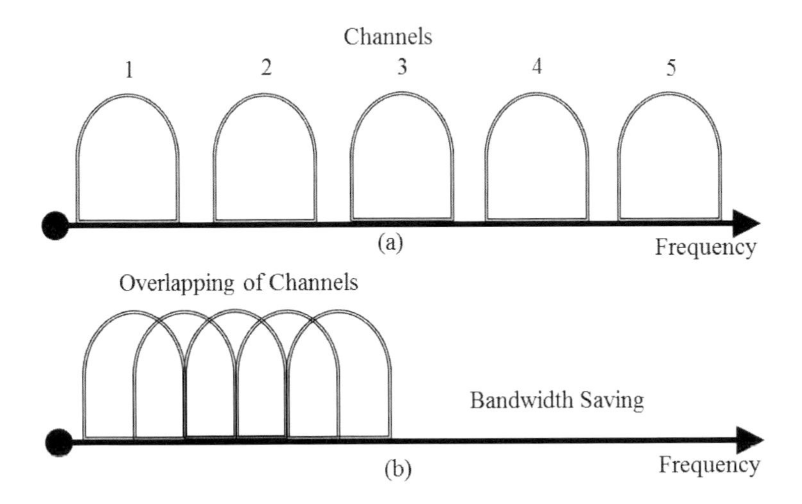

Figure 15.3 Concept of the OFDM (a) FDM (b) OFDM.

(CP) is introduced in each OFDM symbol. This CP is always chosen to be at least equal to the maximum delay spread of the channel. Furthermore, OFDM modulation and demodulation can easily be implemented using Discrete Fourier Transform (DFT) [8].

OFDM has already been employed in a number of communication standards including Digital Audio Broadcasting (DAB), Wireless LAN standards and IEEE 802.16 Broadband Wireless Access System [8]. OFDM is also the chosen modulation technique in next generation wireless standards such as IEEE 802.16j, IEEE.802.16m and 3GPP LTE-A where relaying systems are also being incorporated [1].

15.1.4 Why Resource Allocation?

The efficient allocation of available resources has already been proven effective in improving the performance of different communication systems. Traditional resources include bandwidth, power, time and now antennas in MIMO networks as well. In cooperative communication resource allocation also includes the selection of relay nodes. Since there are many entities involved to achieve a desired goal in cooperative networks, the resource allocation problem becomes even more important.

15.2 COOPERATIVE PROTOCOLS

Many cooperation techniques known as protocols have been proposed based on the concept of relaying. Some of these are amplify & forward [9], decode & forward [7], compress & forward [10], demodulation & forward [11], adaptive relay protocol [12] and coded cooperation [13]. All of these are somehow variations of the two basic protocols, namely, amplify & forward (A&F) and decode & forward (D&F). Here we briefly discuss these two protocols.

15.2.1 Amplify and Forward

Amplify & Forward is the simplest cooperative protocol that was proposed in [9] and more precisely analysed in [7]. In amplify & forward, the scaled version of the received signal at relay is forwarded to the destination as shown in Figure 15.4 (a). The destination receives two versions of the same signal, one from the source and the other from the relay, thus achieving spatial diversity. The performance of this protocol has been widely investigated. The advantages of this protocol are simplicity and low cost. But the major disadvantage is the noise amplification, as the relay receives a noisy version of the signal and forwards this without any alternation/filtration. Obviously this cooperation highly depends on the condition of channel between the source and the relay. The processing of

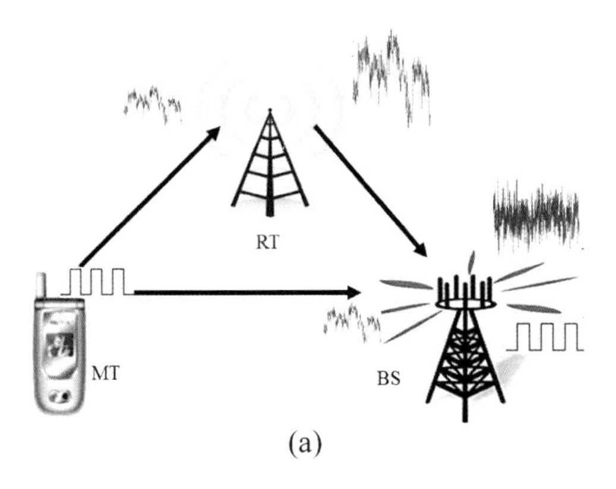

(a)

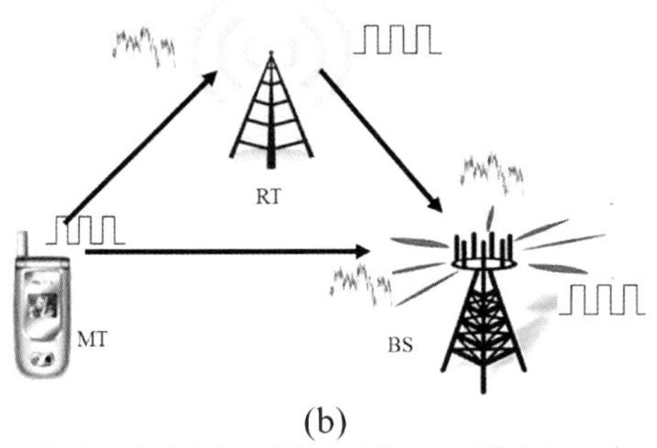

(b)

Figure 15.4 Cooperative Protocols (a) Amplify and Forward (b) Decode and Forward.

sampling, amplifying, and retransmitting analogue values is another potential challenge in amplify and forward relaying protocol [14].

15.2.2 Decode and Forward

Decode & Forward is another commonly used protocol that was also investigated in [7] and [6]. In some literature this is also addressed as detect and forward protocol [15]. In this type of protocol, relay attempts to decode the received signal to get the original data bits as shown in Figure 15.4 (b). After this decoding, the data bits are encoded and transmitted to the destination. This type of protocol looks more practical and also eliminates the noise which is amplified in A&F protocol. Thus, this method can considerably outperform A&F. But the problem with this protocol is that, if decoding errors occur at the relay due to deep fading in the link from the source to the relay, the relay will transmit these incorrect bits to the destination, leading to error propagation and even worse performance [5].

15.3 COOPERATIVE RELAYING

Due to the practical half-duplex nature of devices, there are two types of cooperative relaying proposed in the literature, namely one-way relaying (OWR) and two-way relaying (TWR).

15.3.1 One-way Relaying

In the first phase, MT broadcasts its signal and it is received by both the relay and the BS. In the second phase, the relay transmits the signal to the BS. Then, the same two phases are repeated in reverse for the BS to send data back to the MT as shown in Figure 15.5. In this type of relaying, there is a loss of half of the spectral efficiency as compared to full duplex relaying. Full duplex relaying, in which a relay is able to transmit and receive on the same frequency at the same time, is practically more complex. Therefore, from a practical point of view, half duplex relaying is preferred over full duplex operation even with this loss of spectral efficiency [16].

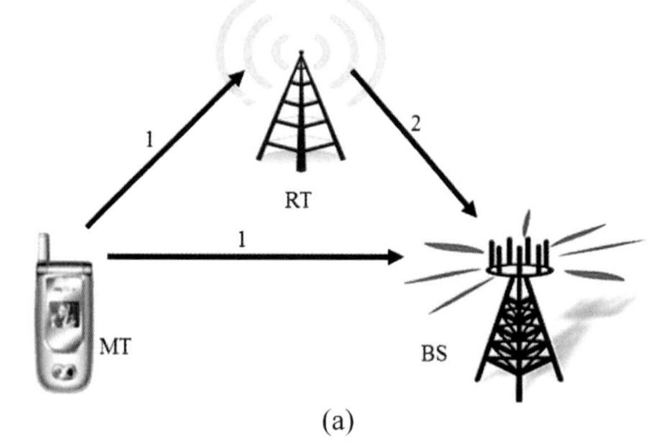

(a)

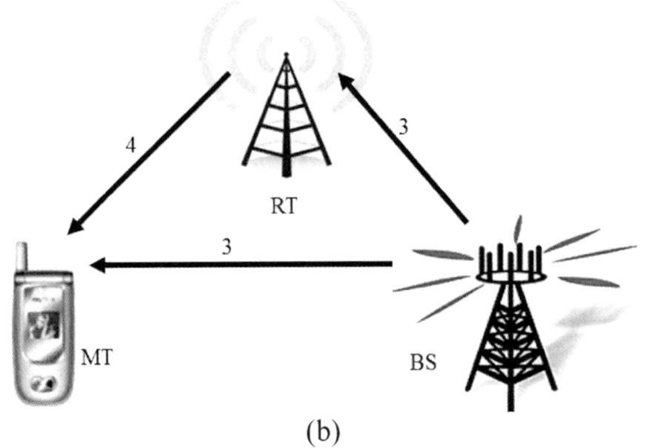

(b)

Figure 15.5 One way Cooperative Relaying (a) MT-to-BS (b) BS-to-MT.

15.3.1.1 AF-OWR

Here we describe the basic mathematical model for AF one-way relaying. Considering that S_1 is the transmitted OFDM symbol with unit power from the BS. The signal received at the relay and MT in the first time slot can be expressed as Y_{br} and Y_{bm}, respectively

$$Y_{br} = \sqrt{P_b}\, S_1 h_{br} + n_r \tag{15.1}$$

$$Y_{bm} = \sqrt{P_b}\, S_1 h_{bm} + n_{m1} \tag{15.2}$$

where P_b is the transmission power of the BS and h_{br} is the channel gain of the BS-RT link. While n_r and n_{m1} is the Additive White Gaussian Noise (AWGN) with variance σ^2, at the relay and mobile, respectively. The received signal is amplified by the relay with a scaling factor β_r as proposed in [7], before retransmitting to the destination

$$\beta_r = \sqrt{\frac{P_r}{|h_{br}|^2 P_b + \sigma^2}} \tag{15.3}$$

where P_r is the transmission power of the relay. The mobile receives the amplified signal from the relay in thc sccond time slot that is given as

$$Y_m = \beta_r Y_{br} h_{rm} + n_{m2} \tag{15.4}$$

Using (15.1) we obtain

$$Y_m = \beta_r \sqrt{P_b} S_1 h_{br} h_{rm} + N_m \tag{15.5}$$

The last term N_m in (15.5) represents the sum of two noise terms, the amplified version of AWGN received from the relay $\beta_r n_r \ h_{rm}$, and n_{m2} at MT, while h_{rm} represents the channel gain of the RT-MT link.

The received Signal-to-Noise ratio (SNR) of each link can be calculated using (15.2) and (15.5). For simplicity, we assume that all noise variances are identical. Let γ_{b-m}^{AF} and γ_{b-r-m}^{AF} represent the SNRs for BS-MT and BS-RT-MT link, respectively

$$\gamma_{b-m}^{AF} = \frac{P_b |h_{bm}|^2}{\sigma^2} \tag{15.6}$$

$$\gamma_{b-r-m}^{AF} = \frac{\beta_r^2 P_b |h_{br}|^2 |h_{rm}|^2}{(\beta_r^2 |h_{rm}|^2 + 1)\sigma^2} \tag{15.7}$$

By using (15.3) in (15.7), we obtain

$$\gamma_{b-r-m}^{AF} = \frac{P_b |h_{br}|^2 P_r |h_{rm}|^2}{(P_b |h_{br}|^2 + P_r |h_{rm}|^2 + \sigma^2)\sigma^2} \tag{15.8}$$

Using Maximum Ratio Combining (MRC) technique at the destination, the received instantaneous SNR is the sum of SNRs of the two links [17]

$$\begin{aligned} \gamma_m^{AF} &= \gamma_{b-m}^{AF} + \gamma_{b-r-m}^{AF} \\ &= \frac{P_b |h_{bm}|^2}{\sigma^2} + \frac{P_b |h_{br}|^2 P_r |h_{rm}|^2}{(P_b |h_{br}|^2 + P_r |h_{rm}|^2 + \sigma^2)\sigma^2} \end{aligned} \tag{15.9}$$

The instantaneous capacity achieved for AF-OWR can be expressed as

$$R_m^{AF-OWR} = \frac{1}{2}\log_2\left(1+\gamma_m^{AF}\right) \tag{15.10}$$

The factor ½ appears here due to the half-duplex operation of relays. It means that relays transmit and receive in two different time slots.

15.3.1.2 DF-OWR

In DF protocol, relay is capable of decoding the signal. Therefore there is no noise amplification. If we assume that relay decodes the signal correctly, then the instantaneous received SNR for relaying link is equal to the minimum of SNRs of two links, BS-RT and RT-MT [7] and is given as

$$\gamma_{b-r-m}^{DF} = \min\left\{ \left(\frac{P_b\left|h_{br}\right|^2}{\sigma^2}\right), \left(\frac{P_r\left|h_{rm}\right|^2}{\sigma^2}\right) \right\} \tag{15.11}$$

While the SNR for direct link is same as shown in (15.6), therefore received SNR at MT after two time slots is equal to the sum of (15.6) and (15.11)

$$\gamma_m^{DF} = \min\left\{ \left(\frac{P_b\left|h_{br}\right|^2}{\sigma^2}\right), \left(\frac{P_r\left|h_{rm}\right|^2}{\sigma^2}\right) \right\} + \frac{P_b\left|h_{bm}\right|^2}{\sigma^2} \tag{15.12}$$

The instantaneous capacity achieved for DF-OWR can be expressed as

$$R_m^{DF-OWR} = \frac{1}{2}\log_2\left(1+\gamma_m^{DF}\right) \tag{15.13}$$

15.3.2 Two-way Relaying

To overcome the spectral loss in OWR, two types of TWR have been proposed in the literature [18]. The first type assumes that no direct link is available between MT and BS and only a relay link is available for transmission. Therefore, two time slots are required to complete the exchange of information between the MT and BS. The second type takes into account the direct link, and requires three time slots in order to complete the exchange of information [19]. These two types of AF-based TWR are known as Analogue Network Coding (ANC) protocol and Time Division Broadcast (TDBC) protocol, respectively [20].

15.3.2.1 Analogue Network Coding (ANC) Protocol

In ANC protocol as shown in Figure 15.6 (a), both MT and BS transmit their signal to the relay during the first time slot. A relay receives the combined signal of MT and BS due to broadcast nature of wireless channel. The relay amplifies this combined signal and then retransmits to both MT and BS in the second time slot.

15.3.2.2 Time Division Broadcast (TDBC) Protocol

In TDBC protocol as shown in Figure 15.6 (b), MT transmits its signal to both BS and relay during first time slot, while BS transmits its signal to both MT and relay in second time slot. In third time slot the relay amplifies the combined signal of BS and MT and retransmits to both MT and BS.

In both cases, the received signal at MT and BS also consists of their own transmitted signal known as self-interference signal. With the knowledge of channel and its own signal, this self-interference signal can be subtracted from the received signal [18].

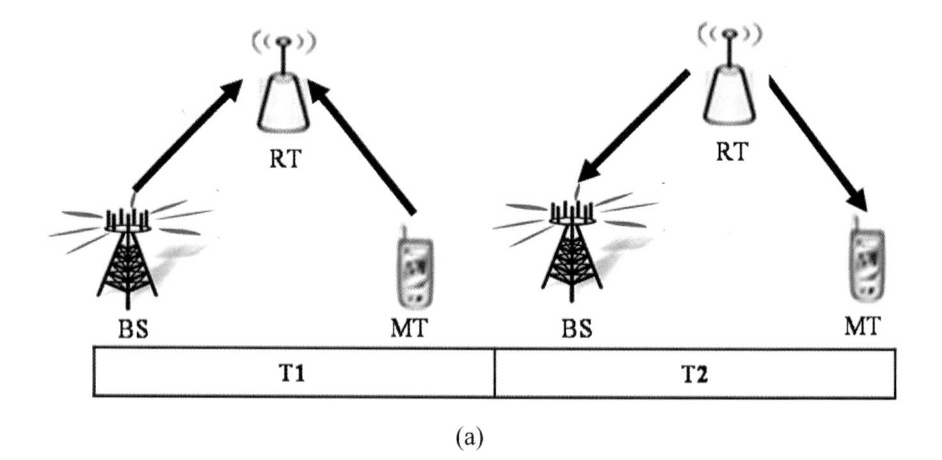

(a)

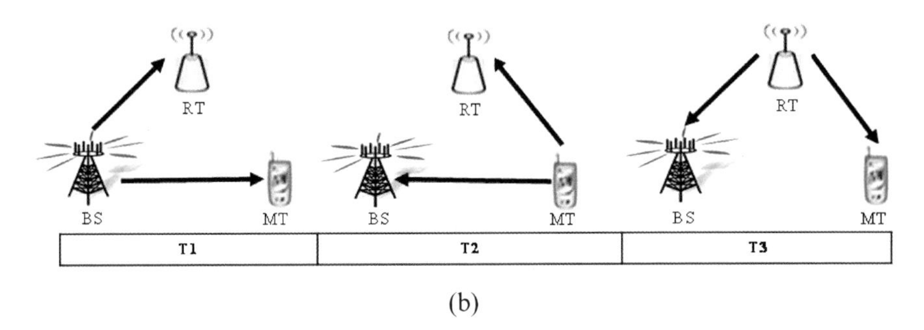

Figure 15.6 Two-Way Cooperative Relaying (a) TWR-ANC (b) TWR-TDBC.

15.3.2.3 TWR with ANC Protocol

Here we describe the basic mathematical model for AF-TWR and assume that there is no direct link available between the MT and BS. Considering that S_1 and S_2 are the transmitted OFDM symbols of the BS with transmitted power P_b and MT with transmitted power P_m, respectively. The combined signal received Y_r^{twr} at the relay can be expressed as

$$Y_r^{twr} = \sqrt{P_b} S_1 h_{br} + \sqrt{P_m} S_2 h_{mr} + n_r \tag{15.14}$$

Where h_{br} and h_{mr} are the channel gains of the BS-RT link and the MT-RT link, respectively. The received combined signal is amplified by the relay with scaling factor β_r^{twr} as proposed in [18]

$$\beta_r^{twr} = \sqrt{\frac{P_r}{|h_{br}|^2 P_b + |h_{mr}|^2 P_m + \sigma^2}} \tag{15.15}$$

By using (15.14) and (15.15) the received signals at MT and BS can be represented as, respectively

$$Y_m = \underbrace{\beta_r^{twr} \sqrt{P_b} S_1 h_{br} h_{rm}}_{\text{Required Signal}} + \underbrace{\beta_r^{twr} \sqrt{P_m} S_2 h_{mr} h_{rm}}_{\text{Self-Interference}} + N_m \tag{15.16}$$

$$Y_b = \underbrace{\beta_r^{twr} \sqrt{P_b} S_1 h_{br} h_{rb}}_{\text{Self-Interference}} + \underbrace{\beta_r^{twr} \sqrt{P_m} S_2 h_{mr} h_{rb}}_{\text{Required Signal}} + N_b \tag{15.17}$$

The terms N_m and N_b in (15.16) and (15.17) represent the noise terms that contain both the amplified version of AWGN components received from the relay and the AWGN component at the receiving node itself. In these equations, the received signal at both the MT and BS also contains their own transmitted signals, known as self-interference signals. With the knowledge of the channel gains and its own signal, this self-interference can be subtracted from the received signal.

Let γ_m^{ANC} and γ_b^{ANC} represent the SNRs at MT and BS respectively:

$$\gamma_m^{ANC} = \frac{\left|\beta_r^{twr} h_{br} h_{rm}\right|^2 P_b}{\left(\left|\beta_r^{twr} h_{rm} \sigma_r\right|^2 + \sigma_m^2\right)} \tag{15.18}$$

$$\gamma_b^{ANC} = \frac{\left|\beta_r^{twr} h_{mr} h_{rb}\right|^2 P_m}{\left(\left|\beta_r^{twr} h_{rb} \sigma_r\right|^2 + \sigma_b^2\right)} \tag{15.19}$$

For simplicity of derivations we assume that all noise variances are identical and respective channel gains are reciprocals (i.e. $h_{xy} = h_{yx}$ and $\sigma_m^2 = \sigma_r^2 = \sigma_b^2 = \sigma^2$). By using the following substitutions,

$\frac{|h_{br}|^2}{\sigma^2} = \frac{|h_{rb}|^2}{\sigma^2} = X$, $\frac{|h_{mr}|^2}{\sigma^2} = \frac{|h_{rm}|^2}{\sigma^2} = Y$, $\frac{|h_{bm}|^2}{\sigma^2} = \frac{|h_{mb}|^2}{\sigma^2} = Z$ the received SNR γ_m^{ANC} given in (15.18) can be written as:

$$\gamma_m^{ANC} = \frac{P_b X P_r Y}{P_b X + (P_r + P_m) Y + 1} \tag{15.20}$$

and γ_b^{ANC} given in (4.19) can be written as:

$$\gamma_b^{ANC} = \frac{P_m Y P_r X}{(P_b + P_r) X + P_m Y + 1} \tag{15.21}$$

The achievable rates from BS to MT and from MT to BS are

$$R_m^{ANC} = \frac{1}{2}\log_2(1+\gamma_m^{ANC}) \tag{15.22}$$

$$R_b^{ANC} = \frac{1}{2}\log_2(1+\gamma_b^{ANC}) \tag{15.23}$$

The factor 1/2 shows that transfer of information between two nodes is completed in two time slots. The instantaneous sum-rate of TWR-ANC protocol can be given as:

$$\begin{aligned} R_{sum}^{ANC} &= R_m^{ANC} + R_b^{ANC} \\ &= \frac{1}{2}\log_2(1+\gamma_m^{ANC}) + \frac{1}{2}\log_2(1+\gamma_b^{ANC}) \end{aligned} \tag{15.24}$$

15.3.2.4 TWR with TDBC Protocol

The received SNRs at MT and BS using TDBC protocol can be calculated in the same way as calculated for ANC protocol earlier. Let γ_m^{TDBC} and γ_b^{TDBC} represent the SNRs at MT and BS respectively for TDBC protocol:

$$\gamma_m^{TDBC} = \frac{P_b X * P_r Y}{P_b X + (P_r + P_m)Y + 1} + P_b Z \tag{15.25}$$

$$\gamma_b^{TDBC} = \frac{P_m Y * P_r X}{(P_b + P_r)X + P_m Y + 1} + P_m Z \tag{15.26}$$

Therefore the instantaneous sum-rate of TWR-TDBC protocol can be given as:

$$R_{sum}^{TDBC} = \frac{1}{3}\log_2(1+\gamma_m^{TDBC}) + \frac{1}{3}\log_2(1+\gamma_b^{TDBC}) \tag{15.27}$$

The factor 1/3 shows that transfer of information between two nodes is completed in three time slots.

15.4 RESOURCE ALLOCATION

In multi-user, multi-carrier systems, different sub-carriers experience different and independent channels for different users. Some of the sub-carriers may experience deep fading for some users and hence are not suitable at that time instant. Therefore, allocation of sub-carriers to different users according to channel conditions, will improve system performance in multipath frequency selective fading environment. Multi-user diversity can also be exploited by adaptive allocation of these sub-carriers [5].

15.4.1 Sub-carrier and Resource block

In Long Term Evolution (LTE) system, Resource block (RB) is the minimal unit to be allocated. A single RB consists of twelve consecutive OFDM sub-carriers [21].

OFDM uses a large number of sub-carriers having smaller bandwidth for multi-carrier transmission. In the frequency domain, the spacing between the sub-carriers is 15 kHz. The resource blocks have a total size of 180 kHz in the frequency domain and 0.5ms in the time domain. Each user is allocated a number of resource blocks in the time–frequency grid. The allocation of resource blocks to users depends on the scheduling mechanisms in the frequency and time dimensions [22].

15.4.2 Resource Block Allocation

There are different schemes, available for the allocation of sub-carriers/resource blocks. Here we discuss three most common schemes.

15.4.2.1 Round Robin Scheme (RRS)

In round robin scheme (RRS) known as user-oriented, RBs are allocated in cyclic way to each MT. In other words RBs are equally allocated to all mobiles, regardless of the channel conditions. It is the most natural fair algorithm to allocate resources, but it does not provide any fairness in term of quality of service (QoS) because it does not exploit channel conditions.

15.4.2.2 Maximum SNR Scheme (MSS)

In maximum SNR scheme (MSS) known as network oriented, RBs are allocated to the MTs which have maximum received SNRs. This scheme optimizes the overall system throughput but fairness between MTs is not considered. In this way, any MT which has bad received SNR due to poor channel conditions may not be allocated any RB.

15.4.2.3 Proportional Fairness Scheme (PFS)

Proportional Fairness Scheme (PFS) shows an intermediate solution for the above two schemes. This scheme considers both fairness in term of individual performance and overall system performance. This scheme shows a good trade-off between overall system efficiency and fairness in term of individual user efficiency.

15.4.3 Power Allocation

Power is always a critical resource in wireless networks. System performance can be improved by efficient use of this resource. Power optimization generally extends the life-time of networks where user's equipment is battery operated. Power allocation in multi-user multi-carrier cooperative systems becomes a more challenging and important problem due to the increased number of entities involved.

15.5 SOLVED EXAMPLES

Example 15.1: In this example we compare the performance of AF-OWR protocol and DF-OWR protocol in terms of achievable capacity in fading channels using MATLAB simulation. It is assumed that only one RT is cooperating with each MT. There are 96 sub-carriers which produce eight RBs that are available for these two MTs. The RBs are allocated in simple round robin fashion in both protocols. The instantaneous data rate achieved can be calculated using (15.10) and (15.13) respectively. If there is M number of MTs and K number of resource blocks, the overall throughput can be calculated for AF and DF protocol using (15.28) and (15.29), respectively.

$$R^{AF-OWR} = \sum_{m=1}^{M} \sum_{k=1}^{K} R_{m,k}^{AF-OWR} \tag{15.28}$$

$$R^{DF-OWR} = \sum_{m=1}^{M} \sum_{k=1}^{K} R_{m,k}^{DF-OWR} \tag{15.29}$$

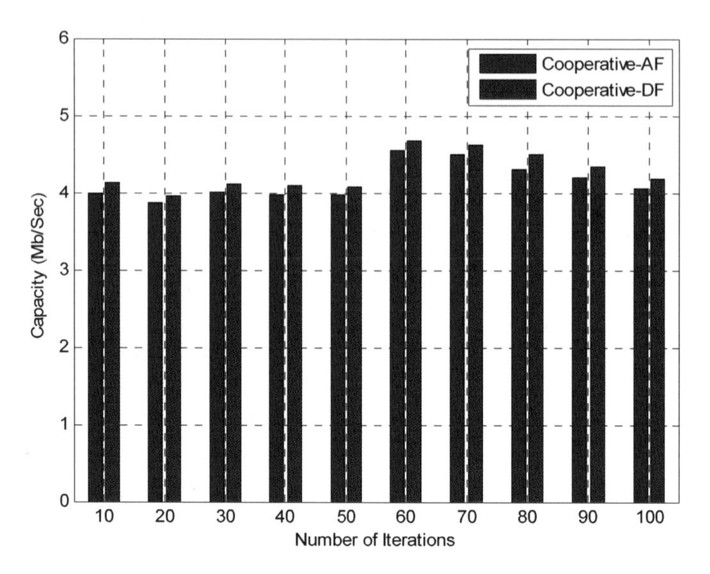

Figure 15.7 Achievable System Capacity with Direct Link (M=2).

Figure 15.7 shows the overall achievable throughput for AF-OWR and DF-OWR protocols at equal power allocation. The performance is observed at different number of iterations at received SNR = 20dB of direct link. It is clear that there is significant increase in capacity of DF protocol over AF protocol.

Example 15.2: In this example, firstly we develop an optimization problem to apply PFS resource block allocation and secondly, we compare the performance of three resource block allocation scheme presented in this chapter with AF-OWR protocol.

Problem Formulation: The system consists of a single BS, R number of RTs and M number of MTs. There is K number of RBs available in the cell. The maximization of system throughput is an objective function with minimum rate requirement for each MT.

The total achievable throughput of the system is given by (15.28). Here we define a variable $\alpha_{m,k}$ as sub-carrier allocation index such that

$$\alpha_{m,k} = \begin{cases} 1 & \textit{if } \text{k}^{\text{th}} \textit{ RB is allocated to mobile } m \\ 0 & \textit{otherwise} \end{cases}$$

The optimization problem for resource allocation can be formulated as

$$R_S^{AF-OWR} = \max \sum_{m=1}^{M} \sum_{k=1}^{K} \alpha_{m,k} R_{m,k}^{AF-OWR} \tag{15.30}$$

Subject to:

$$\alpha_{m,k} \in \{0,1\}, \quad \forall\ k,\ m \tag{15.30a}$$

$$\sum_{m=1}^{M} \alpha_{m,k} = 1, \quad \forall\ m \tag{15.30b}$$

$$\sum_{k=1}^{K} \alpha_{m,k} R_{m,k}^{AF-OWR} \geq R_{\min} \quad \forall\ m \tag{15.30c}$$

The first two constraints (15.30a) and (15.30b) in this optimization problem are that each RB can only be assigned to only one MT to avoid intra cell interference. The constraint (15.30c) is the minimum data rate requirement of each MT.

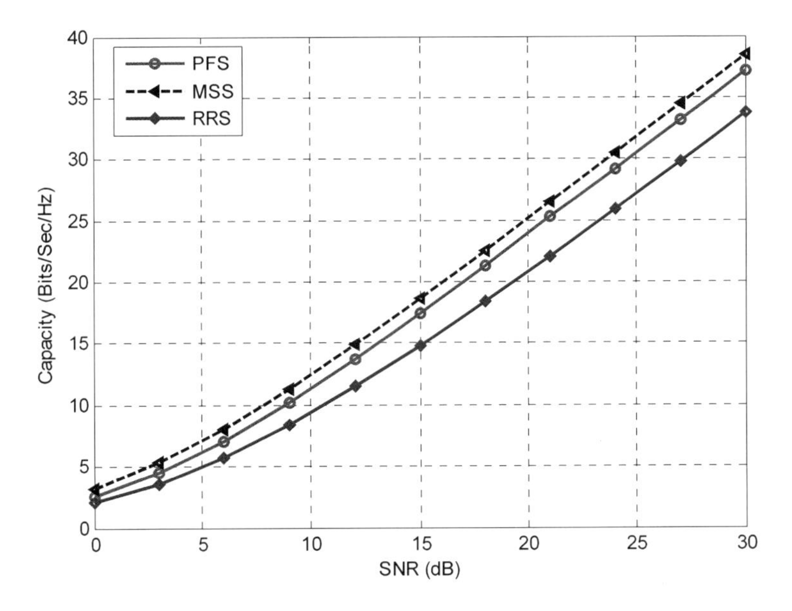

Figure 15.8 System Capacity without Direct Link versus SNR (M=2, K=8).

The optimization problem as described in (15.30) is developed and resource block allocation is made in three steps.

1. In the first step, priority of MT is determined. The MT with highest received SNR is first selected for the allocation of RB.

2. In the second step, RBs are allocated to different MTs to satisfy their minimum data rate (R_{min}) requirement. The MT with highest priority is allowed to select the best RB from available RBs which maximize its rate. This process continues till either all MTs have achieved R_{min} or all the RBs are allocated to MTs. This step guarantees that MTs with higher data rate requirement or with good channel conditions are not allowed to use all the resources at the expense of others.
3. This step aims to maximize the overall throughput of the system after achieving R_{min} for all MTs in the previous step. If RBs are still available, these are allocated to the MTs which maximize the system throughput.

Simulation: Figure 15.8 shows the overall achievable system capacity for these three resource block allocation schemes against received SNR. It is shown that PFS always achieves higher overall system throughput as compared to RRS and achieves very close performance to MSS. The individual pair performance against SNR is shown in Figure 15.9. It is shown that in MSS, one of the pairs always gets much higher rate as compared to others. While in PFS proportional

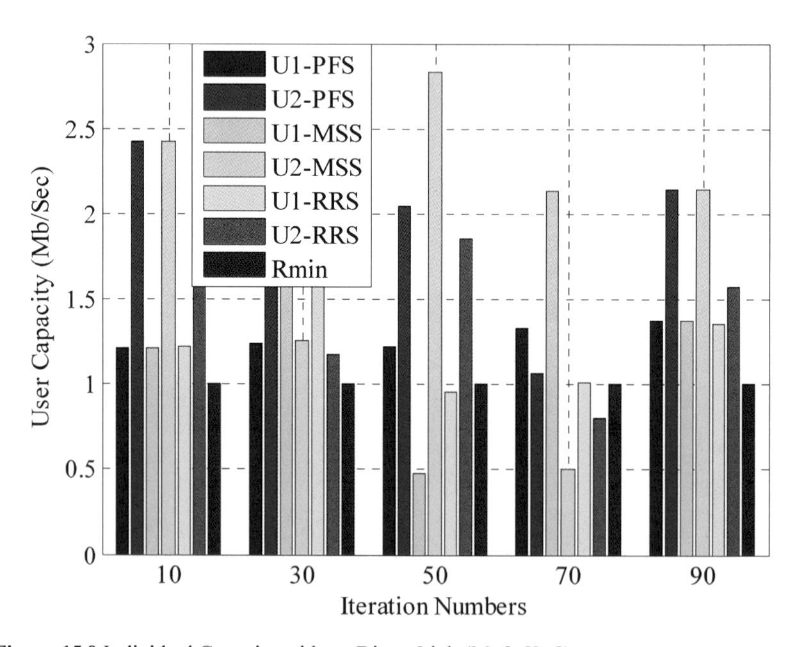

Figure 15.9 Individual Capacity without Direct Link (M=2, K=8).

fair rate is ensured for both pairs. A good trade-off between overall rate and individual rate is observed in PFS. The fairness between all MTs is achieved in PFS, while MSS and RSS are unable to provide fairness in terms of R_{min}. The good trade-off between throughput and fairness is observed in the PFS allocation scheme.

Example 15.3: In this example we discuss the impact of power optimization on the performance of AF protocol and DF protocol.

After the allocation of all RBs as explained in the example 15.2, the discrete variable in the form of RB allocation has been fixed and the power is allocated to each entity by optimizing the objective function:

$$\max \ R_S^{AF-OWR} \tag{15.31}$$

Subject to:

$$(P_b^k + P_r^k) \leq P_t^k \tag{15.31a}$$

$$(P_b^k , P_r^k) > 0 \tag{15.31b}$$

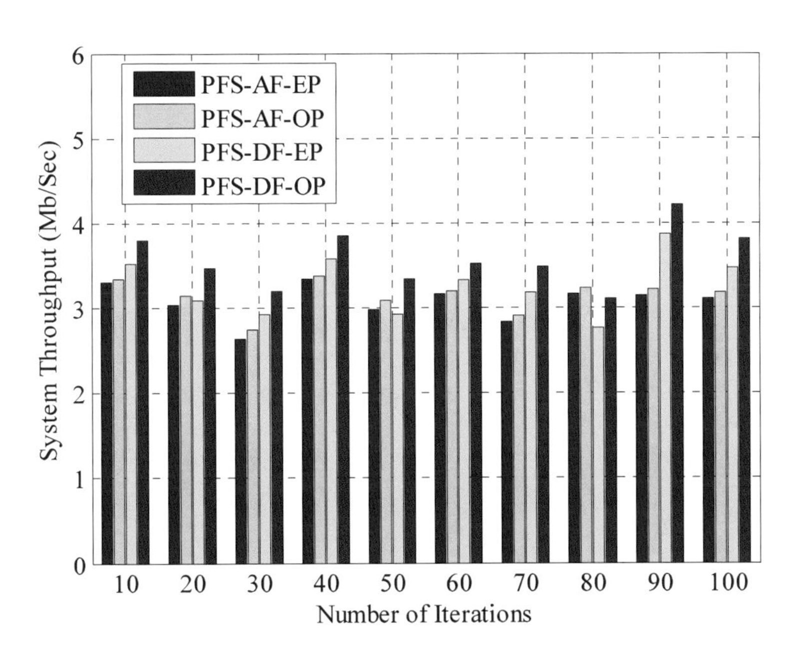

Figure15.10 Comparison of PFS-EP and PFS-OP for AF and DF Systems (M=2, K=8).

Now in this problem, only one variable in the form of power is present. This problem can be solved as simple convex optimization problem. To get optimized values of power for source and relay we apply MATLAB optimization toolbox command "*fmincon*" which is designed to find the minimum of a given constrained nonlinear multi-variable function.

Figure 15.10 presents the comparison of PFS scheme with equal power allocation (PFS-EP) and PFS scheme with optimized power allocation (PFS-OP) for both AF and DF systems. We observe that the PFS-OP achieves a significant throughput performance gain over the PFS-EP.

15.6 CONCLUSIONS

Cooperative communication with multi-carrier system is a promising design technology for next generation wireless networks. This chapter has provided an introduction of cooperative communication systems. Brief descriptions of common relaying techniques and basic protocols have also been discussed. In addition, in this chapter, we have studied the resource allocation problem in multi-carrier cooperative networks. Three common resource block allocation schemes are discussed and presented.

References

[1] K. Loa, et al., "IMT-advanced relay standards [WiMAX/LTE Update]," IEEE Communications Magazine, vol. 48, pp. 40-48, 2010.

[2] K. J. R. Liu, Cooperative Communications and Networking: Cambridge University Press: Cambridge University Press, 2009.

[3] E. C. V. D. Meulen, "Three-Terminal Communication Channels," Advances in Applied Probability, vol. 3, pp. 120-154, 1971.

[4] T. Cover and A. E. Gamal, "Capacity theorems for the relay channel," IEEE Transactions on Information Theory, vol. 25, pp. 572-584, 1979.

[5] M. Abrar, et al., "Cooperative diversity versus antenna diversity in wireless communication systems," in 4th International Conference on New Trends in Information Science and Service Science (NISS) 2010, pp. 260-263.

[6] A. Sendonaris, et al., "User cooperation diversity. Part I. System description," IEEE Transactions on Communications, vol. 51, pp. 1927-1938, 2003.

[7] J. N. Laneman, et al., "Cooperative diversity in wireless networks: Efficient protocols and outage behavior," IEEE Transactions on Information Theory, vol. 50, pp. 3062-3080, 2004.

[8] R. Prasad, OFDM for wireless communications systems: Artech House, Bostan, London, 2004.

[9] J. N. Laneman, et al., "An efficient protocol for realizing cooperative diversity in wireless networks," in IEEE International Symposium on Information Theory, 2001, p. 294.
[10] G. Kramer, et al., "Cooperative Strategies and Capacity Theorems for Relay Networks," IEEE Transactions on Information Theory, vol. 51, pp. 3037-3063, 2005.
[11] T. Wang. and G. B. Giannakis, "High-Performance Cooperative Demodulation With Decode-and-Forward Relays," IEEE Transactions on Communications, vol. 55, pp. 830-830, 2007.
[12] L. Yonghui and B. Vucetic, "On the Performance of a Simple Adaptive Relaying Protocol for Wireless Relay Networks," in IEEE Vehicular Technology Conference, 2008. VTC Spring 2008, pp. 2400-2405.
[13] M. Janani, et al., "Coded cooperation in wireless communications: space-time transmission and iterative decoding," IEEE Transactions on Signal Processing, vol. 52, pp. 362-371, 2004.
[14] A. Nosratinia, et al., "Cooperative communication in wireless networks," IEEE Communications Magazine, vol. 42, pp. 74-80, 2004.
[15] M. Benjillali and L. Szczecinski, "A simple detect-and-forward scheme in fading channels," IEEE Communications Letters, vol. 13, pp. 309-311, 2009.
[16] P. Jingjing and T. See, "Rate performance of AF two-way relaying in low SNR region," IEEE Communications Letters, vol. 13, pp. 233-235, 2009.
[17] Marvin Kenneth Simon and M.-S. Alouini, Digital communication over fading channels, Second ed.: John Wiley & Sons, Inc., Honboken, New Jersey, 2005.
[18] R. Boris and W. Armin, "Spectral efficient protocols for half-duplex fading relay channels," IEEE Journal on Selected Areas in Communications, vol. 25, pp. 379-389, 2007.
[19] A. Agustin, et al., "Protocols and Resource Allocation for the Two-Way Relay Channel with Half-Duplex Terminals," in IEEE International Conference on Communications, 2009. ICC '09. 2009, pp. 1-5.
[20] J. MinChul and K. Il-Min, "Relay Selection with ANC and TDBC Protocols in Bidirectional Relay Networks," IEEE Transactions on Communications, vol. 58, pp. 3500-3511, 2010.
[21] "3rd Generation Partnership Project, Technical Specification Group RadioAccess Network; Physical Layer Aspect; 3GPP TS 36.201 version 10.0.0."
[22] Erik Dahlaman, et al., 3G Evolution: HSPA and LTE for Mobile Broadband: Academic Press, Oxford, UK, 2007.

Chapter 16.

Channel Allocation

Feng-Ming Yang and Wei-Mei Chen

Department of Electronic Engineering, National Taiwan University of Science and Technology
wmchen@mail.ntust.edu.tw

This chapter provides a general overview of Channel Allocation using OFDMA technique in Worldwide Interoperability for Microwave Access (WiMAX) system. In order to analyze the traffic class queue model, the conventional Markov model is implemented here. The chapter therefore lays the foundation for discussions to select suboptimal subcarriers and processes the operations of Modulation and Coding Scheme (MCS) based on channel quality information.

16.1 INTRODUCTION

The Channel Quality Indicator (CQI) value was evaluated at the MAC layer by using CINR reports during frame time, which lead to a flexible priority scheduling algorithm that guarantees throughput for five class traffic services. The ertPS traffic with limited bandwidth considered the piggyback policy on the packet header. The allocated bandwidth was insufficient to serve the traffic load, the MS requests an additional bandwidth by piggybacking its amount. By adjusting the transmission rate, the bandwidth allocator in this study was efficiently regulated and fluctuated markedly, causing the inaccessible bandwidth with ertPS traffic.

The dynamic manner in which the packets are changed to match the channel conditions helps to increase the number of streams on a frame-by-frame basis. On Channel Quality Information CHannel (CQICH), the channel quality information is transmitted over Adaptive Modulation and Coding (AMC) architecture. Based on the bandwidth allocation and request mechanism with piggyback policy, BW-

4G Wireless Communication Networks: Design, Planning and Applications, 319-336,

REQs use extended piggyback request fields in the IEEE 802.16 standard. The communication between MSs and the BS occurs through two directions, that is, uplink and downlink. As packets travel through a BS, a number of packet losses occur for various traffic types and suffering the inaccessible bandwidth after bandwidth allocation. Let T be the transmission stream type of packet. Let μ (in the unit of frame time) be the bandwidth granted to all of users when they are served, which is the parameter to fulfill the satisfaction of the bandwidth requirement of application.

16.2 SUBCARRIER PERMUTATION

WiMAX provides higher transmission rates and mobility when mobile users change dynamically. The system channel quality varies according to the Carrier to Interference and Noise Ratio (CINR). Subcarrier permutation is responsible for reducing the average interference and providing a subcarrier allocation mechanism. The distributed subcarrier permutation can be divided into the Partial Usage of Sub-Shannels (PUSC) model and Full Usage of Sub-Channels (FUSC) in an IEEE 802.16e OFDMA system. The cluster-based PUSC model randomly selects subcarriers and assembles them into a sub-channel. The design goal of distributed permutations is to reduce interference among users. When subcarrier permutation is distributed pseudo-randomly, each of these users obtains a sub-channel with the same overall quality. However, differential service flows are not immediately guaranteed to users. Subcarrier permutation should access subcarriers in the order determined by the users' service flow.

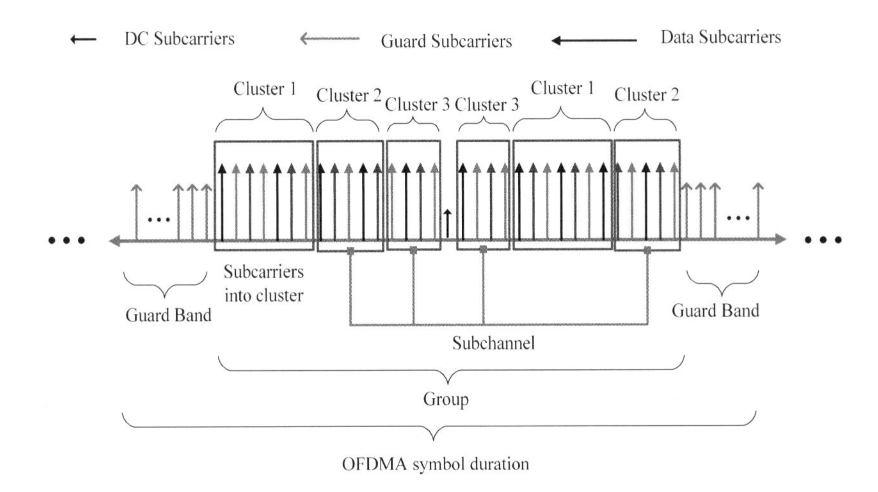

Figure 16.1 Subcarriers permutations with PUSC mode.

The PUSC model is a type of subcarrier permutation defined in IEEE 802.16e OFDMA PHY [1]. The PUSC type of subcarriers permutations is used for both the downlink and uplink sub-frame in terms of OFDM symbol duration. The subcarriers are distributed pseudo-randomly per sub-channel, and multiple users access multiple subcarriers. The set of used subcarriers is partitioned into logical clusters, and the group gathers there clusters. The term N_S is the number clusters within OFDMA symbol duration, including OFDMA symbol time and a symbol guard time. Figure 16.1 shows the PUSC model and illustrates how subcarriers are successfully assembled into the sub-channel. Let the serial number of subcarriers and sub-channels be x and y, respectively. Denote the physical location of subcarriers within the sub-channel as $SCH(x,y)$, which can be obtained by

$$SCH(x,y) = N_G \times N_{SCH}(x,y) + \left(p\left[N_{SCH}(x,y) \bmod N_G\right] + Base\right) \bmod N_G \tag{16.1}$$

where the basic permutation sequence is denoted as $p[w]$ ($0 \leqq w \leqq N_S$). The term N_G is the number of sub-channels in the group. Base is a random variable, which can be obtained from DL-MAP.

$$N_{SCH}(x,y) = \left(x + N_{CLU} \times y\right) \bmod N_{SCA} \tag{16.2}$$

N_{SCA} is the number of subcarriers of sub-channel. N_{CLU} is the number of subcarriers in each cluster. Upon determining the physical location of subcarriers, the PUSC model estimates the adjacent subcarrier permutation for uplink and downlink.

16.3CHANNEL QUALITY INFORMATION CHANNEL

The Channel Quality Indicator (CQI) value was evaluated at the MAC layer by using CINR reports during frame time, which lead to a flexible priority scheduling algorithm that guarantees throughput for five class traffic services.

In the TDD mode, a deterministic signal is transmitted from one MS to the BS. Each MS delivers the deterministic signal to the BS for estimating channel sounding. The BS computes the MS of the uplink channel response to use the downlink closed loop data transmission in broadband mobile TDD channels [2]. In the frame structure, the MS determines the duration time of the channel sounding zone according to uplink map (UL-MAP) information. The sounding signal is transmitted for uplink in the channel sounding zone. The sounding channel is only defined for the uplink quality of signal, not for the received signal quality downlink [3]. The IEEE 802.16e standard proposes that the Downlink Carrier to Interference and Noise Ratio (DL-CINR) report its operation, which enables MS to determine the channel states. The BS obtains the DL-CINR channel report according to the REP-RSP message on Channel Quality Information CHannel (CQICH). The REP-RSP message is sent by MS in response to the REP-REQ message from BS to report the DL channel status.

The CQICH allocation request header is sent by MS to request the allocation of CQICH. The PDU of the CQICH allocation contains the CQICH allocation request header without data payload. The BS and the MS identify the capabilities of the modulation and coding of both entities and the BS allocates CQICH for MSs by using a CQICH control information element (CQICH Control IE) for periodic CINR reports. The MS sends a REP-RSP message, keeping the Modulation and Coding Scheme (MCS) level fashion, to the BS by triggering a band AMC operation. When the BS is ready to trigger the transition to Band AMC mode or to update the CINR reports, it sends a REP-REQ message to the MS. After the MS receives the REP-REQ message, it replies with a REP-RSP message. For the report on REP-RSP message, the MS reports the CINR value to the BS for the various frequency bands with MCS configuration [4]. The reported effective CINR feedback is consistent with the MCS level. The CQI may be obtained by the optimal CINR report to meet the specified target error rate before the next scheduled CINR report. The REP-REQ message is used to specify the new effective CINR over the CQICH. Figure 16.2 illustrates the messages of the CINR feedback for construction of a connection between one MS and one BS. The single-bit header type (HT) field distinguishes the generic MAC header formats. The encryption control (EC) field specifies whether the payload is encrypted. The type and feedback type fields are used to recognize MAC header types and feedback types, respectively. The BS utilizes handover to improve link quality on the fast base station switching (FBSS) mechanism. The anchor BS may change the data flow to another BS depending on the BS selection scheme [5]. The FBSS indicator (FBSSI) predefines the request CQICH of MS during FBSS handover. The BS verifies the Preferred-Period field, which determines the CQICH allocation period of MS within the frame duration. The CID-MSB and CID-LSB messages indicate the MS basic CID for data transmission using the most significant bit (MSB) and least significant bit (LSB) addresses, respectively. The header check sequence (HCS) block records a Cyclic Redundancy Check (CRC) of the MAC header to verify an error regarding a combination of the fixed length physical layer header.

After the CQICH Control IE with index, the CINR report is received; the BS multicasts the REP-REQ messages, and subsequently allocates uplink bandwidth to all MSs of the cell. When MS switches to the new anchor BS, CQICH is not pre-allocated to the MS prior to the anchor BS, and the MS monitors the MAP from the new anchor BS and waits for the CQICH allocation request [6][7]. The MS does not receive a REP-REQ message within the switching period, and requests the new anchor BS to allocate a CQICH channel by transmitting a CQICH allocation request header. If the new anchor BS receives a CQICH allocation request, the BS allocates a CQICH for the MS. The BS enables measurement of the uplink channel quality and determines the channel response. The REP-RSP message reports the measurement of channel quality between the MS and BS after receiving the second REP-REQ message from the MS. The CINR value helps the MS to determine the channel state during channel estimation. Subsequently, the MS determines the connection and BS selects a suitable AMC according to the REP-REQ message. This study assumes that the BS has perfect knowledge of the channel quality information of each user based on the CQICH. The principle is to select the highest channel quality connection that uses fewer slots. The term SBS and ABS are the type of serving BS and anchor BS connection with MS, respectively. Let A be the set of all anchor BS on a cell. Denote T_b as the decision type for serving BS or anchor BS connection, which can be defined as follows:

$$T_b(L,B)=\begin{cases} SBS, & if\left\lceil \dfrac{L}{B_{SBS-MS}} \right\rceil \geq \left\lceil \dfrac{L}{B_{ABS_j-MS}} \right\rceil \\ ABS, & if\left\lceil \dfrac{L}{B_{SBS-MS}} \right\rceil < \left\lceil \dfrac{L}{B_{ABS_j-MS}} \right\rceil \end{cases} \quad \forall j \in A, \tag{16.3}$$

where L is the average packet size, $B_{X\text{-}Y}$ is the transmission block size of the selected AMC between node X and Y, ABS* is the selected BS node and is defined as

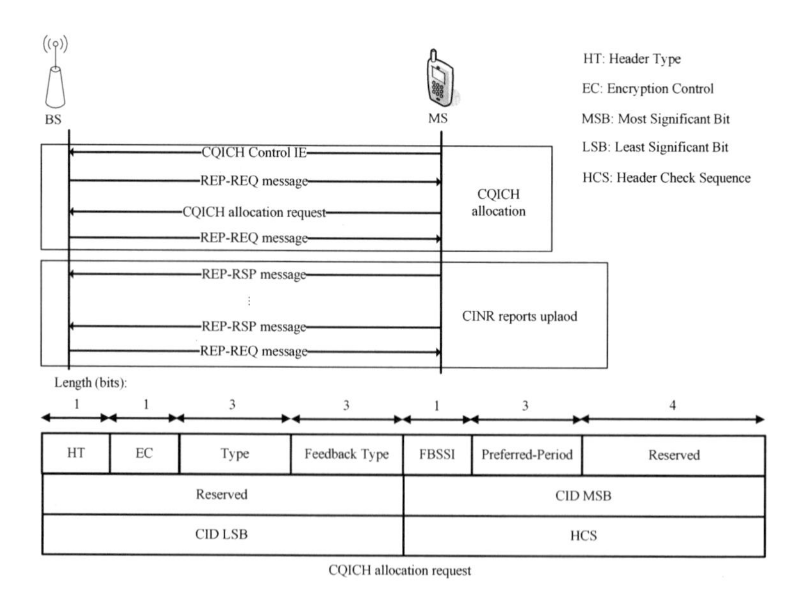

Figure 16.2 Construction of a connection between one MS and one BS.

$$ABS^{*} = \arg\min_{j}\left\{\left\lceil \frac{L}{B_{ABS_j - MS}} \right\rceil\right\} \tag{16.4}$$

16.3.1 Bandwidth Allocation and Request Mechanism

The methods that are suggested in the WiMAX standard to send their BW-REQs are divided into two categories, that is, request-based and polling-based. The request-based bandwidth request model dynamically changes the uplink burst profile. All requests for bandwidth must carry the MAC header and payload. The BW-REQ is transmitted during any uplink allocation, except during any initial ranging interval. The MSs send BW-REQs during contention periods in the UL frame. Due to the possibility of collisions, the contention-based bandwidth requests are aggregated by the BS buffer. The design goal of the request protocol is for MSs to periodically aggregate bandwidth requests. The polling-based bandwidth request is focused on how to effectively allocate bandwidth to various traffic classes. The remaining bandwidth, if any, is allocated to the lower traffic classes that still have traffic in the BS buffer. This chapter also allows a polling-based bandwidth request on data frame transmissions for traffic class. The bandwidth request opportunity size model is a type of subcarrier permutation defined in IEEE 802.16e OFDMA PHY. The subcarriers are distributed pseudo-randomly per sub-channel, and multiple users access multiple subcarriers. The set of used subcarriers is partitioned into logical clusters, and the group gathers their clusters [8].

The coding rate of ertPS traffic is adaptive according to BW-REQ. If the transmission rate must be changed, explicit BW requests are sent by piggybacked BW-REQ headers. Otherwise, the coding rate equals the rate that was used in the last BW request. Figure 16.3 illustrates the BW request mechanism of ertPS when the rate changes and the BW granting scheme [9]. Although the bandwidth allocation of ertPS is similar to that of UGS, some allocated bandwidth is inaccessible because the coding rate changes frame by frame. The inaccessible bandwidth (H^{ertPS}) is expressed as follows:

$$H^{ertPS} = \int_0^{T_f} h^{ertPS}(t)dt = \int_0^{T_f} \sum_{i=1}^{N_c} t\times\left(R_i/E_i - R_{i+1}/E_{i+1}\right)dt \tag{16.5}$$

where N_c denotes the total number of coding rate variations. R_i is the ith coding rate, and E_i is the number of packets in R_i coding rate. T_f is the duration of the frame. Piggybacking, as defined in the IEEE 802.16 standard, is a method that is used by MSs to transmit BW-REQs, and is optional and not able to transmit all types of bandwidth requests.

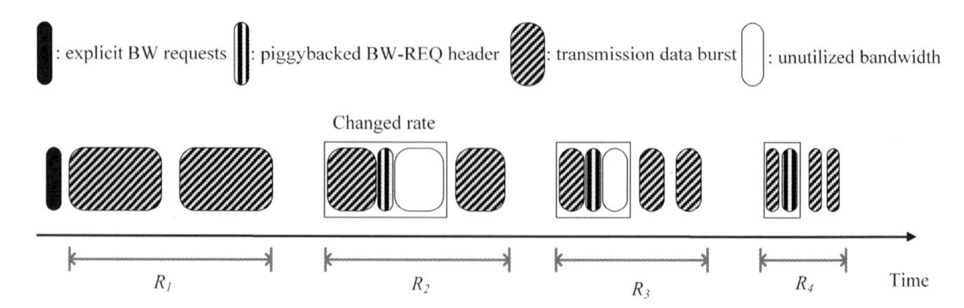

Figure 16.3 Example of bandwidth request mechanism in ertPS.

16.3.2 Channel Sounding

The channel sounding technique is recommended for a mobile environment in which the distance between the BS and MS is dynamic. Several researchers have proposed channel sounding systems and channel sounders, including time-domain [10-11] and frequency-domain [12] channel sounding. The sounder can optimize the design of multipath propagation modes. However, with its limited transmitted power and complex hardware overhead constraints, the sounder cannot achieve operation efficiency by reducing hardware overhead.

Although the Channel Quality Information CHannel (CQICH) contains the most information in the field of feedback mechanism, the effect of subcarriers remains unclear [13]. The IEEE 802.16e standard proposes that the channel sounding mechanism can know that channel states of sub-channels. In the Time Division Duplex (TDD) mode, a deterministic signal is transmitted from one MS to the BS. In the TDD mode, each MS must deliver the deterministic signal to the BS for estimating channel sounding. In the frame structure, the MS knows the duration time of channel sounding zone according to UL_MAP information. The sounding signal is transmitted for uplink in the channel sounding zone. The BS can then sound all of the assignation subcarriers from MSs. To support a mobile environment, the BS can also notify the MS of an uplink channel through periodic channel sounding. The BS can separate the multiple sounding waveforms, and each MS can also occupy all subcarriers during the sounding allocation. The BS computes the MS of UL channel response to use the DL closed loop data transmission in broadband mobile TDD channels [14].

Figure 16.4 shows that management messages in TDD mode consist of the preamble, Frame Control Header (FCH), DL -MAP, UL-MAP, ranging sub-channel, and channel sounding zone. The preamble symbol uses the BS_ID and segment value to identify the subcarrier permutation. The FCH predefines data patterns, each with its own modulation type. The BS can encode the FCH message, which must be decoded by the MS. The DL-MAP and UL-MAP messages indicate the allocations for data transmission in the downlink and uplink directions, respectively. The ranging sub-channel block records the indexes of the sub-channels specified in the UL-MAP message to identify the

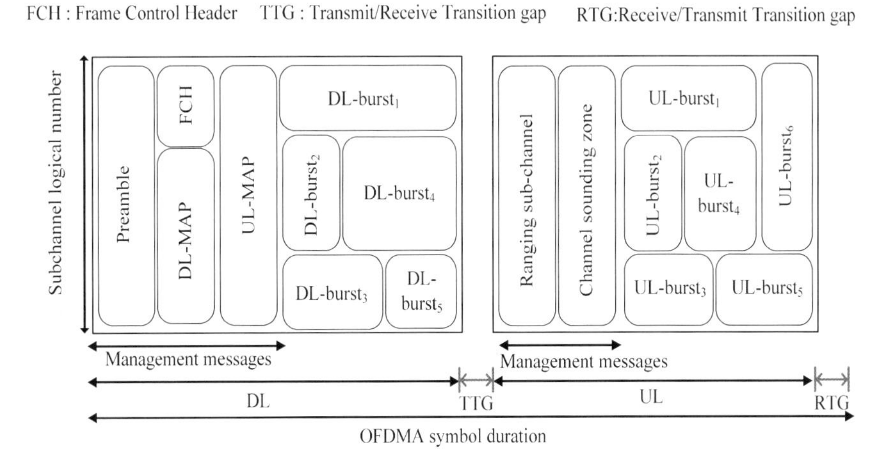

Figure 16.4 Channel sounding zone in TDD mode structure.

ranging codes used for modulation. The channel sounding zone is a burst in one or more OFDMA symbol times at the UL sub-frame. The channel sounding signals are transmitted from the MS to BS through the channel sounding mechanism. The BS makes it possible to measure uplink channel quality and determine the channel response between the BS and MS. This chapter assumes that the BS has perfect knowledge of each users' channel quality information based on the sounding signals. If the uplink and downlink channel are synchronous, the BS estimates the uplink channel state for each MS according to its channel sounding signal. The CINR value helps the MS determine the channel state during channel estimation.

16.4 CROSS-LAYER SUBCARRIER PERMUTATION MECHANISM

The channel state part of one frame is C_j/S_i. Denote the stream type transmission traffic application as TR. Accounting for the packet delay time, the following equation can schedule the stream type as

$$TR = \arg\max_i \left\{ \frac{C_j \times \mu(i)}{S_i \times T_i} \right\} \tag{16.6}$$

By solving the above equality, we can easily obtain the stream type transmission traffic application for the user. The delay time can be minimized with the optimal values, which can be obtained by S_i and C_j parameters. However, the solution of the optimization equation (16.6) is not explicit because S_i and C_j are different layer in MAC and physical layer, respectively. Moreover, the length of frame is not equal to that of packet. The parameters of the MAC and physical layer are non-synchronous. Thus, it is necessary to determine the relationship between the MAC and physical layer and give a cross-layer suboptimal subcarrier allocation solution.

Each frame may have different sub-frames, and the stream types of these sub-frames may have different processing priorities. However, the cross-layer suboptimal subcarrier allocation scheme can divide one frame into sub-frame in the several phases. This study assumes that each stream type i ($i \in \{1,2,3,4\}$) has its own sub-frame. These streams arrive at the MAC layer with a Poisson distribution at a certain rate. The processing time for each stream type within the frame requires k_s phase of service. The mechanism of frame allocation process then becomes a queue scheduling problem for $M/E_K/1$ with a non-preemptive system. The packets in each stream type are based on Poisson distribution at the rate of λ_i, where the total arrival rate of all stream types is λ, and $\lambda=\Sigma\lambda_i$. The phase allocation rate μ is replaced by $\mu\times K$. The expected system waiting time for the average of all packets with stream types in the frame allocation process can be obtained [15] by

$$W_s = \frac{K+1}{2K}\frac{\rho}{\mu(1-\rho)} \tag{16.7}$$

where $\rho=\lambda/\mu$. The proposed allocation process considers that the 802.16e system class agrees with the $M/M/1$ model in allocating frame process at the MAC layer. The WiMAX environment includes our stream types with different frames.

Given K, the expected system waiting time of the frame allocation process should be minimized. Figure 16.5 shows how the choice of parameter ρ affects the system waiting time with different numbers of phases. Increasing the phase of operation decreases the system waiting time. When possible, phase allocation frames are combined if the stream type packet is allocated for the sub-frame. This is because WiMAX defines different stream types. However, the IEEE 802.16e environment has a limited number of phases.

This chapter presents a power allocation scheme for subcarriers based on the channel state to realize an OFDMA system. Water filling power adaptation is the optimal power allocation for subcarrier because channel state is known in the physical layer [16]. A subcarrier without transmission is the result of interference between two subcarriers, as a subcarrier is not transmitted if the receiver cannot receive it correctly. Hence, the cross-layer suboptimal subcarrier allocation scheme adopts the power constraint for water-pouring. If the subcarrier cannot satisfy the power constraint, it should reallocate subcarriers for stream data. Each user occupies one or more sub-channels and subcarriers are grouped from sub-channel in an OFDMA system. Let the total number of users and the total number of subcarriers be N_s and U_s, respectively. The term $P(p,q)$ represents the transmission power for the qth subcarrier and the pth user.

For the distributed subcarrier permutations, the pilot subcarriers are allocated first, and the remaining subcarriers are partitioned into data subcarriers. These subcarriers are distributed pseudo-randomly to average the interference over the available bandwidth. Moreover, with previous allocation bandwidth (or transmission data rate) at the MAC layer, the subcarrier allocation of the cross-layer suboptimal subcarrier allocation scheme is more intelligent in selecting subcarriers.

Example: Given K, the expected system waiting time by frame allocation process should be minimized. It is observed that increasing the phase of operation is lower system waiting time. Combine phase allocation frame when possible if the stream type packet is allocated for sub-frame because the WiMAX defines different stream types. However, the number of phases the IEEE 802.16e environment has is limited. How to choice of parameter ρ affects the system waiting time with different number of phases?

Solution: IEEE 802.16e defines different QoS class types. Multi-phase allocation frame has lower system waiting time if the stream type packet is allocated for sub-frame. It is also observed that increasing the parameter ρ is higher system waiting time.

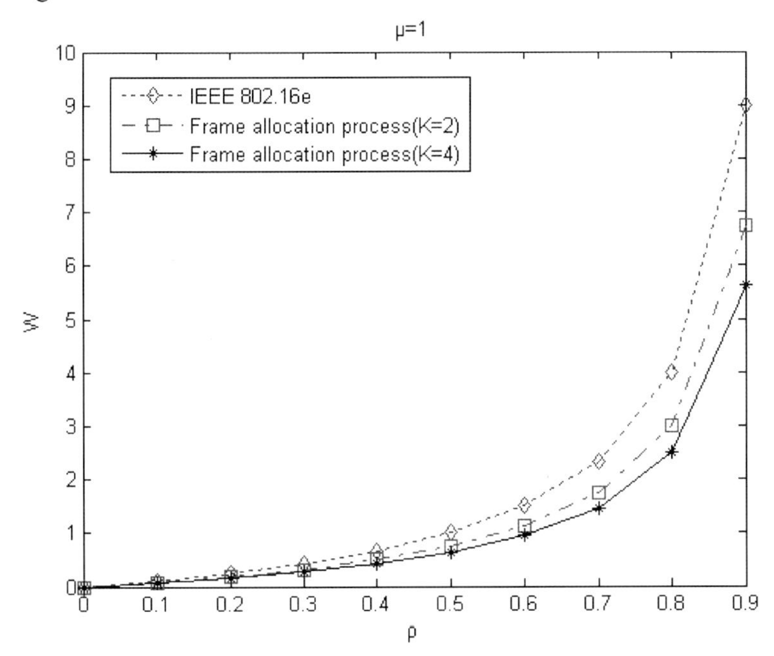

Figure 16.5 The system waiting time as a function of ρ for different values of K.

16.5QUEUEING MODEL FOR TRAFFIC CLASS

The following queuing model principles were considered to conform to QoS requirements and to design an efficient service architecture based on delayed latency of real time services and channel state. From the CQICH allocation method, the channel quality information is relayed back to the CQICH reports through the AMC channel controller within BS. The AMC channel selector also determines the Modulation and Coding Scheme (MCS) to maximize user throughput. The IEEE 802.16 medium access control (MAC) protocol defines several bandwidth request allocation mechanisms and five types of traffic characteristics based on priority, including unsolicited grant service (UGS), extended real-time polling service (ertPS), real-time polling service (rtPS), non-real-time polling service (nrtPS), and best effort (BE) [16]. Figure 16.3 illustrates the schematic diagram of the system model. The BS scheduler determines the uplink or downlink packets scheduling. Traffic of the same type from a wired and wireless network is aggregated into a traffic class queue following a single Poisson arrival process in the BS buffer. All type of packets are classified and aggregated according to service flows. The UGS traffic is assigned a fixed departure rate to meet its stringent delay requirements. The ertPS traffic changes the allocated bandwidth dynamically, depending on the traffic characteristics. The rtPS traffic type guarantees a specified delay constraint. The nrtPS type is the delay-tolerant service, and the BE data stream does not have any specific QoS requirements. The system assigns initial bandwidth for various types of traffic as λ^{UGS}, λ^{ertPS}, λ^{rtPS}, λ^{nrtPS} and λ^{BE}, based on the requested bandwidth of UGS and ertPS, the required minimum bandwidth of rtPS and nrtPS, and the queue length of BE traffic, respectively. The allocated bandwidth of design parameter for the various traffic classes are μ^{UGS}, μ^{ertPS}, μ^{rtPS}, μ^{nrtPS} and μ^{BE}. The allocated bandwidth $\mu^{UGS} + \mu^{ertPS} + \mu^{rtPS} + \mu^{nrtPS} + \mu^{BE} \leq \mu$, where μ is the upper bound available bandwidth within the BS. The average packet allocated bandwidth by five traffic class is less than the available bandwidth within the BS, which implies that the theoretical queueing model can serve as an upper bound to all of traffic.

Model Assumptions: The assumptions of the availability channel model are provided as follows:

1. The arrival of the packets to the BS buffer follows the Poisson process with the arrival rate λ, and the service time for each user to complete a packet is exponentially distributed with parameter μ.
2. The state n represents the number of packets within the frame duration.
3. Assume that N homogeneous users are running in parallel channels during a single WiMAX frame time, and each channel has a fixed coding rate r_k $(1 \leqq k \leqq N)$.
4. Let r_{max} and r_{min} be the MCS of the maximum coding rate and the minimum coding rate, respectively. (where $r_{max} = r1 > r2 > r3 > \ldots > rN = rmin$.)

5. Denote the probability mass function for K to obtain the value of k by $g(k) = \Pr(K = k)$, where $k = 1, 2, \ldots, N$.
6. The service rate is μ_k, which is exponential to the coding rate r_k and $g(k) = \mu_k / \mu$.

For the first, second and third assumptions, the packets of the BS buffer are assumed to follow a Poisson process, which may be explained as either in the operational phase or in a steady state. Their assumptions of independent channels in the BS are an efficient approximation to reality because the packets that are served by the various channels are uncorrelated. The service time, in accordance with the exponential distribution, has also been widely accepted. A frame is generally allowed to contain packets for various users rather than for a single user. Assumption 4 for channel quality information and queue state information is suggested to achieve high throughput while balancing the fairness among multiple users. The packet departure time with a higher coding rate is ahead of those with a lower coding rate channel. Hence, a higher coding rate channel can serve a lower number of waiting packets in a process time. The fifth and sixth assumptions of the actual number of packets in any departure module is also a random variable K, which can obtain any positive integral value less than N with probability $g(k)$. If μ_k is the departure rate of batches of size K, then $g(k) = \mu_k/\mu$, where μ is the composite departure rate of all batches and is $\sum_{i=1}^{N} \mu_i$.

The conventional Markov model was implemented to analyze the traffic class queue model. The Markov model was constructed for the packets queue. The process of the packets queue assumes that N channel servers are working. The state n represents the number of packets in the BS buffer queue. According to the assumption, the packet completing rate is μ and the probability distribution is $g(k)$ for the number of departure packets in the channel of coding rate r_k. Subsequently, the transition completing rate from state n to state n-k is $\mu g(k)$. Denote P_n as the steady probability for the system remaining at state n. It is easy to derive P_n by solving the following Chapman–Kolmogorov:

$$\begin{cases} (\lambda+\mu)P_n = \lambda P_{n-1} + \mu \sum_{i=1}^{N} P_{n+i} g(i), & if\ n \geq 1 \\ \lambda P_0 = \mu P_1 & , if\ n = 0 \end{cases} \tag{16.8}$$

For each state n, there is also (16.8), which can be rewritten in operator notation as

$$\left[\mu \left(\sum_{i=1}^{N} g(i) D^{i+1} \right) - (\lambda+\mu)D + \lambda \right] P_n = 0, \ where\ n \geq 0, D > 0 \tag{16.9}$$

We determined the characteristic equation $f(r)$ by using the boundary condition ΣP_n=1, then

$$f(r) = \mu\left(\sum_{i=1}^{N_q} g(i) r^{i+1}\right) - (\lambda+\mu)r + \lambda = 0 \tag{16.10}$$

Thus, the optimization solving for the designing problem can be obtained by

$$r_s = \min\{r \mid 0 < r < 1, \textit{such that } f(r) = 0\} \tag{16.11}$$

The mean queuing length in the system (that is, L_q), which should be smaller than the BS buffer length, can be calculated from $L_q = r_s^2/(1 - r_s)$. Let W_q be the mean of queuing time of the system, which can be obtained by Little's formula, expressed as

$$W_q = \frac{r_s}{\lambda(1-r_s)} - \frac{1}{\mu}. \tag{16.12}$$

Specifically, (16.12) computes the delay time of the queue by considering the packet arrivals and the packet departures. The BS has notable flexibility in controlling the downlink and uplink by the bandwidth allocator. Let T_r and T_d be the size of the request slots and data slots (both in frames), respectively. We defined bandwidth efficiency as the ratio of the average time utilized by the MSs for data transmissions in a frame to the total time allocated for N_r request slots and N_d data slots in a frame. The probability of receiving request messages in a frame is a binomial distribution. Let N_s denote the average number of MSs that receive bandwidth grant from the BS, which can be computed from:

$$N_s = \sum_{j=0}^{N_r} \min(j, N_d) \binom{N_r}{j} \tau^j (1-\tau)^{N_r - j},\ j \in [0, N_r] \tag{16.13}$$

where τ is the probability that a MS transmits a data burst in a frame. The ratio of the average number of slots that are used to transmit data to the total number of slots allocated to transmission bursts for the requesting mechanism in a frame is the bandwidth efficiency (η), which is obtained as:

$$\eta = \frac{N_s \times T_d}{N_r \times T_r \times N_d \times T_d} \tag{16.14}$$

Similarly, we defined throughput of a BS (denoted by θ) as the average number of bits transmitted from MSs to the BS in one cell. The throughput of a BS is markedly dependent on the physical channel coding rate profile. Subsequently, the throughput θ of a channel can be computed by

$$\theta = \frac{\left(\lambda^{UGS} + \lambda^{ertPS} + \lambda^{rtPS} + \lambda^{nrtPS} + \lambda^{BE}\right)}{\sum_{i=1}^{N} g(i)\mu_i} \times N \times \eta \quad (16.15)$$

Example: Suppose that a WiMAX system must build a channel model. For numerical example, the total number of channels during a single WiMAX frame time is N=4. The expected packet completing rate of the BS is μ=10. To study the effect of traffic load on delay time, we defined traffic load (ρ) as follows

$$\rho = \frac{\left(\lambda^{UGS} + \lambda^{ertPS} + \lambda^{rtPS} + \lambda^{nrtPS} + \lambda^{BE}\right)}{\mu} \quad (16.16)$$

Every frame is assumed to be able to transmit and receive four distinct channel coding rates. As illustrated in Figure 16.6, the service packet ratios among the four channel coding rates are 1:2:3:4, 1:1:1:1 and 4:3:2:1. Assuming that the expected deadline for most packets access delay is d, we set a constraint in that the expected waiting time T_w for the last request in the one full channel is no longer than the expected deadline, that is,

$$T_w = \frac{N_q}{N \times \mu} \leq d \quad (16.17)$$

Subsequently, the limitation of the queue length N_q is satisfactory. The longer the limitation of the queue length, the value of N_q can be dynamically set as

$$N_q = \lceil d \times N \times \mu \rceil \quad (16.18)$$

Although we acknowledge the dynamic relationship of (16.18) in this example, other relationships can also be implemented according to the requirements of real conditions. Please demonstrate the effect of the choice of traffic load ρ on the system waiting time with various numbers of channel coding rates.

Solution: Given $g(k)$, the expected system waiting time of the frame allocation process is minimized. Increasing the number of high channel coding rate operation

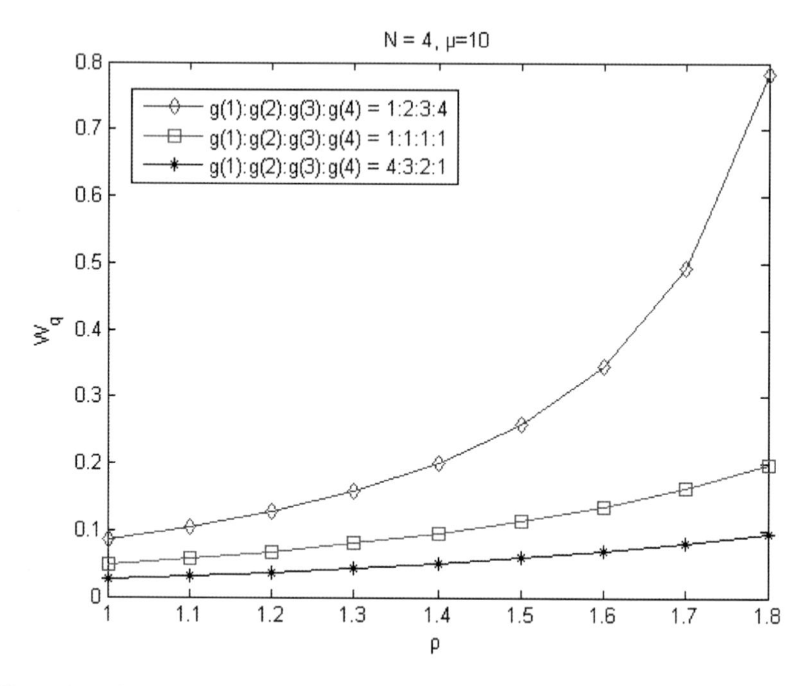

Figure 16.6 The system waiting time as a function of ρ in different values of g(k).

decreases the system waiting time. When possible, various channel coding rates allocation stream types are combined if the packets are allocated for the frame. This is the main reason that the various locations of users have various channel coding rates. However, the high channel coding rate accesses a large number of slots, causing starvation for the lower channel coding rate applications in the IEEE 802.16e environment.

16.6 CONCLUSIONS

This chapter has provided a general overview of WiMAX systems including the channel allocation models for a Point to Multipoint (P2MP) architecture which are analyzed. In the IEEE 802.16e OFDMA system, the efficient method of bandwidth allocator to guarantee QoS performance is by reserving the bandwidth of the BS from the received BW-REQ. A number of scheduling algorithms consider the requirements from all users and the available channel resources. The disadvantage of IEEE802.16e standard is that it cannot fulfill the actual QoS bandwidth requirements of several service flows that are requested by mobile users. By PUSC model in IEEE 802.16e OFDMA system, the popular way to constitute a sub-channel is picking up the subcarriers from the whole transmission spectrum using pseudo-random method. The disadvantage of this method is that it cannot fulfill the special requirements of several service flows requested by mobile users. Further, channel conditions vary as the MSs moving. The efficiencies of channel allocation can also be defined differently as shown in a later chapter of the book.

16.7 PROBLEMS AND SOLUTIONS

1. Define the essential properties of the following types of WiMAX system:
   a. Channel Quality Indicator(CQI)
   b. Carrier to Interference and Noise Ratio (CINR)
   c. Partial Usage of Sub-Shannels (PUSC)
   d. Full Usage of Sub-Channels (FUSC)
   e. Channel sounding
   f. Adaptive Coding and Modulation (ACM)
2. Give two reasons why WiMAX system might use OFDM/OFDMA instead of single carrier being modulate.
3. If packets arrive according to a Poisson distribution with a mean of 4/ms, the mean service time is 0.16ms (with a variance of 25 ms^2). The cross-layer suboptimal subcarrier allocation scheme can divide one frame into sub-frame in the several phases. The BS wish to know, how many packets are waiting for transmission.
4. Consider the queuing model for traffic class. If the packets of the BS buffer are assumed to follow a bulk input without MCS, would any changes be needed to the balance equation? If so, what?
5. IEEE 802.16e supports five service classes. Which service class is the best choice for sending uncompressed video?

References

[1] Stiakogiannakis IN, Kaklamani DI. IEEE802.16e — WiMAX: Performance analysis of Partial and Full Usage of Sub-channels under fractional frequency reuse. Proceedings of Wireless Technology Conference (EuWIT 2009), 28-29 September 2009; 41–44.

[2] Ghosh A, Wolter DR, Andrews JG, Chen R. Broadband Wireless Access with WiMAX/802.16: Current Performance Benchmarks and Future Potential. IEEE Communications Magazine 2005; 43(2):129–136.

[3] Fallah YP, Agharebparast F, Minhas M, Alnuweiri H, Leung VCM. Analytical modeling of contention-based bandwidth request mechanism in IEEE 802.16 wireless networks. IEEE Transactions on Vehicular Technology 2008; 57(5): 3094–3107.

[4] He J, Guild K, Yang K, Chen HH. Modeling contention based bandwidth request scheme for IEEE 802.16 networks. IEEE Communications Letters 2007; 11(8): 689–700

[5] Cicconetti C, Lenzini L, Mingozzi E, Eklund C. Quality of Service Support in IEEE 802.16 Networks. IEEE/ACM Transactions on Networking 2006; 20(2):50–55.

[6] Qiang N, Vinel A, Yang X, Turlikov A, Tao J. "Investigation of Bandwidth Request Mechanisms under Point-to-Multipoint Mode of WiMAX Networks. IEEE Communications Magazine 2007; 45(5): 132-138.

[7] Ma M, Lu J, Fu CP. Hierarchical scheduling framework for QoS service in WiMAX point-to-multi-point networks. IET Communications 2010; 4(9): 1073-1082.

[8] Stiakogiannakis IN, Kaklamani DI. IEEE802.16e — WiMAX: Performance analysis of Partial and Full Usage of Sub-channels under fractional frequency reuse. Proceedings of Wireless Technology Conference (EuWIT 2009), 28-29 September 2009; 41–44.

[9] Sabry A, El-Badawy H, Shehata K, Ali A. A Novel Resource Allocation Technique for VBR Video Traffic in the Uplink over WiMAX Networks. International Conference on Information and Multimedia Technology, 16-18 December 2009; 442–448.

[10] Zwick T, Beukema TJ, Nam H. Wideband channel sounder with measurements and model for the 60 GHz indoor radio channel. IEEE Transactions on Vehicular Technology 2005; 54(4):1266–1277.

[11] Kivinen J. 60-GHz wideband radio channel sounder. IEEE Transactions on Instrumentation and Measurement 2007; 56(5):1831–1838.

[12] Siamarou AG, Al-Nuaimi M. A Wideband Frequency-Domain Channel-Sounding System and Delay-Spread Measurements at the License-Free 57- to 64-GHz Band. IEEE Transactions on Instrumentation and Measurement 2010; 59(3):519–526.

[13] Choi JM, Lee H, Chung HK, Lee JH. Sounding Method for Proportional Fair Scheduling in OFDMA/FDD Uplink. IEEE 65rd Vehicular Technology Conference, 22–25 April 2007; 2732–2735.

[14] Vook FW, Zhuang X, Baum KL, Thomas TA, Cudak MC. Signaling Methodologies to Support Closed-Loop Transmit Processing in TDD-OFDMA. IEEE C802.16e-04/103r2, July 2004.

[15] Gross D, Harris CM. *Fundamentals of Queueing Theory* (3nd edn). Wiley: New York, USA, 1998:116–130.

[16] Cicconetti C, Lenzini L, Mingozzi E, Eklund C. Quality of Service Support in IEEE 802.16 Networks. IEEE/ACM Transactions on Networking 2006; 20(2):50–55.

Chapter 17.

Wireless Systems Operating Underground

Ryszard J. Zielinski, Michal Kowal, Slawomir Kubal and Piotr Piotrowski

Institute of Telecommunications, Teleinformatics and Acoustics, Wroclaw University of Technology, Wroclaw, Poland
ryszard.zielinski@pwr.wroc.pl

17.1 INTRODUCTION

Modern technologies used in underground mining require broadband wireless communication between miners, mining machines and fixed-line telecommunications infrastructure. These broadband wireless systems are going to be used for remote control purposes, for data transmissions between sensors and data acquisition center and for voice communications. The broadband wireless systems can be used during standard operation of the mining enterprise as well as during rescue operations. For either of these purposes it is convenient to have wireless Real Time Location Systems (RTLS) integrated with the telecommunication infrastructure. The chapter presents the current status of wireless underground communications and the latest research results pertaining to the possibility of WLAN, WiMax usage in these types of environments.

Further investigations concentrate on OFDM based systems. This type of radio interface should work properly in multipath propagation environments, which are typical for underground excavations. Research include propagation phenomena and the possibility to improve radio range.

Finally, based on the experimental results, we present hierarchical communication systems.

17.2 WIRELESS SYSTEMS OVERVIEW (STATUS 2008 – 2009)[11]

Some proposals of wireless systems suitable for underground communication and which were implemented worldwide (particularly in Poland) are presented below. All data listed in the next chapters are as they were published in manufacturers'

4G Wireless Communication Networks: Design, Planning and Applications, 337-374,

documentation. Therefore, one should be careful in evaluating range, bandwidth, throughput, capacity, services and the scale of technology implementation and costs, based solely on this data.

17.2.1 Medium frequency (MF) communication systems

This frequency range is used by telecommunication systems developed and distributed by the Conspec Company. They provide data and full duplex transmission either for fixed and/or mobile users. [4]. The main concept of the system is shown in Figure 17.1

Users are equipped with radio terminals with loop antennas, which receive and transmit signals between the terminal and electric wires running along the tunnels (which are the true transmission medium). Repeaters can be used to provide better coverage and extend the range of the system. This system is working with FM modulation in the frequency range between 280 and 520 kHz.

Terminals operate on two frequencies: F1- used for local and F2- used for long-range communications. Terminals always use local frequency F1 to receive information. For transmission directly to the recipient we use the local frequency F1, however in order to communicate with a recipient, who cannot be reached directly, we use the long-range F2 frequency. In this operation, the terminal communicates with a network of repeaters on F2 frequency, which then retransmits information to the remote terminal using the earlier mentioned F1 receiving frequency.

To carry a signal through tunnels, the system can use various conducting construction and infrastructure elements, such as phone lines, electric wires, metal parts like pipes etc. Nevertheless, such concept has some drawbacks, like signal loss via earth wire and severe disturbances from other systems and machines. Because of this, the best way to obtain a reliable system is to provide its own transmission network with dedicated cooper wires.

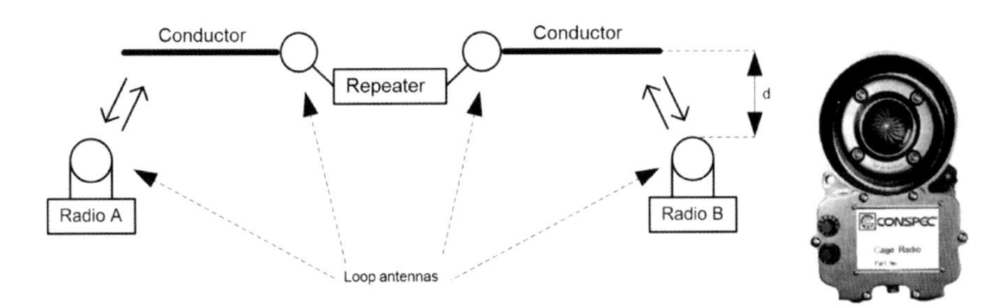

Figure 17.1 A concept of system working on medium waves.

System range strictly depends on loop antenna size and its distance from the transmission medium. Typically, for terminals 2-3 meters away from the medium, the range is about 300 meters. Taking into account an average size of a mine, this is quite close, therefore this is why repeaters constitute the core network

Systems based on medium frequencies are less prone to disturbances compared to systems working using higher frequencies [2]. They are also immune to diffraction loss at tunnel curves, with has a great influence on ultra high frequency propagation. Undoubtedly such approach is also cheaper in comparison to e.g. leaky cable solutions. On the other hand, because of the narrow band, system usage is restricted to voice and low-rate data transmissions. What is also important, it is impossible to implement services like user or equipment localization.

Despite these disadvantages, Conspec has successfully implemented such medium wave systems in countries like USA, Canada or Australia.

17.2.2 Leaky cable

The main concept of a leaky cable system is based on the use of a coaxial cable with degraded shielding parameters (usually by applying a slit in the shield). Such cable installed along a tunnel becomes a kind of a long antenna, which is used to transmit and receive radio signals. For example, systems made by Varis/Becker, using VHF, operate in frequencies ranging from 145 to 185 MHz. Downlink is allocated between 145 – 160 MHz and it is divided into 16 voice channels with a 6 MHz wide channel for data transmission [10]. Uplink starts from 170 to 185 MHz and also has 16 voice channels, however it is only 2MHz wide, as the rest of the uplink bandwidth is used for data transmission for services like CCTV.

The weakened coaxial cable shield leads to considerable growth of attenuation of propagated signal. For example, in leaky cable VHF systems, attenuation is about 4 – 5 dB/100m and for UHF systems it is 7-8 dB/100m. This is the main reason why leaky cable systems have amplifiers installed every few hundreds meters. Usually such amplifiers have automatic gain control and are able to perform auto-diagnostic procedures.

The main part of a leaky cable system is the base unit, having parts responsible for system maintenance and management. The base unit provides functions for connecting radio terminals through to the telecommunication exchange, demodulate video signals, provide system diagnostics and supply power to the amplifiers. Being the most important part of the system, the base unit is located aboveground.

Leaky cable technology offers a wide functionality for service providers [10]. Such system allows terminals to operate at distances exceeding 30 m from the cable. Terminal range can be even further extended two or three times by usage of special antennas called slope antenna, mounted on cable ends and places where range has to be extended or where there is high probability of cable damage.

Nevertheless, implementation of these systems is expensive due to high infrastructure costs (like cables, amplifiers, base unit). Another great disadvantage of this technology is poor flexibility and low immunity to damage. What is more, it is impossible to implement localization and broadband services and Line of Sight (LOS) conditions have to be assured for proper operation of the system. In spite of all this, leaky cable is still the most popular technology for communication in mining environments with usage examples all over the world.

17.2.3 TTE Systems

An interesting group of systems for underground communications uses TTE technology, which acronym is derived from the words "Through-The-Earth". These systems allow users located underground to communicate through a one or two-way link with receivers located on the earth's surface.

The principle of this technology is based on a lower attenuation of the

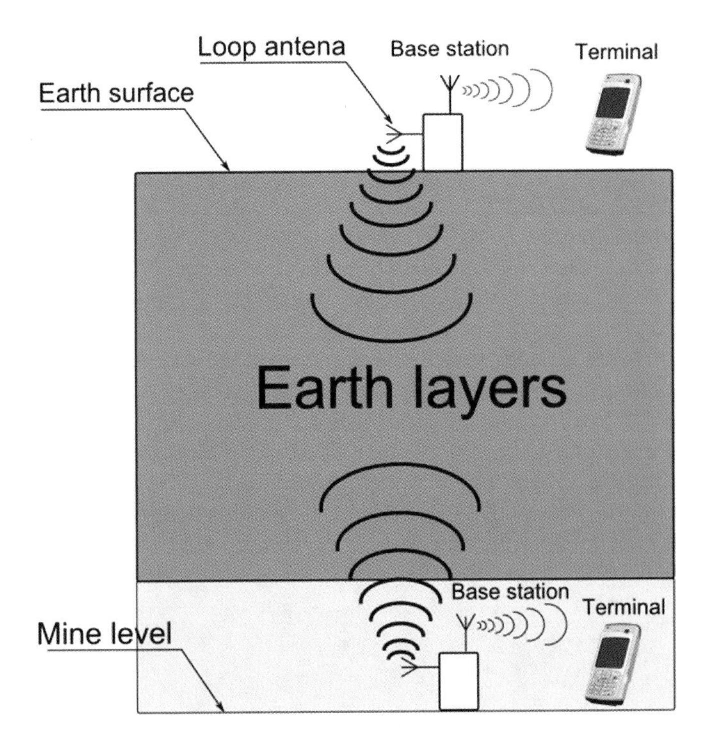

Figure 17.2 Concept of the TTE system made by Telemag.

magnetic component at very low frequency (few kHz) EM wave in the earth layers. Attenuation is so low that it allows for a radio signal to pass through ground, even at a distance of several hundred meters. This distance depends on many factors, the main ones pertaining to electrical parameters of the geological layers [6].

There are a number of systems of this type, but the most advanced solution is called Telemag, made by Transtek, commissioned by the U.S. government [1]. It allows two-way transmission of voice in half – duplex mode at frequencies between 3 and 8 kHz, using SSB (Single-Side Band) modulation. Concept of the TTE system made by Telemag is shown in Figure 17.2.

To establish a connection between the earth surface and the mine level, two magnetic loop antennas are used. In order to obtain the best signal level these antennas must be placed in a straight line. The base stations act as signal relays between user terminals and loop antennas.

The system was successfully tested in coal and limestone mines, where the distances between the earth surface and mine level depths were about 90m. The maximum theoretical distance is around 600m.

The services offered by these systems are limited to voice communications only, whilst the great difficulties related to mobility of loop antennas limit Telemag applications, where it is being used only as a supplement to other, more universal technologies.

It is very likely that in the near future there will be more complex systems based on TTE technology. For example, Gamma Services International has developed the TramGuard [6] system, whose main goal is to minimize risks associated with miners being to close to mining machinery. The mining conditions (noise, dust, insufficiency lighting, etc.) create a great danger and often these are the main cause of accidents. In parallel to TramGuard, MinerTrack is being developed to locate and identify miners.

The main idea of both systems is to transmit information using Magnetic Field Generators (MFG) that operate at frequencies around 3kHz and communicate with the earth's surface. Miners will be equipped with portable Personal Alarm Devices (PAD) through which they can send various types of information (identification number, text messages etc.). The message will be transferred to the nearest MFG, while it continuously transmits data to the surface. PAD will also include signal receiving functionality. Furthermore, mining machines will be equipped with emergency stop mechanisms, used whenever the system detects any miner in dangerous proximity to the machine.

17.2.4 WLAN Networks

The relative low cost and easy installation made WLAN networks common in most sectors of industry. Not surprisingly, they started to be used also in mines.

Currently on the market, there are several solutions based on 802.11 networks, for example, the ImPact system manufactured by Mine Site

Technologies [8]. Wireless connectivity is achieved through Wi-Fi access points, using the 802.11 b/g standard, operating at 2.4 GHz with bit rate of 11 Mbps. Access points are connected to a switch through wired Ethernet (copper and fiber). Usage of TCP/IP stack makes such networks very flexible when it comes to implementation of various services. Users are equipped with mobile terminals for voice services (based on VoIP), SMS and voicemail. Video cameras and sensors supervising the mine (monitoring concentration of methane, oxygen, etc.) can also be connected to the network.

There is also a possibility to implement user identification and tracking using TRACKER units, based on RFID tags (Radio Frequency Identification). An RFID antenna is a small loop antenna working at 433 MHz (older version TRACKER modules) or communicating directly with Wi-Fi access points at 2.4 GHz. Each user is equipped with such antenna, which transmits to the access point a unique identifier (ID), and other information, e.g. battery status, to a maximum distance of about one hundred meters. RFID is available in two versions. The first solution is a separate device equipped with a 4.5 V battery, worn on the belt, or attached to the mining machine. Another very common solution is to integrate the device within the mining lamp battery, located on the mining helmet, this solution is known as Integrated Communications Cap Lamp (ICCL).

17.2.5 Mesh networks (802.11s)

An interesting WLAN implementation solution in mine environments relates to the 802.11s standard. These networks are based on mesh topology, where network nodes are able to dynamically establish connections with each other on ad-hoc basis. By measuring quality parameters, the routing protocol determines which path to choose. Currently 802.11s standard is not officially approved, but some companies are conducting extensive research to implement such systems.

An example of a system, which can be used in the mining environment, is BreadCrumb, manufactured by Rajant. BreadCrumb is based on the 802.11b standard and operates at a frequency of 2.4 GHz. Access Points are portable and powered by batteries [1]. Maximum range in a straight corridor is about 450 meters (for voice) but maintaining connectivity requires line of sight, due to large signal attenuation. Breadcrumb is approved by the U.S. Department of MSHA (Mine Safety and Health Administration). First installation of the system took place in the 2008 in the Beaconsfield gold mine (Australia).

The main advantage of mesh network topology is that it is more failure proof compared to networks using cable infrastructure. Thanks to dynamic choice of transmission routes, even in the event of a disaster (e.g. collapse of excavation) communication between a node and the rest of the network could be maintained

17.2.6 DOTRA System

This is a system developed by Inova Technical Innovation Center from Poland, which enables wireless connectivity on the excavation sites, production halls and other areas of the mine [7]. One base station is able to handle a total of approximately 100km of mine corridors. DOTRA is a trunking system based on MPT 1327 standard (with analog voice and digital control channels) which means that channels are allocated dynamically for the duration of each call. After the call, the channel is released and can be used by another user.

Users can communicate through terminals having unique identification numbers and in addition to voice can also send and receive short text messages. There is also a mode for conversation with a group of terminals (group call) and for broadcast communications. System also allows for establishing emergency calls and prioritizing them and offers an option for unconstrained calls to the PSTN network. The set of different DOTRA terminals is presented in Figure 17.3.

DOTRA system is also able to transmit low-resolution video images, for example from CCTV cameras monitoring certain areas of the mine. The system can also remotely control and monitor mine equipment, like: sensors, water pumps, mining machines, etc. DOTRA is currently used in Polish copper mines of KGHM Polska Miedz S.A.

Figure 17.3 Set of different DOTRA terminals [7]

17.2.7 MultiCOM

The construction and operating principle of MultiCOM is very similar to leaky cable systems. In this system the signals are transmitted in the 150 - 174 MHz band, which is divided into 32 channels for voice and data transmissions. The remaining bandwidth is distributed between 16 channels for video transmissions (uplink only).

Users of the system can talk with each other, using portable radios, such as MRS INTEL10 and MRS INTEL 20. It is also possible to make calls to PSTN and send SMS messages. The transmission of video signals is performed in real time. In order to maintain sufficient image quality, there are video amplifiers installed every 350 meters.

MultiCOM is equipped with a monitoring system called MULTITAG, which allows for localization assessment and identification of miners or machinery, in the area covered by the network. Each user is equipped with a small MULTITAG card, which is secured to mining lamp and is powered from the lamp's battery.

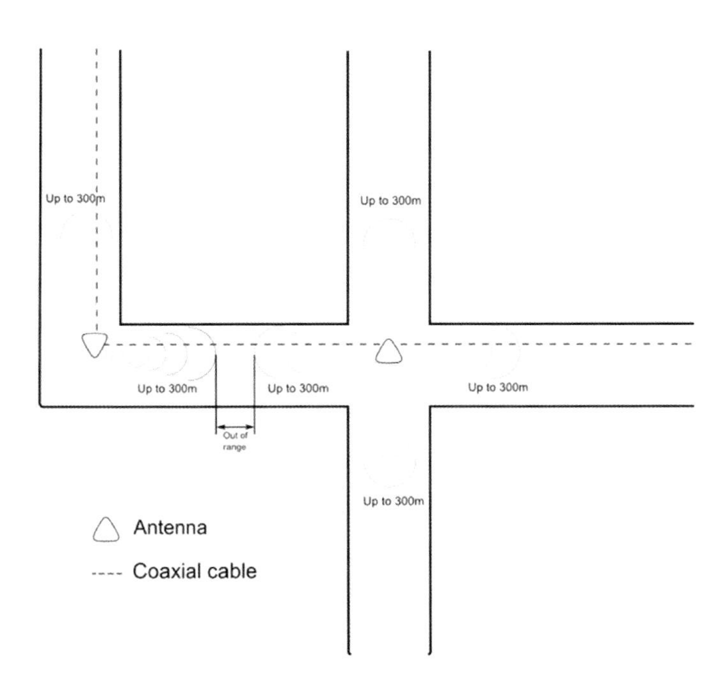

Figure 17.4 DAS system architecture (source [9]).

These cards send signals, which include an identification number and is received by specially designed readers, placed along the leaky cable. This allows for assessment of precise location of the miner at given times. The received data is sent to the main computer, where it is presented in graphical form and stored in a database.

The system has vast functionality related to control of equipment and automation of multiple mine operation processes. With the MULTIDATA module it is possible to take full or partial control over transport systems, water pumps operation, etc. The communication between the network and the devices is conducted via RS 232 serial communications modems. In the same way the system can be connected to sensors monitoring levels of toxic/explosive gases, humidity or temperature.

17.2.8 SIAMnet

This system is based on a solution, known as DAS (Distributed Antenna System) , see Figure 17.4. The backbone network consists of a coaxial cable. However, in contrast to leaky cable systems, this coaxial cable is fully shielded. It is adapted for data transmission in the 600 - 900 MHz band. In this system antennas are placed only at locations where users require connectivity to the network [9].

Typically, these are omnidirectional antennas (Figure 17.5a), the operating frequency is 800 MHz and, depending on the type of excavation and the structure of its walls, it has a maximum range of up to 300 meters. If necessary, for example when communication is required at a distance greater than standard antenna provides, it is also possible to install directional antennas (Figure 17.5b).

In order to amplify the signal there are compact amplifiers placed along the network, which are supplied with power through the coaxial cable. Compared with the technique of leaky cable, this system requires much less amplifiers due to lower propagation loss.

The system allows for duplex voice operation, both in conventional or trunking mode. Terminals communicate with the network at a frequency of 860 MHz, using digital modulation. Different mode connections (private conversation, conference, emergency call, broadcast) are established in the headend through a relay device (Voice Repeater). It is also possible to add a module for connections to the PSTN network, as well as a dedicated relay device (Crossband Repeater) to connect the system to existing UHF/VHF radio systems already present in the mine.

The system has eight data channels. Each channel offers a data rate of 1500 kbps. To operate in this mode two types of radio modems are used: BMRM (Master Base Radio Modem) and SRM client modems (SIAMnet Radio Modems). One radio channel is used by one BMRM device and any number of SRM, which communicate with BMRM through three ports: two of which are asynchronous (115 kbps), while the third is synchronous with a high data rate for services such as LAN. Scheme of data transmission in SIAMnet is shown in Figure 17.6.

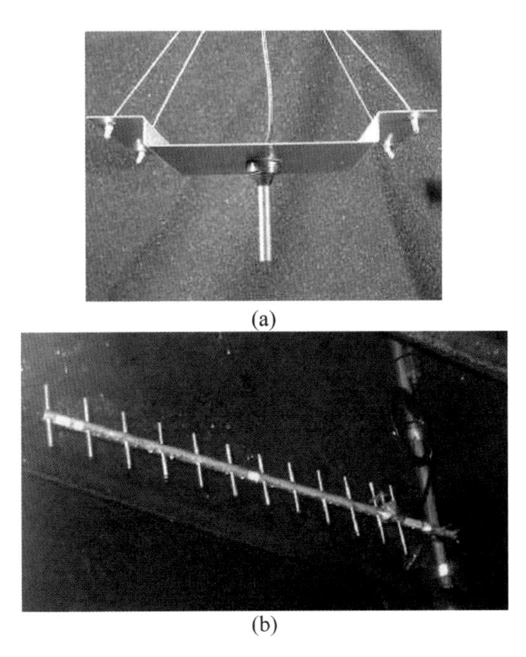
(a)

(b)

Figure 17.5 Antennas used in SIAMnet: a) typical omnidirectional, b) directional Yagi antenna [9].

SIAMnet for broadband data transmission uses DOCSIS 1.1 standard (DOCSIS 2.0 and EuroDOCSIS 2.0 are also supported). It is worth mentioning that the very popular among cable television subscribers broadband Internet

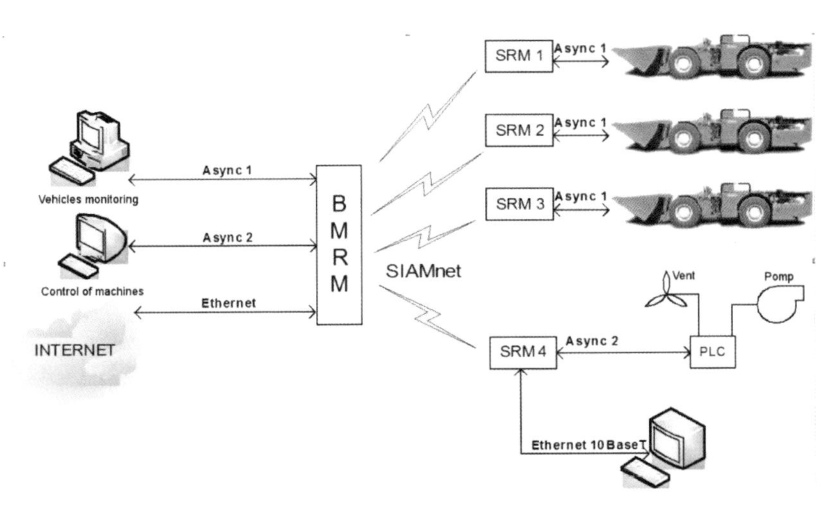

Figure 17.6 Data transmission in SIAMnet.

access service also operates according to the DOCSIS standards. The heart of this module is a unit called CMTS (Cable Modem Termination System), through which client modems communicate with the network. SIAMnet allows for installation of six CMTS in the headend. This configuration results in a throughput of 250 Mbps for the downlink and 150 Mbps for the uplink. In this solution, there is no problem with LANs, even one-kilometer underground, or with connecting Wi-Fi access points.

17.2.9 System Kopalnianej Telefonii Bezprzewodowej (SKTB)

SKTB, which is Polish mine's wireless telephony system, uses the radio interface based on the DECT standard. It allows voice communication and data transmission (small packages) within the different scale networks as well as integration with the mine dispatcher network [5].

The radio resources are shared with Time-Frequency Division Multiplexing (TFDM). Ten radio channels are located in the band 1880 - 1900 MHz, each divided into 24 time slots. With a single base station it theoretically enables communication for 120 duplex channels. The mean transmit power is 10mW. Base stations are located, depending on the structure of the area, every 50 – 300 meters, where the maximum distance from the access module is limited to 4km. A large number of base stations in the specified area provide a high system capacity, several times higher compared to mobile networks. Increasing coverage is accomplished using radio transmitters.

The basic system architecture contains one access module interacting with sixteen base stations, which are further supported by several dozen radio transmitters. It ensures maximum number of logged users equipped in radio terminals equal to 64. System capacity can be increased by attaching additional access modules to the base stations. One base station allows for four simultaneous voice connections; it also offers connection handover, so the users can move between base station ranges.

The radio terminal in SKTB is a personal radio broadcasting device OSTAR. It provides voice connections, alarm connections or dispatcher reports, as well as allowing for text messages and transmission of other data. Terminals periodically communicate with the base station to update information, for example about their location.

17.3 REQUIREMENTS FOR WIRELESS COMMUNICATION SYSTEMS OPERATING IN A MINE

Wireless networks operated in a mine should ensure provisions for special multimedia services. The major characteristics of such services include bandwidth demand, real-time operation and sensitivity to delay variations. The set of requirements for networks transmitting voice and video are different than for data transmission networks, especially for network used for real time video transmissions. The basic problem is to fulfill quality of service requirements QoS, which are different for different services [15]. The requirements for services and applications should be well known and these are defined by users. Users are not interested in the way the given service is implemented in the network but rather in the observed final effect of service operation. Considering this, the quality of service requirements should comply with all of aspects, taking into account the end user point of view. They should concentrate on issues concerning user perception rather than phenomena appearing in the transmission network [12],[13],[14],[17],[18].

Here are the most important quality parameters for users:

- Delay – time needed to perform a given service, calculated from the moment the user requests the services to the moment the user receives requested information when the service is already active. Depending on the type of service the delay can influence user satisfaction in a variety of ways.
- Delay variation – especially important in transport layer of packet switched network. It describes variation of time for incoming successive packets. Services very sensitive to delay variation must use packet-buffering techniques. These allow for elimination or reduction of delay variations noticed by the user by increasing the initial delay.
- Information loss – has direct influence on final information quality presented to the user, regardless if it is voice, picture, video or data transmission services. Information loss is not only bit or packet distortion but also the effects related to quality reduction caused by coding or compression (it is used for more effective transmission channel utilization – voice codecs and video codecs).

Depending on the type of service the requirements for service performance are also changing. Basic service classes relate to providing audio, video or data information. Five quality levels were developed for audio services [16]. The basic types of audio services are as follows:

- Conversational voice
  The delay has a great influence on conversational voice quality. It results in two observable effects. The echo phenomenon, which can highly reduce the quality of transmitted voice especially for delays around dozens of milliseconds, this effect is reduced by echo cancellers. For delays around several hundred of milliseconds the second phenomenon can be noticed, which influences conversation dynamics – delays in respondent answers.
  The human ear is also very sensitive to rapidly changing delay variations (jitter), which cause voice vibration. For that reason variations

of incoming packet times must be removed through special buffers (de-jitterizing buffer) for all voice services in packet networks. The information loss requirements conform to the voice signal distortion tolerance characteristics of a human ear. For connectionless transmissions the main source of reduced voice quality relates to performance of codecs operating in packets loss conditions.

- Voice messaging
  Requirements related to information loss are the same as for conversational voice and depend on voice codec. The main difference is in the delay. Voice messaging is much less sensitive to delay, because there is no direct information exchange. Based on research made in regular telecommunication networks, delays of even around few seconds are still acceptable.

- Streaming audio
  These services foresee delivery of audio signals with quality better than phones. Therefore the requirements on packet loss are much more stringent, but considering that there is no interactive character to them, the requirements related to delay can be more lenient.

General classification of video services contains six quality levels [16] and their representation for different services. The main services are:

- Videophone
  Videophone services could be performed in systems with duplex transmissions. They are connected with audio and video signal transmissions with conversational quality. Therefore the requirements for them are lower as for conversational voice services, additionally the system must provide synchronization of audio and video streams. It can be said, based on human eye properties, that connectionless systems permit loss of some part of information. For example, well operating MPEG-4 codecs ensure good video signal quality for packet loss lower than 1%.

- Streaming video
  There is no interaction required related streaming video, therefore the requirements for these services are similar to streaming audio.

From the user point of view the basic requirement for data transmission services is errorless delivery. Delay and delay variations have less meaning, except for data transmission services related to real time localization procedures. For some services it is necessary to assure some level of synchronization for different data streams. For example, during a multimedia session, the audio signal should be

synchronized with text appearing on the whiteboard. Different applications impose requirements related to the time between a request for information is made to the time it is presented on the user terminal. There are a lot of data transmission services, for examples: web browsing, bulk data transfer, E-commerce, command/control services, still image transfer, e-mail, instant messaging and others.
The last group includes background applications. Their requirements are limited to errorless information delivery. Requirements for earlier mentioned services are shown in Table 17.1.

Service requirements are mapped into parameters, which describe properties of transmission system in the backbone and access domain. For connection techniques the dominant parameter is link throughput, for connectionless techniques the rest of QoS parameters also become important.

Table 17.1 QoS requirements for audio, video and data transmission services

Service type	*Application*	*Typical data amount*	*Basic parameters and values*		
			Delay (one way) [ms]	*Delay variation [ms]*	*Information loss level*
Audio	conversation	4-64kbps	<150 <400 (borderline value)	< 1	< 3% (PLR)
	Voice messaging	4-32kbps	< 1s (playback) < 2s (recording)	< 1	< 3% (PLR)
	streaming	16-128kbps	< 10s	< 1	< 1% (PLR)
Video	videophone	16-384kbps	< 150 <400 (borderline value)		< 1% (PLR)
	streaming	16-384kbps	< 10s		< 1% (PLR)
Date	Interactive games	< 1kB	< 200		zero
	Telnet	< 1kB	< 200		zero
	E-mail (access to server)	< 10kB	preferred < 2s accepted < 4s		zero
	E-mail (between servers)	< 10kB	to several minutes		zero
	Fax	~ 10kB	< 30s/page		<10-6 (BER)

PLR (Packet Loss Rate), **BER** (Bit Error Rate)

It is very difficult to find direct relation between these requirements and the parameters of IP network. For TCP/IP networks (packet switched) it is impossible to define QoS parameters in the same way as for connection switched networks or for ATM networks. The basic problem with QoS in TCP/IP network is that these networks had been created to enable transmission in all conditions without any quality guarantee. User applications in terminals should be responsible for the quality of transmitted data. Even small modification of this kind of transmission procedures in order to guarantee even very limited quality by the network, needs to implement some negotiation mechanisms. The possible set of parameters one can modify cannot be easily associated with the classical concept of parameters describing quality of service.

17.4 ASSESSMENT OF WIRELESS SYSTEMS IN MINE UNDERGROUND ENVIRONMENTS

Research was performed in the copper mine in Lubin at the depth around 600m below ground surface. Two typical mine environments have been chosen: long straight corridor with a length around 700m and a grid of corridors covering an area of around 300m x 100m. Both environments were shown in Figure 17.7, which also indicate the location of base stations.

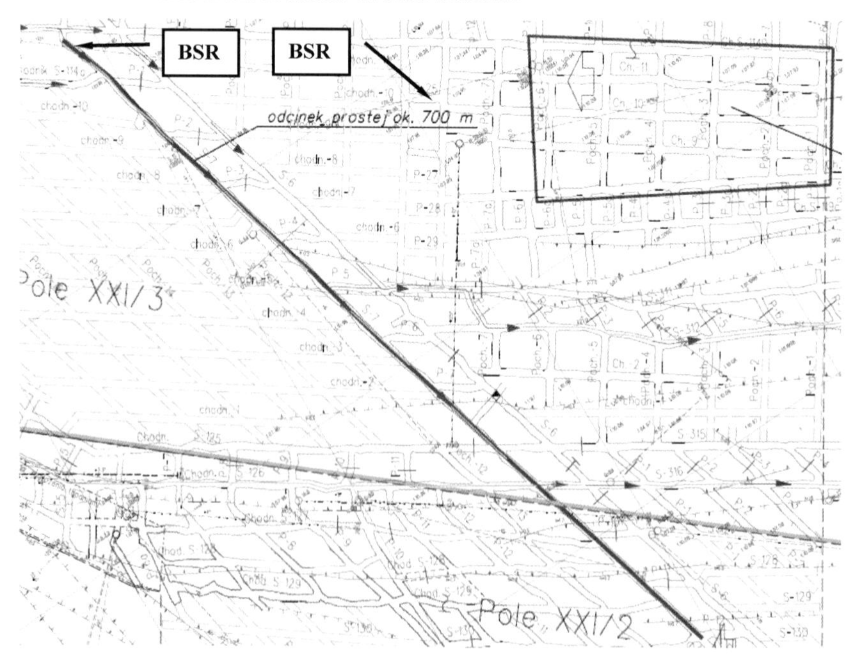

Figure 17.7 Plan of the mine area under research.

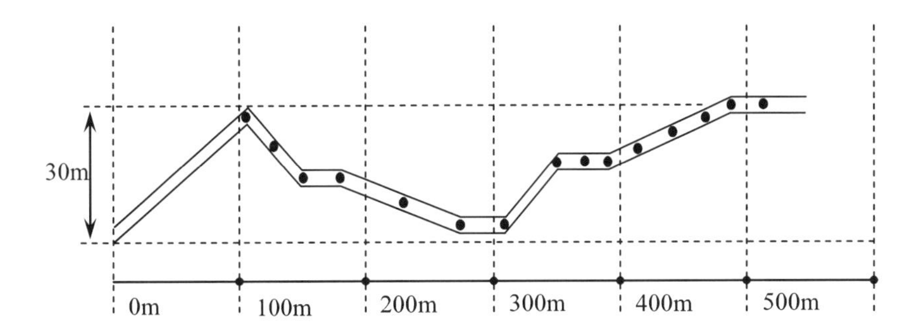

Figure 17.8 Cross-section of the straight corridor.

All tunnels forming the grid were generally flat and on the same level, whilst the straight corridor did rise and fall along its length. Approximate cross-section of the corridor with marked measurement points is shown in Figure 17.8. The indicated lengths were measured along the corridor floor using a laser rangefinder.

Height of the straight corridor was around 4m and the width was around 5m (Figure 17.9a). The grid corridors were a little lower, at around 3m, having the same width as the straight tunnel (and Figure 17.9b).

17.4.1 Measurement system

Measurements configuration and methodology had to be adapted to the environment in view of very damp soil (there were puddles and deep mud in many places). All systems had to be tested in the same way using the same tools to achieve comparable results. Additionally, the necessity to use external power sources (buffer power supplies) for system and measurement devices determined that point-to-point architecture should be used. Configuration of the test set is shown in Figure 17.10.

Figure 17.9. a) View on the straight corridor, b) part of grid area.

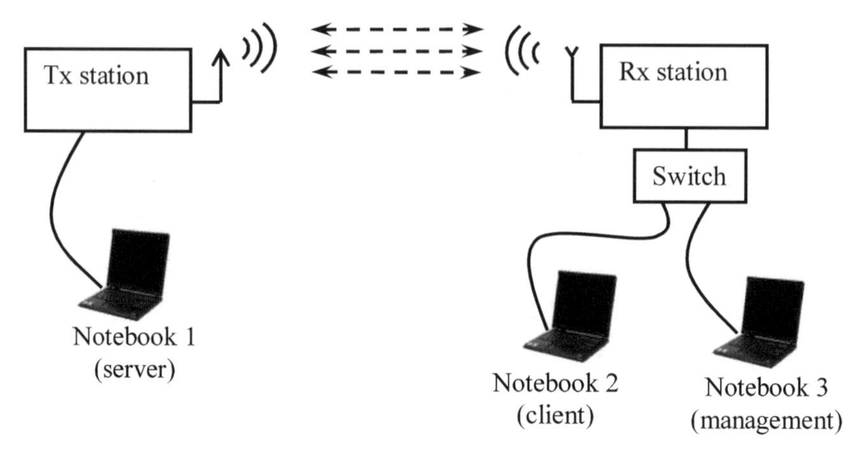

Figure 17.10 Scheme of the test set.

Research was performed using three notebooks (Celeron 1.86GHz, 1GB RAM, operating system – Centos). As a transmitter, depending on the System Under Investigation (SUI), WiMAX systems used a base station whilst the WLAN systems used an access point. The receiver consisted of a user terminal (WiMAX systems) or client card (WLAN systems). The transmitter (Tx station) was located in a fixed point at a height around 3m and around 0.5m from the wall (see Figure 17.11). The receiver (Rx station) was placed on a car at a height around 2m (in the tunnel center) and it was moved along the corridor.

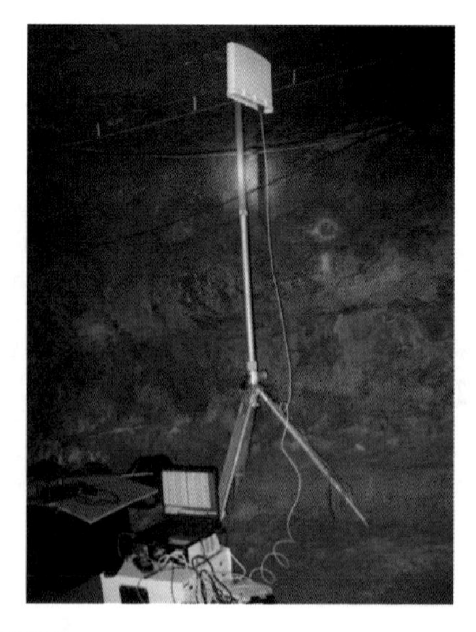

Figure 17.11 Base station set-up.

Measurements were performed using three applications: rude/crude, iperf and ping. The first two were used for throughput measurements in selected test points; the third - ping was used to measure transmission delay. Iperf was used during assessments in the grid area, where the time of a single test was set to 60s and the results were recorded every second. The final throughput results were then calculated based on these sixty individual measurements [27].

17.4.2 Systems under investigation

All experiments in the mine environment were performed using two WiMAX systems: Airspan operated in the 1.5GHz band and Axxcelera operated in the 3.5GHz band, as well as a system with IEEE 802.11b/g radio interface operated in the 0.9GHz band (Ubiquiti + Mikrotik) and a system using IEEE 802.11n (Linksys). The selected systems, having quite typical radio parameters, allowed for assessment of their suitability for operation in the specified environment. The technical parameters of SUI are shown in Tables 17.2, 17.3 and 17.4.

Table 17.2 Basic parameters of Axxcelera modules

Parameter	*Axxcelera CPE*	*Axxcelera AP*
Frequency Band	3572-3600 & 3472-3500 MHz	3572-3600 & 3472-3500 MHz
Air Interface Standard	802.16-2004	802.16-2004
Architecture	Point to Multipoint (PMP)	Point to Multipoint (PMP)
Duplex Scheme	Full Duplex FDD	Full Duplex FDD
FDD Duplex Spacing	100 MHz typical	
RF Channel Sizes	7 MHz, 3.5 MHz and 1.75 MHz	14 MHz, 7 MHz, 3.5 MHz and 1.75 MHz
Modulation Types	Dynamic Adaptive: 64QAM, 16QAM, QPSK, BPSK	Dynamic Adaptive: 64QAM, 16QAM, QPSK, BPSK
Coding Rates	1/2, 2/3 and 3/4	1/2, 2/3 and 3/4
Net Throughput	40 Mbps (7 MHz channel size)	35 Mbps (7 MHz, FDD)
Transmit Power	+20dBm	+20dBm; +27dBm (optional)
Receive Sensitivity	-100dBm	-100 dBm; -112 dBm (Uplink Sub-channelization)
Antenna Beam Width	15 degrees	60 degree
Antenna Gain	19 dBi	16.5 dBi

Table 17.3 Basic parameters of Airspan modules

	AirSpan CPE	*AirSpan AP*
Frequency Band	1426.5 – 1524 MHz	1426.5 – 1524 MHz
Air Interface Standard	Adaptive TDMA, as defined by IEEE802.16-2004	Adaptive TDMA, as defined by IEEE802.16-2004
Architecture	Point to Multipoint (PMP)	Point to Multipoint (PMP)
Duplex Scheme	TDD	TDD
RF Channel Sizes	5 MHz, 3.5 MHz and 1.75 MHz	5 MHz, 3.5 MHz and 1.75 MHz - optimized for 5MHz
Modulation Types	64QAM, 16QAM, QPSK, BPSK	64QAM, 16QAM, QPSK, BPSK
Coding Rates	1/2, 2/3 and 3/4	1/2, 2/3 and 3/4
Transmit Power	+24dBm	+27dBm
Receive Sensitivity	-104dBm for 1.75MHz, -100dBm	-104dBm for 1.75MHz, -100dBm
Antenna Beam Width	Azimuth - 60°, Elevation - 30°	60°
Antenna Gain	10.5 dBi	10.5 dBi

Table 17.4 Radio link and device parameters of Ubiquiti Networks SuperRange9 system

Ubiquiti Networks SuperRange9 with Mikrotik board	
Radio Operation	Proprietary 900MHz based on 802.11b/g CCK/OFDM
Data Rates	1Mbps-11Mbps (802.11b), 6-54Mbps (802.11g)
Channels	5/10/20 MHz
900MHz Frequency Channel Support	907MHz-5/10MHz, 912MHz-5/10/20MHz, 917MHz-5/10MHz, 922MHz-5/10/20MHz,

17.4.3 Measurement results

All of the systems were tested in the straight corridor and in galleries forming the grid. The research results are shown below, separately for the two characteristic mine areas.

17.4.3.1 Long straight corridor

The long straight corridors are typical for this kind of environment. They compose the main communication ducts used by excavation vehicles. The preliminary measurements were performed using the ping tool to estimate range (checking the connection possibility between transmitter and receiver) and select the test points and theirs density. This assessment used the Airspan system, which according to theoretical expectations should ensure the highest range. Results of this test specified a preliminary range of around 500m. The test point density was verified experimentally during research – the more throughput variations in a specified area the more test points. Based on this, the beginning section of the corridor includes less test points than the end. Additionally, in some places the conditions in the mine (deep and wide puddles, steep inclines with soft soil) did not allow measurement performance, because it was not possible to stop the car there. The throughput (using rude/crude), delay (using ping) and received power (using management application) were measured during research. The research results are shown in Figure 17.12. The test results for 802.11n system differ significantly from the results obtained for other systems and showing them together would make the data unclear, therefore they are presented separately in Figure 17.13.

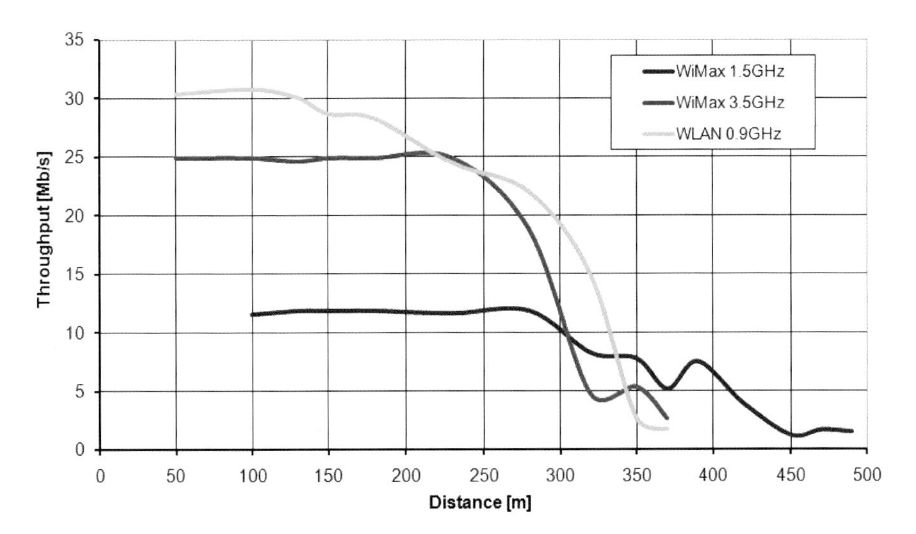

Figure 17.12 Throughput for wireless systems as a function of distance.

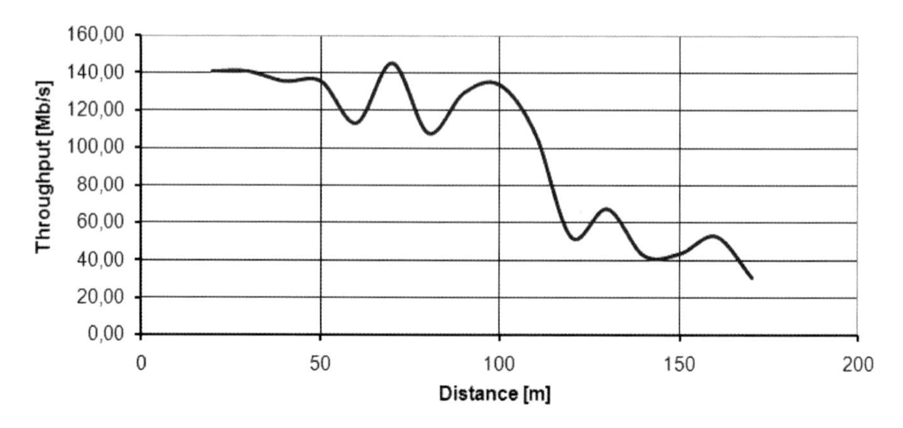

Figure 17.13 Throughput for 802.11n system as a function of distance.

The assessment results show that Airspan achieved the best performance. The main priority consisted of achieving the highest possible range, whilst maintaining some minimum throughput value. The Axxcelera and Mikrotik systems allowed for higher throughput, however after reaching around 350m the connection was lost. All of the three systems worked with comparable Tx power. According to expectations, the highest range was achieved for the Airspan system. This system operates in a lower band compared to Axxcelera, therefore the rough surface scattering has lower influence, whilst it also works with a different Tx protocol compared to MikroTik. Achieving a range of around 500m in these

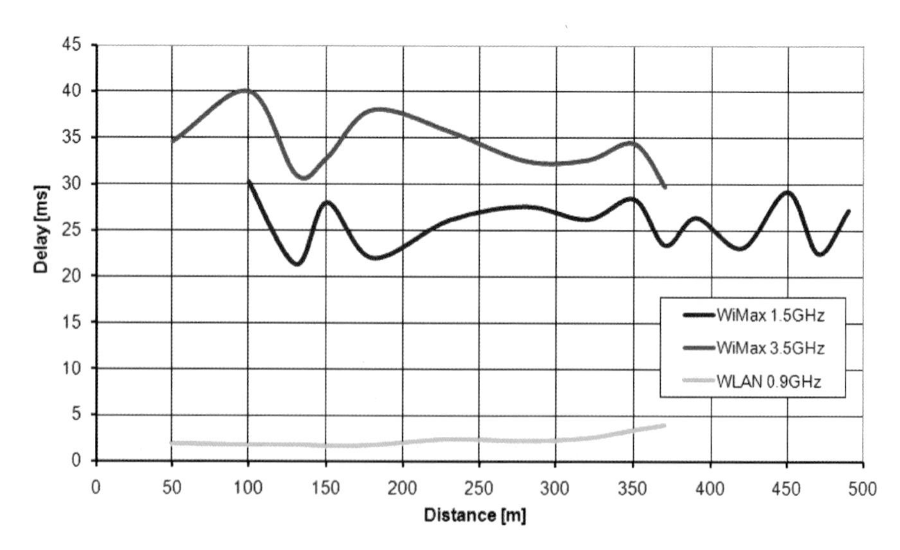

Figure 17.14. Delay as a function of distance in straight corridor.

conditions is very promising for future potential use in underground environments. The range of the 802.11n system dropped rapidly compared to others, but this can be attributed to lower Tx power and lower Tx and Rx antenna gains. However, very high throughput is not the key objective in this case. Throughput around 1Mbps (achieved by Airspan at the boundary region of range) is a value sufficient for successful data transmission for different services, e.g. voice transmission. The higher distance from transmitter and lower throughput are consistent with theoretical assumptions. Increasing the distance causes a drop in Rx signal power, therefore modulation used for information transmission is adopted to one more resistant to distortion. On the other hand, modulation with lower state numbers allows for transmission of information with a lower throughput.

Delay measurement results show that it is quite stable along the corridor. The ping application, used to test delay, measures it during transmission from transmitter to receiver and back to transmitter. Lowest delay was achieved for Mikrotik and 802.11n, which could be attributed to the different transmission protocol compared to WiMAX systems. Delay of around 20ms still allows for implementation of wideband services e.g. voice transmissions and services for which fast reaction is needed e.g. control, monitoring or alarms. The results of delay measurements for SUI in straight corridor are shown in Figures 17.14 and 17.15.

In each case the received power was measured using the managing application of each system. It should be noted, that this power highly depends on devices parameters, which have influence on the link energy balance. Unfortunately these parameters were different for each system e.g. gain of Tx and Rx antennas. However, the noted drop of received power along the corridor is about the same in each case. Results of RSSI (Received Signal Strength Indication) measurements for all of the systems are shown in Figure 17.16.

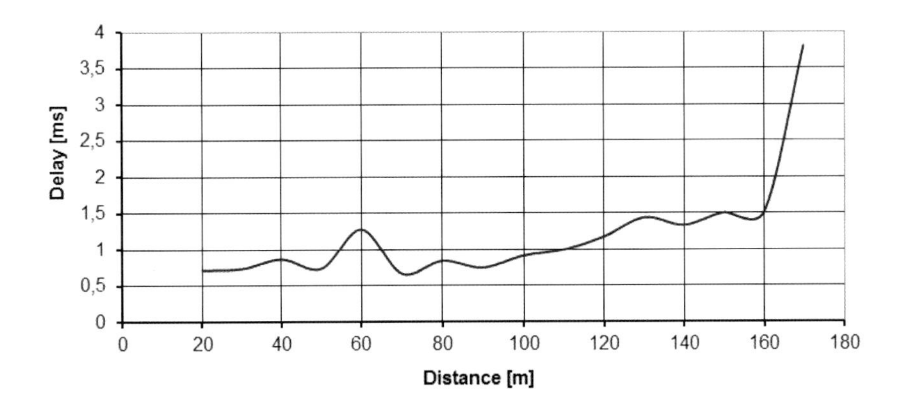

Figure 17.15 Delay as a function of distance for 802.11n system.

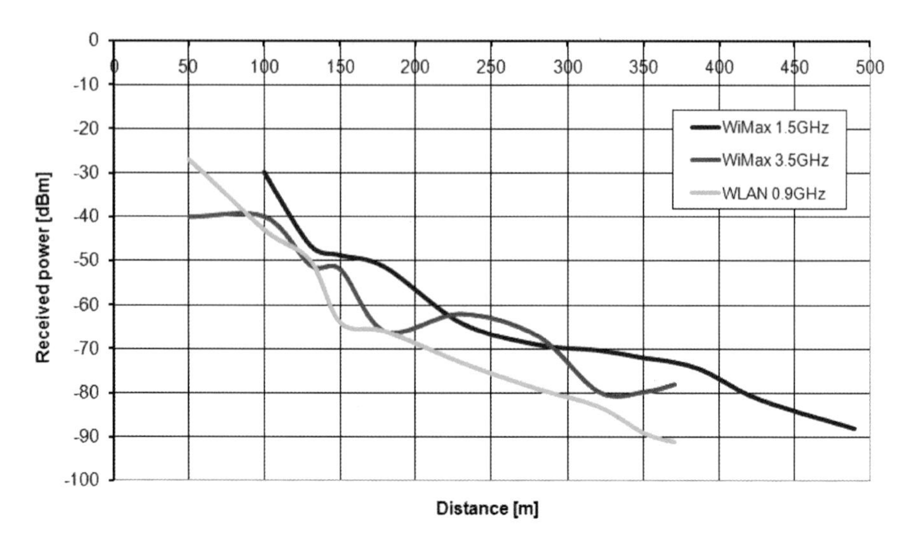

Figure 17.16 Received power as a function of distance.

Systems were set to automatically selected type of modulation depending on the received power. For the Airspan system one could note 64QAM (Rx power to around -69dBm), 16QAM (Rx power to around -72dBm), QPSK (Rx power to around -75dBm) and BPSK modulations. Modification of modulation type had direct influence onto the achieved throughput value.

17.4.3.2 Grid of corridors

The system of corridors is arranged in a grid as shown in Figure 17.17. This is the second characteristic environment encountered in underground mines. Arrangement of corridors is related to the method of excavation of copper ore. The whole length between Tx and the 19-th point is about 311m and the distance between adjacent transverse corridors is about 50m. During the first part of the research, system measurements were performed in a straight line Tx-19, which provided the optical visibility (LOS). At the end of the corridor the drop in signal power level was approximately 4dB (WiMAX System) and 30dB (MikroTik system) in comparison to the values recorded near the base station. However, it had no effect on the performance of the systems, where maximum efficiency was achieved along the whole corridor system.

In the second part, the tests were performed at intersections of corridors, at points as marked in Figure 17.17. Between points 8 and 9 there was a rubber curtain used to control air circulation in the mine, therefore this area included a higher concentration of measurement points.

All examined systems achieved maximum efficiency at the end of the long corridor. However, with increased distance from the straight line Tx-19 into the grid, one could notice a drop in throughput, until a total loss of connectivity

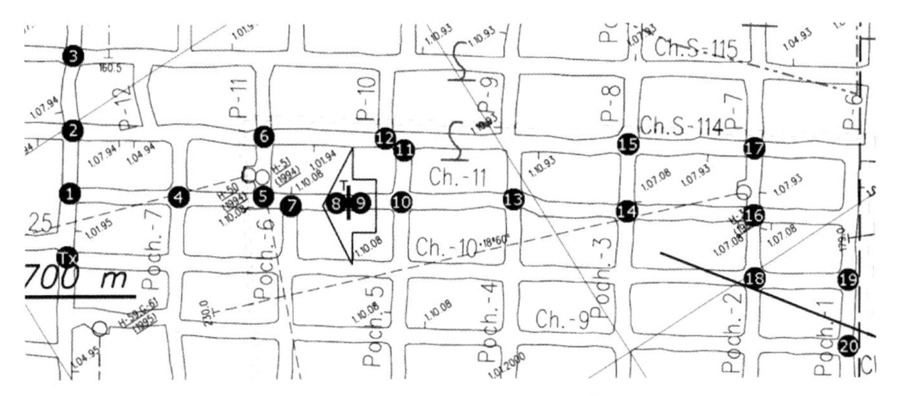

Figure 17.17 Part of grid of corridors with marked measurement points.

between the end points of transmission. It is worth noting that in case of measurement points 14, 15 and 16 only the Airspan system allowed transmission, see Figure 17.18. This phenomenon is related to the propagation of electromagnetic waves and their reflection from the walls. The operation frequency of this system was much lower than the Axxcelera system, therefore the wavelength was also longer. As expected, the greater the wavelength the less it is susceptible to scattering on wall roughness. The lack of MikroTiK system coverage in the above points resulted from proprieties of the transmission protocol and power differences in the energy balance. Wave scattering from the rough surface did not have a big influence here, also because of low frequency.

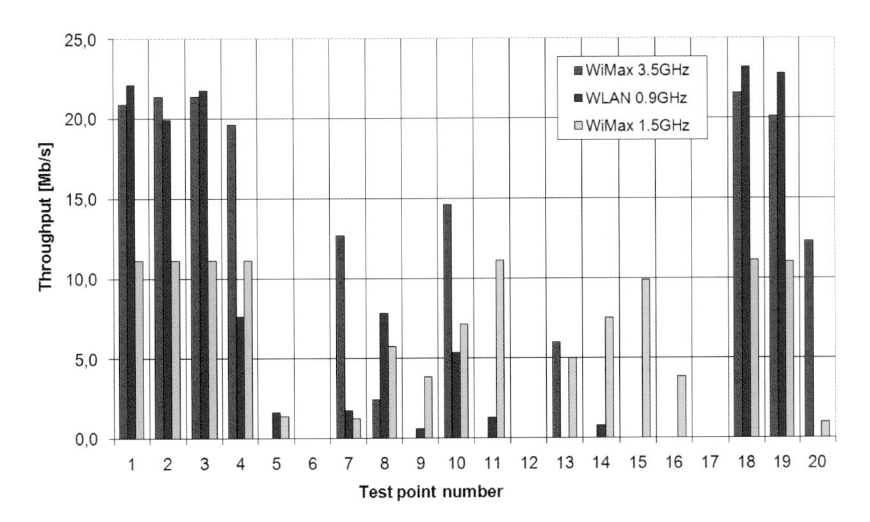

Figure 17.18 System throughput in subsequent measurements.

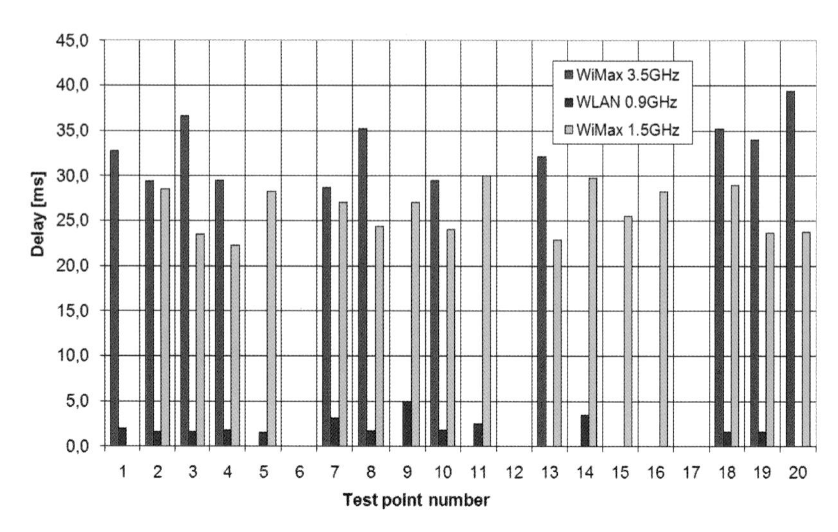

Figure 17.19 Delay in subsequent measurement points in grid of corridors.

The delay test results from the grid points provide similar conclusions as for the straight corridor and are shown in Figure 17.19. Differences in delay measurements at the subsequent measurement points for every system are small. WiMAX systems have significantly higher delay than the MikroTik (WLAN) system, due to their operation principles and the transmission protocol.

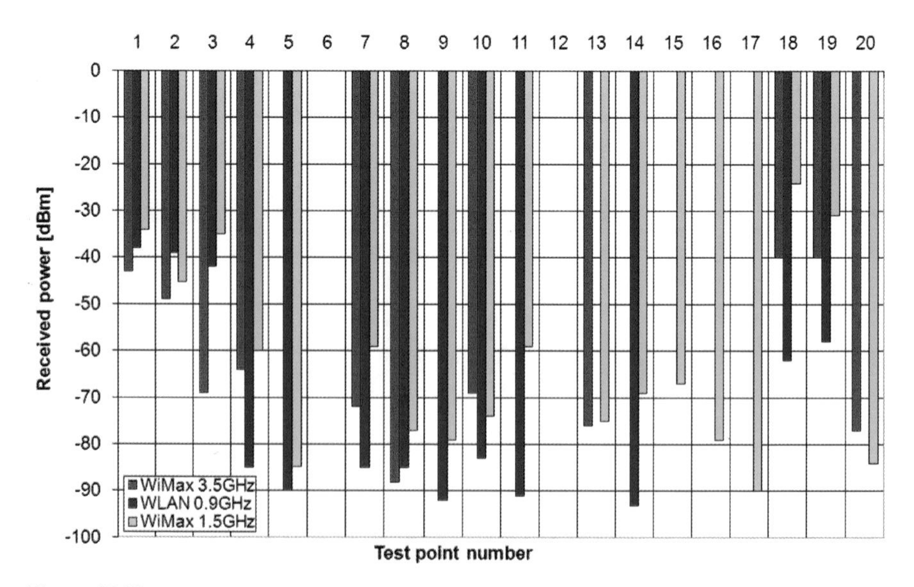

Figure 17.20 Received power level in subsequent measurement points in the grid of corridors.

The measurements of received power in the grid of corridors were conducted the same as in the straight corridor. Test results relate to throughput of the systems and are presented in Figure 17.20. The higher level of the received signal corresponds to the higher order modulation and it is associated with higher throughput. The results were also examined from electromagnetic wave propagation point of view. Additional tests were also carried out and are described in literature [28, 29].

17.4.4 Summary of the mesurements

The measurement results are very interesting. They show that wireless systems operating on the basis of widely used standards can also be used in different conditions than for what they had been designed. According to the coverage criterion it has been proven that the best was the system based on WiMAX using 1.5 GHz. In the straight corridor system it achieved the highest range - about 500m, whilst the others stopped working after 370m. Again, in the grid of corridors this system also achieved the best range. With a few exceptions in the whole test area it allowed for communication, leaving far behind the other investigated systems. Although the throughput of the system at certain points was only about 1Mbps, this is still sufficiently fast to allow implementation of the most services provided in packet networks. The measured delays of about 30ms are greater than in wired networks, still they do not affect the quality of services.

Wireless networks based on WiMAX 1.5GHz system can be used to provide communications in areas where for various reasons it is not cost-effective to mount leaky cable, or when the throughput offered by leaky cable is insufficient. System based on the WiMAX standard can also be used as a backbone network for implementing simultaneous voice, video and telemetry data services. The WLAN system can be used as the access network. Of course proper communication in such difficult conditions increases security of people working in such areas. Performed measurements show that system based on the WiMAX standard can be successfully used for communications in underground mines. Construction and implementation of such solutions in this very difficult propagation environment is very promising for further work.

17.5 TEST RESULTS OF INTERACTION BETWEEN WLAN AND WIMAX AIRSPAN SYSTEMS

Studies related to the interaction between WLAN and WiMAX systems were limited to measurement of the transmission parameters of both systems during simultaneous transmission in a reverberation chamber. Because both systems work in different frequency bands (AirSpan using 1.5GHz and WLAN using 2.4GHz) their simultaneous work should not compromise the quality of transmission if the out of band emissions of transmitters of both systems is low - acceptable to the receivers of each of the systems.

The tests were performed using three laptops. Two of them work as data sources for terminals of all systems (the Iperf client was running on the terminals), whilst the third was the data streams receiver running Iperf server.

The assessment was performed for four scenarios (a-d) related to different settings of both system devices in the chamber. Different amounts of electromagnetic wave absorbers were placed in different places in the chamber. The data rate was observed for all systems every second during constant and slow stirrer rotation from 0 to 360 degrees. The tests were performed for several scenarios including different access point placements. The test setup and the four scenarios (a-d) of device and absorbers settings are shown in Figure 17.21. An average throughput for all scenarios for both system is compared in Table 17.5. Figure 17.22 and Figure 17.23 show some example of data rates for each scenario as function of position (angle) stirrer, for the WiMAX and WLAN systems.

17.5.1 Conclusions

The propagation environment in the reverberation chamber is very demanding for wireless data transmission systems due to the large number of wave reflections from the walls of the chamber. This extremely difficult environment makes it possible to test the efficiency of transmission techniques used in radio devices. Only systems with extraordinary properties can successfully operate in such conditions.

Tests performed in the chamber confirmed the excellent properties of the WiMAX system in difficult propagation conditions. Although the achieved transmission rates differ significantly from the maximum, it is important that the system was able to transmit data for each configuration and each scenario measured in the reverberation chamber.

The system configuration had important influence on the measurement results. Test performed in an anechoic chamber shown that for proper operation the system requires a significant reduction of the received signal level. This is why during each test absorbers shielded at least one device. This configuration decreases the level of reflected waves from the walls. Unfortunately, this was the only possible solution for devices with integrated antennas. Surprisingly good results were obtained for the WLAN system. However, it is easy to notice that

Table 17.5 Average throughput in reverberation chamber WiMax and WLAN systems during their simultaneous operation

System	*Averages data rates*			
	Scenario a [Mbps]	*Scenario b [Mbps]*	*Scenario c [Mbps]*	*Scenario d [Mbps]*
AirSpan	0,966	0,880	0,815	0,869
WLAN	8,453	10,601	10,708	13,105

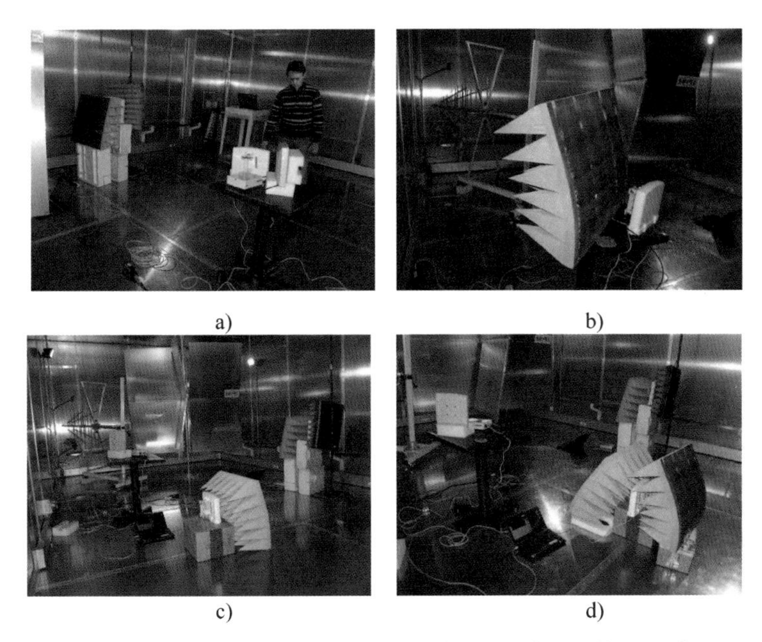

Figure 17.21 Test setup in the reverberation chamber during tests of mutual impact of WLAN and WiMAX systems onto their data rates in four different scenarios (a-d).

these were obtained with more absorbers mounted in reverberation chamber. This is clearly visible in testing of the compatibility of both systems. With each successive scenario, more absorbers were added and at the same time higher transmission speed was obtained. But it was possible only when using the latest techniques of transmission, which provide bandwidth of more than 6 Mbps. These techniques are resistant to adverse events occurring in a radio channel.

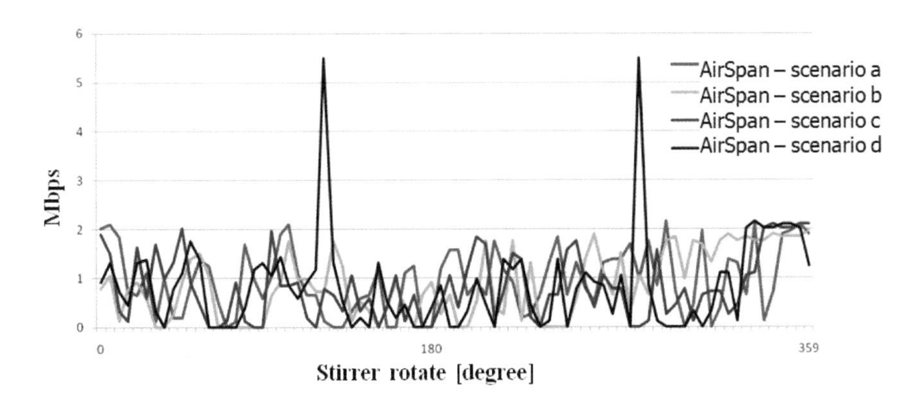

Figure 17.22 The data rate of WiMAX AirSpan system vs stirrer angle measured in the reverberation chamber – in the presence of WLAN system operation.

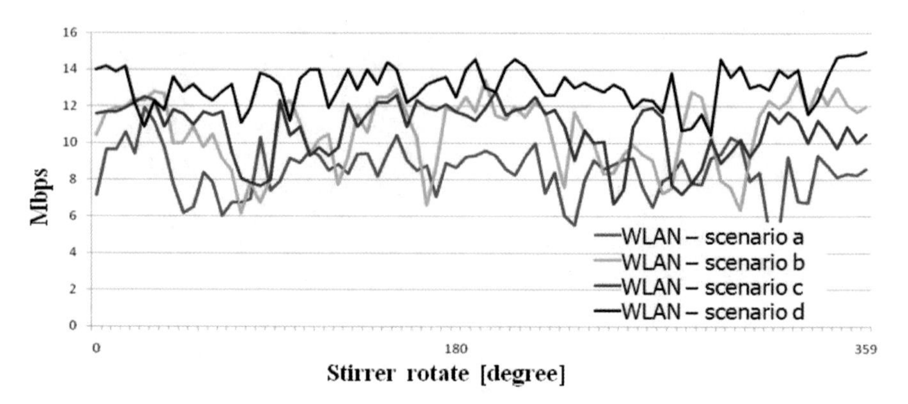

Figure 17.23 The data rate of WLAN vs stirrer angle measured in the reverberation chamber – in the presence of WiMAX AirSpan system operation.

17.6 COOPERATION OF WIMAX AND WLAN SYSTEMS

Systems based on WiMAX standard, due to the operation principle, are used for the construction of backbone networks. They offer a lower bit rate than WLAN networks or other low range systems, but are able to transmit data over longer distances. In mine environments in long straight corridors WiMAX devices will operate as a backbone network, whilst the WLAN system will be used in the hall, where spoil is produced. This is why proper cooperation between these two systems is very important. The general architecture of a hybrid system, using WiMAX and WLAN systems is shown in Figure 17.24.

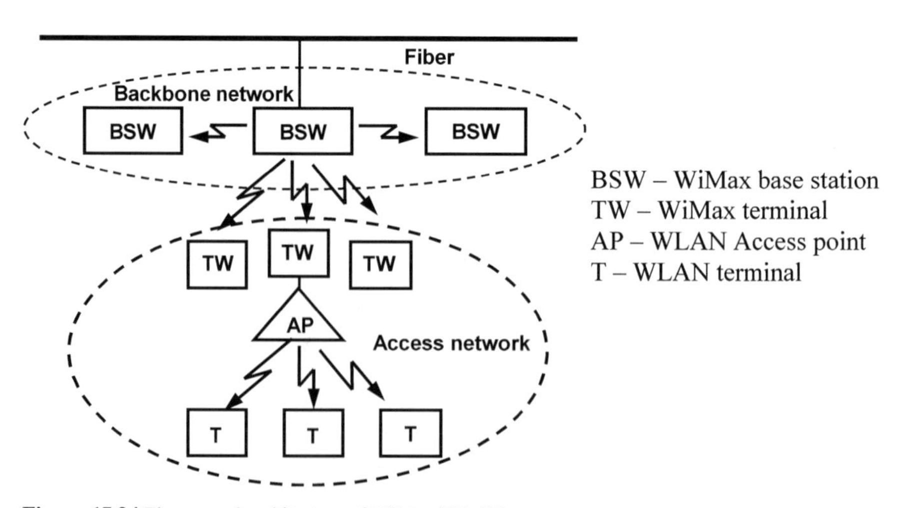

Figure 17.24 The general architecture of WiMax/WLAN systems.

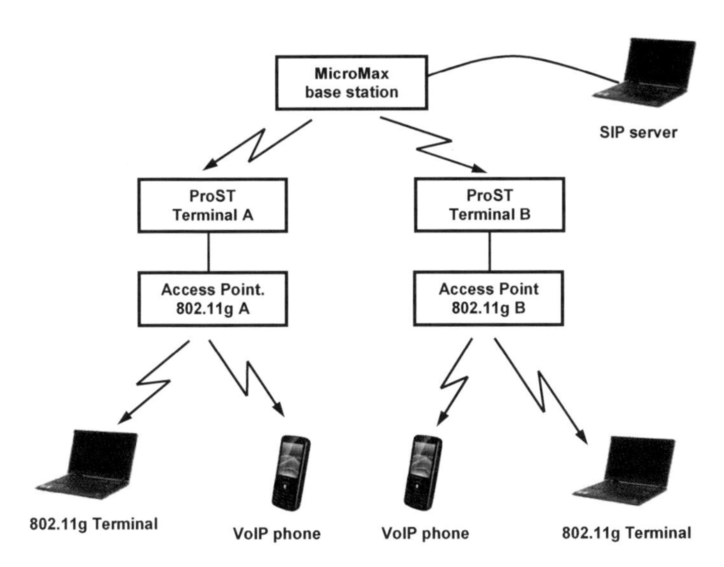

Figure 17.25 Hybrid WiMax/WLAN network.

The hybrid WiMAX/WLAN system consists of WIMAX Base Stations (BSW), which cooperate with WiMAX terminals. The WLAN access points are connected to WiMAX terminals by cable and provide services for WLAN terminals. In this configuration it is possibility to make a direct connection between WLAN terminals and the external wired network, through the access and the backbone networks.

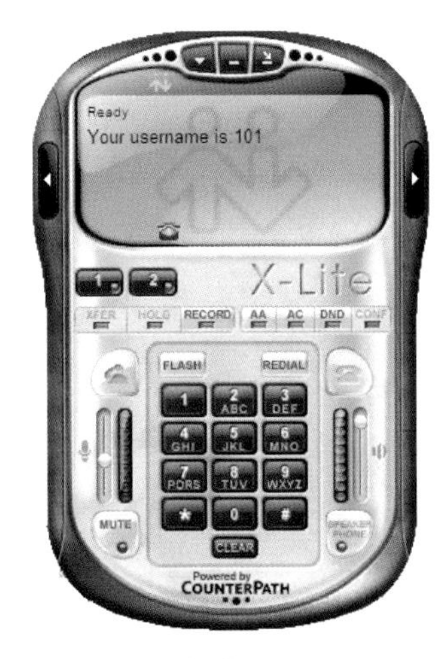

Figure 17.26 Main window of the X-Lite application.

The structure of the hybrid system, based on Airspan's solution (WiMAX 1.5 GHz) and IEEE 802.11g (2.4 GHz WLAN) is shown in Figure 17.25. This configuration was set up in a mining excavation in order to study possibilities of cooperation between systems and test implementation of services. A test network segment uses a MicroMax base station and two Airspan terminals ProST, two 802.11g access points, three laptops equipped with microphones and cameras and two Nokia E51 phones.

The presented network segment was used to test the possibility for implementation of real-time voice and video services. Each laptop, including the laptop with the SIP server was equipped with a microphone and camera. An application allowing calls over the IP protocol was run on the computer – for this we have chosen the free X-Lite program (Figure 17.26).

This application allows video and voice calls using the SIP protocol, which can also be recorded. To make a call a user has to enter the recipients number using the keypad and click on the green handset. Configuration is very simple and it is limited to filling out the following fields in the settings:

- SIP server IP address (e.g. 10.0.0.11),
- user name (in this case 101),
- password (101).

The tests included also two Nokia E51 phones (simple GSM phone). These phones were equipped with software allowing for SIP calls. Nokia E51 phones are able to establish a network connection with 802.11b/g access points. When the phone is connected to a network, it is possible to make a call using the built-in SIP client. Before making such call it is necessary to configure the SIP server IP address, user name and password in the phone.

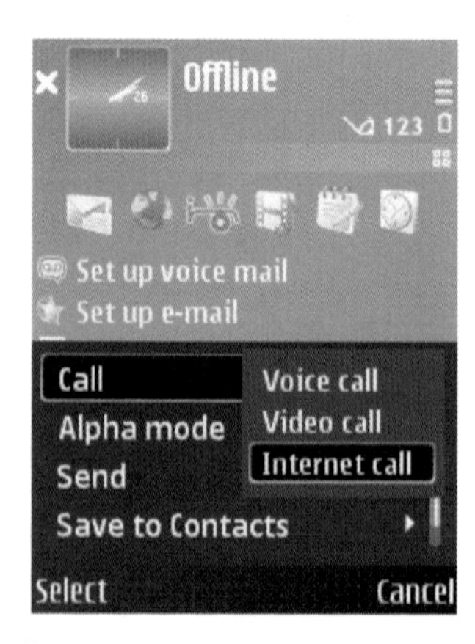

Figure 17.27 Making the SIP call.

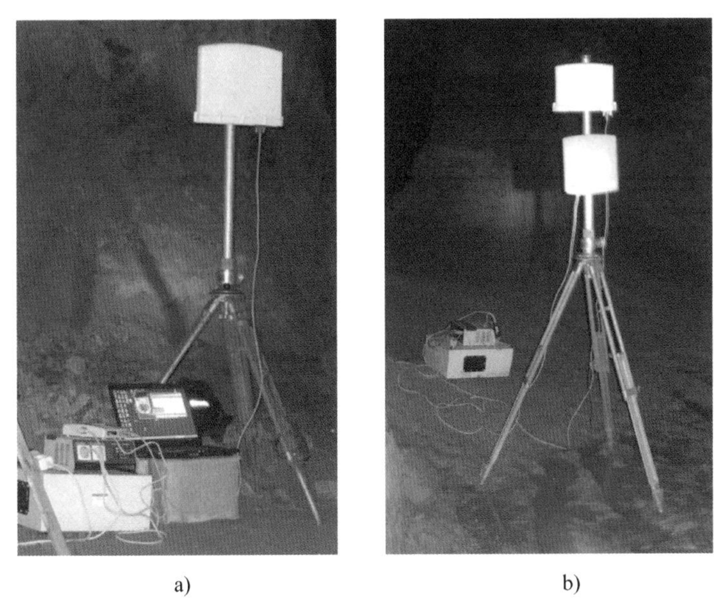

a) b)

Figure 17.28 a) MicroMAX of Airspan system connected to the laptop with SIP server, b) ProST terminal of Airspan system (at the top) and 802.11g access point.

The SIP call is carried out in the same way as a GSM call. The user has to enter the number from the keyboard and from a popup menu can choose to perform an Internet call (Figure 17.27).

Photos of WiMAX devices in the underground mine environments are presented in Figure 17.28a, b and Figure 17.29.

Figure 17.29 ProST terminal of Airspan system (on the right) and 802.11g access point.

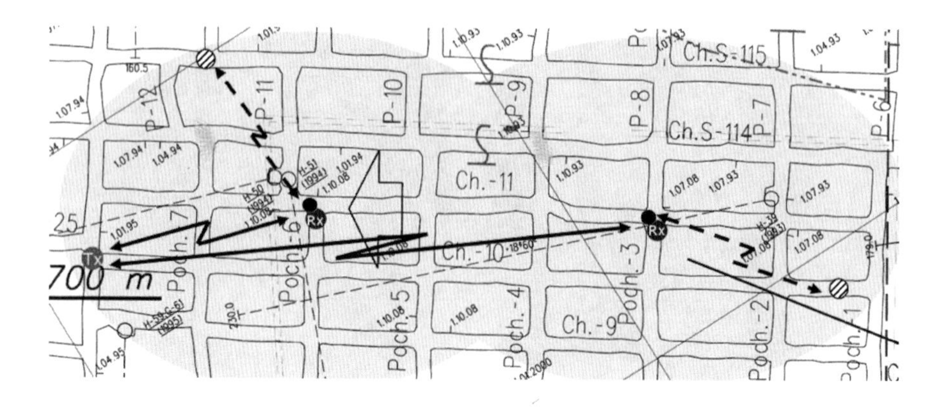

Figure 17.30 The hybrid wireless network test setup in the mine corridors.

During the tests ProST devices connected to 802.11g access points were set at points where earlier a high signal level from MicroMAX base station has been measured. The distribution of the backbone (WiMAX) and access network (WLAN) units is shown in Figure 17.30.

Black solid lines in Figure 17.30 indicate WiMAX connections (backbone) between the base station BSR (Tx point) and client ProST (Rx points) units. According to Figure 17.25, 802.11g access points were connected (black points in Figure 17.30) to ProST client units. Black dashed lines are wireless WLAN connections between access points and laptops or mobile phones playing the role of WLAN terminals. Gray ovals indicated the predicted range of the access network, which was defined on the basis of previous measurements.

We have tested the possibility of establishing data connections and voice or video transmissions between WLAN terminals and between WLAN terminals and a laptop connected to the base station. It was found that the laptop equipped with an antenna having a higher gain than the one mounted in the phone is able to stay connected at a greater distance from access point. Specific ranges achieved for both: the laptop (gray) and phone (black) are shown in Figure 17.31.

Assessment of implemented services has shown that the previously measured delay does not affect the quality of voice conversations. The quality of voice was very good even at the borderline connectivity range. The voice call was dropped at the same time the access point connection was lost. Additionally, wherever tests of voice call were conducted, we have also evaluated video calls, voice and video streaming transmissions. Results of these tests also proved positive. The video stream transmitted through the network was smooth and had good quality. It should be noted, that images transmitted by industrial cameras can be used in the mine - they need less bandwidth than the images sent by web cameras used during the tests.

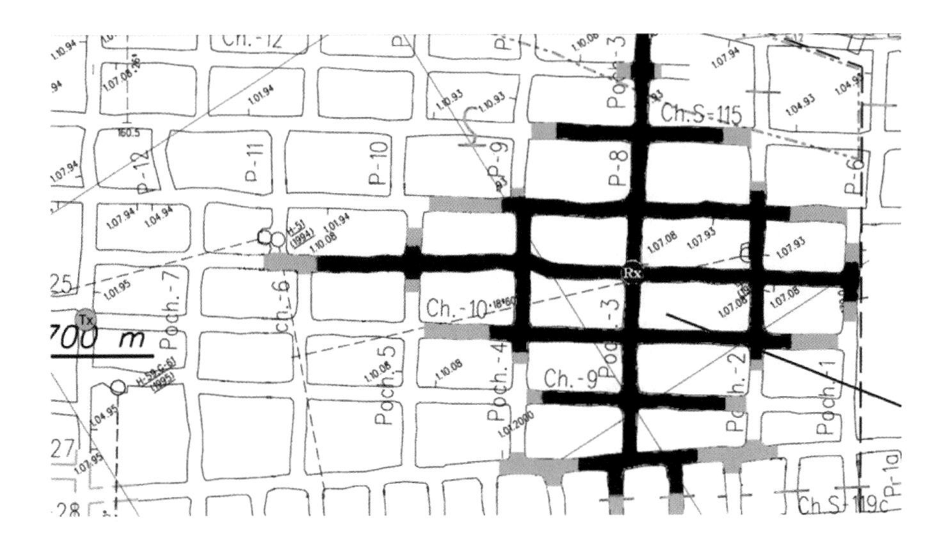

Figure 17.31 Range of the access network.

When two terminals were connected to two different access points the data was transmitted through a total of four wireless links. They had to pass through two wireless links between WLAN terminals and the WLAN access points and two wireless links between WiMAX terminals and the WiMAX base station. Delay through the whole path was about 60ms, but it did not affect the quality of call.

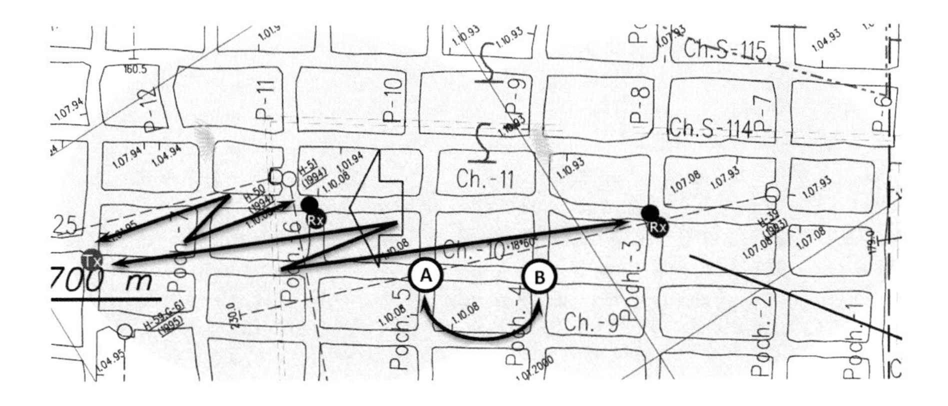

Figure 17.32 Call handover.

An important phenomenon observed during the experiments was handover during the switching from the range of access point A to the range of access point B. The access points were set in such a way that their ranges overlap in a small area (Figure 17.32). During a call, the SIP server holds the voice call when a terminal leaves the coverage area of access point A and when the terminal comes into range of access point B the SIP server automatically resumes the call. The whole operation takes about half a minute and there is no need to re-connect the call.

The assessment results for the hybrid wireless system consisting of a backbone network using WiMAX 1.5 GHz standard connected to a WLAN IEEE 802.11g network are very interesting. They prove that it is possible to provide wireless voice calls, real time streaming video and data transmissions in a range exceeding 100m from the WLAN access points. It is possible to extend the network range to reach designated points of the mine where spoil is produced. This solution is competitive with leaky cable, which is not able to support broadband.

17.7 SUMMARY

The assessment results show that wireless transmission technologies, currently available on the market, are sufficient to build wireless communication systems in underground mines. They are ready to provide telephone services, real time voice and image transmissions and fast data transmission. It was found that for these types of environments and for the expected applications, the best suited are WiMAX systems operating in the 1.5 GHz band. This type of system has proved to have the best resistance to multipath phenomenon. This is why the obtained range was close to 500m, whilst the bandwidth did not drop below 1 Mbps. Because of its size, the WiMAX system end terminals can only be used in mine machines. Therefore, in terms of flexibility and a multi servicing architecture, hybrid networks are proposed. In such networks the WiMAX terminals are only the intermediary stations in the provision of advanced telecommunications services. In the proposed structure WLAN access points are connected to WiMAX terminals. WLAN access points are operating in the 2.4 GHz band and also are using the OFDM technique (IEEE 802.11g/n WLAN). The study shows that they can provide coverage from 150m to about 200m. Their main advantage is low price and widespread availability. Miniature terminals, cards with PCMCIA and USB interfaces and devices having built-in terminals (for instance mobile phones, PDAs, miniature computers, etc.), are widely available on the market. The study showed that the WiMAX system operating in the 1.5 GHz and the 2.4 GHz WLAN can cooperate well in the same area. Due to the shorter range of WLAN systems hybrid network nodes should be properly deployed.

In the mine a phenomenon was observed that electromagnetic wave penetration is weaker in the transverse corridors to the direction of wave propagation. Further measurements were therefore made along with analysis of application of passive repeaters. Additionally a localization system was tested using a chirp signal. The results of this research are very promising, but further description of the above subjects is beyond the scope of this paper.

Experiments performed in the real underground mine environment have shown the (current) best market solutions. The tested devices were typical, the same as used in terrestrial systems. Therefore they should be modified in order to meet the conditions and requirements of underground mining. The modification should involve:

- development of suitable dust-proof and watertight cases,
- development of intrinsically safe (Ex) power supplies,
- modification of hardware and software associated with limitation of system capacity and security (in the underground mine such large system capacity and very advanced security systems are not needed),
- development of simple antennas resistant to mechanical damage.

References

[1] Chirdon D., Barkand T., Damiano N., Dolina K., Dransite G., Hill J., Retzer P., Shumaker W. "*Report of findings: ECT Committee underground communication and tracking system tests at CONSOL* Energy Inc., McElroy Mine", MSHA Technical Support, June 2006.

[2] Grayson L., Beauchamp M., Bumbico A. "*Improving mine safety and training*: Establishing U.S. Global Leadership", MSTTC Commision Report, University of Missouri, December 2006.

[3] Grohman M. "Undergroung cells (in polish)", Polska Miedź, nr 10, s. 7, October 2006.

[4] www.conspeccontrolsinc.com/products/radio/

[5] www.emag.katowice.pl

[6] www.gsmining.com/TGminertrack.htm

[7] www.inova-cit.home.pl

[8] www.minesite.com

[9] www.siamtec.com/SIAMnet/

[10] www.varismine.com/support/

[11] Kowalczyk W.: *The model of hierarchical wireless system in the underground environment* (in polish), MSc thesis, (supervisor: prof. Ryszard J. Zielinski), Wroclaw Univ. of Tech., Electronics Dept, 2009.

[12] ITU-T G.902 Recommendation

[13] www.imtc.org

[14] www.h323forum.org

[15] ITU SG12 – Contribution 37,:"*Draft New Recommendation G.QoSRQT – End-User Multimedia QoS Categories*", 8/2001.

[16] ITU-T Recommendation F.700, "*Framework Recommendation for multimedia services*",11/2000.
[17] ITU-T Recommendation Y.1540, "*Internet Protocol Data Communication Service – IP Packet Transfer and Availability Performance Parameters".*
[18] Draft New Recommendation Y.1541, *Network Performance Objectives for IP-Based Services*, 5/2001.
[19] EN 301 199, *„Digital Video Broadcasting; Interaction channel for Local Multi-point Distribution Systems* (LMD).
[20] Więckowski T. W., Zieliński R. J., Dąbrowski A., Lewicki F.,: *"Multimedia wireless systems (in polish),* STM 2000 Conference, Łódź, March 2000.
[21] ETSI TR 101 274: *„Transmission and Multiplexing (TM); Digital Radio Relay Systems (DRRS); Point-to-multipoint DRRS in the access network; Overview of different access techniques"*, v.1.1.1, 1998-06.
[22] ETSI EN 301 080,:*"Transmission and Multiplexing (TM); Digital Radio Relay Systems (DRRS);Frequency Division Multiple Access (FDMA); Point-to-multipoint DRRS in the bands allocated to the fixed service in the range from 3 GHz to 11 GHz*, v.1.1.1, 1999-02.
[23] ETSI EN 301 753,*"Fixed Radio Systems; Point-to-Multipoint equipments and antennas; Generic harmonized standard for Point-to-Multipoint digital fixed radio systems and antennas covering the essential requirements under Article 3.2 of the Directive 1999/5/EC"*, v.1.1.1, 2001-03.
[24] Siemens, information materials.
[25] *„Broadband Wireless Point-to-Multipoint System OnDemand 10 GHz",* Lucent Technologies, PL/00-WR 469, August 2000.
[26] Lucent Technologies, information materials.
[27] IETF RFC 2678, *IPPM Metrics for Measuring Connectivity*, September 1999.
[28] Staniec Kamil, "*Modeling the influence of walls shapes on radio signal properties in mines", in proc. IEEE International Symposium on Antennas and Propagation & USNC/URSI, Charleston, SC USA, June 1-5 2009. Piscataway, NJ. IEEExplore 978-1-4244-3647-7/09*
[29] Staniec Kamil, *„Radio wave propagation in real mine environment" (in Polish), Przegląd Telekomunikacyjny, Wiadomości Telekomunikacyjne. 2009, vol. 82, no. 6, pp. 482-485*

Chapter 18.

Virtual MIMO Using Energy Efficient Network Coding in Sleep Apnoea Monitoring Systems

Abdur Rahim[1] and Nemai C. Karmakar[2]

Depertment of Electrical & Computer Systems Engineering, Monash University Clayton, Melbourne, Victoria 3800, Australia
[1]abdur.rahim@monash.edu, [2]nemai.karmakar@monash.edu

In wireless sleep apnoea monitoring system, signal propagation paths may be affected by fading because of reflection, diffraction, energy absorption, shadowing by the body, patients mobility and the surrounding environment. MIMO has emerged as a significant technique to combat such fading and to provide enhanced channel capacity. However, for a small sensor node in WBAN systems, it is difficult to install multiple antennas on a sensor due to size limitations. Hence, to take full advantage of the MIMO system, multi-sensor cooperation using energy efficient network coding thus forming virtual MIMO for WBAN is proposed for sleep apnoea monitoring system in this chapter.

18.1 INTRODUCTION

Life is becoming fast and complex due to anxieties, unrest, failure to meet work demands, keeping a balance between work and family, and long working hours. These factors created depressions among people. Sleep Apnoea is a breathing disorder caused by frequent closure of the upper airway during sleep. The symptoms of sleep apnoea include extremely loud heavy snoring, depression, morning headaches, fatigue, learning and memory difficulties and falling asleep while driving or at work [1]. The three different forms of sleep apnoea, for example, central sleep apnoea (CSA), obstructive sleep apnoea (OSA), and mixed sleep apnoea (MSA) constitute 0.4%, 84% and 15% of cases respectively [2].

4G Wireless Communication Networks: Design, Planning and Applications, 375-400,

Researchers have been working on wireless sleep apnoea monitoring system for many years to avoid uncomfortable sleep in an unfamiliar sleep laboratory in traditional PSG-based wired monitoring systems. Using such wireless monitoring system, the patients are able to sleep at own home and their natural sleeping patterns can be monitored without any disturbance due to wires. In existing wireless sleep apnoea monitoring systems, physiological signals such as Electrocardiography (ECG), Electromyography (EMG), Electroencephalography (EEG), Electrooculography (EOG), etc. can be sent wirelessly to the remote base station using the Single Input Single Output (SISO) environment under the Wireless Body Area Networks (WBAN) scenario. In a WBAN system, a variety of sensors are attached to the body, on clothing or even implanted under the skin, where the patient experiences greater physical mobility and they are no longer required to stay in the hospital. WBAN devices have typical communications ranges of up to 5 metres. WBANs applications are divided into three classes; medical applications, applications for assisting persons with disabilities, and entertainment applications [3].

There are several challenges in sleep apnoea monitoring systems under WBAN for transmitting physiological data wirelessly to the remote base station. These include attenuation, fading, multipath effect, delay spread and so on. A wireless signal is severely sensitive to fading. Particularly in WBAN, propagation paths can experience fading due to energy absorption, reflection, diffraction, shadowing by the body, body postures [4], body movement, polarization mismatch and scattering of electromagnetic signals due to the body and the dynamic environment condition [5]. In WBAN system, the other reason for fading is multipath due to the environment around the body. Multipath is the result of the original signal reaching the receiver at different times within a specific transmission time slot. The transmitted signal can take several paths to reach the receiver, that is, directly, after being diffracted or after it has been reflected off another object [6]. For communication between two sensors on the human body, transmitted signals can reach at the receiver in several ways [7], such as-propagation through the body, diffraction around the body and reflections off nearby scatters and then back toward the body. Multi path causes a challenge for any wireless communication system and results in additional complexity of the system design. Moreover, in WBAN, attenuation can be caused due to transmission path length, obstructions in the signal path, and multipath effects. The Multiple Input Multiple Output (MIMO) system uses multiple antennas at the transmitter and receiver to combat such fading as well as to produce significant capacity and diversity gains over conventional SISO systems using the same bandwidth and power in highly fading environment [8]. To the best of our knowledge, no research exists yet related to MIMO-based sleep apnoea monitoring systems. Hence, the application of MIMO in Sleep Apnoea monitoring system will be of great interest for dynamic WBAN.

The high data rate and reliable transmission between body-worn wireless sensors in patient monitoring systems, sports and entertainment and military applications demand the use of MIMO for the WBAN channels [9]. Furthermore, for a small sensor node in WBAN systems, it is difficult to install multiple antennas on a sensor due to size limitations. Hence, to take full advantage of the MIMO system, multi-sensor cooperation (virtual MIMO) using energy efficient network coding for WBAN is proposed for sleep apnoea monitoring. Because, sleep Apnoea patient have various levels of mobility, for example, walking, wheel chairing, eating, sitting, twisting, turning, etc., thus lead to dynamic environment in WBAN. To the best of our knowledge, energy efficient network coding for virtual MIMO-based cooperative communication in WBAN has not been used yet. The application of energy efficient network coding in such highly fading or dynamic environment will be a great interest. Hence the energy efficient network coding for transmitting physiological data from various sensors to the base station to achieve cooperative diversity in sleep apnoea monitoring has been proposed.

In this chapter, Virtual MIMO using energy efficient Network Coding in sleep apnoea monitoring system is proposed. The rest of the chapter is organized as follows: In section 18.2 and 18.3, Network coding technique in WBAN is represented. Section 18.4 shows the cooperative communication using network coding for WBAN channels and section 18.5 represents the sensor node development in WBAN. Finally section 18.6 concludes the paper.

18.2 NETWORK CODING TECHNIQUES

Network coding is a well-designed and novel technique which improves network throughput and performance. The concept of network coding was first introduced by Ahlswede et al. in their paper [10] in order to achieve improved capacity of multicast connections. The principal idea of Network Coding is that each node in the network combines previously received packets and forwards the combination to neighbouring nodes, rather than storing and forwarding individual packet. This leads to fewer transmissions and thus helps to save transmission and reception energy.

18.2.1 Network coding techniques for very highly fading or dynamical environment

One of the research challenges in wireless sleep apnoea monitoring system is that the fading effect should be reduced in dynamic environment. Dynamic environments refer to the networks where the structure, topology, and demands may fluctuate in a short time scale as compared to the data transfer. The main benefit of network coding in a wireless environment is in the situations where the topology dynamically changes [11].Network coding has been used in various fading channel such as two way relaying networks over Rayleigh fading channels and wireless fading channel. We know that Rayleigh fading is a highly fading or dynamic channel in wireless communication. WBAN is also a dynamic environment, therefore network coding will provides a substantial benefit to combat fading for such dynamic environment.

Application of network coding for wireless applications in general has been investigated in the more recent literature. Sundararajan et al. in [12] mentioned that the theory of network coding promises significant benefits in network performance, especially in lossy networks and in multicast and multipath scenarios. Wireless networks suffer from interference and, in some cases, considerable delay. The works related to practical implementations of network coding in wireless networks has been studied in [13-15]. Network coding has been shown to improve throughput and reliability in such practical settings. Network coding algorithms for dynamic networks have been studied in [16-18]. Ho et al. [17] showed that the network coding approach provides substantial benefits in dynamically varying environments. In addition, Network Coding is currently used in highly dynamic topologies such as content-distribution networks [19], peer-to-peer networks [20] ,and wireless networks [13]. Network coding has been used in various fading channel such as two way relaying networks over Rayleigh fading channels [21] and wireless fading channel [22]. We know that Rayleigh fading is a highly fading or dynamic channel in wireless communication. WBAN is also a dynamic environment due to following reasons:

- In most medical conditions, doctors recommend patients to move or walk, as much as they can tolerate, in order to improve their health. Hence Sleep Apnoea patient have various levels of mobility, for example, walking, wheel chairing, eating, sitting, twisting etc., thus lead to dynamic environment in WBAN.
- Based on the posture and the positions of the antennas, part of the body may shadow the line-of-sight (LOS) path. Movement of the body changes the orientation of the antennas, and the shadowing conditions [23], thus create dynamic scenario. Moreover, several studies [24-26] indicated that body movement creates a dynamic WBAN channel.
- •WBAN nodes experience different channel condition, which can vary dynamically on a time-scale within the same order of magnitude of the data transmission time [27].Moreover, in case of WBAN communication, propagation paths can be affected by fading due to energy absorption, reflection, diffraction, shadowing by the body, body postures, and the surrounding environment [4],[5].Hence all these factors contribute to make a dynamic WBAN system.

Based on the discussion in this section, we can conclude that network coding can be used in highly fading scenario or dynamic environment. Since WBAN also creates a dynamic scenario, therefore network coding will provides a substantial benefit to combat fading for such dynamic environment. Furthermore, in the proposed research of wireless sleep apnoea monitoring system for WBAN, the network is often assumed to be small, with limited number of sensor nodes.

Nonetheless, even in larger dynamic networks, network coding can be useful. Hence network coding will provide a significant improvement in terms of energy efficiency and throughput. However, to the best of our knowledge, there is no significant study using network coding in dynamic WBAN system. Hence the application of energy efficient network coding in WBAN to achieve diversity via cooperative communication (Virtual MIMO) scenario in order to combat fading will pose significant research challenge.

18.3 NETWORK CODING TECHNIQUES IN WBAN

Network coding is a novel technique which improves network throughput and performance. It is an interesting method for reliability improvement in WBAN. The fewer number of transmission in network coding schemes help to save transmission and reception energy. Hence the applications of network coding in WBAN will offer significant benefit to monitor patient's physiological data.

18.3.1 Power Efficient Network coding techniques for dynamic WBAN system

In WBAN, sensors mounted on body are required to operate for extended periods of time without much heat dissipation. Hence energy efficiency is especially important. The consumption of energy of WBAN sensor node can be divided into three domains such as sensing, wireless data communication and data processing. Out of theses, majority of energy consumption is due to data communication.

Shi et al. [28] used Network coding in WBAN for energy efficient data transmission. The following description based on [28] shows that Network coding is capable to reduce total transmission energy required in WBAN. Consider a

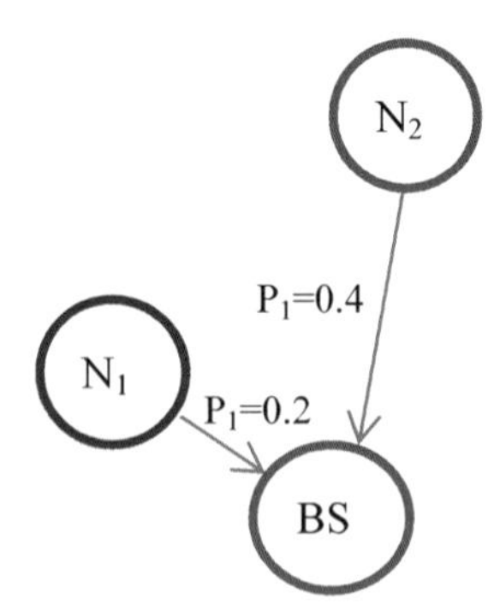

Figure 18.1 2 nodes star topology [28].

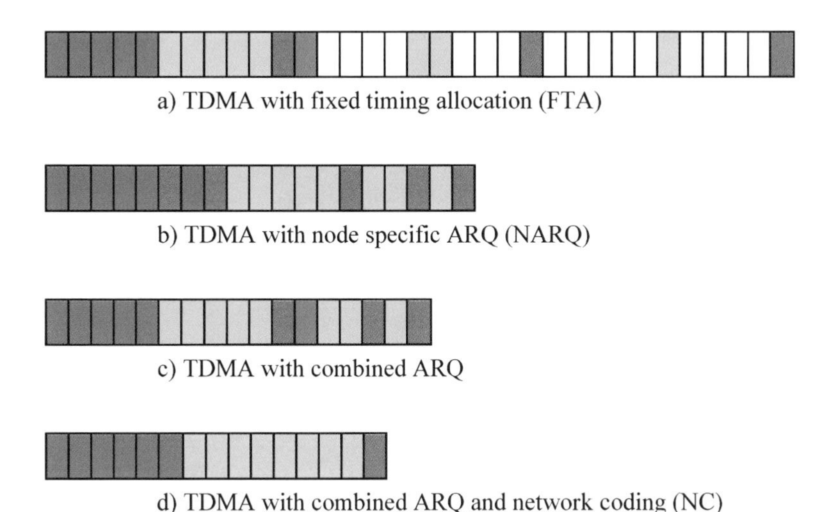

Figure 18. 2 Network coding benefit in WBAN in terms of energy efficiency [28].

two-sensor star network shown in Figure 18.1 with nodes N1 and N2, each trying to directly upload 5 packets to a BS through the same frequency band. In the link layer, assume the packet erasure probabilities are time invariant, at 0.2 and 0.4 respectively. Figure 18.2 shows instances of four different possible communication schemes, all based on time division multiple access (TDMA) with automatic repeat requests (ARQ) [28].Here blue and green cells represent the packets of N1 and N2 respectively and orange cells represent the acknowledgement (ack).

Define a transmission round to be the transmission of data packets by one or more sensor nodes, followed by a broadcasted ack packet. Both nodes wake up at the end of a transmission round to listen to the ack, which contains retransmission requests and schedules for the next round.

(a) Fixed Timing Allocation (FTA): In this scheme (Figure 18.2a), each node is allocated 5 slots per round, and both wake up at the end of each round to receive the broadcasted ack.
(b) Node-specific ARQ (NARQ): Here (Figure 18.2b) each node transmits until all of its packets are received successfully. The ack packet contains retransmission requests for the actively transmitting node and scheduling information for both nodes.
(c) Combined ARQ (CARQ): Both nodes are allocated specific transmission periods each round, with a combined ack packet broadcasted at the end in this case (Figure 18.2c) [28].
(d) Combined ARQ and network coding (CARQ-NC): Each node linearly combines its 5 data packets prior to transmission (Figure 18.2d). Since each coded packet represents an additional degree of freedom rather than

a distinct data packet, more than 5 coded packets can be sent to compensate for anticipated losses [28].

For evaluation of energy use, it is assumed that every data packet transmission and every ack packet reception consumes an equal amount of E units of energy. Table 18.1 compares the total energy required for the schemes shown in Figure 18.2 and also the throughput. Excluding ack periods and time during which nodes are sleeping, all schemes require 12E in data transmission. On the other hand, the energy used for ack reception varies significantly across the different schemes.

According to Table 18.1, it is seen that Combined ARQ and network coding (CARQ-NC) Is the most energy efficient compare with other schemes. The very similar results can be obtained if more sensor nodes are added. Hence network coding is energy efficient in WBAN and it will provide significant benefit by creating cooperative communication in WBAN to reduce fading effect.

Another simple example of how network coding reduces traffic in a broadcast scenario can be seen in Figure 18.3 [29]. Here a source node *S* wants to transfer two bits b_1 and b_2 to both destination *D*1 and *D*2 through the network. Each link in the network is assumed to be error free with capacity of 1. When store-and-forward is used, each intermediate node replicates what it receives and then forwards it to neighborhood nodes. With store-and-forward switching, the network throughput of the butterfly network is dictated by the bottleneck node *C* [29]. Since the capacity of the link between *C* and *D* is one, the node *C* transmits one bit at a time. In this way, 10 transmissions are required to complete data transfer. However, when network coding is applied, the bottleneck node *C* can mix incoming data b_1 and b_2 and compute the exclusive-or (XOR) of the two. Since *D*1 (*D*2) knows both $b_1(b_2)$ and $b_1 \oplus b_2$ it can also figure out $b_2(b_1)$ by taking the XOR of $b_1(b_2)$ and $b_1 \oplus b_2$.

Table 18.1 Comparison of Completion energy per accepted data packet, packet delivery energy per throughput rate [28]

	Total transmission energy, E_{TX}	Energy spent on listening to acknowledgement packets, E_A	Total completion energy per accepted data packets, E_{tot}	Throughput, η	Pkt delivery E per thrput rate
a) FTA	14E	4E	10E/5	10/33	330E/50
b) NARQ	14E	10E	12E/5	10/19	228E/50
c) CARQ	14E	4E	10E/5	10/17	170E/50
d) CARQ-NC	14E	2E	8E/5	10/15	120E/50

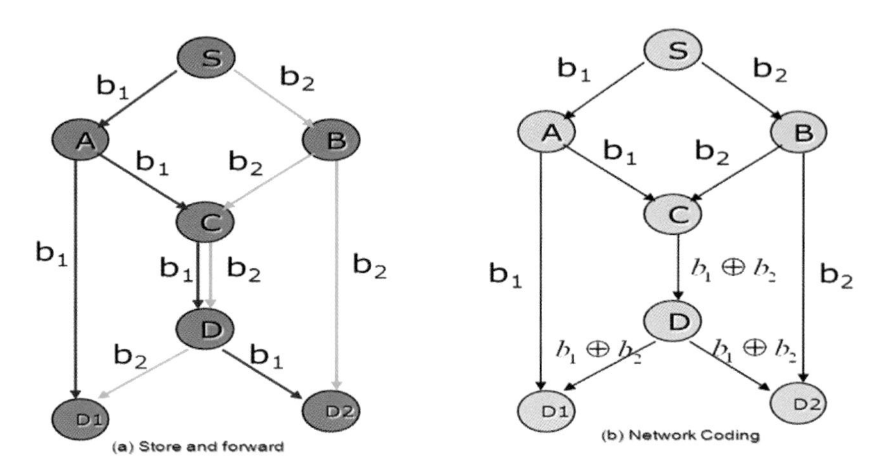

Figure 18.3 Example of network coding (butterfly network) [29].

With network coding, only 9 transmissions are needed. In this case, the number of transmitted bits is reduced and, hence, network coding saves bandwidth and improves energy consumptions.

The discussion in this section represents that network coding is energy efficient in WBAN and also it can be able to handle data from large numebr of sensors, even though in WBANs, the network is often assumed to be small, with limited number of sensor nodes.

18.3.2 Network Coding enhances throughput

Network coding has recently emerged as a promising transmission technology to improve spectral efficiency and system throughput. In this section I shall represent how network coding provides throughput improvement by several network coding scheme. Consider a three nodes traditional network shown in Figure 18.4(a), where nodes *A* and *B* exchange information via relay *R*. Let x_a denote the frame initiated by node.

A who first sends x_a to *R*, and then *R* relays x_a to node *B*. After that, *B* sends x_b in the reverse direction. Hence a total of four time slots are needed for the exchange of two frames in opposite directions.

Straightforward Network Coding Scheme

Reference [30] outline the straightforward way of applying network coding in the three-node wireless network. Network coding can be divided into two generic schemes such as digital and analogue. Digital network coding (DNC) refers to coding at the packet level, meaning that the network coding will XOR the bits of the packets to be encoded. This scheme is also known as Straightforward Network

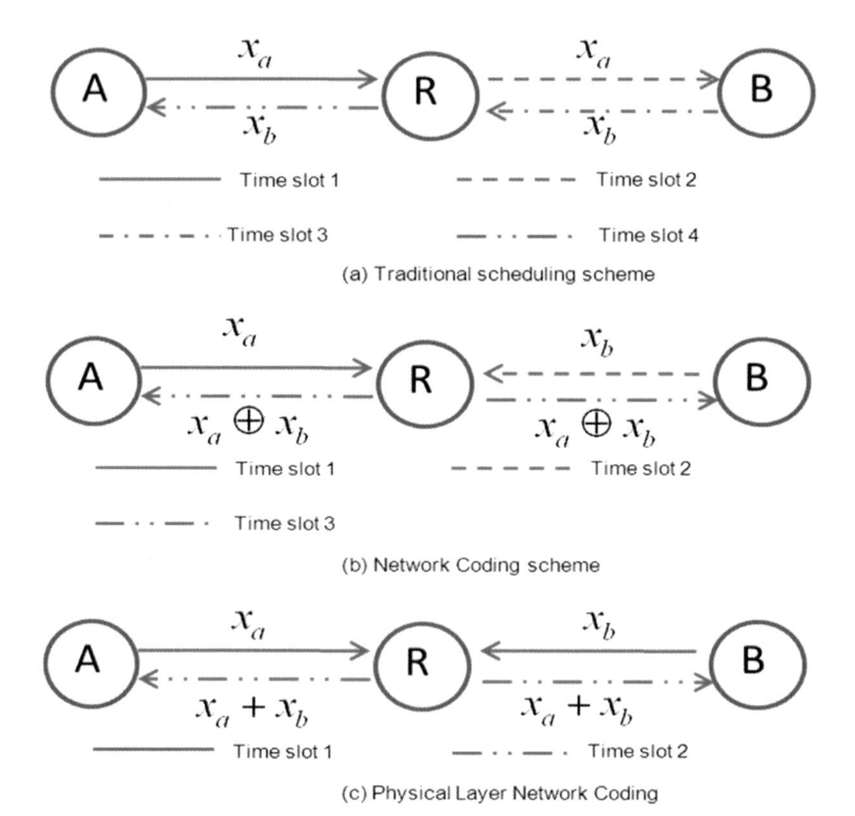

Figure 18.4 Several Network coding schemes.

Coding Scheme as shown in Figure 18.4(b). DNC is implemented in the network layer, where relay node combines bits and forwarding. In digital network coding, senders transmit sequentially, and relays mix the content of the packets and broadcast the mixed version as shown in Figure 18.4(b). First, *A* sends x_a to *R* and then *B* sends frame x_b to *R*. After receiving x_a and x_b, *R* encodes frame as $x_r = x_a \oplus x_b$ .where $\oplus$ denote bitwise exclusive OR operation being applied over the entire frames of x_a and x_b . *R* then broadcasts the encoded frame x_r to both *A* and *B*. When *A* receives x_r, it extracts x_b from x_r using the local information x_a as follows: $x_a \oplus x_r = x_a \oplus (x_a \oplus x_b) = x_b$. Similarly, *B* can extract x_a .Hence a total of three time slots are needed, for a throughput improvement of 33% over the traditional transmission scheduling scheme.

Physical Layer Network coding (PLNC)

Physical-layer network coding (PNC), as proposed is also known as Analog Network Coding (ANC). Analogue network coding refers to coding at the signal level where senders transmit simultaneously. In this scheme, further improvement can be achieved, because to implement the same function as Straightforward Network Coding Scheme, only two time slots are needed as shown in Figure 18.4(c). This scheme provides 100% throughput improvement. The comparison of various schemes is represented in Table 18.2. Therefore, it is observed that straightforward network coding (Figure 18.4b) provides 33% throughput improvement whereas physical layer network coding (Figure 18.4c) provides 100% throughput improvement compare with traditional network (Figure 18.4a).

18.4 VIRTUAL MIMO-BASED COOPERATIVE COMMUNICATION IN WBAN USING ENERGY EFFICIENT NETWORK CODING

Generally, the human body is not a friendly environment for wireless communication. It is partially conductive and consists of materials of different dielectric constants, thickness, and characteristic impedance [31]. Therefore, the human body can significantly influence the behaviour of propagation and lead to high losses. Furthermore, the movement of the body can lead to significant signal fluctuations. In addition, according to the discussion in section 18.21, it is seen that WBAN is a dynamic environment. Diversity is an important tool to combat fading in such dynamic environment. More specifically cooperative communication (virtual MIMO) is an efficient way to attain diversity for small sensor nodes in WBAN. Network coding has an extensive potential for improving throughput, reliability, and robustness of wired and wireless networks [32]. In several papers [33],[34-37], Network coding has been applied to implement cooperative communication schemes. In [38], Shengli et al. [38] investigated cooperative wireless network based on physical layer network coding. However, it has not been applied yet to achieve cooperative diversity in WBAN to the best of our knowledge. Hence the principles of network coding can be applied to implement cooperative communications in WBAN to combat fading.

Table 18.2 Comparison of various network coding schemes

Time slot	Traditional	Network coding	Physical Layer Network coding(PLNC)
1	$A \xrightarrow{X_a} R$	$A \xrightarrow{X_a} R$	$A \xrightarrow{X_a} R$, $B \xrightarrow{X_b} R$
2	$R \xrightarrow{X_a} B$	$B \xrightarrow{X_b} R$	$R \xrightarrow{X_a+X_b} A$ and B
3	$B \xrightarrow{X_b} R$	$R \xrightarrow{X_a \oplus X_b} A$ and B	
4	$R \xrightarrow{X_b} A$		

Cooperative communications (virtual MIMO) is a strategy where users not only transmit their own encoded information but also relays re-encoded versions of other users' information to a common base station. As I mentioned before, a wireless signal is severely sensitive to fading. Particularly in WBAN communications, propagation paths can experience fading due to energy absorption, reflection, diffraction, shadowing by the body, body movement, body postures [4], polarization mismatch and scattering of electromagnetic signals due to the body and the surrounding environment [5]. In order to counter these effects, one strategy is to increase the diversity order of the communications channel by providing several independent copies of the same transmitted signal. Many approaches such as space, frequency and polarization diversity can be used to combat the fading effect in WBAN. However, in such diversity technique, multiple antennas are needed at the transmitter and receiver. Due to small sensor nodes in WBAN system, it is difficult to place multiple antennas. Cooperative diversity, (also discussed in [39, 40]) is an alternate way to attain the performance benefits of MIMO transmissions without using multiple antennas on a single sensor node in WBAN. Hence cooperative communication (virtual MIMO) using energy efficient network coding is another way to attain diversity by cooperation among sensor nodes in WBAN.

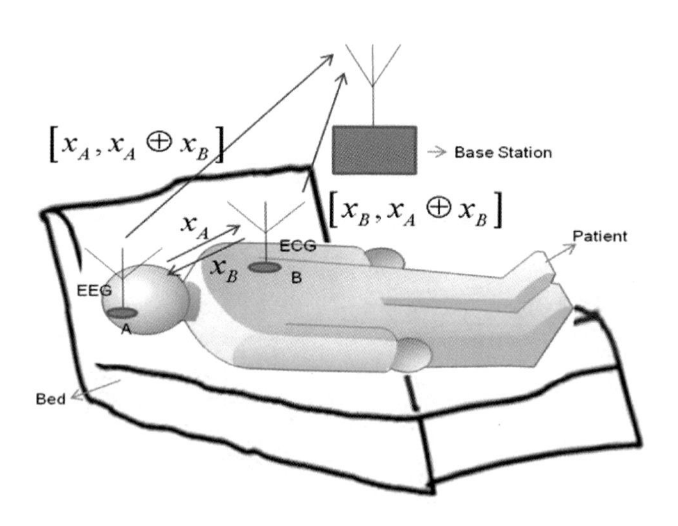

Figure 18.5 Virtual MIMO in WBAN using network coding.

18.4.1 System model for sensor cooperation (virtual MIMO) using network coding

Sensor cooperation has emerged as a promising spatial diversity technique for multi-sensor wireless system. In Figure 18.5, channels between sensor nodes, from sensor node to Base Station (BS), and from relay to BS are considered to follow block-fading Rayleigh distribution [41]. Let h_{ij} represents the channel gain from Sensor Node (SN) i to j. h_{iR} is the channel from SN S_i to relay node R. h_{iB}, and h_{RB} indicate the channel from SN S_i and relay node to destination, respectively. Each node has a power limitation of P_i. Half duplex constraint mode has been considered. Further, it has been considered that each sensor node takes a random binary message $(x_A, x_B) \in \{-1, 1\}$ and sends it to BS [41].

During first phase of transmission, signal received at S_1 from S_2 is [41]

$$y_A = x_B h_{BA} + n_{sA} \tag{18.1}$$

and the signal received at S_2 from S_1 is

$$y_B = x_A h_{AB} + n_{sB} \tag{18.2}$$

After two phases of transmission, Base station receives [41]

$$\begin{bmatrix} y_{b1} \\ y_{b2} \\ y_{bc} \end{bmatrix} = \begin{bmatrix} h_{b1} & 0 & 0 \\ 0 & h_{b2} & 0 \\ 0 & 0 & h_{b1} + h_{b2} \end{bmatrix} \begin{bmatrix} x_A \\ x_B \\ x_A \oplus x_B \end{bmatrix} + \begin{bmatrix} n_{b1} \\ n_{b2} \\ n_{bc} \end{bmatrix} \tag{18.3}$$

y_{b1} and y_{b2} represent the signals received at base station from both of the SNs during first and the second time slots, and y_{bc} shows the signal received during cooperative phase, i.e., during third time slot, from both of the SNs. Partner SNs take hard decisions on the received bits, and the base station uses maximum-likelihood (ML) decoder, to detect the bits x_A and x_B [41].

18.4.2 Investigation of sensor cooperation with network coding

Network coding is an efficient and low cost option for the implementation of cooperative communications [42]. It is assumed that users (sensors) cooperate with each other only in the presence of reliable inter-user channel conditions as shown in Figure 18.5. When data from sensor A and sensor B cannot both be correctly recovered at the base station, a system outage occurs. For both sensor cooperation schemes (with or without network coding), assume sensor A and sensor B transmit with BER p_a and p_b respectively during the first time slot. In the second time slot, transmission from the two cooperating sensors have BERs p_{Ra} and p_{Rb}, respectively. The outage probability for conventional cooperation can be expressed as [43]:

$$P_{s1} = p_a p_{Rb}(1-p_b)(1-p_{Ra}) + p_b p_{Ra}(1-p_a)(1-p_{Rb}) + p_a p_b . p_{Ra} p_{Rb} \tag{18.4}$$

For the network-coded cooperation scheme, The outage probability [43],

$$P_{s2} = (1-p_b)p_a \cdot p_{Ra} p_{Rb} + (1-p_a)p_b \cdot p_{Ra} p_{Rb} + p_a p_b \tag{18.5}$$

Consider $\{p_a, p_b, p_{Ra}, p_{Rb}\} \sim p \sim \dfrac{1}{SNR}$

Equation (18.5) and (18.4) can then be simplified according to [43]

$$P_{s1} = p^2(2-2p+3p^2) \sim \frac{1}{SNR^2} \tag{18.6}$$

$$P_{s2} = p^2(1+2p-2p^2) \sim \frac{1}{SNR^2} \tag{18.7}$$

From equation (18.4) and (18.7), [43]

$$\Delta(P_s) = P_{s1} - P_{s2} = p^2(1-4p+5p^2) > 0, \forall_p \in (0,1)$$

It is observed that both schemes provide a diversity order of 2. Also it has been proved that network coding has lower system outage probability. Following the analysis above, the system outage probability of a generalized network for N sensor node can be written as [43]:

$$P_s = \frac{1}{\binom{2N}{N+1}} p_s^N (1-p_s)^N + \sum_{k=N+1}^{2N} p_s^k (1-p_s)^{2N-k}$$

$$= \frac{1}{\binom{2N}{N+1}} p_s^N (1-p_s)^N + O(p_s^{N+1}) \sim O\left(\frac{1}{SNR}\right)^N$$

which, theoretically, provides a diversity order N. According to outage probability calculation in this section, it is seen that network coding provides significant diversity gain for N sensor nodes.

18.5 WBAN SENSOR NODE DEVELOPMENT

In order to create cooperative communication (Virtual MIMO) scenario in WBAN, several sensor nodes are required to place at various position on body. The following factors are considered in various parts of a sensor node for creating virtual MIMO in WBAN scenario.

1. Half Duplex sensor node: The transmission between sensor nodes is considered as half duplex mode.

   a. Tx/Rx Control: The node should be capable to operate either in transmit mode or receive mode.
   b. Synchronization between node to node: The WBAN sensor networks require a way to synchronize among the nodes.

2. Sensing Data

   a. Data Acquisition: It is the process of sampling signals and converting the resulting samples into digital numeric values that can be manipulated by a computer. The components of data acquisition systems include sensors, Analog to Digital Converter (ADC), Digital to Analog Converter (DAC).
   b. Data Storing: A data storage device is a device for storing information.

3. Data Transmission: The transceiver is required to transmit data.
4. Data Encoding
   a. Applying network coding on sent and received data

5. Data Receiving: The transceiver is used to receive data

a. Node to Node handshaking: Handshaking refers to communication between two nodes that establishes the parameters for continued communication.
b. Storing Received data at memory

A WBAN sensor node is mainly divided into two parts as shown in Figure 18.6. These are: Sensor data processing, Zigbee and netwok coding. Various hardware components of these two parts of a sensor node for collecting ECG signal are described below.

A. Amplifier:

Amplifier is an essential component of sensor data processing part. The electrical signals of heart (ECG) obtained from the patient's body are of very low strength (approximately 2.5mV [44]), low frequency & mixed with noise. The instrumentation amplifier amplifies the low strength signal. The two signals entering the differential amplifier are subtracted to cancel the common noise present in the signal.

B. High Pass Filter and Amplifiers:

High Pass Filter is used to block DC offset present in the signal. If DC is not getting blocked at this level it will increase with gain of amplifiers connected after DC blocker, will induce noise [44]. We want to get higher gain.

C. Notch filter:

Notch filter eliminates the 50 Hz supply noise picked up by body. It consists of one low pass filter and one high pass filter.

E. Analog to Digital Converter:

Analog to digital converter is needed for converting analog ECG signals into digital in order to interface with the microcontroller. In our case, 12- bit Analog to Digital converter is selected in order to get good accuracy.

F. Microcontroller:

The core of the WBAN sensor node is the microcontroller which performs tasks, processes data and controls the functionality of other components in the sensor node. Microcontrollers are the most suitable choice for sensor node and for embedded systems [45] because of their flexibility to connect to other devices, programmable and low power consumption. A general purpose desktop microprocessor is not a suitable choice for sensor node because of higher power consumption compare with the microcontroller [46].We have chosen MSP430 series microcontroller which is an ultra-low-power 16-bit RISC mixed-signal processors from Texas Instruments (TI).

The MSP430FG4618/F2013 experimenter's board is a comprehensive development target board that can be used for numerous applications. The MSP-EXP430FG4618 kit comes with one MSP430FG4618/F2013 experimenter's board shown in Figure 18.7 [47] and two AAA 1.5 V batteries. This experimenter's board is based on the Texas Instruments ultra-low power

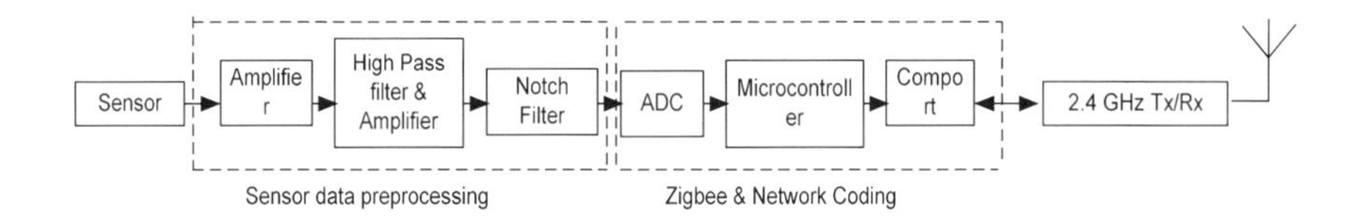

Figure 18. 6 A WBAN sensor Node.

Figure 18.7 MSP430FG4618/F2013 Experimenter's Board (Copyright © Texas Instruments Incorporated MSP430FG4618/F2013 Experimenter's Board [47]).

MSP430 family of microcontrollers. The MSP430FG4618 and the MSP430F2013 microcontrollers are resided on this board. We want to use microcontroller MSP430FG4618 which are microcontroller configurations with two 16-bit timers, a high-performance 12-bit A/D converter, dual 12-bit D/A converters, three configurable operational amplifiers, one universal serial communication interface (USCI), one universal synchronous/asynchronous communication interface (USART), DMA, 80 I/O pins, and a liquid crystal display (LCD) driver with regulated charge pump [47]. This microcontroller is used for controlling analog to digital converter operation as well as to write program of Network coding. Programming of the microcontroller is done in C code.

G. Com port :

The data acquired from memory is converted into serial data & transmitted through Com1 port to transceiver module.

H. CC2420 Zigbee development Kit:

The CC2420MSP430ZDK generally comprising of the CC2420 transceiver and the ultra-low power MSP430FG4618 MCU purchased from Texas Instruments [47]. The MSP430 Experimenters board shown in Figure 18.7

offers the user with functionality such as LCD, capacitive touch-pad, push buttons, buzzer, prototyping space, RS232 communication interface, JTAG Programming Interfaces etc [47]. The CC2420MSP430ZDK is supported by IAR Embedded Workbench for MSP430. IAR Embedded Workbench is used to program/debug the on-board MSP430 devices with the developed code. Programming and debug is done using JTAG1 and JTAG2 which provide interfacing to the MSP430FG4618 and MSP430F2013 respectively. The CC2420MSP430ZDK consists of one CC2420EMK, two MSP430FG4618/F2013 Experimenter Board and one 1 x Programmer (MSP-FET430UIF). The CC2420EMK evaluation kit includes two CC2420EM modules and antennas [47].

MSP-FET430 Flash Emulation Tool (FET)

The flash emulation tool allows the application development on the MSP430 MCU. There are two debugging interfaces, USB and parallel port, respectively named MSP-FET430UIF (Figure 18.8) and MSP-FET430PIF [47]. These two debugging interfaces are used to program and debug the MSP430 in-system through the JTAG interface.

Transceiver

Sensor nodes mostly use of Industrial, Scientific and Medical (ISM) band which gives free radio, huge spectrum allocation and global availability. The various choices of wireless transmission media are Radio Frequency (RF), optical communication (laser), and infrared. Lasers require less energy, however, it requires direct line of sight for communication and it is also sensitive to atmospheric conditions. Infrared, like lasers, need no antenna but are limited in its broadcasting capacity [48].Whereas RF based communication is the most relevant that fits to most of the Wireless Sensor Network (WSN) applications. WSN's use the license free communication frequencies between about 433 MHz and 2.4 GHz [49]. Transceivers are used in sensor nodes which provide the functionality of both transmitter and receiver. The operational states of the transceiver are Transmit, Receive, Idle, and Sleep. As mentioned before, we are using CC2420 transceiver to implement virtual MIMO. The CC2420 is a true single-chip 2.4 GHz

Figure 18. 8 MSP-FET430UIF flash emulation tool (Copyright © Texas Instruments Incorporated MSP-FET430UIF flash emulation tool [47]).

IEEE802.15.4 compliant RF transceiver providing the PHY and some MAC functions. The CC2420 is controlled by the TI MSP430 microcontroller through the SPI port and a series of digital I/O lines and interrupts [47]. An experimental setup of CC2420 with MSP430FG4618/F2013 Experimenter board is shown in Figure 18. 9 at our lab.

18.5.1 Virtual MIMO in WBAN

The sensor node described in the previous section will be used to form virtual MIMO in WBAN. In virtual MIMO, two or more users (sensors) sending messages to a common base station thus form partners to help each other on information transmission. When user 1 (sensor 1) communicates to the BS, the relay (partner user 2) also receives the messages due to the broadcasting property of the wireless medium. Therefore, the information messages from user 1 are transmitted to the BS through two fading paths: one direct path and one indirect path through relay (user 2). For two sensors cooperative communication scenario as shown in Figure 18.5, consider sensor A and sensor B has physiological signals x_A and x_B respectively. Sensor A broadcasts its fresh information (x_A) in the first time slot

Figure 18. 9 TI CC2420 transceiver on MSP430FG4618/F2013 Experimenter board.

which is received by adjacent sensor and the base station due to broadcast nature. In the second time slot, same thing is performed by sensor *B*. Finally in the third time slot, both sensors send refinement information part of their message which is same for both the sensors. Figure 18.10 [43] shows the exchange of information in different time slots between two sensor nodes and BS. This message is formed by taking XOR operation of sensors' own message with the message received from other cooperative sensor. Hence with sensor cooperation-based network coding, the information transmitted from both sensor *A* and sensor *B* can be retrieved correctly, even if the direct uplink fails. For instance, if the base station fails to decode x_A, however x_B and $x_A \oplus x_B$ both arrive correctly, the base station can recover x_A from $x_A = x_B \oplus (x_A \oplus x_B)$.
At the base station side, it detects and saves the messages received by both users during the first phase. In the second phase of transmission by users, it gets the XOR copy of both of the messages and uses this to get the second copy of the message for each user. By this way, multiple copies of the same signal are sent to the BS to attain diversity. This network coding will be implemented at the microcontroller at the sensor node as shown in Figure 18.1 and the data will be decoded at the base station side as shown in Figure 18.11.

A transmits x_A, B and BS will listen	A transmits x_B, A and BS will listen	A and B transmits $x_A \oplus x_B$, BS will listen

→ Time

Figure 18. 10 Exchanging information in different time slots.

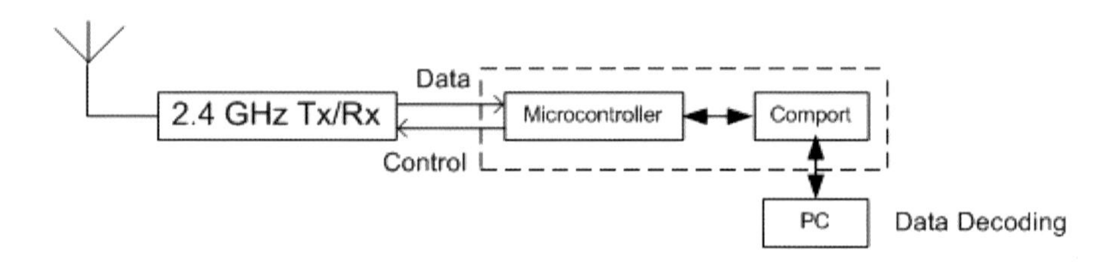

Figure18.11 Base Ststion.

18.6 CONCLUSION & FUTURE WORK

According to above discussion in this chapter, several points can be proved. Firstly, In WBANs, the network is often assumed to be small, with limited number of sensor nodes. Nonetheless, even in larger networks, network coding can be useful depending on the application as discussed. Thus Network coding can be used to handle data from large number of sensors. Secondly, in order to reduce the fading effect in dynamic WBAN channel, network coding can be useful which is shown in various literatures for dynamic environments. Thirdly it has been proved that network coding is energy efficient for WBAN. Finally, network coding can be used to create cooperative communication scenario in WBAN for small and energy constrained sensor nodes. After all, it is feasible to apply Network coding in WBAN to form cooperative communication (virtual MIMO) for transmitting physiological data to the base station. We have provided theoretical description of energy efficient network coding for creating virtual MIMO in WBAN for sleep apnoea monitoring systems. Currently we are working toward the implementation of the sensor node. Also we are working to incorporate all the hardware mentioned before in order to form a virtual MIMO in WBAN. In addition, our concern is that a key requirement of physical layer network coding (PNC) is synchronization among nodes which has been addressed in other wireless networks such as ad hoc and sensor networks etc. However, unlike them, WBAN sensor nodes pose a number of unique challenges such as very low permitted transmission energy by on-body sensors due to their close proximity to human body, low transmission range compared to traditional sensor network and low processing capabilities at each body sensors. Hence, symbol level time, carrier frequency and carrier phase synchronization issue in PNC for WBAN need further research. Therefore this issue will be investigated in dynamic WBAN system.

18.6.1 Acknowledgement

This research is supported by the Regni Health Pty Ltd and Australian Research Council (ARC) linkage project grant (LP0776796: Radio Frequency Wireless Monitoring in sleep apnoea).We would also like to thank Yang Yang and AKM Azad for helping to choose the components. Spccial thanks to Ray Cooper for purchasing components.

18.7 PROBLEMS AND SOLUTIONS

In order to ensure whether a patient has sleep apnoea, doctors usually recommend patients to having sleep test in laboratories. Researchers have been working on wireless sleep apnoea monitoring system for many years to avoid uncomfortable sleep in an unfamiliar sleep laboratory in traditional PSG-based wired monitoring systems. Using such wireless monitoring system, the patients are able to sleep at own home and their natural sleeping patterns can be monitored without any disturbance due to wires [50]. From the existing research on wireless sleep apnoea monitoring systems, it is known that physiological signals can be sent wirelessly to the remote base station only in the Single Input Single Output (SISO) environment under the WBAN scenario. However, a wireless signal is severely sensitive to fading. Particularly in wireless body area network (WBAN) communications, propagation paths can experience fading due to energy absorption, reflection, diffraction, shadowing by the body, body postures [4], body movement, polarization mismatch and scattering of electromagnetic signals due to the body and the surrounding environment [5]. The Multiple Input Multiple Output (MIMO) system uses multiple antennas at the transmitter and receiver to combat such fading as well as to produce significant capacity and diversity gains over conventional SISO systems using the same bandwidth and power [8]. Hence, the application of MIMO in Sleep Apnoea monitoring system will be of great interest for WBAN. However, for small sensor nodes in WBAN, due to size constraints, cooperative communication-based virtual MIMO system has emerged recently as an important research area. Moreover, Sleep Apnoea patient have various levels of mobility, for example, walking, wheel chairing, eating, sitting, twisting, turning which lead to dynamic environment in WBAN. As for a wireless body area network, movements and the low power constraints of sensor nodes can lead to severe losses. The behavior of network coding in such environments is an active area of research. To the best of our knowledge, energy efficient network coding for cooperative communication in dynamic WBAN has not been used yet. Hence, in this research, the energy efficient network coding for transmitting physiological data from various sensors to the base station to achieve cooperative diversity in WBAN has been proposed. Moreover, multi-sensor cooperation with energy efficient network coding will be developed to investigate the performance of the Sleep Apnoea monitoring system.

To the best of our knowledge, energy efficient network coding for cooperative communication in WBAN has not been used yet. Therefore virtual MIMO-based cooperative communication using network coding can be applied in dynamic WBAN system to combat fading. In this chapter, we have demonstrated that WBAN is a highly fading dynamic environment. We have also proved that network coding is an energy efficient technique. Furthermore, we have described several network coding scheme and their throughput, thereby demonstrated that Network coding enhances throughput. In addition, we have shown that Network coding is reliable and it is capable to handle large amount of data from various sensors. Also it is shown that Network coding provides significant diversity gain.

References

[1] Obstructive Sleep Apnoea [Online]. Available: http://www.med.monash.edu.au/medicine/alfred/research/sleep/ob-apnoea.html

[2] Sleep Apnoea [Online]. Available: http://www.tititudorancea.com/z/sleep_apnea.htm [Accessed: 15 Dec 2010]

[3] P. Yunjoong, P. Sang Kyu and L. Ho Yong, "Performance of Wireless Body Area Network over on-human-body propagation channels," in *Sarnoff Symposium, 2010 IEEE*, 2010, pp. 1-4.

[4] Kamya Yekeh Yazdandoost and K. Sayrafian-Pour. Channel model for Body Area Network [Nov 2010] [Online]. Available: https://mentor.ieee.org/802.15/dcn/08/15-08-0780-12-0006-tg6-channel-model.pdf [Accessed: Nov 30, 2010]

[5] I. Khan, "Diversity and MIMO for body-centric wireless communication channels," PhD thesis, School of Electronics, Electrical, & Computer Engineering, University of Birmingham, UK, September 2009.

[6] S. Suthaharan, "Space time coded MIMO-OFDM systems for wireless communications: Signal detection and channel estimation," Master Thesis, National University of SIngapore, 2003.

[7] Jae-Myeong Choi, Heau-Jo Kang and Y.-S. Choi, "A Study on the Wireless Body Area Network Applications and Channel Models," in *Second International Conference on Future Generation Communication and Networking*, 2008.

[8] Abdur Rahim and N. C. Karmakar, "Measurement of correlation coefficient for dynamic WBAN channels in sleep apnoea monitoring system," in *6th International Conference on Broadband Communications & Biomedical Applications(IB2COM11)*, Melbourne, Australia, November 21 - 24, 2011.

[9] I. Khan and P. S. Hall, "Experimental Evaluation of MIMO Capacity and Correlation for Narrowband Body-Centric Wireless Channels," *Antennas and Propagation, IEEE Transactions on,* vol. 58, pp. 195-202, Jan 2010.

[10] R. Ahlswede, C. Ning, S. Y. R. Li and R. W. Yeung, "Network information flow," *Information Theory, IEEE Transactions on,* vol. 46, pp. 1204-1216, 2000.

[11] C. Fragouli, J. Widmer and J. Y. le Boudec, "On the Benefits of Network Coding for Wireless Applications," in *Modeling and Optimization in Mobile, Ad Hoc and Wireless Networks, 2006 4th International Symposium on*, 2006, pp. 1-6.

[12] J. K. Sundararajan, D. Shah, Me, x, M. dard, S. Jakubczak, M. Mitzenmacher and J. Barros, "Network Coding Meets TCP: Theory and Implementation," *Proceedings of the IEEE,* vol. 99, pp. 490-512, 2011.

[13] S. Katti, H. Rahul, H. Wenjun, D. Katabi, M. Medard and J. Crowcroft, "XORs in the Air: Practical Wireless Network Coding," *Networking, IEEE/ACM Transactions on,* vol. 16, pp. 497-510, 2008.

[14] Sachin Katti, Dina Katabi, Hari Balakrishnan and M. Medard, "Symbol-level Network Coding for Wireless Mesh Networks," in *SIGCOMM'08*, Seattle, Washington, USA., August 17–22, 2008, .

[15] P. A. Chou, Y. Wu and K. Jain, "Practical network coding," in *Allerton Conference on Communication, Control, and Computing, Monticello, IL*, October 2003.

[16] Z. Fang and M. Medard, "Online Network Coding for the Dynamic Multicast Problem," in *Information Theory, 2006 IEEE International Symposium on*, 2006, pp. 1753-1757.

[17] T. Ho, M. Medard and R. Koetter, "An information-theoretic view of network management," *Information Theory, IEEE Transactions on,* vol. 51, pp. 1295-1312, 2005.

[18] T. Ho, B. Leong, M. Medard, R. Koetter, C. Yu-Han and M. Effros, "On the utility of network coding in dynamic environments," in *Wireless Ad-Hoc Networks, 2004 International Workshop on*, 2004, pp. 196-200.

[19] C. Gkantsidis and P. R. Rodriguez, "Network coding for large scale content distribution," in *INFOCOM 2005. 24th Annual Joint Conference of the IEEE Computer and Communications Societies. Proceedings IEEE*, 2005, pp. 2235-2245 vol. 4.

[20] W. Mea and L. Baochun, "Network Coding in Live Peer-to-Peer Streaming," *Multimedia, IEEE Transactions on,* vol. 9, pp. 1554-1567, 2007.

[21] L. Wei, L. Jie and F. Pingyi, "Network Coding for Two-Way Relaying Networks Over Rayleigh Fading Channels," *Vehicular Technology, IEEE Transactions on,* vol. 59, pp. 4476-4488, 2010.

[22] B. Du and J. Zhang, "Physical-layer network coding over wireless fading channel," in *Information, Communications and Signal Processing, 2009. ICICS 2009. 7th International Conference on*, 2009, pp. 1-5.

[23] Jun-ichi Takada, Takahiro Aoyagi, Kenichi Takizawa, Norihiko Katayama, akehiko Kobayashi, Kamya Yekeh Yazdandoost2, Huan-bang Li and R. Kohno. Static Propagation and Channel Models in Body Area [Online]. Available:
http://www.ap.ide.titech.ac.jp/publications/Archive/COST2100_TD%2808%29639%280810Takada%29.pdf [Accessed: 30 Nov 2010]

[24] A. Fort, C. Desset, P. deDoncker, P. Wambacq and L. v. Biesen, "An ultra-wideband body area propagation channel model-from statistics to implementation," *IEEE Transactions on Microwave Theory and Techniques,* vol. 54, pp. 1820-1826, April 2006.

[25] D. Smith, L. Hanlen, J. Zhang, D. Miniutti, D. Rodda and B. Gilbert, "Characterization of the dynamic narrowband on-body to off-body area channel," in *IEEE International Conference on Communications ICC '09*, June 2009, pp. 1-6.

[26] D. Smith, L. Hanlen, D. Miniutti, Z. Jian, D. Rodda and B. Gilbert, "Statistical characterization of the dynamic narrowband body area channel," in *Applied Sciences on Biomedical and Communication Technologies, 2008. ISABEL '08. First International Symposium on*, 2008, pp. 1-5.

[27] Xiaomeng Shi, Muriel M´edard and D. E. Lucani, "Network Coding for Energy Efficiency in Wireless Body Area Networks," Massachusetts Institute of Technology2010.

[28] X. Shi, Muriel M´edard and D. E. Lucani, "When Both Transmitting and Receiving Energies Matter: An Application of Network Coding in Wireless Body Area Networks," in *Network Coding Applications and Protocols Workshop NC-Pro 2011*, Valencia, Spain, May 2011.

[29] J. Kim, "Performance Analysis of Physical Layer Network Coding," PhD Thesis, Electrical Engineering: Systems, The University of Michigan, 2009.

[30] Y. Wu, P. A. Chou and S. Y. Kung, "Information Exchange in Wireless Networks with Network Coding and Physical Layer Broadcast," Aug. 2004.

[31] Li Huang, Maryam Ashouei, Firat Yazicioglu, Julien Penders, Ruud Vullers, Guido Dolmans, Patrick Merken, Jos Huisken, Harmke de Groot, Chris Van Hoof, and a. B. Gyselinckx, "Ultra-Low Power Sensor Design for Wireless Body Area Networks: Challenges, Potential Solutions, and Applications," *International Journal of Digital Content Technology and its Applications,* vol. 3, 2009.

[32] M. A. R. Chaudhry, S. Y. El Rouayheb and A. Sprintson, "Efficient Network Coding Algorithms for Dynamic Networks," in *Sensor, Mesh and Ad Hoc Communications and Networks Workshops, 2009. SECON Workshops '09. 6th Annual IEEE Communications Society Conference on*, 2009, pp. 1-6.

[33] X. Lei, T. Fuja, J. Kliewer and D. Costello, "A Network Coding Approach to Cooperative Diversity," *Information Theory, IEEE Transactions on,* vol. 53, pp. 3714-3722, 2007.

[34] L. Xiao, T. Fuja, J. Kliewer and D. Costello, "A network coding approach to cooperative diversity," *IEEE Transaction on Information Theory,* vol. 53, pp. 3714-3722, Oct. 2007.

[35] S. Fu, K. Lu, Y. Qian and M. Varanasi, "Cooperative network coding for wireless ad-hoc networks," *Proc. IEEE Globecom,* pp. 812-816, Nov. 2007.

[36] X. Bao and J. Li, "Adaptive network coded cooperation (ANCC) for wireless relay networks: matching code-on-graph with network-ongraph," *IEEE Transaction on Wireless Communication,* vol. 7, pp. 574-583, Feb 2008

[37] Z. Han, X. Zhang and V. H. Poor, "High performance cooperative transmission protocols based on multiuser detection and network coding," *IEEE Transaction on Wireless Communication,* vol. 8, pp. 2352-2361, May 2009 May 2009.

[38] F. Shengli, L. Kejie, Z. Tao, Q. Yi and C. Hsiao-Hwa, "Cooperative wireless networks based on physical layer network coding," *Wireless Communications, IEEE,* vol. 17, pp. 86-95, 2010.

[39] A. Sendonaris, E. Erkip and B. Aazhang, "User Cooperation Diversity Part I and Part II," *IEEE Transaction on Communication,* vol. 51, pp. 1927-1948, Nov. 2003.

[40] J. N. Laneman, D. N. C. Tse and G. W. Wornell, "Cooperative diversity in wireless networks: Efficient protocols and outage behavior," *Information Theory, IEEE Transactions on,* vol. 50, pp. 3062-3080, 2004.

[41] Gordhan Das Menghwar and C. F. Mecklenbr¨auker, "User Cooperation versus Multiple-Access-Channel with Dedicated-Relay using Network Coding," in *Third Mosharaka International Conference on Communications, Computers and Applications*, Amman, Jordan, 2009.

[42] G. D. Menghwar, W. Shah and C. F. Mecklenbrauker, "Throughput and outage for block-Markov encoding implementation with network coding for cooperative communications," in *Wireless Communication, Vehicular Technology, Information Theory and Aerospace & Electronic Systems Technology, 2009. Wireless VITAE 2009. 1st International Conference on,* 2009, pp. 414-418.

[43] C. Yingda, S. Kishore and L. Jing, "Wireless diversity through network coding," in *Wireless Communications and Networking Conference, 2006. WCNC 2006. IEEE*, 2006, pp. 1681-1686.

[44] A.B. Kanwade, S.P. Patil and D. S. Bormane, "Wireless ECG Monitoring System," *Computer Scienec & Electronic Journal,* vol. 3, 2010.

[45] P. Shah, "Wireless Sensor Networks," Bachelor of Technology, Nirma University Institute of Technology, Ahmedabad, Oct 2010.

[46] Sensor Node – Eus-1001 Tester Manufacturer – Xdectm Ecu [Online]. Available: http://www.ecurepairs.net/ecu/sensor-node-eus-1001-tester-manufacturer-xdectm-ecu/ [Accessed: 5 Nov 2011]

[47] Texas Instruments [Online]. Available: http://www.ti.com/ [accessed: Nov 2011]

[48] O.K. Boyinbode and K. G. Akintola, "Transforming Nigeria Health Care System through Wireless Sensor Networks," *The Pacific Journal of Science and Technology,* vol. 10, pp. 310-323, May 2009.

[49] Sensor Node [Online]. Available: http://en.wikipedia.org/wiki/Sensor_node [Accessed: 22 Nov 2011]

[50] Y. Yang, "Wireless Passive Radio Frequency-Based Monitoring System for Sleep Apnoea Diagnosis," PhD confirmation report, Monash University, 2010.

Chapter 19.

Electromagnetic Interference Immunity Enhancement Techniques and Applications

Hung Q. Nguyen, Jim Whittington and John Devlin

Department of Electronic Engineering, La Trobe University, Bundoora, Victoria, Australia
qh4nguyen@students.latrobe.edu.au, j.whittington@latrobe.edu.au, j.devlin@latrobe.edu.au

In previous chapters, the benefit of electromagnetic propagation in communication was introduced. Electromagnetic radiation is the primary principle behind the operation of transmitters and receivers in radios, televisions, satellites, navigation systems and radars, to name a few. However, electromagnetic radiation also is the core problem behind many of the signal integrity issues in electronic and electrical devices and systems. This chapter explores the negative effects of electromagnetic radiation on electronic systems. The chapter first introduces fundamentals of electromagnetic fields, then switches to classifying sources of electromagnetic interference. Following this, various electromagnetic interference reduction techniques are presented. A case study is then provided and discussing electromagnetic interference assessment methods and immunity mechanisms applied in a real application. The case study focuses on the efforts made in order to enhance electromagnetic interference immunity for a distributed clock system in the multi-transceiver digital TIGER-3 radar. Experimental results carried out during tests are presented to verify the immunity techniques. The main objective of this chapter is to provide a reference for students, engineers, and researchers in order to understand principles of electromagnetic induction and to consider its effects during the design phase of a project.

4G Wireless Communication Networks: Design, Planning and Applications, 401-426,

19.1 INTRODUCTION

Electromagnetic propagation has become an integral of modern technology. In the last decades, electromagnetic applications have exploded quickly with significant achievements in wireless communication over a very wide frequency band. Therefore, it is important that electronic products have to comply with interference limits to ensure *electromagnetic compatibility* (EMC). EMC is defined as the ability of an electronic equipment to successfully operate in its intended electromagnetic environment, and to not generate interference to the environment [21]. It is a major challenge in design of electronic and electrical devices and systems to achieve and improve EMC. Thus, an understanding the mechanisms of electromagnetic emissions and coupling is very important. It is, hence, helpful to firstly review a few general concepts related to electromagnetic radiation.

The foundation of understanding of *electromagnetic fields* (EMF) dates back to the early 1800s. Hans Christian Oersted discovered that when he switched on a current from a "voltaic pile", a magnetic compass nearby moved, confirming a direct relationship between electricity and magnetism [18]. Further study determined that electric currents create magnetic fields, which are circular around the currents, as shown in Figure 19.1.

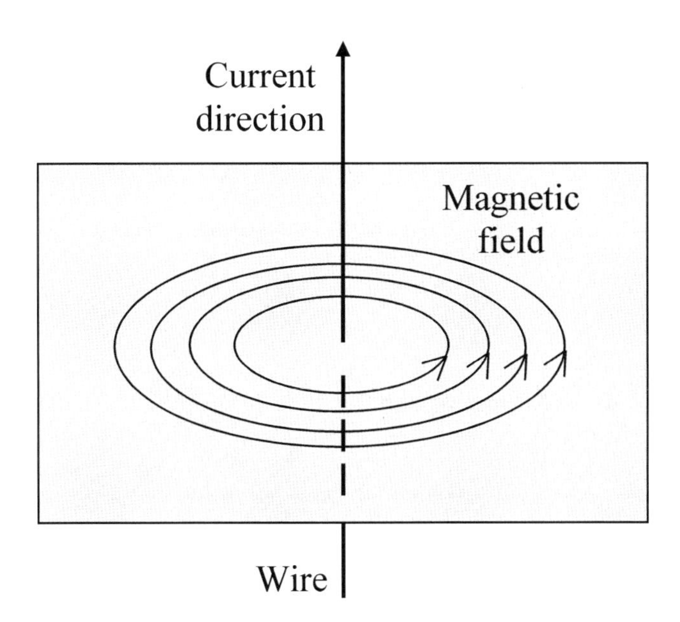

Figure 19.1 Magnetic field of the electric current

Mathematically, there have been various documents analysing basic electromagnetic theory, Faraday's law and Maxwell's equations of electromagnetic induction. These classical theories can be seen in [7, 26, 22, 6]. In light of the aim of this chapter, only the basic relationship between electronic and magnetic fields is discussed. The density of the magnetic field (B) generated by an electric current can be calculated as

$$B = \frac{KI}{R} \tag{19.1}$$

where, B is strength of the magnetic field (T); I is the electric current (A); R is the distance from the wire (m); K is a medium constant.

The magnetic field intensity (H) related to magnetic field with the permeability of the material is denoted as

$$H = \frac{B}{\mu} \tag{19.2}$$

where, H is the magnetic field intensity (A/m); μ is the magnetic permeability of a material (H/m).

In differential and integral form, the magnetic filed intensity generated by a current is commonly expressed in vector notation as

$$dH = \frac{Idl \times \hat{r}}{4\pi R^2} \tag{19.3}$$

$$H = \int \frac{Idl \times \hat{r}}{4\pi R^2} \tag{19.4}$$

where, $\hat{r}$ is a unit vector in the direction of the magnetic field; dl is the differential length of the current;

On the other hand, whenever there is a change of flux, an EMF is generated. Faraday's law of EM induction states that the EMF generated in a conducting loop in which there is a changing magnetic flux is equal to the negative rate of change of flux [4]

$$\varepsilon = \frac{\Delta\Phi_B}{\Delta r} \tag{19.5}$$

where, ε is electromotive force (V); Φ_B is magnetic flux (Wb).

As discussed previously, EMF is the combination of an electric field and a magnetic field running perpendicular to each other. With the above analysis of relationship between electric and magnetic fields, it is clear that EMFs can be coupled to each other. Any change in an electric field on one wire causes a change of current in adjacent wires. The changing of current on one wire also causes a changing of a magnetic field that is perpendicular to the electric field. This changing of the magnetic field can cause an electromotive force, hence, a current flowing in adjacent wires. Therefore, EMFs can be affected by each other by electric coupling and magnetic coupling. Where the electric field is capable of inducing a voltage in a wire, a magnetic field is capable of inducing a current [24]. These coupling mechanisms are discussed more details in the Section 19.3.

When an electronic circuit is negatively affected by an EMF, it is called *electromagnetic interference* (EMI). In other words, EMI is defined as any electromagnetically conducted or radiated voltages or currents that cause malfunctions, performance degradation in electronic and electrical systems [28, 1]. The effects of EMI can range from performance degradation, like interference on radio and television, to critical disruption that can threaten life, such as, the wrong operation of monitoring medical equipments or loss of control in aircraft.

Nowadays, with the increase in clock frequency; the high speed of analog-to-digital converters (ADC), and digital-to-analog converters (DAC), and the high density of integrated circuits, the immunity to EMI is more essential yet more challenging overcome, especially at high frequencies [14]. It is vitally important to understand the mechanisms that generate electromagnetic fields and the mechanisms by which they interfere with electronic devices. This offers designers more solutions to deal with EMI at the beginning stage of system design rather than adopt it at the very final phase of product development. In this chapter, realistic approaches are discussed to suppress the EMI performance in electronic circuits. These approaches range from printed circuit board (PCB) design techniques, such as grounding, differential signalling; to signal processing techniques, such as pulse shaping mechanisms.

19.2 EMI CLASSIFICATION

EMI can be classified in a variety of ways. In [20, 23], Nitsch and Sabath discuss the classification of EMI effects in regard to system level assessment. They focus on a comparison of different manifestations of high-power electromagnetic (HPEM) effects in different systems. In these publications, EMI is characterized by physical mechanism, duration and criticality. The classification by physical mechanism allows to analysis of susceptibility data focused on the physical mechanism, causing the observed effect. The duration effect provides users with information on how long the desired function will be disturbed. While the criticality classification provides the operational impact and the essential information on the functionality of the system. This type of EMI classification is very useful to assess the impact of HPEM.

Alternatively, one can characterize EMI by its bandwidth. Following this, Giri et al. first categorises EMI as intentional and natural EMI [12, 11]. Then they focus on classification of intentional EMI, which is further sub-classified to spectral attributes including narrow band, moderate band, untramoderate band and hyperband. These bands are defined based on the ratio between the high and low frequency of the EM spectrum.

Another way is to classify EMI is due to its duration effect, where EMI can be continuous interference or impulse noise. The continuous EMI generally arises from sources that continuously generate energy, such as man-made transmitters. While impulse noise may be man-made or naturally occurring including lightning, pulse radars, and switching systems.

The aim of this chapter is to provide fundamental information to improve EMI immunity in electronic system design. Thus, the EMI is classified according to the source by which it is generated. The electromagnetic environment consists of natural EMFs and man-made fields.

19.2.1 Natural electromagnetic interference

Natural EMFs can originate from many sources, such as cosmic noise and electrical discharges in the atmosphere. The spectrum of natural EMFs is extreamly small compared to man-made fields. For example, Vecchia et al. state that the main part of spectral energy of lightning discharges pulses is below 100 kHz [27]. In [29], Willett et al. also note that the spectral energy of natural EMFs decays as $1/f^2$ in the 0.2 - 20 MHz frequency band and much faster above that band. At higher radio frequencies, above 30 MHz, the natural EMFs mainly originate from the sun and the extraterrestrial microwave background radiation from space [3]. Solar magnetic storms are common causes of EMI. Its activity tends to occur in cycles that peak in frequency and intensity every 11 years. This peak EMI radiation may damage electronic and electrical equipment over a vast region. As reported by Kappenman [16], the solar storm in March 1989 disrupted the entire power grid in Quebec, leaving 6 million customers without power for 9 hours.

19.2.2 Man-made electromagnetic interference

The everyday use of electronic devices emitting radio frequency has significantly increased in the last decades. Sources generating EMF can be found across the world. These sources include household devices, such as computing devices, wireless routers, microwave ovens, televisions, power transmission lines; to high power transmitters, such as cellular mobile networks, radar transmitters, just to name a few. All of these sources could be radiated or conducted; and hence, may interrupt, obstruct, or otherwise degrade or limit the effective performance of electronic circuits. Man-made EMI can be sub-characterised to intentional and unintentional radiation emitters. Intentional EMI sources are those originating from devices whose primary function depends on radiation operation. These sources include radar, satellite and communication transmitters. EMI also can be induced intentionally, as in some forms of electronic warfare [1]. While unintentional EMI sources are emitted from devices that transmit radio frequencies, although their primary function is not to radiate energy. Switching power supplies, transmission power cables and electric motors can be considered as sources of unintentional EMI.

19.3 EMI COUPLING MECHANISMS

In order to reduce performance degradation caused by EMI, understanding the methods that EMI couples to electronic devices is very important. Generally, there are two principle ways in which EMI can be coupled from sources to devices, radiation and conduction [5]. In the other words, EMI penetrations are the coupling between adjacent circuits since parasitic voltages and currents are propagated via interconnections. Coupling through the radiation of electromagnetic fields is either in the form of emission within circuits or in the form of external fields to the circuits. Conducted penetrations may consist of EMI conduction through power, control and communication cables; whilst radiated penetrations come from wireless transmitters, such as radars, radio stations and even from spaceborne emitters.

19.3.1 Conduction

Conducted EMI is the noise that travels along electrical conductors, wires and electronic components. There are two coupling types of EMI conduction: common-mode and differential-mode. Common-mode is caused by injection of the EMI to ground plane, inputs, outputs and cables, since these cables and traces act as antennas. Katzir [17] notes that common-mode EMI can be generated by the switching of a parasitic capacitance in the components of power circuits, i.e. transistors, diodes, and transformers. This results in unwanted current flowing to the ground of the circuit. Whilst the current loops formed by the ground plane and signal tracks cause differential-mode noise [25]. This differential-mode noise is unavoidable and needs to be minimized.

Most conducted EMI problems are caused by power supplies. The main reason is that electrical grid spreads in a vast region, it is vulnerable to radiation energy. For example, EMI can induce into the electrical grid via main station, power lines and devices connected to it.

EMI also can be generated internally in digital circuits. Since DC currents may be interrupted due to the operation of digital elements. Such sporadic changes in currents cause EMI noise which is conducted to other circuits via power or interconnection tracks. Some components such as crystal oscillators, clock generators are apparent EMI sources. As clock speeds have increased higher and higher frequencies, plus these clocks tend to be distributed throughout a large system, EMI from these devices has a high potential for coupling into signal tracks or I/O ports. This type of EMI source significantly contributes to overall EMI.

19.3.2 Radiation

Radiated EMI is noise that propagates through the air in forms of magnetic fields or radio waves. This EMI induces voltages and currents in other circuits, and hence, can degrade their performance. Unlike common impedance coupling, there is no requirement of a conduction path. Radiated EMI is also known as radio frequency interference (RFI). This type of coupling can be further classified into near-field and far-field EMI. For the far-field radiation, EMI propagates outward as electromagnetic waves through free-space. As unintentional antennas can be formed by PCB traces, internal and external cables, radiated EMI maybe directly induce into circuits. The near-field EMI is also known as cross-talk. This coupling is not a direct connection as conducted EMI. Cross-talk can be coupled if there is any mutual impedance between two circuits located very close to each other. Through this mutual impedance, which can be inductive or capacitive, quick changes of voltages or currents in one circuit can induce a voltage or current to the circuit next to it. Capacitive cross-talk is more common than inductive, since inductive cross-talk is caused by magnetic couplings. A current flowing on a signal generates a magnetic field surrounding it. This magnetic field, in turn, induces a current in a parallel signal. While capacitive cross-talk directly couples to the voltages of adjacent signals. Radiated EMI has a more significant affect on electronic and electrical devices, and has become unavoidable with our increasing dependence on wireless equipment.

After understanding the fundamentals of EMI coupling mechanisms, in the next sections, EMI reduction techniques are discussed, followed by a case study illustrating their application.

19.4 EMI REDUCTION TECHNIQUES

The intrinsic goal of EMI immunity enhancement is for a circuit to reduce and suppress EMI emitted locally from the circuit or radiated from the external environment, so as to guarantee the performance of the system. There are a number of methods for dealing with EMI. In the past, shielding was the most prevalent method used to decrease EMI. Shielding is a method in which the circuit is isolated from the electromagnetic environment. Another method is grounding, in which the inductance levels in interconnections are minimized. A third method involves the use of EMI filters, with inductors and capacitors placed appropriately according to the circuit impedance. One powerful technique to reduce EMI is the use of spread spectrum modulation for signalling, which spreads the radiated energy over a wider frequency range. This section covers techniques that are simple to apply, yet provide effective EMI reduction. Solutions which relate to system design and board layout are discussed first, followed by solutions that can be easily employed later in the product design cycle.

19.4.1 Grounding

Many EMI problems can be solved by having proper grounding, which is usually a straightforward proposition. With proper grounding, EMI caused by electronic and magnetic flux coupling or common impedance coupling can be reduced. To achieve good grounding, inductance levels in interconnections should be minimized [15].

A key technique to reduce EMI is to keep the ground return path as short as possible and minimize the current loop area between a current path and its return path. This can be explained by the relationship between EMI radiation or absorption, current, loop area and frequency in the following equation

$$EMI(V/m) = KIAf^2 \tag{19.6}$$

where, K is constant of proportionality; I is the electric current (A); A is the current loop area (m^2); f is frequency (MHz);

Another way is to reduce inductance in a circuit is to use power and ground planes. Typically, planes provide the lowest inductance path for power supplies. Therefore, using planes is one technique to control ground bouncing and power instability. Although it is important to note that splitting ground and power planes should be avoided, as doing so may create different reference voltages.

Separating digital logic grounds from analog grounds also helps to reduce noise coupling. The technique becomes more critical as high resolution ADCs and DACs are used. For example, if a 14-bit DAC are used, effective operation requires at least 84 dB of noise isolation. Therefore, noise induced by stray digital current flowing to analog ground region must be considered. Otherwise, this noise might cause a different voltage between the analog ground pin and the digital ground pin of the DAC which brakes the noise isolation requirement. In order to prevent this problem, the analog region and digital region must be separated. Also, only one common connection point directly under ADC/DAC should be used to connect the two regions. Importantly, this connection must be kept as short as possible so as minimize voltage difference between the analog and digital pins.

19.4.2 Differential signalling

Differential signals are usually carried over a twisted pair, as such, external noise, including crosstalk and electromagnetic noise is uniformly induced on both signal wires. The receiving circuit determines the signal level based on the voltage difference between the two wires, thus, reduces the impact of external noise. Therefore, differential signalling is less susceptible to noise than single-ended signalling.

19.4.3 Capacitive decoupling

Decoupling capacitors help to minimize return current loops, as they allow high frequency currents on the power planes to reach ground. Otherwise, these currents may return to ground via I/O cables or power connectors, creating larger return loops. Thus, using decoupling capacitors is a very cheap, yet effective solution for many EMI problems. To achieve proper effect, decoupling capacitors should be placed as close as possible to the power pins of associated devices. Although the bypass frequency depends on the value and type of the decoupling capacitor, so the choice of capacitor(s) must be carefully considered. For example, tantalum capacitors are more effective at higher frequencies than aluminium electrolytic capacitors [15].

19.4.4 EMI filtering

EMI filters constructed from capacitors and inductors are very useful to eliminate high frequency noise. EMI filters not only protect from external noise being induced into circuits, but also prevent internally generated noise being radiated, which could affect other systems. Determining the configuration and the values of capacitors and inductors depends on the frequency of EMI noise that needs to be filtered and the impedance of the node that requires the filter. A high-impedance node requires a predominately capacitive load, while a low-impedance node

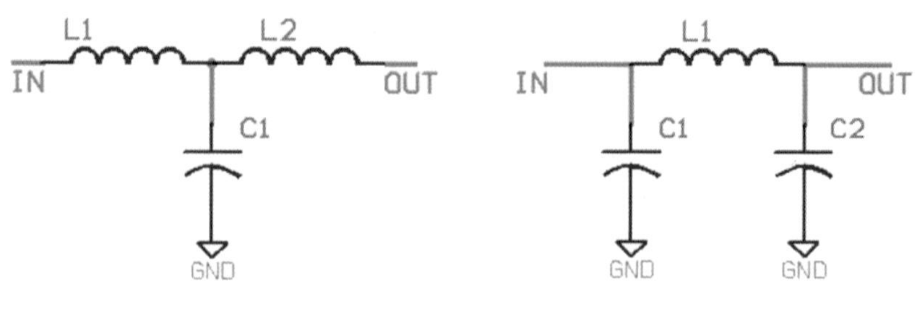

Figure 19.2 T-circuit and PI-circuit lowpass filters.

requires a predominately inductive load. These configurations can be in forms of T-circuits, PI-circuits or L-circuits. For example, Figure 19.2 depicts the T-circuit and PI-circuit, which are suitable for low-impedance and high-impedance loads, respectively.

19.4.5 EMI suppressing with ferrites

As discussed previously, common-mode noise caused by EMI energy coupled to power planes and ground pins is almost exclusively transferred via input/output cables at frequencies lower than 133 MHz. In this case, the EMI energy is capacitive coupled to the shield. Hence, shielding has very little effect, even with differential signalling [1]. To attenuate high frequency common-mode EMI, ferrites with high permeability are used. Since the impedance of ferrites depends on frequency, at high frequency they act more like a lossy transformer, and hence, can be used to suppress radiation.

19.4.6 Spectrum spreading

RF radiation is the main EMI source that significantly contributes to environment interference. The more powerful the RF signal of a given frequency, the more severely it interferes with other devices. Therefore, one can reduce the interference by spreading the spectrum energy of the signal over a wider frequency band. The spectrum spreading technique is primarily applied to square-wave signals. Square-wave signals include both the fundamental frequency and odd multiples of the fundamental frequency, with energy contained in both the fundamental and the harmonics. The most severe radiation involves the fundamental, the third, and fifth harmonic of the signal. When spreading at a suitable degree, harmonic components are buried under the background noise, thus, EMI generated by these harmonics are reduced. Katrai et al. note that by using spread spectrum, radiation can be lowered from 7 to 20 dB depending on the degree of modulation [25].

19.4.7 Shielding

The last EMI minimizing technique discussed in this section is shielding. To protect electronic devices from EMI, circuits, especially radiating devices shielded with metal cases is often used. Electromagnetic shielding is also used to prevent EMI from radiating its energy out of the shielded device. In order to protect large areas from electromagnetic radiation, shielding paints are an another choice. The electro-conductive coatings can protect people and sensitive high-technology equipment against EMI for whole buildings, such as, in nurseries and hospitals. This guarantees that important medical data is derived correctly and will not be altered by EMI.

For an effective shielding, gaps must be sealed properly. Ideally, there should be no gap larger than 1/10 of a wavelength of the fastest operating frequency. Furthermore, the shielding material and thickness also play important roles. As denoted in Equation 12.3, magnetic field intensity is inversely proportional to the magnetic permeability of the material. This means that materials with high permeability provide better shielding performance. The shielding thickness chosen depends on the EMI strength and its frequency. Since the higher frequency and the stronger EMI, the deeper the EMI penetrates the skin depth of a shield.

19.5 CASE STUDY - EMI IMMUNITY ENHANCEMENT FOR DISTRIBUTED CLOCK SYSTEM IN DIGITAL HF RADAR

In this section, EMI suppression techniques applied to a real application, the digital TIGER-3 radar system, are discussed. Firstly, the TIGER-3 radar is briefly introduced to provide basic understanding of the system. Thereafter, EMI immunity enhancement methods including filtering, differential signalling, magnetic transformer usage, pulse shaping and lastly, shielding are presented. Accordingly, experimental results are provided to demonstrate the EMI immunity improvement.

19.5.1 TIGER-3 system architecture

The Tasman International Geospace Environment Radar (TIGER) is a part of the Super Dual Auroral Radar Network (SuperDARN) which is an international network of HF radars dedicated to observe the auroral and polar cap ionosphere [13, 9, 10]. Existing TIGER operations consists of a dual HF radar operating over the 8-20 MHz frequency band. The main antenna array consists of 16 log-periodic antennas. The beam formed by this array scans over 52° of azimuth in 16 steps separated $\approx 3.25^o$ by using a phasing network. The beam widths are 4° at 10 MHz, 3° at 14 MHz and 2° at 18 MHz [10]. The phasing network consists of delay lines, which produce relative time delays between the transceivers as appropriate for the selected beam direction. These time delays vary with gain settings, component ageing and environment changes, hence significant calibration and maintenance is required. These limitations will be reduced in TIGER-3 Radar, currently under development, where FPGA technology is being used to implement multi-channel transceivers and control radar timing.

The TIGER-3 radar is fundamentally a digital system, which consists of 20 *Transceivers*, a *Main Computer* and a *Timing Hardware Server* (Figure 19.3). The radar operates in 8 - 18 MHz frequency range. Each *Transceiver* connects to a twin terminated folded dipole (TTFD) antenna [8] and transmits pulses at 2.4 kW with a duty cycle of up to a 6.5% . The radar performance provides a range of more than 5000 km and a field of view (FoV) of up to 90 degrees.

The digital *Transceivers*, realised on Xilinx Virtex-5 SX50T devices receive control data from the *Main Computer* via a *Gigabit Control Network* connection. This data is processed to extract operational parameters for the set-up of each *Transceiver*. The operational parameters include phase offset, transmit frequency, beam number etc. Received data streams from the *Transceivers* are sent back to the *Main Computer* through the Ethernet interface, where they are processed and merged to form the standard SuperDARN data set [19].

The *Timing Hardware Server* (THS) is implemented on an FPGA Virtex-5 FX70T device. *Timing signals* including a common *system clock* and a *start pulse* signal are generated by the THS to coordinate the operation of the whole system. For the system to work, transmission and reception operation of *Transceivers* are clocked by the tightly synchronised clock. Also, to synchronise pulse transmission on all *transceivers*, a common *start pulse* is used to indicate the start of every pulse sequence; whilst on reception, each transceiver must wait for a specified period of time after the *start pulse* to start sampling backscatter signals. The TIGER-3 radar is being developed as an "all digital" radar with 20 integrated digital transceivers, as such, accurate coordination of all 20 transceivers is essential for the generation of transmit signals, and, collection and merging of receive data to form a standard SuperDARN data set. Accordingly, the system clock frequency must be highly stable and be tightly synchronised. In order to achieve this, a clock synchronisation method to coordinate the operation of entire system using a highly stable, accurate common clock source distributed to each transceiver has been implemented. In the next part, the hardware architecture of the distributed clock system is presented.

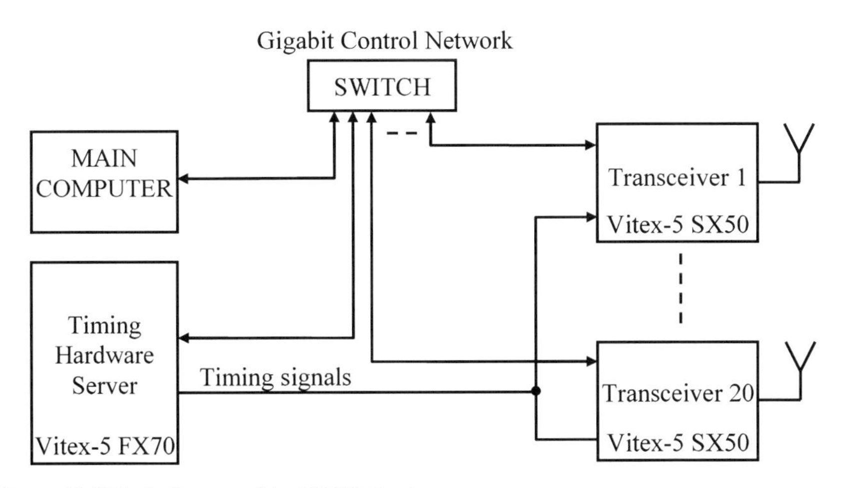

Figure 19.3 Block diagram of the TIGER-3 radar system.

19.5.2 Clock synchronisation hardware architecture

In order to synchronise 20 transceivers for the radar system, the THS board generates a common clock and performs the synchronisation strategy for all transceivers. With correct distribution this will ensure that all transceivers operate at the same frequency within the specified phase alignment. Since a common clock is provided throughout the system, only relative phases between transceivers are important. The THS is implemented on a Xilinx Virtex-5 device which provides very flexible, high performance clocking features. Two of these features are the Digital Clock Managers (DCMs) and Phase-Locked Loops (PLLs). The DCMs are able to synthesise high frequency clocks with low jitter and to correct clock duty cycle. While the PLLs can be used as a clock jitter filter and can also serve as a frequency synthesiser in conjunction with DCMs for a wide range of frequencies [30].
The hardware for the proposed technique is illustrated in Figure 19.4. Instead of utilising a highly accurate and stable crystal oscillator in each transceiver, which would be expensive and yet provide no guarantee of exact frequency and phase lock. A master clock is generated from a single oven controlled crystal oscillator (*OCXO*) circuit which offers frequency stability and low clock jitter. A dedicated *Clock Synthesiser* chip then synthesises the master clock to 125Mhz, which is the required clock frequency within the radar system. The synthesised clock works as the clock source in the THS. In the FPGA fabric, a DCM is used to correct the clock duty cycle and filter jitter. Based on the *Control Data* sent from the *Main Computer*, the *start pulse* is generated and synchronised to the common clock used to operate and synchronise the *Transceivers*. Since transceivers and THS are physically located in different racks, the *Clock Distributor* then distributes these *timing signals* utilising equal length CAT-6 cables to each of the 20 *Transceivers*. On the transceiver side, where Virtex-5 SX50T devices are deployed, the *timing signals* are reconstructed using PLLs in conjunction with DCMs.

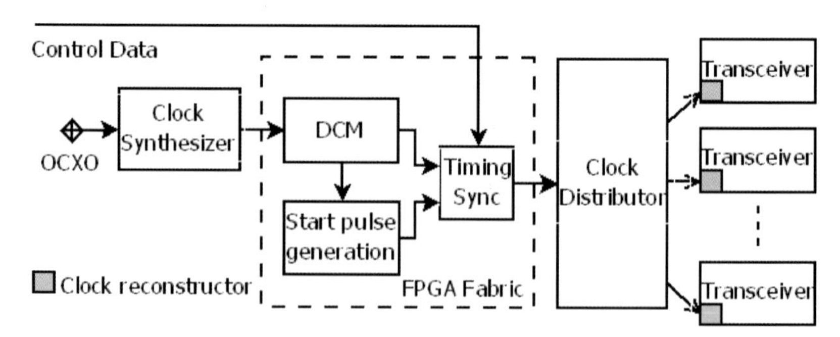

Figure 19.4 Block diagram of the clock synchronisation system.

On each transceiver, there is also a locally generated clock created via a standard 125 MHz crystal. The *local clock* is used for functions that do not require synchronisation, such as, Ethernet data transferring and general operation monitoring. This allows basic operation of the transceiver for troubleshooting in case of loss of the synchronised clock.

As each transceiver is designed to transmit at 2400 Watts full power, EMI radiated by the whole system is extremely strong. This EMI could inject back into the system and cause spurious start pulses and clock distortion. Therefore, a vital requirement is that timing signals must be immune to EMI. The focus of the remainder of this section is on addressing the issue of correct timing signal generation and EMI immunity to ensure correct system operation.

19.5.3 Experimental models and EMI noise assessment

To determine the seriousness of any EMI and to minimize EMI affects, accurate information including magnitude and frequency needs to be examined. Generally, EMI can be analysed in frequency-domain and time-domain. For the experiments, a spectrum analyser was employed to assess EMI in the frequency-domain. As the synchronised clock is distributed to all transceivers, EMI can be injected to the clock system in any section of clock path. For a thorough clock examination, the EMI noise assessment must be taken at the far-end of the distribution system. Accordingly, experiments are performed at the transceivers. Furthermore, on transmission, the synchronised clock is fed to DDSs to synthesise the transmit signal at RF frequency. Thus, EMI immunity of the clock system can be examined in frequency-domain based on synthesised RF signal analysis.

In the time-domain, EMI does not interfere with the clock consistently on every clock cycle, as such, EMI impact cannot be analysed by simply examining the clock signal waveforms at any point in time. Therefore, a custom monitoring circuit was built to assess EMI effects over a long period of time.

19.5.3.1 Frequency-domain

An intrinsic requirement for the radar is the maintenance of the quality of the transmit signals of transceivers. Hence, for our purposes it is important to test the clock quality at the end of the clock paths. The experimental model for these measurements is depicted in Figure 19.5. The synchronised 125MHz clock *Sync. Clock* is fed to a DDS which is responsible for generating the carrier frequencies within the 8-20MHz range. Two 14-bit Digital-to-Analog Converters (DACs) translate the digital output from the DDS to analog signals at RF frequencies. The analog signal from the DAC1 is amplified by a power amplifier and then fed to a dummy load. While the quality of DAC2 output waveform is examined using an HP 4395A spectrum analyser.

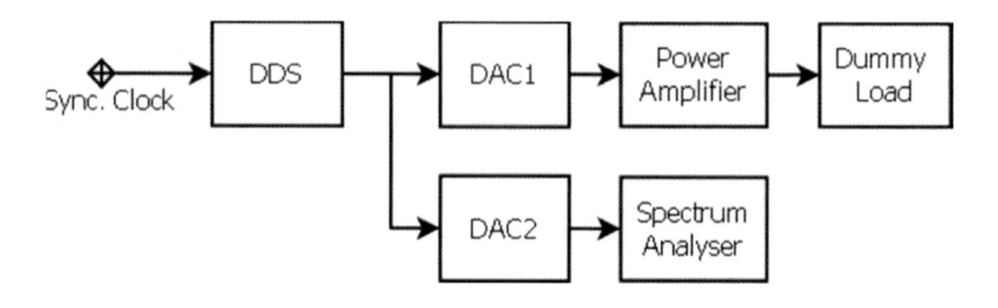

Figure 19.5 Noise suppression assessment block diagram.

19.5.3.2 Time-domain

Since, the synchronised clock runs through a vulnerable circuit with long cables, there more EMI noise on the clock line compared to the local clocks in the transceivers. In order to analyse the affect of EMI on the clock system, a clock monitoring circuit was constructed to compare the synchronized clock sent from THS and a local clock on each transceiver. The circuit monitors the difference in the number of clock cycles every second between the local clock and the distributed clock generated by THS. As the circuit is built between clock domains, metastability must be considered. The hardware block diagram of the circuit built by fabric on Xilinx SX50T device is shown in Figure 19.6.

The *Loc. clock counter* counts the number of local clock cycles and generates the *Enable* signal, which activates the *Syn. clock counter* to count the distributed clock in every second. The *Subtracter* calculates the difference in the number of clock cycles from *Loc. cycles* and *Syn. cycles* sent from the two counters. This difference is sent to the Main Computer to be plotted on the fly during the experiment. In order to avoid metastability problem, all data must be

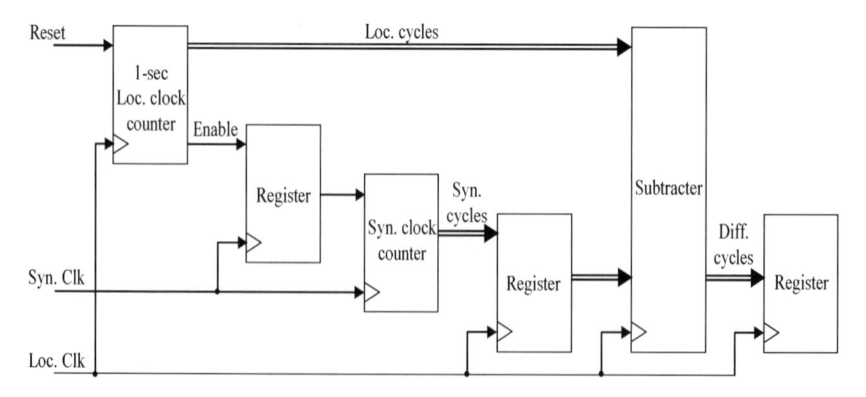

Figure 19.6 Clock monitoring circuit block diagram.

synchronized to capturing clock before usage. To achieve this, signals and data generated from local clock domain must be latched in distributed clock domain before being processed in the distributed clock domain and vice verse. Specifically, the *Enable* signal generated by the *Loc. clock counter* is registered in the distributed clock domain before being used by the *Syn. clock counter*. In addition, data from the *Syn. clock counter* is latched in the local clock domain by *Register* before it is sent to the *Subtracter*.

The application of these experimental models and their results are analysed during discussion of each EMI reduction technique in the next section.

19.5.4 EMI immunity enhancement techniques

19.5.4.1 Switched-mode power supply filtering

Switching regulators can be considered as source of EMI, as voltages are switched at a high frequency ranged from 50 kHz to 1 MHz. In fact, most conducted EMI within switch-mode power supplies originates from the main switching MOSFETs, transistors, and output rectifiers. Therefore, lowpass filters should be added to the output of switched-mode power supplies to suppress switching frequency noise.

For the TIGER-3 transceiver power supply, Figure 19.7 indicates the results of the experiment, which is set up as in the Section 19.3.1, for suppressing the switching noise. These experimental results prove noise immunity improvement

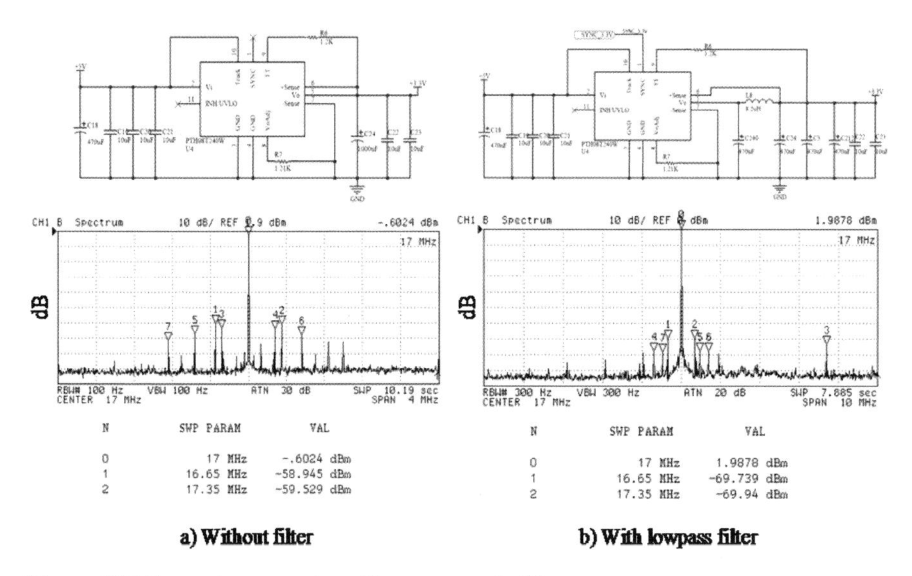

a) Without filter **b) With lowpass filter**

Figure 19.7 Harmonic suppression with power supply filter.

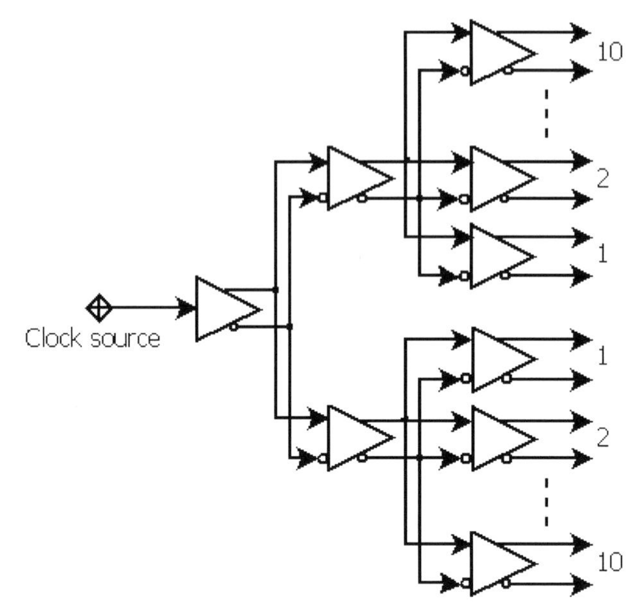

Figure 19.8 Clock distribution tree.

after the insertion of a PI-circuit lowpass filter to a switched-mode power supply.

The output signals of the transceiver at 17 MHz are measured by the HP 4395A spectrum analyser. The results show the harmonics around the operational frequency. In this discussion, only the two strongest harmonics indicated by markers number 1 and 2 at frequencies 16.65 MHz and 17.35 MHz, respectively, are considered. These harmonics are mixed from 17 MHz and the 350 KHz switching frequency which is generated by the 3.3 V regulator. It is clear that the harmonics are decreased by more than 10 dB after the filter is added to the output of the regulator.

19.5.4.2 Differential signalling

Since differential signalling is less susceptible to noise than single-ended signalling, the technique provides a good mode for high frequency clock transmission. For the TIGER-3 radar, the clock and start pulse generated by the THS are fed to the Clock Distributor which is responsible for distributing the signals to the transceivers. A common general approach to clock distribution is the use of buffered trees, as shown in Figure 19.8. To improve noise performance, differential timing signals are used. To achieve this, each signal utilizes a precision fanout clock driver with two banks of 10 Low Voltage Differential Swing (LVDS) output pairs for the 20 transceivers. This enables the precisely aligned, ultra low-skew signals to be driven to the transceivers.

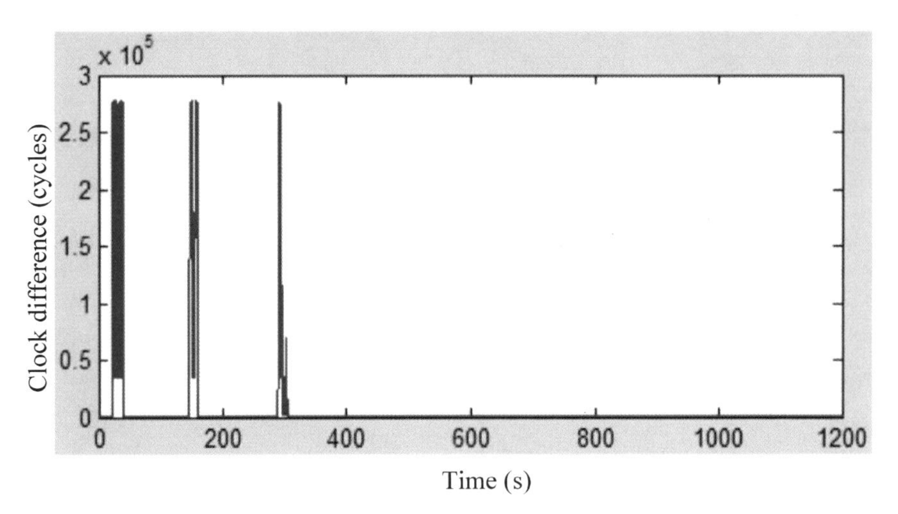

Figure 19.9 Clock missing due to EMI.

19.5.4.3 EMI suppressing with magnetic transformers

To assess EMI immunity improvement through application of magnetic transformers in the clock distribution system. Experiments were firstly performed without using magnetic transformers. These results are then compared with that of the scenario in which the transformers are employed.

Figure 19.9 depicts the result of the clock monitor program without magnetic transformers. The results show that the clock is missing due to EMI noise, clearly

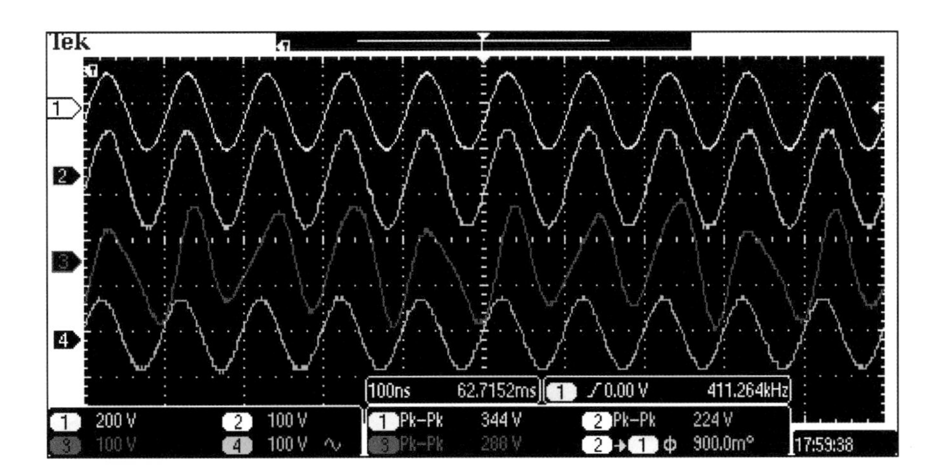

Figure 19.10 RF waveforms distortion due to clock missing.

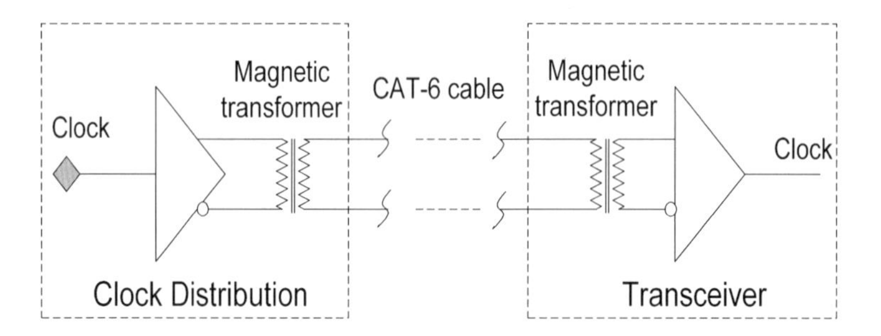

Figure 19.11 EMI suppression with magnetic transformers for clock.

notable by some periods of time, clock differences are approximately 2.8x 10^5 clock cycles; whilst the numbers are normally around 1755 (which is shown clearly in Figure 19.12). In this experiment, the distributed clock was distorted at very low transmitted power of 600 W with the output RF voltage was 288 V_{p-p} at 10 MHz. This can be seen in Figure 19.10 which is a snapshot of an oscilloscope screen with the experiment was set up as in Section 19.5.3.2. The RF waveforms were probed at the dummy loads of 4 transceivers using the distributed clock. The deformation of the third (from top down) RF waveform demonstrates the impact of the clock distortion.

In order to achieve better EMI immunity performance in the TIGER-3 clock system, magnetic transformers are employed at both ends of clock transmission cables.

A CAT-6 cable with 2 twisted pairs connects the timing signals from the Clock Distributor to each transceiver, one pair for the differential clock, another pair for the differential start signal. In order to suppress the common-mode EMI emitted on the cables, the 2 twisted pair CAT-6 cables are AC coupled through magnetic transformers on both sides. Figure 19.11 shows the circuit for the clock pair. Each receive pin pair is differentially terminated into an external 100 Ω resistor to match the cable impedance. To save board space and reduce component count, RJ-45 connectors with integrated magnetics are used.

In Figure 19.12, it is shown that the magnetic transformer circuitry enables the clock system to function at full capacity, without clock loss. A snapshot of the clock monitor shown in Figure 19.12 (a) indicates the clock difference fluctuating around 1755 clock cycles. The maximum clock difference between 2 adjacent measurements is only one cycle. This is explainable as the two frequencies are not exactly the same and will drift slightly with time. In other words, the fluctuation is not due to missing clock cycles. The correct formations of RF waveforms at 4 transceiver outputs in Figure 19.12 (b) also demonstrate the EMI immunity improvement. During this measurement, the radar system was operating at approximately 1.8 kW per transceiver, with a maximum peak-to-peak voltage of the outputs was 860 V.

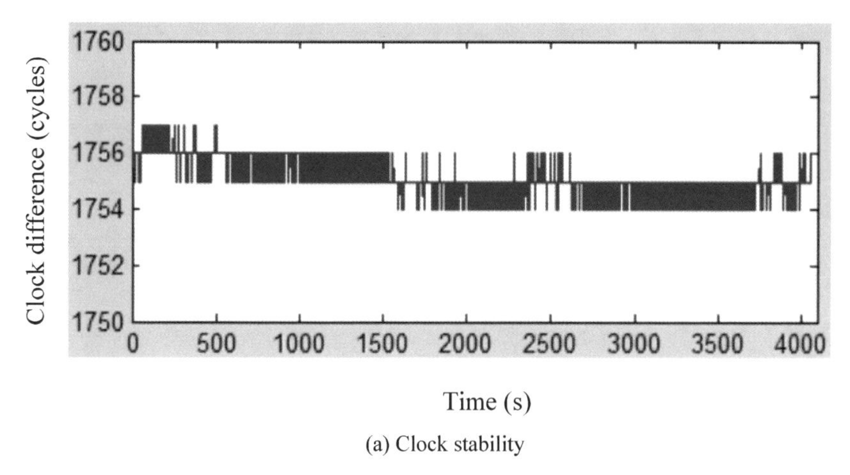

(a) Clock stability

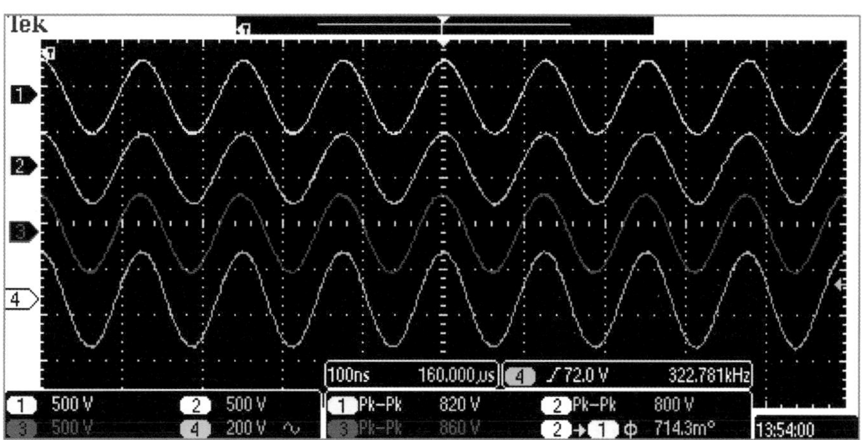

(b) RF waveforms at 4 transceiver outputs

Figure 19.12 Clock system immune from EMI with magnetic transformers usage.

19.5.4.4 Harmonic suppression with pulse shaping technique

For the TIGER-3 system, in order to maintain a well-behaved frequency spectrum, output transmit pulses require a smooth envelope. To some extent, the output pulse waveforms are evaluated for spectral purity with slower transition edges. Bienvenu et al. developed a pulse shaping technique for that purpose [2]. The TIGER-3 radar transmits a sequence of seven or eight 100 μs or 300 μs pulses depending on the operational modes. Figure 19.13 depicts the implemented full eight-pulse waveform at the outputs of four transceivers. When pulsing the output

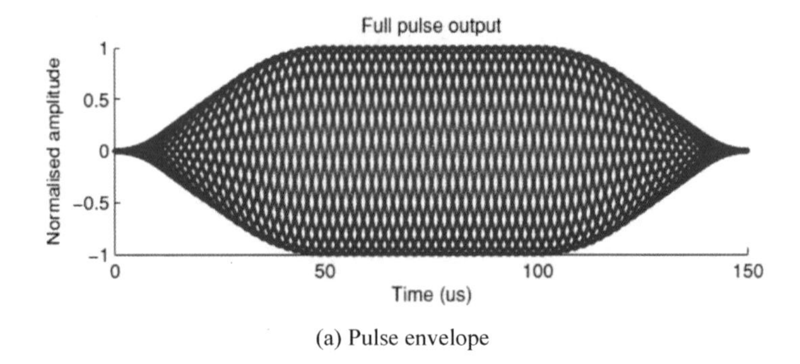

(a) Pulse envelope

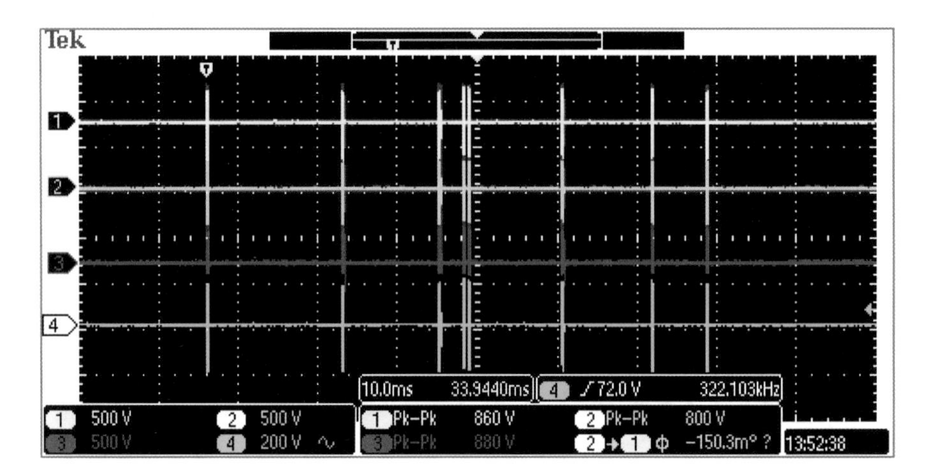

(b) Pulse sequence

Figure 19.13 TIGER-3 output signal waveform.

signal, the rise and fall characteristics of the envelope alter the output frequency spectrum.

Figure 19.14 compares spectrum of a rectangular pulse and the implemented pulse using the pulse shaping technique. The TIGER-3 aims to meet spectrum requirements with sidelobes to be lower than -70 dB at 70 kHz from the centre frequency. These requirements are fulfilled by the implemented pulse, while the rectangular envelope is clearly not able to meet the requirements. Moreover, as unwanted harmonics, especially the 3rd and 5th harmonics, are suppressed, EMI generated by them is reduced. For example, if the radar transmits at 12.5 MHz, then the 5th harmonic is 62.5 MHz, which is one with half of the clock frequency. It is thus highly possible that this harmonic could interfere the synchronised clock, should the harmonic energy be high. Consequently, the pulse shaping mechanism helps reduce EMI.

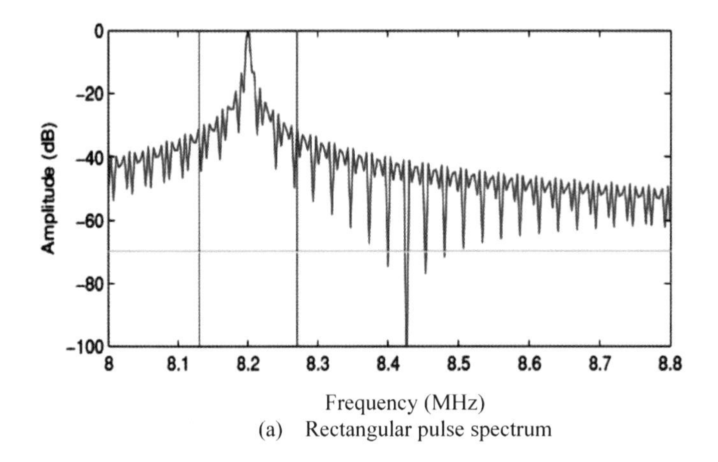

Frequency (MHz)

(a) Rectangular pulse spectrum

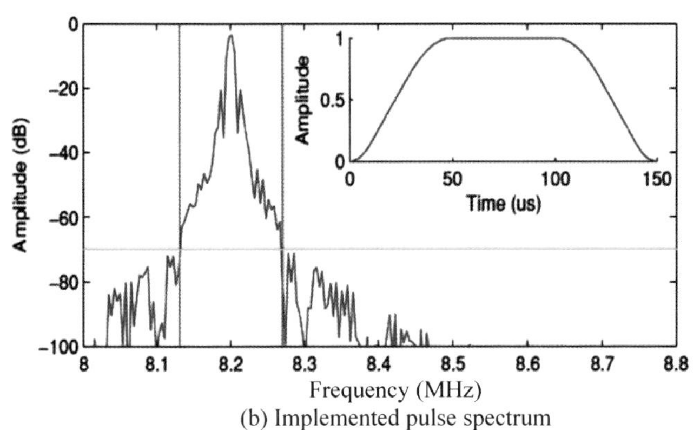

Frequency (MHz)

(b) Implemented pulse spectrum

Figure 19.14 Spectrum comparison.

19.5.4.5 Shielding

In the TIGER-3 radar system, each transceiver sits in a specifically designed aluminium case. Moreover, sensitive circuits are on separately shielded metal boxes. For example, RF circuit, power amplifier and power supply are electromagnetically shielded in different cases. In addition, cords and cables connecting the peripherals, such as timing signals and data signals, are shielded to keep unwanted RF energy from entering or leaving the circuits.

19.6 SUMMARY

In this chapter negative aspects of electromagnetic radiation, resulting as EMI, have been discussed. Prior to providing solutions for EMI suppression, a background on electromagnetic fields has been presented. In which, basics of electronic and magnetic field theory were briefly introduced. Then EMI was classified according to source by which it is generated. These sources include man-made and natural EMI. Among them, man-made EMI has larger spectrum and can spread up to several GHz. On the other hand, spectral energy of natural EMI decays very quick above 20 MHz. EMI coupling mechanisms were also discussed. As the understanding of these greatly assists in dealing with EMI issues from the very first step in system design, and further effectively troubleshooting EMI problems once the hardware has been constructed. Next, the chapter focused on EMI reduction techniques. Finishing with a case study discussing the application of EMI reduction techniques applied to the distributed clock system in the TIGER-3 radar.

Accurate, stable and EMI immune clock synchronisation is essential to the operation of the new digital TIGER-3 radar. An external clock synchronisation with EMI immunity enhancement approach is implemented using FPGA technology. The solution focuses on addressing the EMI with range from simple methods, such as shielding and grounding; to more complicated techniques, such as pulse shaping mechanism. Experimental results indicate that with the usage of those techniques, EMI immunity is of sufficient quality to allow a correct operation of the TIGER-3 radar.

References

[1] Donald J. Arndt. Demystifying Radio Frequency Interference: Causes and Techniques for Reduction. Trafford, 2009.

[2] B. Bienvenu and N. V. Vu and J. Whittington and J. C. Devlin. Variable-length quadratic envelope pulse shaping and automatic gain control for HF radars. 6th IEEE International Conference of Broadband Communications and Biomedical Applications (IB2COM 2011), Melbourne, Australia, 2011.

[3] Bernard F. Burke and Francis Graham-Smith. An introduction to radio astronomy. Cambridge University Press, Third edition, 2010.

[4] Rob Chapman and Keith Burrows and Carmel Fry and Doug Bail and Alex Mazzolini and Jacinta Devlin and Henry Gersh. Heinemann physics 12. Malcolm Parsons, Third edition, 2009.

[5] Christos Christopoulos. Handbook of engineering electromagnetics, chapter Electromagnetic Compatibility, pages 347-376. Marcel Dekker, Inc, 2004.

[6] P. C. Clemmow. The Plane Wave Spectrum Representation ofElectromagnetic Fields. Oxford University Press and IEEE Press, 1996.

[7] Robert E. Collin. Field theory of guided waves. John Wiley & Sons, Inc.,

Second edition, 1991.
[8] E. Custovic and H. Q. Nguyen and J. C. Devlin and J. Whittington and D. Elton and A. Console and H. Ye and R. A. Greenwald and D. A. Andre and M. J. Parsons. Evolution of the SuperDARN antenna: Twin Terminated Folded Dipole Antenna for HF systems. 6th IEEE International Conference of Broadband Communications and Biomedical Applications (IB2COM 2011), Melbourne, Australia, 2011.
[9] Dyson, P.L. and Devlin, J.C. The TIGER Radar - An Extension of SuperDARN to sub-auroral latitudes. :9-31, 2000.
[10] Dyson, P.L. and Devlin, J.C. and Parkinson, M.L. and Whittington, J.S. The Tasman international geospace environment radar (TIGER) - current development and future plans. Radar Conference, 2003. Proceedings of the International, pages 282 - 287, 2003.
[11] Giri, D.V. Classification of intentional EMI based on bandwidth. American Electromagnetics (AMEREM) 2002, Annapolis, MD, 2002.
[12] Giri, D.V. and Tesche, F.M. Classification of intentional electromagnetic environments (IEME). Electromagnetic Compatibility, IEEE Transactions on, 46(3):322 - 328, 2004.
[13] R. A. Greenwald and K. B. Baker and R. A. Hutchinsa and C. Hanuis. An HF phased-array for studying small-scale structure in the high-latitude ionosphere. Radio Sci., number 1, pages 63--79, 1985.
[14] IEC 61000-4-1. Electromagnetic compatibility (EMC) - Part 4-1: Testing and measurement techniques. Technical report, International Electrotechnical Commission, 2006.
[15] Intel. Design For EMI. Application note, Intel Corporation, 1999.
[16] John Kappenman. A Perfect Storm of Planetary Proportions. Technical report, IEEE spectrum, 2012.
[17] Katzir, L. and Singer, S. Reduction of common-mode electromagnetic interference in isolated converters using Negative feedback. Electrical and Electronics Engineers in Israel, 2006 IEEE 24th Convention of, pages 180 - 183, 2006.
[18] Roberto De Andrade Martins. Resistance to the Discovery of Electromagnetism: Orsted and the Symmetry of the Magnetic Field. :245-265, 2003.
[19] H. Q. Nguyen and N. V. Vu and J. Whittington and E. Custovic and B. Bienvenu and J. Devlin. Noise immunity enhancement for a distributed clock system in digital HF radar. 6th IEEE International Conference of Broadband Communications and Biomedical Applications (IB2COM 2011), Melbourne, Australia, 2011.
[20] Nitsch, D. and Sabath F. Electromagnetic Effects on Systems and Components. American Electromagnetics (AMEREM) 2006, Albuquerque, New Mexico, 2006.
[21] Clayton R. Paul. Introduction to electromagnctic compatibility. John Wiley & Sons, Inc., Hoboken, New Jersey, Second edition, 2006.

[22] Milica Popovic and Branko D. Popovic and Zoya Popovic. Handbook of engineering electromagnetics, chapter Electromagnetic Induction, pages 123-162. Marcel Dekker, Inc, 2004.

[23] Sabath, F. Classification of electromagnetic effects at system level. Electromagnetic Compatibility - EMC Europe, 2008 International Symposium on, pages 1 -5, 2008.

[24] Matthew J. Schneider. Design Considerations to Reduce Conducted and Radiated EMI. Master's thesis, Purdue University, West Lafayette, Indiana, USA, 2010.

[25] EMI Reduction Techniques. EMI Reduction Techniques. Technical report, Pericom Semiconductor Corporation, 1998.

[26] Jack Vanderlinde. Classical Electromagnetic Theory. Kluwer Academic Publishers, Second edition, 2004.

[27] Paolo Vecchia and Rüdiger Matthes and Gunde Ziegelberger and James Lin and Richard Saunders and Anthony Swerdlow. Exposure to high frequency electromagnetic fields, biological effects and health consequences (100 kHz-300 GHz). International Commission on Non-Ionizing Radiation Protection, 2009.

[28] Violette, J. L. Norman and White, Donald R. J and Violette, Michael F. Electromagnetic compatibility handbook. New York : Van Nostrand Reinhold, 1987.

[29] J. C. Willett and J. C. Bailey and C. Leteinturier and E. P. Krider. Lightning Electromagnetic Radiation Field Spectra in the Interval From 0.2 to 20 MHz. JOURNAL OF GEOPHYSICAL RESEARCH, 95(D12):20,367-20,387, 1990.

[30] Xilinx. Virtex-5 FPGA User Guide. Xilinx Inc., v5.1 edition, 2009.

Chapter 20.

Gbps Data Transmission in Biomedical and Communications Instruments

Jean Raphaël Olivier Fernandez and César Briso Rodríguez

Grupo de Radiocomunicación (GRC), Escuela Universitaria de Ingeniería Técnica de Telecomunicación (EUITT), Universidad Politécnica de Madrid (UPM), Crta. Valencia km. 7, 28031 Madrid, Spain
jean.raphael.olivier.fernandez@gmail.com, cbriso@diac.upm.es

In this chapter, the design of data transmission networks for reducing crosstalk signals and reflection effects is explained. Two examples of such networks are presented with the binary rates of 20 Mbps and 8 Gbps, respectively. Each example concludes by discussing the effects of resistive networks in both systems.

20.1 INTRODUCTION

Next generation of biomedical and test instruments requires the use of internal high–speed serial data transmission to communicate to internal modules. These communication types in large biomedical and aero–spatial instruments [1] can supply a range up to 30 m in point–to–multipoint configuration, with speeds up to Gbps over copper. For example, within the hospital or extended care environment, there is an overpowering need for monitoring of more and more vital body functions. Reliability is a particularly important factor to consider when treating the monitoring of patient's vital signals from sensors. X–rays, MRI/NMR and CT applications require robust, reliable and always available networks to perform important, often critical functions, when every second counts. There are some desirable features that have to be investigated such as: immunity to electromagnetic interference, distribution network robustness, seamless

4G Wireless Communication Networks: Design, Planning and Applications, 427-440,

integration by means of modular solutions, by using components with standard interfaces. In this way, interconnections are simplified.

In this case, data communications have problems of crosstalk, transmission errors and attenuation. These problems have to be overcome with a careful and special design of the distribution network.

Another important configuration is the connection of large instruments to arrays of sensors and other components located few meters away from a central processing unit using a serial bidirectional configuration. In this case, it is necessary to use high–speed differential drivers with low–loss balanced distribution and transmission lines. In the current applications, this information must be transmitted in a point–to–multipoint configuration.

20.1.1 Data transmission network basics

This section presents data transmission network basics.

Main elements of a half–duplex transmission network are differential line driver and receiver for digital transmission over balanced lines, amplifiers that ensure achieving the threshold level, according to serial data standard used, and balanced lines that carry signals (information). Figure 20.1 shows a simple block diagram of a typical half–duplex transmission network.

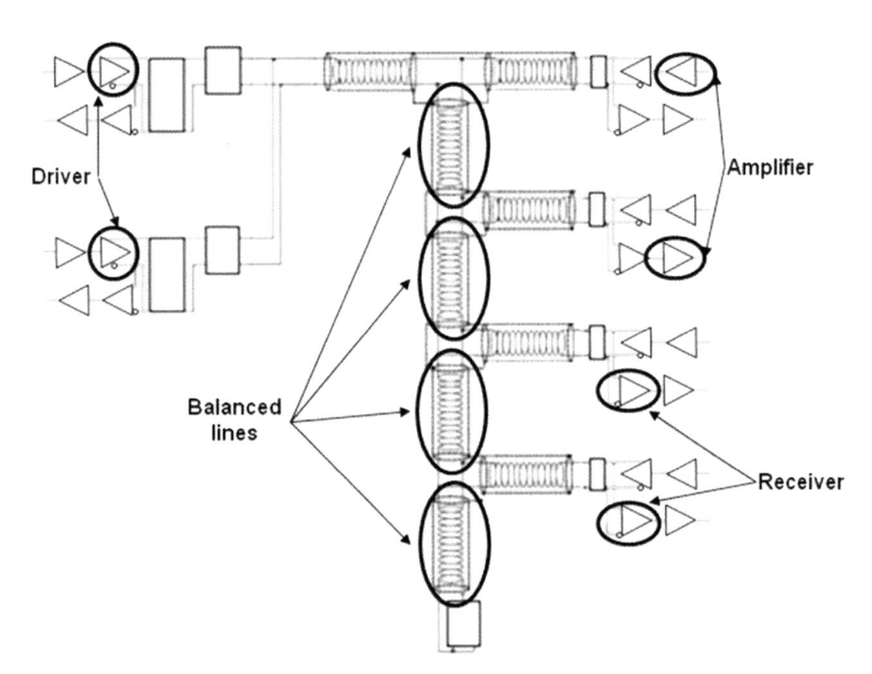

Figure 20.1 Half–duplex transmission network.

20.1.2 MODELING AND SIMULATION OF HIGH SPEED DATA TRANSMISSION NETWORKS

A point–to–multipoint transmission network has been developed. It is capable to work up to Gbps with one transmitter and up to sixteen receivers.

The design is based on the simulation and measurements of a 20 Mbps prototype. The complete network has been modeled and measured to obtain a simulation model able to be used up to Gbps.

In the design process, it is very important to get the optimal transmission and load impedance together with the accurate configuration of the distribution network to minimize crosstalk and reduce transmission errors. First, the impedance matching is treated by finding the loads that make the transmission ideal, avoiding reflections, and optimizing the signal level received. Nevertheless, the crosstalk added in the connectors and transmission lines does inevitably introduce additional loads in the devices, drivers and receivers, which fit the signal level of every section to obtain optimized levels that allow us to settle these problems. In this case, crosstalk is a phenomenon that must be considered [2], since a differential mode voltage level higher than +/– 200mV [3] at the input of a digital device might induce errors. To avoid this, it is necessary to assure that the transmission in the buses does not generate a high enough crosstalk level to produce this effect in other signals.

The following section describes the entire design process of high speed data transmission networks.

20.1.2.1 Point–to–multipoint transmission network

The design has been tested with a 20 Mbps transmission network designed for a research instrument that must communicate to an array of sensors located up to 5 meter away from the central processor.

The equivalent electrical model (using the AWR's Microwave Office software) of the prototype, is shown in Figure 20.2.

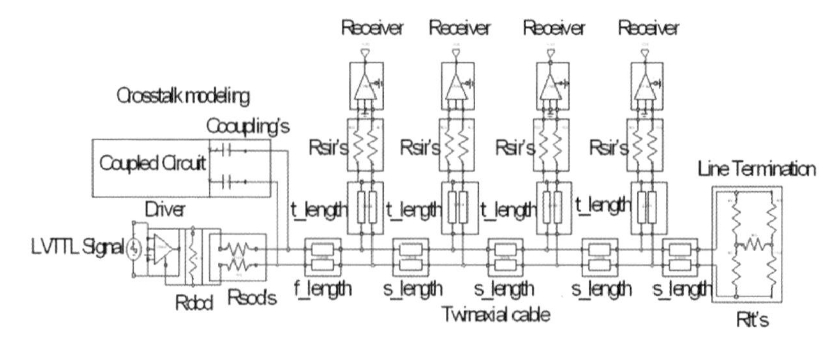

Figure 20.2 Equivalent electrical circuit of the Transmission/Reception System.

This subsystem is composed fundamentally of a HS–26CLV31RH driver, which converts the transmitted signal from LVTTL (a square signal ranged from 0 to 3.3V) at 10 MHz to RS422, and 4 HS–26CLV32RH receivers. The above mentioned active devices have been modeled as operational amplifiers due to their differential nature. Therefore, their principal parameters will be gain, input impedance, output impedance, cut–off frequency and delay, among others. The connections have been made by means of twinaxial cable of different lengths. These transmission lines will be subsequently featured by their characteristic impedance, Z_0, their relative dielectric permittivity, ε_r, and their losses at a reference length and a reference frequency, $f_{scaling\_loss}$.

The first step was modeling of the driver used: a HS–26CLV31RH driver. This procedure was carried out on the basis of the models described in [4] and has been optimized by the accomplishment of measurements in the laboratory. The above mentioned measurements have been carried out by exciting the driver HS–26CLV31RH with a LVTTL signal at several frequencies up to 100 MHz and loading it with a segment of twinaxial cable — its characteristic impedance (Z_0) is 120 Ω — with three different loads (Z_L): 56 Ω, 120 Ω and ∞ Ω.

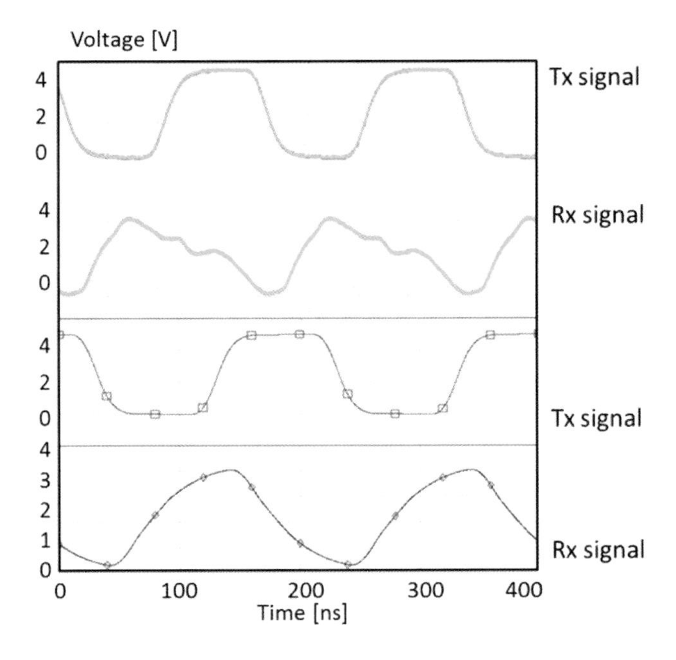

Figure 20.3 Practical measurements vs. simulation (Z_L = 56 Ω).

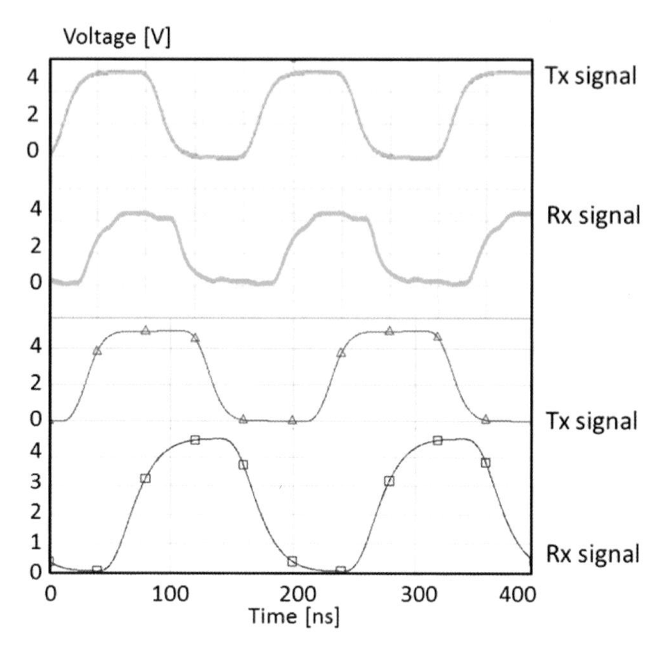

Figure 20.4 Practical measurements vs. simulation (Z_L = 120 Ω).

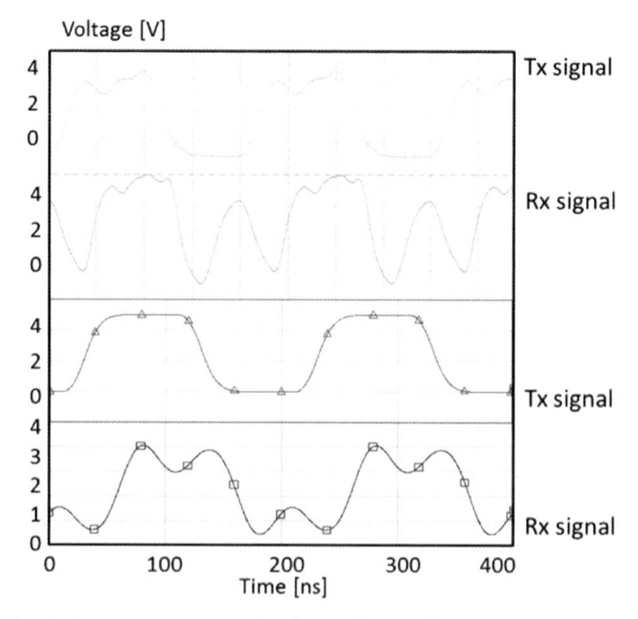

Figure 20.5 Practical measurements vs. simulation (Z_L = ∞ Ω).

Results are shown in figure 20.3, 20.4, and 20.5, respectively. Thus, parameters in the simulation must be fitted for obtaining the input impedance, the output impedance, the gain and the bandwidth of the device.

As we can see in the previous figures, the optimal load is $Z_L = Z_0 = 120\ \Omega$ (impedance matching).

The parallel bus called "Coupled Circuit" allows us to simulate the crosstalk [5]. R_{dod}, R_{sod}, R_{sir} resistors are used as a protection against possible short circuits, as well as to equalize the transmitted and received signals.

The back termination network called "Line Termination" and composed by R_{lt} resistors, serves to make the correct impedance–matching to the system. All simulation parameters are given in Table 20.1.

In the case of driver and receiver parameters, it is important to remark that the specification "Z_{high}= 50 KΩ" simulates that the driver and receiver devices are disabled. In other words, both output pins and input pins remain in a high–impedance state. The crosstalk has been modeled using two parallel transmission lines (adjacent buses constituted by twinaxial cable) measuring the coupling coefficient between the above mentioned lines. To model the simulation circuit we have used a capacitive model of crosstalk and we have computed the capacitor values, $C_{coupling}$, connected between the buses [6]. Therefore, the values of the above mentioned elements are tuned using measurements of the signal level transmitted and coupled. This model can be easily scaled in frequency.

Table 20.1 Equivalent Electrical System Parameters.

Driver				
Gain	Z_{in}	Z_{out}	$f_{cut\text{-}off}$	Delay
1	1 KΩ	8 Ω	100 MHz	10 ns
Receiver				
Gain	Z_{in}	Z_{out}	$f_{cut\text{-}off}$	Delay
1	8 Ω	8 Ω	100 MHz	10 ns
Twinaxial cable segment parameters				
Z_0	ε_r	Loss	$f_{scaling\_loss}$	
120 Ω	2.8	0.15 dB/m	10 MHz	
Twinaxial cable segment lengths				
f_length = 4.2 m	s_length = 1.2 m		t_length = 0.5 m	
Crosstalk modeling				
$C_{coupling}$ = 8.1 pF				
Protection/equalizer/impedance-matching resistors				
R_{dod} = 270 Ω	R_{sod} = 22 Ω	R_{sir} = 27 Ω	R_{lt} = 120 Ω	

20.1.2.2 Analysis of the output impedance of the driver

An important aspect of the design is to equalize transmission in both directions. For this purpose, serial resistors, R_{sod}, are used to attenuate the transmitted power of the drivers. Nevertheless, if we observe the crosstalk effect [7] we note that the exactly opposite situation happens, for what we try to come to a commitment (the crosstalk was diminished by up to 30% without attenuating the useful signal too much). Finally, we must not forget the protection effects against driver short circuits that are contributed by the above mentioned resistors. The graphical representation of this analysis is shown in Figure 20.6 and 20.7.

20.1.3 Modeling and Simulation of a 8–Gbps Transmission Network

From the measurements carried out at 20 Mbps, a new distribution network able to be used up to several Gbps has been developed. An extrapolation method was used, supposing that binary data are transmitted at a rate of 8 Gbps and that the wiring used in the previous prototype was replaced by broadband couplers. Finally, possible aspects were analyzed, as it was done in the high–speed data transmission system, such as the crosstalk [8] or interference between adjacent

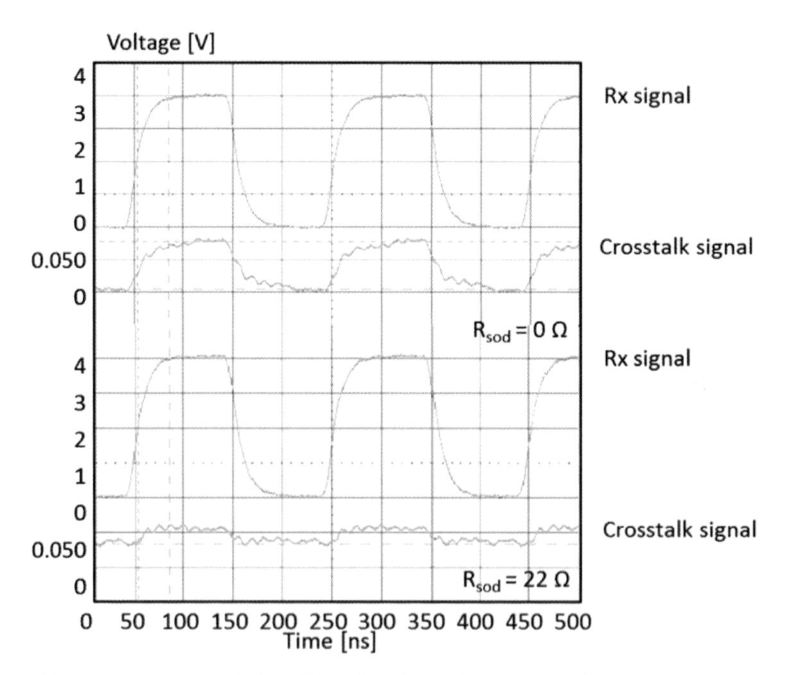

Figure 20.6 Measurements of the effect of serial resistors on received and crosstalk signal (Rsod = 0 Ω and Rsod = 22 Ω).

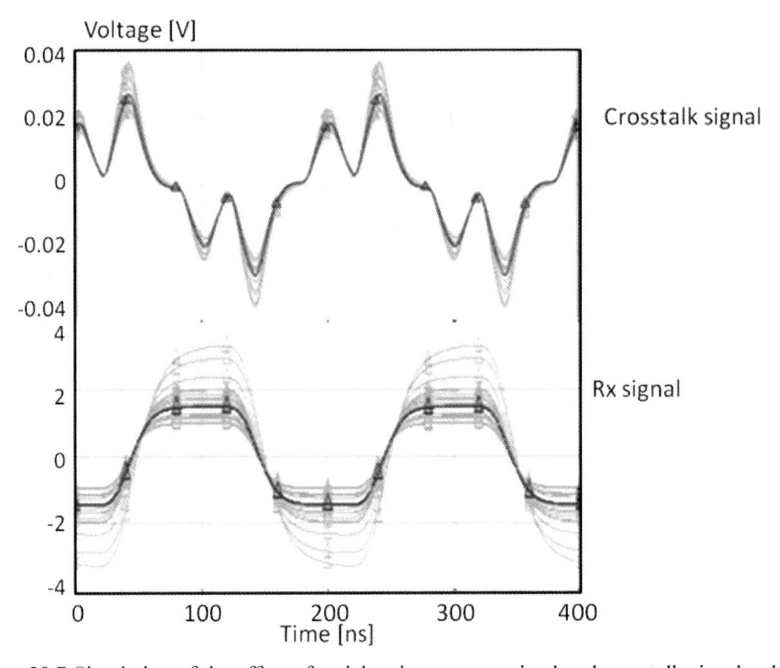

Figure 20.7 Simulation of the effect of serial resistors on received and crosstalk signal voltage level variation depending on R_{sod} and R_{sir} series resistors (from 0 to 44 Ω).

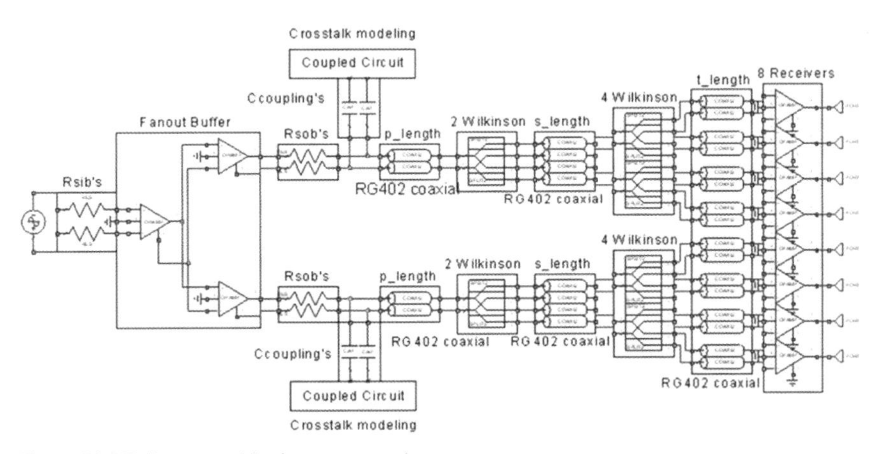

Figure 20.8 Point–to–multipoint system under test.

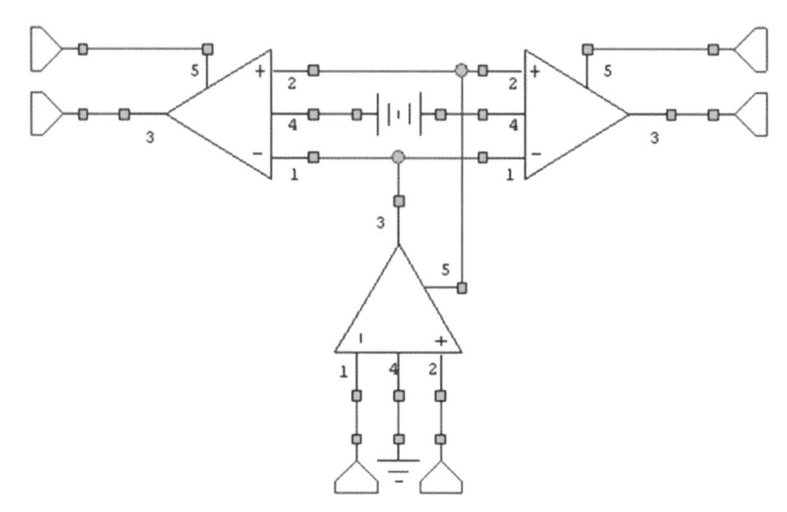

Figure 20.9 HMC720LC3C Gbit Buffer architecture.

buses [9] and the possible introduction of some equalization mechanism related to the devices selected to effectuate the data transmission. In both cases, AWR'S Microwave Office has been used as simulation software.

In the new design, the system matching impedance is Z_0 = 50 ohms and balanced lines have been used. The Figure 20.8 shows the point–to–multipoint data distribution for a 1 to 8 system.

The devices chosen to implement point-to-multipoint data transmission have been the Hittite HMC720LC3C Gb/s buffers. These active devices have been modeled as ultra fast operational amplifiers, due to their differential nature, in association with other operational amplifiers. Therefore, as in the previous case, their principal parameters will be gain, input impedance, output impedance, cut–off frequency and delay. In Figure 20.9 the equivalent electrical circuit of this architecture is shown.

Connections between circuit elements must be made by means of matched pairs of RG402 coaxial cables, to implement balanced lines of different lengths. In the case that is being treated, the total length range of coaxial cables is 5 m. These transmission lines will be once again characterized by their characteristic impedance, Z_0, their relative dielectric permittivity, ε_r, and their losses at a reference length and a reference frequency, $f_{scaling\_loss}$. The signal distribution is made using a power divider network designed using several multioctave stripline 3dB hybrids covering the 4–12 GHz band [10]. These devices have a high thermal stability. In the worst case, the amplitude and phase unbalance are ±0.3 dB and ±2°, respectively, showing advantages over other commercial devices. Moreover, couplers exhibit input and output return loss of 22 dB.

R_{sib}, R_{sob} resistors are used to equalize the transmitted and received signals. The crosstalk has been simulated using a parallel bus called "Coupled Circuit" and a capacitive model, $C_{coupling}$, calculated from the prototype presented in the previous section by means of a linear transformation (20.1).

$$C_{coupling|8Gbps} = \left(C_{coupling|20Mbps} \cdot f_{|20Mbps}\right) / f_{|8Gbps} \tag{20.1}$$

All these parameter values are summarized in the Table 20.2.

The frequency response of the HMC720LC3C Gb/s buffer is shown in Figure 20.10.

20.1.3.1 Analysis of the series resistor influence to the Buffer

In the ideal case, there should be a perfect match at the input of the matching network and the transmission lines discontinuities [11] and, therefore, these factors do not have to be considered. In a real case, where transmission line behavior and interconnection discontinuities must be considered to preserve the integrity of the transmitted and received signals [12], serial resistors that work as a filter are required to equalize the transmitted signal [13]. It is observed that as the transmitted signal level gets lower, buffer series resistance values, R_{sib} and R_{sob} get higher. Moreover, the crosstalk signal progressively diminishes as the signal is equalized [14]. Voltage traces of the Figure 20.11 show equalized signals with R_{sib} = 15 Ω and R_{sob} = 27 Ω.

Table 20.2 Equivalent Electrical System Parameters.

<table>
<tr><td colspan="8">Gb/s Buffer</td></tr>
<tr><td>Gain</td><td colspan="2">Z_{in}</td><td>Z_{out}</td><td>$f_{cut\text{-}off}$</td><td colspan="2">Delay</td><td>Output Return Loss</td></tr>
<tr><td>27 dB</td><td colspan="2">50 Ω</td><td>50 Ω</td><td>14 GHz</td><td colspan="2">120 ps</td><td>10 dB</td></tr>
<tr><td colspan="8">RG402 coaxial cable segment parameters</td></tr>
<tr><td colspan="2">Z_0</td><td colspan="2">ε_r</td><td colspan="2">loss</td><td colspan="2">$f_{scaling\_loss}$</td></tr>
<tr><td colspan="2">50 Ω</td><td colspan="2">2.07</td><td colspan="2">0.01 dB/m</td><td colspan="2">1 GHz</td></tr>
<tr><td colspan="8">RG402 coaxial cable segment lengths</td></tr>
<tr><td colspan="8">f_length = s_length = t_length = 0.85 m</td></tr>
<tr><td colspan="8">Crosstalk modelling</td></tr>
<tr><td colspan="8">$C_{coupling}$ = 0.02 pF</td></tr>
<tr><td colspan="8">Equalizer resistors</td></tr>
<tr><td colspan="4">R_{sib} = 15 Ω</td><td colspan="4">R_{sob} − 27 Ω</td></tr>
</table>

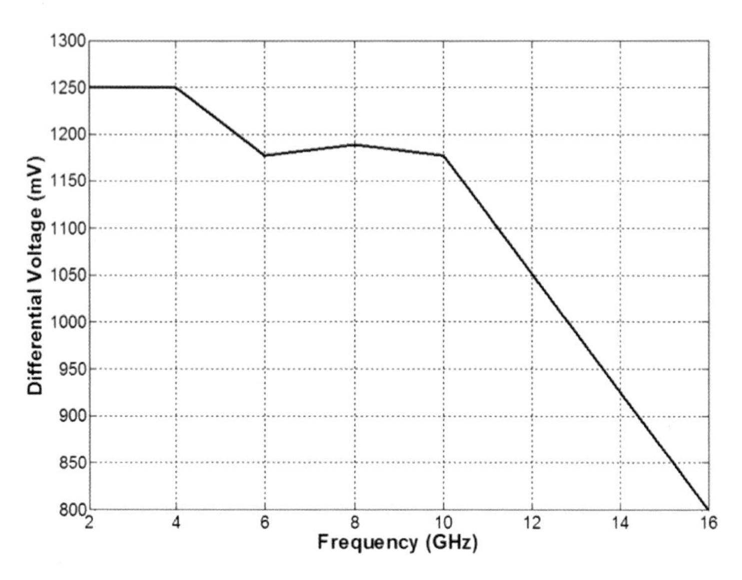

Figure 20.10 Output Differential vs. Frequency.

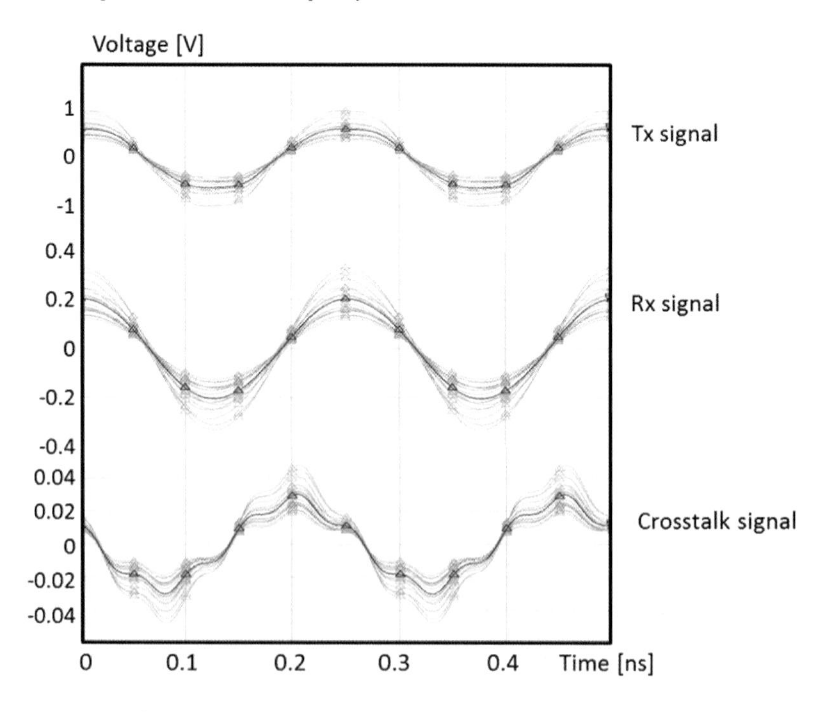

Figure 20.11 Transmitted, received and crosstalk signal voltage level variation depending on resistance values of R_{sib} and R_{sob} resistors.

From the Figure 20.11, it is possible to establish a coupling coefficient between the received voltage and the crosstalk voltage signal, $Coeff_{coupling}$, which is defined as (20.2):

$$Coeff_{coupling} = V_{Rx} / V_{Crosstalk} \tag{20.2}$$

where V_{Rx} and $V_{Crosstalk}$ are the received and the crosstalk voltage signals, respectively. In this case, the $Coeff_{coupling}$ value is approximately 15.85.

All of the values of a Gbps prototype have been calculated supposing that the load impedance is Z_0 = 50 Ω. Under these conditions and assuming a reference temperature of 290 K, the noise power, P_{dBm}, is estimated from the following expression (20.3):

$$P_{dBm} = -174 + 10 \cdot \log_{10}(B) \tag{20.3}$$

where P is expressed in dBm and B is the bandwidth in Hz over which the noise is measured. With a 4 GHz bandwidth required for the transmission, the noise power is –78 dBm.

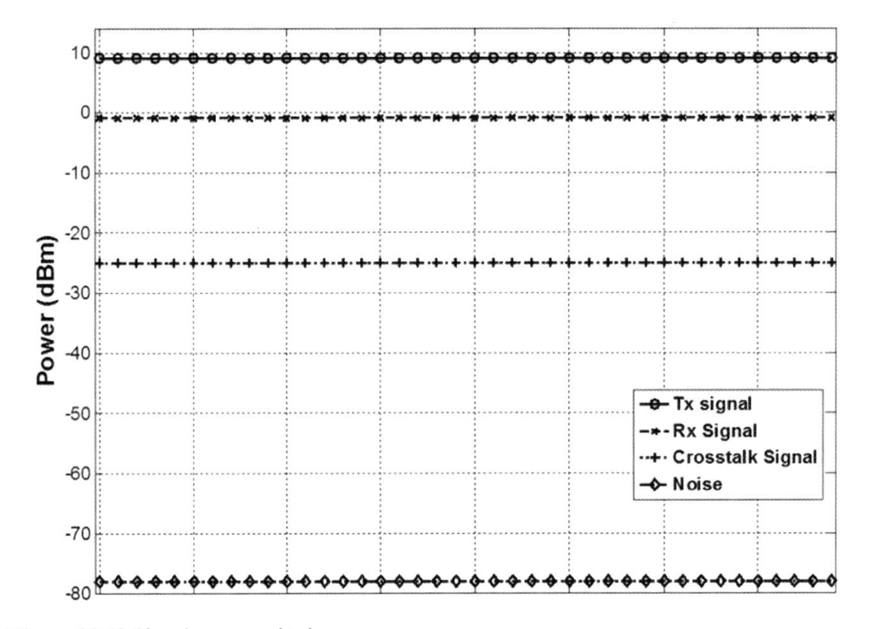

Figure 20.12 Signal–power criteria.

In the test circuit, the output power of the driver is 10 dBm and the crosstalk has been reduced using the serial resistors. This last one is the critical parameter and has been optimized at –25 dBm. Then the received power, in this case with 5 m of cable length, is –1 dBm and the noise power over 4 GHz bandwidth is –78 dBm. The balanced interconnection helps reduce this phenomenon, increase noise immunity, and eliminate ground noise [15], [16].

There is no a general rule about the crosstalk reduction. It will depend on the vulnerability of the transmitted and received signals in the face of these phenomena, without forgetting voltage thresholds of used multipoint data transmission standards. In this case, the crosstalk has been diminished approximately 30% (the signal–to–crosstalk ratio is 24 dB). Therefore, there is a margin to increase the length range of the coaxial cable. Figure 20.12 shows signal–power criteria followed in this design.

The proposed high–speed transmission network has a good behavior for data transmission over copper. It is a simple alternative that can in some cases substitute fiber–optic networks that are more complex for medium and short distances. Thus, disadvantages associated with this technology can be avoided, such as: specialist skills needed, cost of transmission equipment, etc. New high–speed digital drivers allow the development of the proposed distribution network.

The circuit has been optimized reducing crosstalk and matching endpoints (it does not need line termination resistors), and all of them have been simulated and modeled allowing the designer to easily fit the design to new requirements or different components.

References

[1] N. Schneier, "High speed digital design," IEEE Proceedings Aerospace Conference, pp. 257–261, March 1999.

[2] H. You, and M. Soma, "Crosstalk analysis of interconnect lines and packages in high-speed integrated circuits," IEEE Trans. Circuits and Systems, Vol. 37, pp. 1019–1026, August 1990.

[3] National Semiconductor Application Note AN-1031, "TIA/EIA-422-B Overview," National Semiconductor Inc., January 2000.

[4] L. Shi, and B. Guo, "RS485/422 solution in Embedded Access Control System," BMEI' 2nd International Conference on Biomedical Engineering and Informatics, pp. 1–4, October 2009.

[5] W. Chen, S. K. Gupta, and M. A. Breuer, "Analytical models for crosstalk excitation and propagation in VLSI Circuits," IEEE Trans. Computer-Aided Design of Integrated Circuits and Systems, vol. 21, pp. 1117–1131, October 2002.

[6] P. Heydari, and M. Pedram, "Capacitive coupling noise in high-speed VLSI Circuits," IEEE Trans. Computer-Aided Design of Integrated Circuits and

Systems, Vol. 24, pp. 478–488, March 2005.

[7] W. Chen, S. K. Gupta, and M. A. Breuer, "Analytic models for crosstalk delay and pulse analysis under non-ideal inputs," Proceedings Test Conference, pp. 809–818, 1997.

[8] H. You, and M. Soma, "Crosstalk and transient analysis of high-speed interconnects and packages," IEEE Trans. Solid State Circuits, Vol. 26, pp. 319–30, March 1991.

[9] D. S. Gao, A. T. Yang, and S. M. Kang, "Modeling and simulation of interconnection delays and crosstalks in high-speed integrated circuits," IEEE Trans. Circuits and Systems, Vol. 37, pp. 1–9, January 1990.

[10] I. Malo-Gómez, J. D. Gallego-Puyol, C. Díez-González, I. López-Fernández, and C. Briso-Rodríguez, "Cryogenic hybrid coupler for ultra-low noise radio astronomy balanced amplifiers," IEEE Trans. Microwave Theory and Techniques, Vol. 57, pp. 3239–3245, December 2009.

[11] T. Y. Otoshi, "Simulation diagnostics of multiple discontinuities in a microwave coaxial transmission line," IEEE Trans. Microwave Theory and Techniques, Vol. 43, pp. 1310–1314, January 1995.

[12] J. Hauenschild, and H. M. Rein, "Influence of transmission-line interconnections between gigabit-per-second ICs on time jitter and instabilities," IEEE Joumal of Solid-State Circuits, Vol. 25, pp. 763–766, June 1990.

[13] C. Pelard, E. Gebara, A. J. Kim, M. G. Vrazel, F. Bien, Y. Hur, M. Maeng, S. Chandramouli, C. Chun, S. Bajekal, S. E. Ralph, B. Schmukler, V. M. Hietala, and J. Laskar, "Realization of multigigabit channel equalization and crosstalk cancellation integrated circuits," IEEE Journal of Solid-State Circuits, vol. 39, pp. 1659–1670, October 2004.

[14] R. Kaupp, "Waveform degradation in VLSI interconnections," IEEE Joumal of Solid-State Circuits, Vol. 24, pp. 1150–1153, August 1989

[15] F. Broydé, and E. Clavelier, "Crosstalk in balanced interconnections used for differential signal transmission," IEEE Trans. Circuits and Systems I, Vol. 54, pp. 1562–1572, July 2007.

[16] A. J. Rainal, "Crosstalk Transmission properties of balanced interconnections," IEEE Trans. Components, Hybrids, and Manufacturing Technology, Vol. 16, pp. 137–145, February 1993.

Chapter 21.

All-Optical Signal Processing Circuits Using Multimode Interference Structures on Silicon Waveguides

Trung-Thanh Le[1] and Laurence Cahill[2]

[1]*Hanoi University of Natural Resources and Environment, Hanoi, Vietnam*
[2]*Department of Electronic Engineering, La Trobe University, Bundoora, Victoria, Australia*
[1]*thanh.le@hunre.edu.vn;* [2]*l.cahill@latrobe.edu.au*

This chapter provides the background theory of multimode interference (MMI) couplers upon which a class of all-optical signal processing circuits are based. The amplitudes and phases of three kinds of MMI couplers, namely general interference (GI), restricted interference (RI), and symmetric interference (SI) MMIs, are derived. The use of multimode interference (MMI) devices for building basic elements to realize all-optical signal processing circuits is presented. The proposed structures are optimized using analytical and numerical simulations. In addition, an approach for realizing all-optical discrete Fourier transforms using the proposed structures is presented. The approach can be extended to design other larger scale all-optical signal processing circuits. The designs of these MMI based devices use silicon waveguides, which are compatible with existing CMOS technology.

21.1 INTRODUCTION

For many years, optical techniques have been considered for a variety of signal processing tasks such as pattern recognition, the generation of ambiguity surfaces for radar signal processing and image processing applications [1,2]. The major

4G Wireless Communication Networks: Design, Planning and Applications, 441-470,

reason for using an optical signal processor is its high bandwidth advantage over electronic processors. Due to its high throughput, the application of optical signal processing in optical communication systems is a very attractive research area. Photonic signal processing transforms such as the discrete Fourier transform (DFT), discrete cosine transforms (DCT) and discrete wavelet transforms (DWT) are useful for spatial signal processing and optical computing such as spectrum analysis, filtering, and encoding, etc.

Early efforts used lens systems [3, 4], directional couplers [5, 6], and single mode star networks [7] to develop optical signal processing transforms such as the Hadamard transform, the DFT and wavelet filters. However, the systems based on these technologies are usually quite large, lack accuracy and require high precision mechanical placement. In addition, the structure for implementing the transforms based on fibre technology requires bulky crossovers of fibre cables. Recently, the design of DFT and DCT transforms using fibre directional couplers has been presented by Moreolo and Cincotti [8]. In the literature [9, 10], a few transforms such as Hadamard transforms and discrete unitary transformations have employed MMI structures and multimode waveguide holograms. However, these devices were designed for the InP material system. For the device using holograms, a complex fabrication process is required. The presence of holograms within the multimode waveguide tends to introduce additional losses.

Optical signal processing circuits require the basic operations of sum and difference, and exchange. These core functions can be implemented simply by using multimode interference (MMI) structures with some small modifications [11]. MMI devices use the principle of self-imaging within multimode optical waveguides to produce single or multiple images, of an input field, at periodic distances along the waveguide. Such multimode interference couplers have the desirable advantages of low loss, compactness and good fabrication tolerances [12].

The available material systems used for such multimode devices include polymers, silica on silicon and silicon-on-insulator (SOI). The high-index contrast silicon-on-insulator (SOI) platform has attracted much interest due to its potential for miniaturization, improved performance, and compatibility with existing CMOS technology [13].

In this chapter, we show that multimode interference couplers along with phase shifters can become extremely useful elements for building all-optical signal processing circuits. The chapter presents a method of realizing all-optical discrete Fourier transforms using the proposed structures. The photonic circuits are analysed and optimized using the transfer matrix method and the beam propagation method (BPM) [14]. A description of the general theory behind the use of multimode structures to achieve the basic functions of sum, subtraction, exchange and butterfly is presented in Section 21.2. Optimal design of MMI based structures for signal processing along with simulation results are covered in Section 21.3. A brief summary of the results of this research is given in Section 21.4.

21.2 THEORY OF MULTIMODE INTERFERENCE

Optical waveguides are the basic building blocks for fabricating more complex devices, such as directional couplers, MMI couplers and Mach-Zehnder interferometers. The simplest such structure is the dielectric slab waveguide. Electromagnetic wave propagation in a planar slab waveguide will be discussed first, because it is the simplest structure to be analysed from the point of view of its mathematical description, and from it, the general features of more complex waveguide structures can be understood. Maxwell's equations will be used to derive the wave equations for transverse electric (TE) and transverse magnetic (TM) propagation that govern light behaviour in planar waveguides. These wave equations will be solved for the case of symmetric step-index planar waveguides.

21.2.1 DIELECTRIC SLAB WAVEGUIDE

One of the simplest forms of optical waveguide is a three layer planar waveguide (dielectric slab waveguide) as shown in Figure 21.1 [15-18]. W is the width of the waveguide and n_c, n_f and n_s ($n_f > n_s$ and $n_f > n_c$) are the refractive indices of the cladding, the film (core), and the substrate, respectively. Propagation in the z direction is assumed.

Light is confined and guided in the waveguide due to the total internal reflection (TIR) phenomenon. The analysis of light propagation in a dielectric slab waveguide can be carried out by using the ray optics approximation [19]. However, in order to fully describe the fields within a dielectric waveguide, Maxwell's equations need to be used.

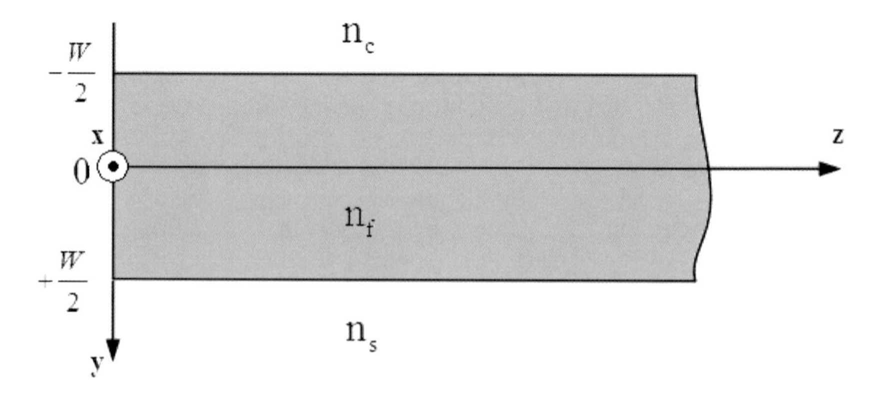

Figure 21.1 A three-layer dielectric slab waveguide

Maxwell's equations describe the interrelation between electric and magnetic fields as well as their interaction with surrounding media. Maxwell's equations for an electromagnetic wave are given by [20]

$$\nabla \cdot \mathbf{D} = \rho \tag{21.1}$$

$$\nabla \mathrm{x} \mathbf{E} = -\mu \frac{\partial \mathbf{H}}{\partial t} \tag{21.2}$$

$$\nabla \cdot \mathbf{H} = 0 \tag{21.3}$$

$$\nabla \mathrm{x} \mathbf{H} = \mathbf{J} + \frac{\partial \mathbf{D}}{\partial t} \tag{21.4}$$

Here, boldface represents a vector quantity, and E, H, and $\mathbf{D} = \varepsilon \mathbf{E}$ represent the electric field, magnetic field and electric displacement, respectively. The symbols ε, μ, ρ represent permittivity, permeability, and free charge density, respectively. The refractive index is $n = \sqrt{\varepsilon_r \mu_r}$, where $\varepsilon_r = \varepsilon / \varepsilon_0$, $\mu_r = \mu / \mu_0$ are the relative permittivity and permeability, respectively. ε_0 and μ_0 are the permittivity and permeability of free space. In this work, the material is non-magnetic. This means that $\mu = \mu_0$ and $n = \sqrt{\varepsilon_r}$.

For isotropic and non-conducting media with no free charge, the wave equation derived from Maxwell's equations is

$$\nabla^2 \mathbf{\Psi} - \mu_0 \varepsilon \frac{\partial^2 \mathbf{\Psi}}{\partial t^2} = 0 \tag{21.5}$$

where $\mathbf{\Psi}$ is either the electric or magnetic field. For the case of the dielectric slab waveguide, the refractive index depends only on a single Cartesian coordinate and the electric and magnetic fields take the form [21]

$$\mathbf{\Psi}(x, y, z, t) = \psi(y) \exp[j(\omega t - \beta z)] \tag{21.6}$$

Here ω is the angular frequency. Given a refractive index distribution n(y) that defines the planar waveguide, the equations for the electromagnetic fields are reduced and this enables exact solutions to be found for the complex field amplitudes $\psi(y)$ as well as for the propagation constant β. Substituting equation (21.6) into the wave equation (21.5) yields the equation describing the spatial dependence of the fields in the three layers of the dielectric slab waveguide

$$\nabla^2\psi + [k^2 n^2(y) - \beta^2]\psi = 0 \tag{21.7}$$

where $k = 2\pi/\lambda$ is the free space wave-number, λ is the optical wavelength, and n is the refractive index. In order to find the propagation modes in a planar waveguide, two independent situations will be investigated. In the first case, the electric field associated with the mode has only the set of field components $\{E_x, H_y, H_z\}$ and its solutions are the TE modes. The second case involves the situation in which the electric field has only field components $\{E_y, E_z, H_x\}$ and the solutions are called TM modes.

In the case of TE propagation along a planar waveguide, the boundary conditions imply continuity of the E_x and H_z fields at the cover-film interface and at the substrate-film interface (Figure 21.1). The following coupled equations are obtained for the TE modes [22, 23]

$$H_y = -(\frac{\beta}{\omega\mu_0})E_x \tag{21.8}$$

$$H_z = j\frac{1}{\omega\mu_0}\frac{\partial E_x}{\partial x} \tag{21.9}$$

$$j\beta H_y + \frac{\partial H_z}{\partial x} = -j\omega\varepsilon_0 n^2 E_x \tag{21.10}$$

In the case of TM propagation along a planar waveguide, E_z and H_x are continuous at the cover-film interface and at the substrate-film interface. Applying these boundary conditions yield the following coupled equations

$$E_y = \frac{\beta}{\omega\mu_0 n^2}H_x \tag{21.11}$$

$$E_z = \frac{1}{j\omega\mu_0 n^2}\frac{\partial H_x}{\partial x} \tag{21.12}$$

$$j\beta E_y + \frac{\partial E_z}{\partial x} = -j\omega\varepsilon_0 H_x \tag{21.13}$$

For simplicity, only the TE modes is considered in the following derivation. The case for the TM modes can be solved similarly by using the corresponding boundary conditions. Equation (21.7) yields three equations for the spatial dependence of E_x in the y direction as follows

$$\frac{d^2E_x}{dy^2}-\gamma_c^2E_x=0,\ y\geq +W/2 \ \text{(cover)} \tag{21.14}$$

$$\frac{d^2E_x}{dy^2}+\kappa_f^2E_x=0,\ -W/2\leq y\leq +W/2 \ \text{(film)} \tag{21.15}$$

$$\frac{d^2E_x}{dy^2}-\gamma_s^2E_x=0,\ y\leq -W/2 \ \text{(substrate)} \tag{21.16}$$

where the three parameters γ_c, γ_s, and κ_f are given by

$$\gamma_c^2=\beta^2-k^2n_c^2 \tag{21.17}$$

$$\kappa_f^2=k^2n_f^2-\beta^2 \tag{21.18}$$

$$\gamma_s^2=\beta^2-k^2n_s^2 \tag{21.19}$$

It is assumed that the dielectric waveguide is symmetrical, which means $n_s = n_c$. By solving equations (21.14), (21.15), and (21.16), the electric field in the cover, film and substrate regions can be expressed as

$$E_x(y)=\begin{cases} Ae^{\gamma_s(y+W/2)}, & y\leq -W/2 \\ Be^{j\kappa_f y}+Ce^{-j\kappa_f y}, & -W/2\leq y\leq +W/2 \\ De^{-\gamma_c(y-W/2)}, & y\geq +W/2 \end{cases} \tag{21.20}$$

where A, B, C, and D are constants, which are determined from boundary conditions. Employing the boundary conditions, the following dispersion equation is obtained [24, 25]

$$V\sqrt{1-b}=\upsilon\pi+2\tan^{-1}\left(\sqrt{\frac{b}{1-b}}\right) \tag{21.21}$$

where $\upsilon=0,1,..,M-1$ represents the mode number. Parameters V (V-number) and b (normalized propagation constant) are defined by $V=kW\sqrt{n_f^2-n_c^2}$, and $b=(n_e^2-n_c^2)/(n_f^2-n_c^2)$, respectively. $n_e=\beta/k$ is the effective refractive index. By solving the dispersion equation (21.21), the effective index of every mode supported by the waveguide can be found.

21.2.2 Multimode Interference Coupler

The most common analytical method for the design of MMI couplers is the guided mode propagation analysis (MPA). In this section, the conventional MPA for the design of MMI couplers will be reviewed and the approximations employed in the method will be highlighted. The three different mechanisms for imaging in MMI couplers are then outlined. Finally, the concept of the transfer matrix method is presented.

21.2.2.1 Approximations in the Guided Mode Propagation Analysis

Consider a step index multimode slab waveguide with a width $W = W_{MMI}$. The waveguide supports a total of M guided modes with modal field profiles $\phi_\upsilon(y)$ and propagation constants β_υ, where $\upsilon = 0,...,M-1$ is the mode number. It is assumed that the radiation modes carry negligible power and therefore can be neglected in this analysis. An input field profile $\psi(y,0)$ at the entrance of the multimode waveguide then can be decomposed into the modal field distribution $\phi_\upsilon(y)$ of all guided modes [26]

$$\psi(y,0) = \sum_{\upsilon=0}^{M-1} c_\upsilon \phi_\upsilon(y) \tag{21.22}$$

where the field excitation coefficient c_υ can be calculated by using the overlap integral

$$c_\upsilon = \frac{\int \psi(y,0)\phi_\upsilon^*(y)dy}{\int \left|\phi_\upsilon(y)\right|^2 dy} \tag{21.23}$$

where * indicates the complex conjugate.

The total field profile $\psi(y,z=L)$ at position z=L then can be approximated as a superposition of all guided modes

$$\psi(y,z=L) = \sum_{\upsilon=0}^{M-1} c_\upsilon \phi_\upsilon(y) e^{-j\beta_\upsilon L} \tag{21.24}$$

where the propagation constant of mode υ for a high index contrast system can be calculated approximately by [27]

$$\beta_\upsilon \approx k_0 n_f - \frac{(\upsilon+1)^2 \pi\lambda}{4 n_f W_e^2} \tag{21.25}$$

Here, n_f is the core refractive index and W_e is the effective waveguide width for the fundamental mode and λ is the operating wavelength. The difference between the propagation constants of the fundamental mode ($\upsilon = 0$) and mode υ can be expressed as

$$\beta_0 - \beta_\upsilon \approx \frac{\upsilon(\upsilon+2)\pi\lambda}{4 n_f W_e^2} \tag{21.26}$$

By defining $L_\pi = \frac{\pi}{\beta_0 - \beta_1} \approx \frac{4 n_f W_e^2}{3\lambda}$, as the beat length of the two lowest order modes (the fundamental and first order modes), equation (21.26) can then be written approximately as

$$\beta_0 - \beta_\upsilon \approx \frac{\upsilon(\upsilon+2)\pi}{3 L_\pi} \tag{21.27}$$

Therefore the propagation constant β_υ is dependent of the square of the mode number υ.

In the MPA analysis, by using the notion of an effective width, the Goos-Hänchen shift can be incorporated into the field description in an approximate way. The mode field amplitudes are then given by [28]

$$\phi_\upsilon(y) = \begin{cases} cos[(\upsilon+1)\pi y / W_e], \; for \;\; \upsilon \;\; even \; and \; |y| \le W_e \\ sin[(\upsilon+1)\pi y / W_e], \; for \;\; \upsilon \;\; odd \; and \;\; |y| \le W_e \\ 0 \; for \; |y| > W_e \end{cases} \tag{21.28}$$

The effective width of the υth mode ($W_{e\upsilon}$) is assumed to be the same as that for the fundamental mode W_e

$$W_{e\upsilon} \approx W_e = W_{MMI} + \frac{\lambda}{\pi}\left(\frac{n_f}{n_c}\right)^{2\xi}(n_f^2 - n_c^2)^{-1/2} \tag{21.29}$$

where n_f and n_c are effective indices of the core and the cladding. ξ =0 for the TE modes and ξ =1 for the TM modes.

A useful expression for the field at position z=L can then be found by substituting equation (21.27) into equation (21.24), yielding

$$\Psi(y,z=L)=e^{-j\beta_0 L}\sum_{\upsilon=0}^{M-1} c_\upsilon \phi_\upsilon(y) exp[j\frac{\upsilon(\upsilon+2)}{3L_\pi}L] \quad (21.30)$$

Thus, the form of the output images will be determined by the modal excitation factor c_υ and the mode phase factor $exp(j\frac{\upsilon(\upsilon+2)}{3L_\pi}L)$. The output field profile $\Psi(y,z=L)$ will be a reproduction of the input field $\Psi(y,0)$ if the parameters are appropriately chosen.

There are three main interference mechanisms. The first is the general interference (GI) mechanism which is independent of the modal excitation. The second is the restricted interference (RI) mechanism, in which excitation inputs are placed at some special positions so that certain modes are not excited. The last mechanism is the symmetric interference (SI), in which the excitation input is located at the centre of the multimode section. These three interference mechanisms will now be investigated in detail.

21.2.2.2 NxN General Interference MMI (GI-MMI)

For an NxN MMI coupler based on the general interference (GI-MMI) mechanism [12], there is no restriction on the placement of the input waveguides. The direct and mirror single images of the input field are ideally formed at distances z that are even and odd multiples of the length $3L_\pi$, respectively.

In addition, multiple images can be formed as well. An NxN general interference MMI coupler has length $L=L_{MMI}=\frac{p}{N}3L_\pi$, where p is a positive integer such that p and N are without a common divisor. In practical designs, the shortest devices are obtained for p=1. The resulting amplitudes from image input i (i=1,..,N) to output j (j=1,..,N) can be given in a compact form

$$A_{ij}=A_{ji}=\sqrt{\frac{1}{N}} \quad (21.31)$$

where $|A_{ij}|^2$ is the normalized powers of the output images. The phases ϕ_{ij} of the equal output signals at the output waveguides can be calculated by [12]

For i+ j: even, $\phi_{ij}=\phi_0+\pi+\frac{\pi}{16}(j-i)(8-j+i)$

and for i+j: odd , $\phi_{ij}=\phi_0+\frac{\pi}{16}(i+j-1)(8-j-i+1)$ (21.32) where the input ports i (i=1, 2,..,N) are numbered from bottom to top and the output ports j (j=1, 2,..,N) are numbered from top to bottom in the MMI coupler. $\phi_0 = -\beta_0 L_{MMI} - \frac{\pi}{2}$ is a constant phase that depends upon the MMI geometry and therefore can be implied in the following calculations.

21.2.2.3 NxN Restricted Interference MMI (RI-MMI)

In an NxN MMI coupler based on the restricted interference (RI) theory, only some of the guided modes in the MMI waveguide are excited by the input field $\psi(y,0)$. This mechanism is implemented by suitable placement of the access waveguides. The theory shows that N-fold images occur at the distance $L_{MMI} = \frac{p}{N} L_\pi$, where p and N have no common divisor. The phases ϕ_{ij} associated with imaging an image input i to an output j in an RI-MMI coupler can be derived as [29]

$$\phi_{ij} = (i^2 + j^2)\frac{\pi}{12} - j\frac{\pi}{2} - \frac{\pi}{2} \text{ for } A_{ij} > 0,$$

$$\text{and } \phi_{ij} = (i^2 + j^2)\frac{\pi}{12} - j\frac{\pi}{2} - \frac{3\pi}{2} \text{ for } A_{ij} < 0 \quad (11.33)$$

where $A_{ij} = \frac{2}{\sqrt{N}} sin[(N-j)i\frac{\pi}{2N}]$ is the associated image amplitude from an image input i to an image output j.

21.2.2.4 1xN Symmetric Interference MMI (SI-MMI)

If an MMI coupler is centre-fed, then ideally only the even symmetric modes are excited. Hence, 1xN MMI couplers can be realized. In general, N-fold images of the input field $\psi(y,0)$ are obtained at distances $L_{MMI} = \frac{p}{N}(\frac{3L_\pi}{4})$. These images are symmetrically located along the y axis with equal spacing W_e / N. The positions y_i (i=0,1,...,N-1) and phases ϕ_i of the output images can be expressed as [29]

$$y_i = (N-1-2i)\frac{W_e}{2N} \quad (21.34)$$

$$\phi_i = -\frac{\pi}{4N}(N-1-2i)^2 \quad (21.35)$$

Table 21.1 Summary of different MMI types

MMI type	*MMI length*	*Condition on the placement of the access waveguide (y)*
General interference (GI)	$\frac{p}{N}3L_\pi$	None
Restricted interference (RI)	$\frac{p}{N}L_\pi$	$y=-\frac{W_e}{6}$, $y=+\frac{W_e}{6}$
Symmetric interference (SI)	$\frac{p}{4N}3L_\pi$	$y=0$

The input waveguide is placed at the centre of the MMI section to implement a 1xN MMI coupler based on the symmetric interference (SI) mechanism. Such a device can also be reversed and forms an Nx1 combiner if the input signals have the appropriate phases.

In summary, the three types of rectangular MMI couplers are summarized in Table 21.1. The coupler lengths and the placement of the access waveguides for producing N outputs are provided. Here, p is an integer.

21.2.2.5 Transfer Matrix of MMI Couplers

The characteristics of an MMI device can be described by a transfer matrix [30, 31]. This transfer matrix is a very useful tool for analyzing cascaded MMI structures.

The phase ϕ_{ij} associated with imaging an input i to an output j in an MMI coupler was given in equation (21.22) for GI-MMI couplers, in equation (21.33) for RI-MMI couplers and in equation (21.35) for SI-MMI couplers. These phases ϕ_{ij} form a matrix $\mathbf{\Phi}$, with i representing the row number, and j representing the column number. Then the transfer matrix of the MMI coupler $\mathbf{M}$ is directly related to $\mathbf{\Phi}$, and the output field distribution emerging from the MMI coupler can be written as

$$\mathbf{b} = \mathbf{Ma} \tag{21.36}$$

where $\mathbf{a} = [a_1\ a_2\ \ldots\ a_N]^T$, $\mathbf{b} = [b_1\ b_2\ \ldots\ b_N]^T$ and $\mathbf{M} = [m_{ij}]_{NxN}$. The superscript T indicates the transpose of a matrix. a_i (i=1,..,N) is the complex field amplitude at input waveguide i and b_j (j=1,..,N) is the complex field amplitude at output waveguide j. Elements of the transfer matrix $\mathbf{M}$ are $m_{ij} = m_{ji} = A_{ij}e^{j\phi_{ij}}$, where A_{ij} is the field amplitude transfer coefficient and ϕ_{ij} is the phase shift when imaging from input i to output j.

21.3 ALL-OPTICAL SIGNAL PROCESSING CIRCUITS USING MMI COUPLERS

In this section we show how simple MMI couplers can implement core signal processing functions. 2x2, 4x4 and 8x8 MMI couplers can be used to realize the basic signal processing functions.

21.3.1 Sum and Subtraction Unit

We start with the 2×2 normalised Haar matrix that is associated with the Haar wavelet filter (or the Daubechies wavelet filter of order M=1) [12]. This matrix can be written as

$$H_2 = \frac{1}{\sqrt{2}}\begin{bmatrix} 1 & 1 \\ 1 & -1 \end{bmatrix} \tag{21.37}$$

The above matrix also represents the first order Hadamard transform. The above filter obviously forms the sum and difference of a 2-point signal [9].

We shall now compare the above matrix with the transfer matrix describing the relation between the input and output fields of a 2x2 restricted interference MMI (RI-MMI) coupler. Consider a 2x2 restricted interference MMI coupler using the silicon channel waveguide structure as shown in Figure 21.2. It consists of a central multimode waveguide with two input and two output single-mode waveguides, where $y_1 = y_2 = \frac{W_{MMI}}{6}$ and $y_3 = y_4 = -\frac{W_{MMI}}{6}$. The width and length of this MMI coupler are W_{MMI} and L_{MMI}, respectively.

In a restricted interference (RI) MMI coupler, suitable placement of the access waveguides means that certain modes are not excited within the multimode region. This results in a reduced length periodicity and hence smaller size MMI couplers. The transfer matrix describing the relation between the input and output

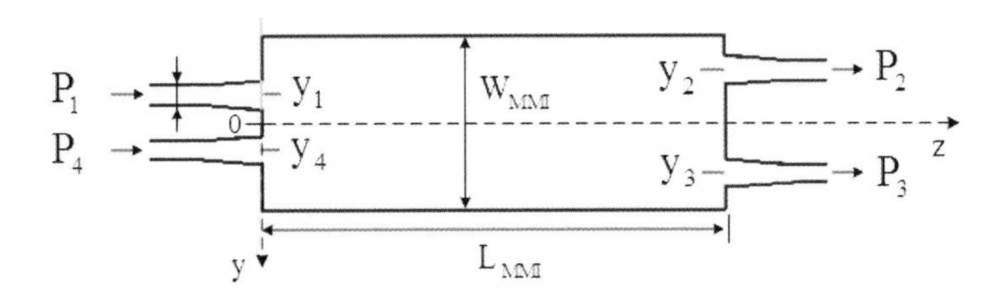

Figure 21.2 Structure of a 2x2 MMI coupler along with the positions of the ports.

electric fields of the above 2x2 restricted interference ideal MMI coupler (RI MMI) with a length of $L_\pi/2$ is

$$\mathbf{M}_{RI} = \frac{e^{j\phi_{RI}}}{\sqrt{2}}\begin{bmatrix} 1 & j \\ j & 1 \end{bmatrix} \tag{21.38}$$

where ϕ_{RI} is a constant phase, $L_\pi = \dfrac{\pi}{\beta_0 - \beta_1}$ is the beat length of the two lowest order modes of the coupler having propagation constants of β_0 and β_1, respectively.

By adding phase shifters at the input and output waveguides for a 2x2 RI MMI coupler, as shown in Figure 21.3 (a) and (b), respectively, the overall transfer function for the coupler has the form

$$\mathbf{M}_\phi = \frac{e^{j\phi}}{\sqrt{2}}\begin{bmatrix} 1 & 1 \\ 1 & -1 \end{bmatrix} \tag{21.39}$$

where φ represents the phase shift due to propagation of the fields along the device. Therefore this element can be used as a sum and subtraction unit.

Similarly, the transfer matrix $\mathbf{M}_{GI}$ for an ideal 2x2 general interference MMI (GI-MMI) coupler of length $3L_\pi/2$ can be expressed as

$$\mathbf{M}_{GI} = \frac{e^{j\phi_{GI}}}{\sqrt{2}}\begin{bmatrix} 1 & -j \\ -j & 1 \end{bmatrix} \tag{21.40}$$

where ϕ_{GI} is a constant phase. By adding phase shifters at the input and output waveguides for a 2x2 GI-MMI coupler as shown in Figure 21.4 (a) and (b),

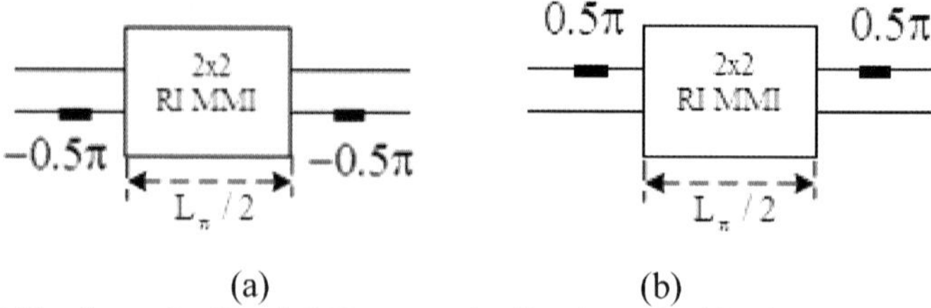

Figure 21.3 The first order (2-point) Haar wavelet filter is realized by the use of 2x2 RI MMI couplers with different positions of phase shifters.

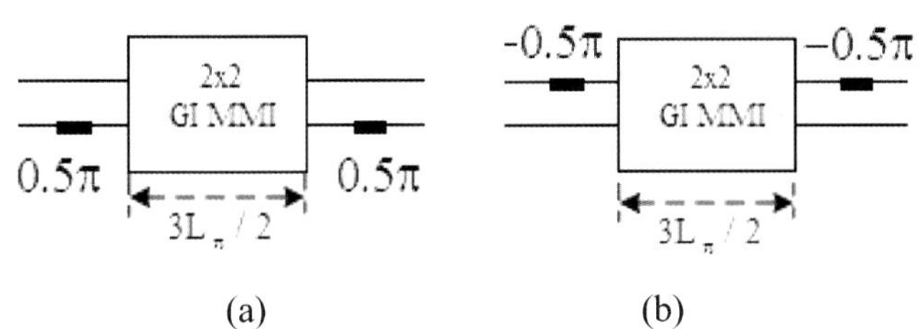

Figure 21.4 The first order (2-point) Haar wavelet filter is realized by the use of 2x2 GI MMI couplers with different positions of phase shifters.

respectively, the overall transfer matrix for the coupler has the form required for a sum and difference unit.

21.3.2 Exchange Unit

Exchange unit can be now formed from an ideal 2x2 RI-MMI coupler of length L_{π}, which has a transfer matrix S_{RI} given by

$$\mathbf{S}_{RI} = je^{j2\phi_{RI}}\begin{bmatrix}0 & 1\\ 1 & 0\end{bmatrix} = e^{j(2\phi_{RI}+\frac{\pi}{2})}\begin{bmatrix}0 & 1\\ 1 & 0\end{bmatrix} \tag{21.41}$$

Note that there will be a phase shift of $2\phi_{RI}+\frac{\pi}{2}$ introduced to both outputs. The exchange unit using a 2x2 RI MMI coupler and its circuit symbol are shown in Figure 21.5(a) and 21.5(b), respectively.

Similarly, the transfer matrix for a 2x2 GI-MMI coupler having a length of 3Lpi is given by

$$\mathbf{S}_{GI2} = -je^{j2\phi_{GI}}\begin{bmatrix}0 & 1\\ 1 & 0\end{bmatrix} = e^{j(2\phi_{GI}-\frac{\pi}{2})}\begin{bmatrix}0 & 1\\ 1 & 0\end{bmatrix} \tag{21.42}$$

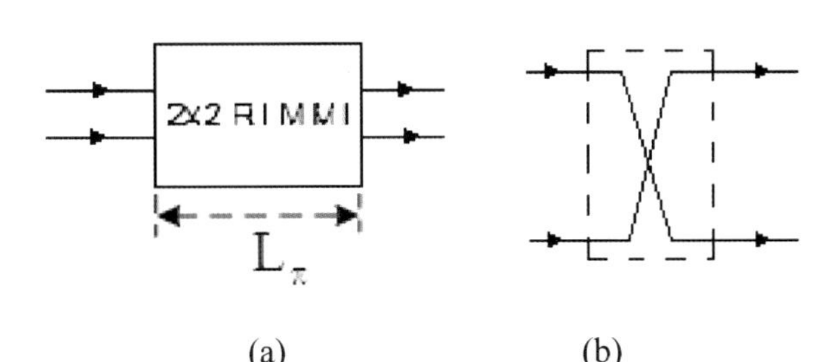

Figure 21.5 (a) Exchange unit (a) using a 2x2 RI-MMI coupler and (b) its circuit symbol.

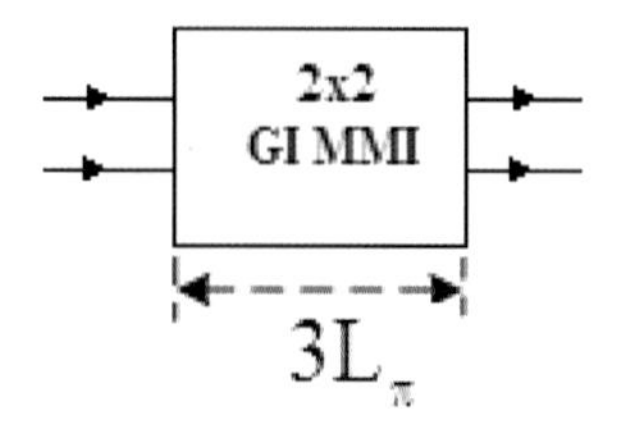

Figure 21.6 An exchange unit realized with a 2x2 GI-MMI coupler at length $3L_\pi$.

Therefore, the exchange unit can be implemented by using a 2x2 GI-MMI coupler as shown in Figure 21.6. In this case a phase shift of $2\phi_{GI}\frac{\pi}{2}$ is introduced.

21.3.3 FFT Butterfly Unit

In this section, the realization of all-optical discrete Fourier transform (DFT) based on the proposed structures is presented.

The DFT decimation in frequency algorithm is performed by division of data at each stage of the fast Fourier transform (FFT) process and by multiplication with weights. The DFT of an input sequence $\{x_n\}\big|_{n=0,\dots,N-1}$ is given by [32]

$$y_k = \sum_{n=0}^{N-1} x_n W^{nk} \tag{21.43}$$

where k=0,…, N-2, $u_n = x_n$ for 0≤n≤N/2, $u_n = x_n + N/2$ for 0≤n≤N/2-1, and $W^{nk} = e^{-j\frac{2\pi kn}{N}}$. Equation (21.43) can be rewritten by

$$y_k = \sum_{n=0}^{N/2-1} (u_n + v_n)W^{2nm} + \sum_{n=0}^{N/2-1} [(u_n - v_n)W^n]W^{2nm} \tag{21.44}$$

The basic butterfly operation for calculating the FFT algorithm is shown in Figure 21.7.

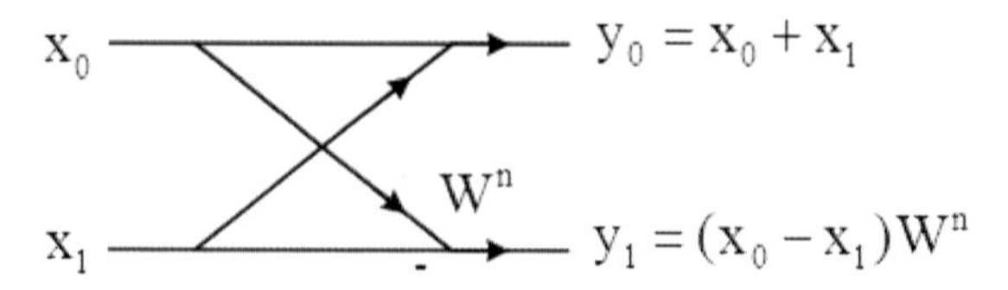

Figure 21.7 Basic butterfly operation for calculating the FFT algorithm.

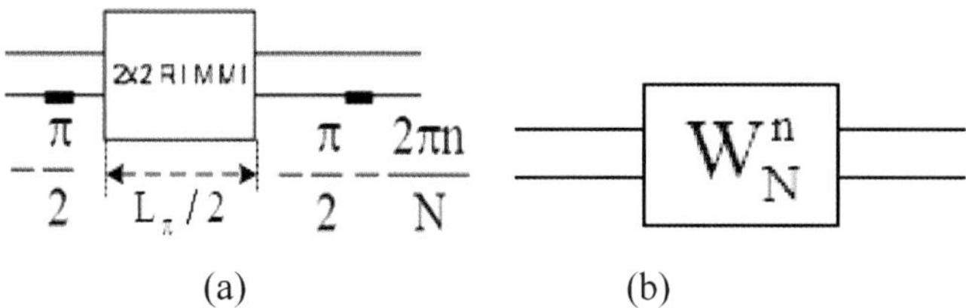

Figure 21.8 2-point DFT butterfly operation (a) realization using a 2x2 RI-MMI coupler and (b) circuit symbol.

This element can be implemented by using a 2x2 RI-MMI coupler cascaded with phase shifters in the input and output waveguides as shown in Figure 21.8(a). The symbol for this element is shown in Figure 21.8(b).

21.4 SIMULATION RESULTS AND DISCUSSIONS

In the present work, silicon on insulator (SOI) technology is used for the design of MMI devices. Silicon photonics [33-35] was chosen because the fabrication of such devices require only small and low cost modifications to existing fabrication processes. SOI technology is compatible with existing complementary metal–oxide–semiconductor (CMOS) technologies for making compact, highly integrated, and multifunction devices [36, 37]. The SOI platform uses silicon both as the substrate and the guiding core material. The large index contrast between Si (n_{Si} =3.45 at wavelength 1550nm) and SiO_2 (n_{SiO_2} =1.46) allows light to be confined within submicron dimensions and single mode waveguides can have core cross-sections with dimensions of only few hundred nanometres and bend radii of a few micrometers with minimal losses. Moreover, SOI technology offers potential for monolithic integration of electronic and photonic devices on a single substrate. A variety of photonic devices have been designed and realized on the SOI platform, including lasers [38-40], high speed optical modulators [41-43], optical switches [44, 45], optical polarization combiners/splitters [46], spot size converters [47], wavelength converters [48], optical amplifiers [49], and optical oscilloscopes [50], etc.

There are two common geometries for practical SOI optical waveguide structures, namely the rib and the channel waveguide (also called the silicon waveguide or photonic wire) used in silicon photonics. The cross-sectional views of these structures are shown in Figure 21.9(a) and (b). Both waveguide structures have their own advantages and drawbacks. To achieve single mode operation, an SOI channel waveguide must have small dimensions, which can be challenging both in terms of coupling to single mode fibre and in fabrication. On the other hand, single mode rib waveguides can have larger dimensions than the channel waveguide. This leads to easier coupling, but the bend radii are larger. Throughout this chapter, a variety of devices for photonic signal processing will

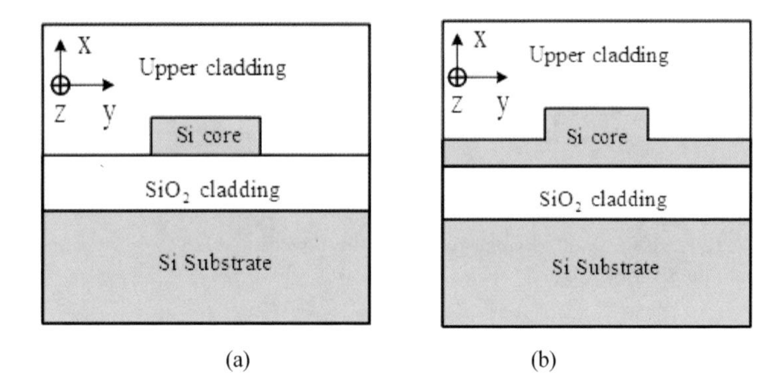

Figure 21.9 Cross-sections of (a) SOI channel and (b) SOI rib waveguides.

be designed and optimised, so both types of SOI waveguide structures may be employed.

In addition, it is well known that the finite-difference time-domain (FDTD) method is a general method to solve Maxwell's partial differential equations numerically in the time domain. Simulation results for devices on the SOI channel waveguide using the 3D-FDTD method can achieve a very high accuracy. However, due to the limitation of computer resources and memory requirements, it is difficult to apply the 3D-FDTD method to the modelling of large devices on the silicon waveguide. Meanwhile, the three dimensional beam propagation method (3D-BPM) was shown to be a quite suitable method that has sufficient accuracy for simulating devices based on SOI channel waveguides [51, 52]. Therefore, the design for devices using silicon waveguides will now be performed using the 3D-BPM. The theory of the 3D-BPM is studied in detail in the literature [53].

21.4.1 Optimal Design of 2x2 MMI Based Structures

The silicon waveguide structure used in the designs is shown in Figure 21.10. An upper cladding region is needed for devices using the thermo-optic effect in order to reduce loss due to metal electrodes. Also, the upper cladding region is used to avoid the influence of moisture and environmental temperature [54].

The parameters used in the designs are as follows: the waveguide has a standard silicon thickness of $h_{co} = 220nm$ and access waveguide widths are $W_a = 0.5\mu m$ for single mode operation. It is assumed that the designs are for the TE polarization at a central optical wavelength $\lambda = 1550nm$.

For a 2x2 RI-MMI coupler, the MMI width is $W_{MMI} = 3(s+W_a)$. From the above results, the minimum MMI width to limit crosstalk between the two access waveguides is $W_{MMI} = 3(s+W_a) = 3.09\mu m$. In these designs, a final taper width of

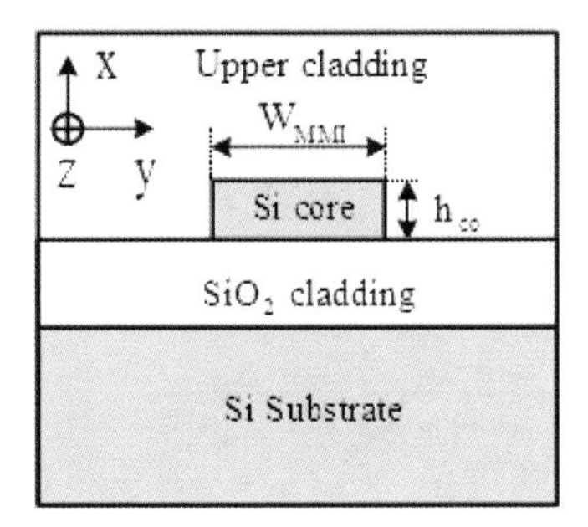

Figure 21.10 Silicon waveguide cross-section used in the designs of the proposed device.

800nm is used and the MMI width is chosen to be $W_{MMI} = 4\mu m$ in order for input and output parallel waveguides to be well separated. This width leads to a separation of s=0.85 μm.

First, the 3D-BPM is used to carry out the simulations for the device having a length of $30\mu m$ for the 2x2 RI-MMI coupler. The aim of this step is to find roughly the positions which result in a power splitting of 50/50, i.e., a 3dB coupler. Then, the 3D-BPM is used to perform the simulations around these positions to locate the best lengths. Our BPM simulations show that the optimised length for obtaining the best performance with the chosen parameters is $20\mu m$. for the 2x2 RI MMI devices. Figure 21.11(a) shows the normalised output powers of this MMI coupler. The field propagations in 2x2 optimised RI- MMI simulated using the BPM are shown in Figure 21.11(b).

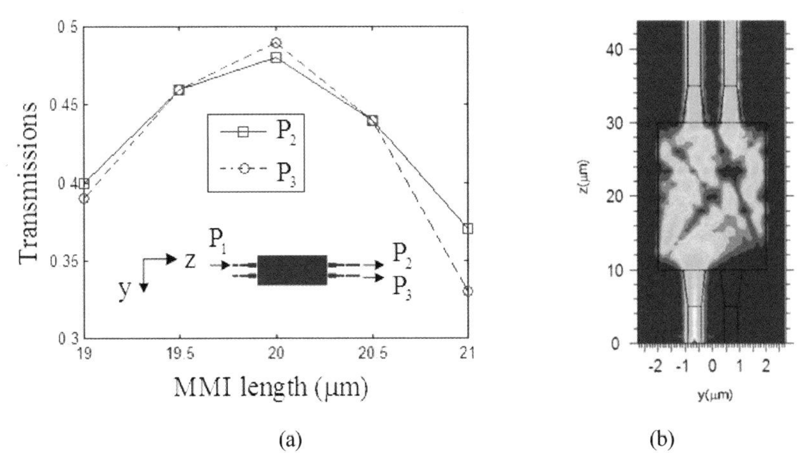

(a) (b)

Figure 21.11 Field propagation in (a) optimised RI-MMI and (b) optimised GI-MMI couplers.

21.4.2 Optimal Design of the Exchange Unit

The exchange unit can be realized using MMI couplers having the appropriate dimensions ($L_{MMI} = 3L_{\pi}/2$ for 2x2 GI-MMI couplers and $L_{MMI} = L_{\pi}/2$ for 2x2 RI-MMI couplers) to form a mirror image. The optimised lengths of the MMI coupler used for the exchange unit is $40\mu m$ for the RI-MMI coupler. The calculated excess losses in both cases are 0.4dB.

The 3D-BPM simulations for the exchange unit having normalized powers of 0.8 and 0.2 presented at input ports 1 and 2 are plotted in Figure 21.12.

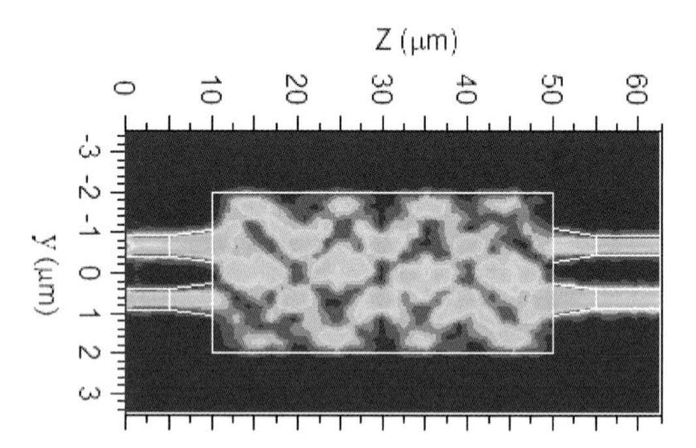

Figure 21.12 The 3D-BPM simulations for an exchange unit based on a 2x2 RI-MMI coupler. The exchange is performed for two input signals having normalized powers of 0.8 and 0.2 at two input ports.

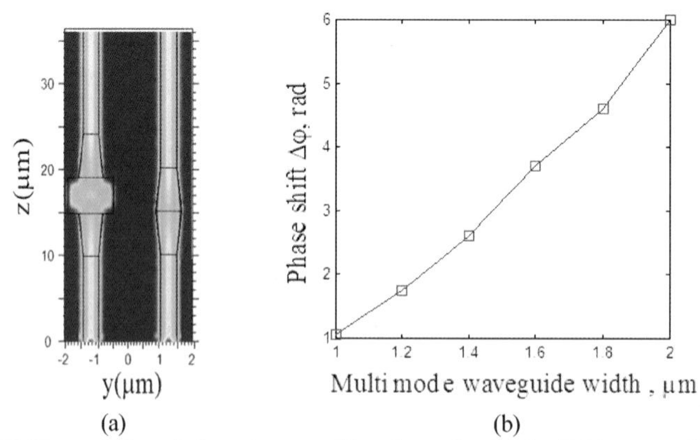

Figure 21.13 Phase shift made by using a multimode section (a) the field propagation through a 1x1 multimode waveguide having a width of 1.5μm compared to a single mode waveguide and (b) phase shifts at different multimode waveguide widths.

21.4.3 Optimal Design of Phase Shifters

The phase shifters incorporated with the MMI structures are particularly important to realize the appropriate functions of all-optical signal processing circuits using 2x2 MMI couplers. It is possible to realize an optical phase shifter by using a curved waveguide section, a wide waveguide, a multimode waveguide, a special patterned waveguide or a heated waveguide based on the thermo-optic effect. Here, we investigate the approach for implementing the phase shifters by using the multimode waveguides [55] due to their advantages of small size, low loss and ease of fabrication with the existing CMOS technology.

The multimode section can be viewed as a 1x1 SI-MMI coupler and the symmetric interference (SI) theory can be used to determine the length L_M to give a single self-image at the end of the section. The width of the multimode silicon waveguide section is chosen to be in the range $1\mu m$ to $2\mu m$ in order to support at least three guided modes. It therefore acts as a small MMI coupler. The overall size of the device is not increased significantly. The length of the phase shifter is chosen to form a single-image at the output. If there are two adjacent waveguides and an additional phase shift is required for one waveguide, then a small MMI with tapered input and output sections can be used as shown in Figure 21.13(a). This figure also shows the 3D-BPM simulation for the fields propagating through a multimode waveguide section having a width of $1.5\mu m$ and through a single mode waveguide as an example. The tapered waveguides are used to reduce losses in this design. The phase shift due to the MMI taper is compensated by inserting a taper into the other waveguide. In practice, the single mode waveguide would not have a tapered section. The phase shift which can be achieved using different widths and optimised lengths of the multimode waveguide section is shown in Figure 21.13(b). It can be seen from this diagram that a particular phase shift can be achieved simply by appropriately choosing the width and length of the multimode waveguide section.

21.4.4 All-optical DFT Using MMI Structures

The 8-point DFT transform can be designed using 8x8 MMI structures. A new signal flow graph for computing the 8-point DFT transform is shown in Figure 21.14. This structure is suitable for realization with MMI couplers.

The structure consists of two basic operation blocks, called DFT-a and DFT-b. Here, the designs of these building blocks using 4x4 and 8x8 MMI couplers are presented in detail.

Design of block DFT-a: Building block DFT-a is plotted in Figure 21.15(a). This structure can be realized by using 8x8 MMI structures with phase shifts located at the input and output ports. The implementation of this device is shown in Figure 21.15(b).

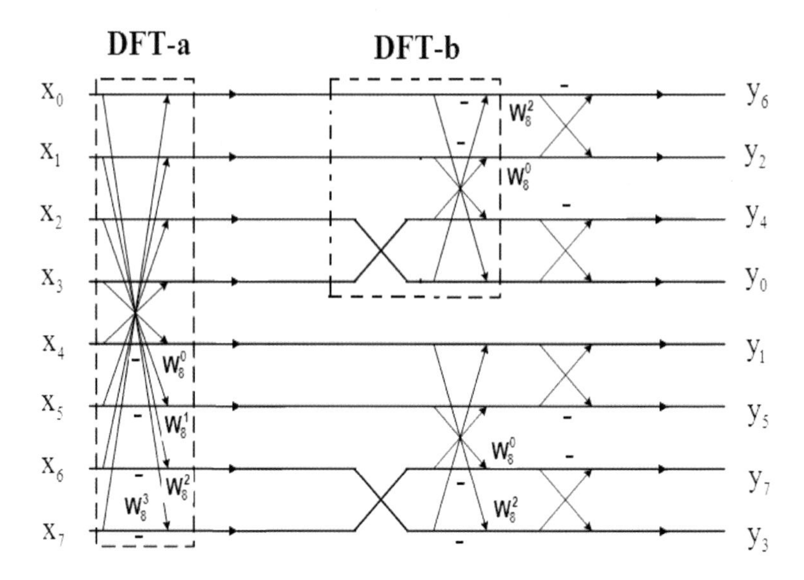

Figure 21.14 A modified structure for implementing the 8-point DFT transform.

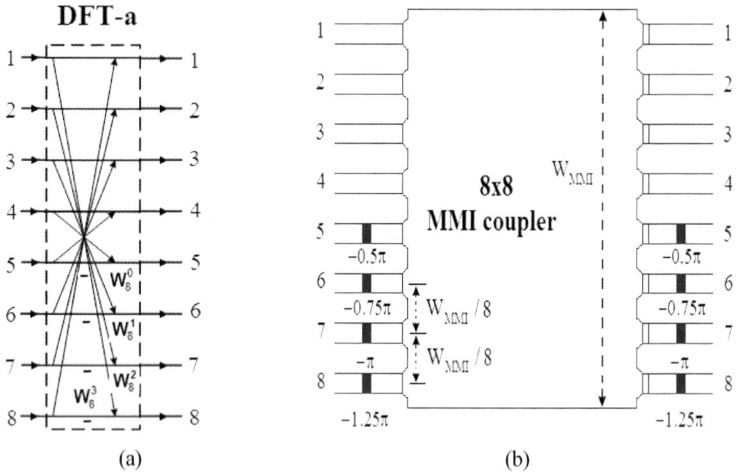

Figure 21.15 DFT building block (a) signal flow graph of DFT-a and (b) implementation using the 8x8 MMI structure.

Design of block DFT-b: It is possible to implement the building block DFT-b operation by using 4x4 MMI, 2x2 MMI structures and phase shifters. The signal flow graph is given in Figure 21.16(a) and its realization is shown in Figure 21.16(b). The 2x2 MMI coupler is used for an exchange unit and can be realized by using a 2x2 GI-MMI or 2x2 RI-MMI coupler.

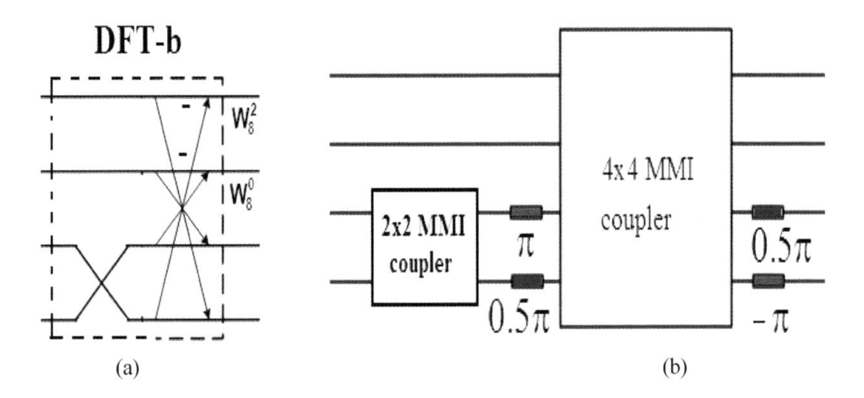

Figure 21.16 Building block DFT-b (a) signal flow graph and (b) its implementation using 2x2 MMI and 4x4 MMI structures.

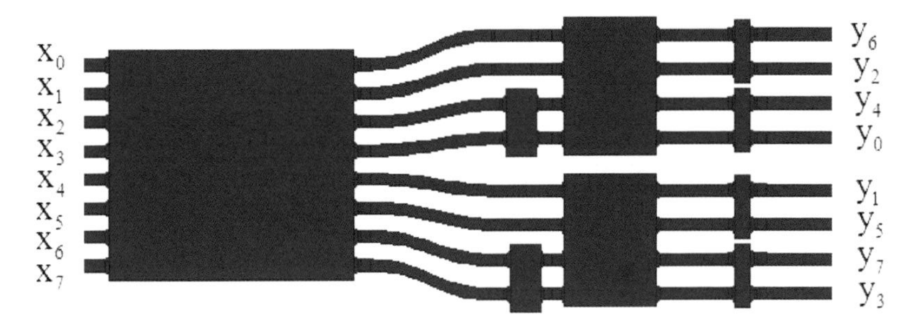

Figure 21.17 Optical circuit for computing the 8-point DFT transform.

By cascading these building blocks, an optical circuit for implementing the 8x8 DFT transform can be created, as shown in Figure 21.17.

In order to realize the 4x4 and 8x8 MMI structures used for the DFT, the positions of input and output waveguides, MMI widths and MMI lengths need to be chosen appropriately. Figure 21.18 shows two 4x4 MMI couplers have the same width W_{MMI} and length $L_{MMI} = \frac{3L_\pi}{2}$ connected together via two phase shifters $\Delta\varphi_1$ and $\Delta\varphi_2$.

A single 4x4 MMI coupler at a length of $L_1 = \frac{3L_\pi}{4}$ is described by the following transfer matrix

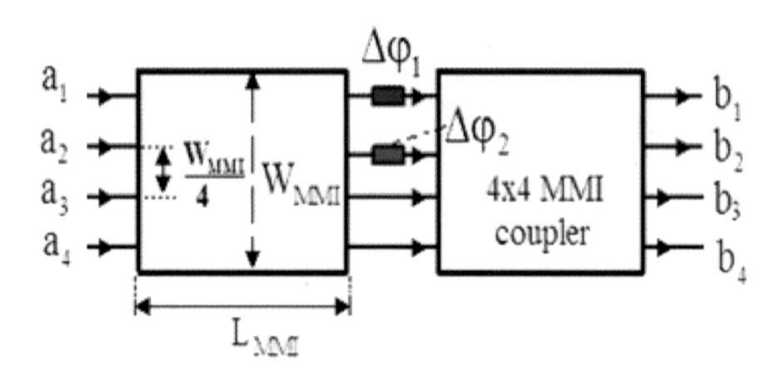

Figure 21.18 4x4 MMI structure.

$$\mathbf{M} = \frac{1}{2}\begin{bmatrix} -1 & -e^{j\frac{3\pi}{4}} & e^{j\frac{3\pi}{4}} & -1 \\ -e^{j\frac{3\pi}{4}} & -1 & -1 & e^{j\frac{3\pi}{4}} \\ e^{j\frac{3\pi}{4}} & -1 & -1 & -e^{j\frac{3\pi}{4}} \\ -1 & e^{j\frac{3\pi}{4}} & -e^{j\frac{3\pi}{4}} & -1 \end{bmatrix} \tag{21.45}$$

If the length is doubled to $L_{MMI} = 2L_1 = 3L_\pi/2$, a new 4x4 MMI coupler is formed and its transfer matrix is

$$\mathbf{S} = (\mathbf{M})^2 = \frac{1}{2}\begin{bmatrix} 1-j & 0 & 0 & 1+j \\ 0 & 1-j & 1+j & 0 \\ 0 & 1+j & 1-j & 0 \\ 1+j & 0 & 0 & 1-j \end{bmatrix} \tag{21.46}$$

The 3D-BPM simulations for a 4x4 MMI coupler having a length of $L_1 = \frac{3L_\pi}{4}$ are shown in Figure 21.19(a) for an signal presented at input port 1 and Figure 21.19(b) for an input signal at port 2. The optimised length of the MMI coupler calculated by using 3D-BPM is $L_1 = 71.70\,\mu m$. The calculated excess loss is 0.35dB for each case.

If two 4x4 MMI couplers with the same length of $L_{MMI} = 3L_\pi/4$ are cascaded together, then a 4x4 MMI coupler having a length of $L_{MMI} = 3L_\pi/2$ is formed. The transfer matrix of this new 4x4 MMI coupler is given by Equation (21.46). The 3D-BPM simulations for this 4x4 MMI coupler are shown in Figure 21.20(a) for the signal at input port 1 and Figure 21.20(b) for the signal at input port 2. The optimised length of each MMI coupler is found to be $L_{MMI} = 141.7\ \mu m$.

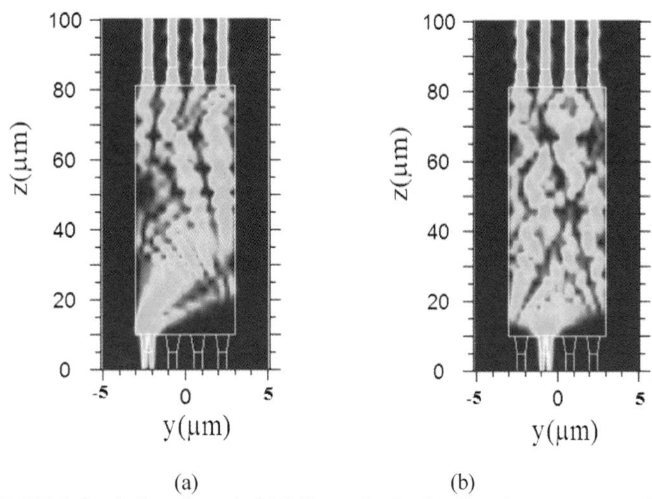

Figure 21.19 BPM simulations for a 4x4 MMI coupler having length $L_1 = 3L_\pi / 4$ with (a) input signal at port 1 and (b) input signal at port 2.

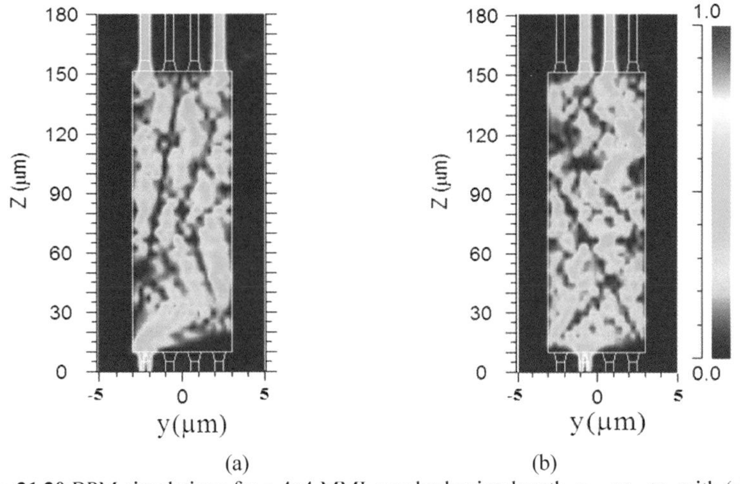

Figure 21.20 BPM simulations for a 4x4 MMI coupler having length $L = 3L_\pi / 2$ with (a) input signal at port 1 and (b) input signal at port 2.

We can see that the BPM simulations predict accurately the transfer matrix expressed by Equation (21.45) and (21.46). The 3D-BPM simulations for optimised designs of 8x8 MMI structures based on the silicon waveguide having a width of $W_{MI} = 9\mu m$ are shown in Figure 21.21. The optimised length calculated to be $L_{MI} \approx 382\mu m$.

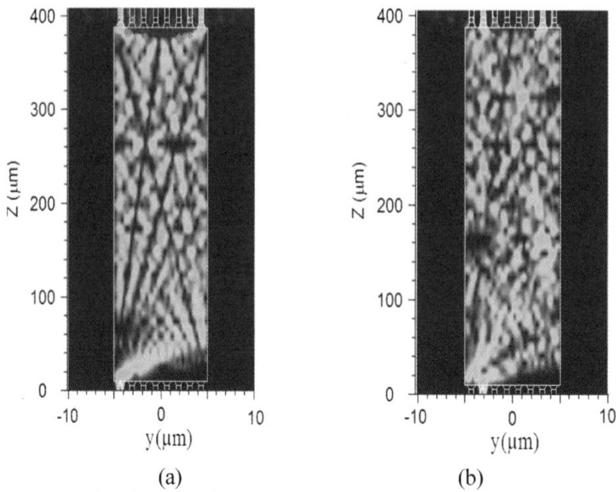

Figure 21.21 3D-BPM simulations of an 8x8 MMI structure for two cases (a) the signal entered at input port 1 and (b) signal entered at input port 2.

It is obvious from the BPM simulation results that the theory predicts adequately the field propagation in the devices. The proposed approach can be extended to design higher order all-optical signal processing circuits such as discrete cosine transforms (DCT), discrete sine transforms (DST) and discrete wavelet transforms (DWT) by cascading the proposed building blocks.

21.5 CONCLUSIONS

This chapter has presented an overview theory of multimode interference couplers and a general theory for the design of all-optical elements using 1x1, 2x2, 4x4 and 8x8 MMI structures based on silicon waveguides, including optical sum and subtraction unit, exchange unit, phase shifters suitable for building large scale and high speed all-optical signal processing circuits. A method of realizing all-optical discrete Fourier transforms based on these building blocks has been presented. The proposed device structures have been optimized for high performance and compactness using the transfer matrix method and 3D-BPM.

21.6 PROBLEMS AND SOLUTIONS

Problems

1. Prove that overlap integral can be calculated by Equation (21.23).

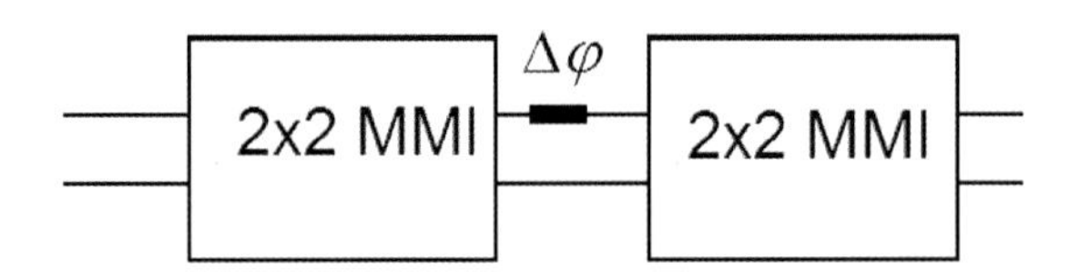

Figure 21.22 Realization of the 2-point Haar transform by using an MZI structure.

2. Prove that the transfer matrix of a 2x2 RI-MMI coupler at the length $L_\pi/2$, can be expressed by $\mathbf{M}=\frac{K}{\sqrt{2}}\begin{bmatrix}1 & j\\ j & 1\end{bmatrix}$; where K is a constant.

3. Prove that the transfer matrix of a 2x2 GI-MMI coupler at the length $3L_\pi/2$ can be expressed by $\mathbf{M}=\frac{K}{\sqrt{2}}\begin{bmatrix}1 & -j\\ -j & 1\end{bmatrix}$; where K is a constant.

4. The Mach-Zehnder Interferometer (MZI) structure is shown in Figure 21.22; where $\Delta\varphi$ is a phase shifter. Prove that the MZI structure can be used as a Haar transform.

References

[1] A. VanderLugt, *Optical signal processing*. New York: J. Wiley & Sons, 1992.

[2] N. B. Le, *Photonic signal processing : techniques and applications*: CRC Press, 2007.

[3] J. W. Goodman, A. R. Dias, and L. M. Woody, "Fully parallel, high-speed incoherent optical method for performing discrete Fourier transforms," *Optics Letters*, vol. 2, pp. 1-3, 1978.

[4] D. G. Sun, N. X. Wang, and L. M. H. e. al., "Butterfly interconnection networks and their applications in information processing and optical computing: applications in fast-Fourier-transform-based opticalinformation processing," *Applied Optics*, vol. 32, pp. 7184-7193, 1993.

[5] A. E. Siegman, "Fiber Fourier optics," *Optics Letters*, vol. 26, pp. 1215-1217, 2001.

[6] G. Cincotti, "Fiber wavelet filters," *IEEE Journal of Quantum Electronics*, vol. 38, pp. 1420-1427, 2002.

[7] M. E. Marhic, "Discrete Fourier transforms by single-mode star networks," *Optics Letters*, vol. 12, pp. 63-65, 1987.

[8] M. S. Moreolo and G. Cincotti, "Fiber optics transforms," presented at 10th Anniversary International Conference onTransparent Optical Networks

(ICTON 2008), Athens. Greece, 22-26 June 2008.
[9] A. R. Gupta, K. Tsutsumi, and J. Nakayama, "Synthesis of Hadamard Transformers by Use of Multimode Interference Optical Waveguides," *Applied Optics*, vol. 42, pp. 2730-2738, 2003.
[10] S. Tseng, Y. Kim, C. J. K. Richardson, and J. Goldhar, "Implementation of discrete unitary transformations by multimode waveguide holograms," *Applied Optics*, vol. 45, pp. 4864-4872 2006.
[11] L. W. Cahill and T. T. Le, "Photonic Signal Processing using MMI Elements," presented at 10th International Conference on Transparent Optical Networks (ICTON 2008), Athens, Greece, 2008.
[12] M. Bachmann, P. A. Besse, and H. Melchior, "General self-imaging properties in N x N multimode interference couplers including phase relations," *Applied Optics*, vol. 33, pp. 3905-, 1994.
[13] L. B. Soldano and E. C. M. Pennings, "Optical multi-mode interference devices based on self-imaging :principles and applications," *IEEE Journal of Lightwave Technology*, vol. 13, pp. 615-627, Apr 1995.
[14] W. P. Huang, C. L. Xu, W. Lui, and K. Yokoyama, "The perfectly matched layer (PML) boundary condition for the beam propagation method," *IEEE Photonics Technology Letters*, vol. 8, pp. 649 - 651, 1996.
[15] D. Marcuse, *Theory of Dielectric Optical Waveguides*: Academic Press, 1991.
[16] A. Yariv, *Optical Electronics*: H R W Series in Electrical and Computer Engineering, 1991.
[17] C. R. Doerr and H. Kogelnik, "Dielectric Waveguide Theory," *IEEE Journal of Lightwave Technology*, vol. 26, pp. 1176-1187, 2008.
[18] A. Ghatak and K. Thyagarajan, *An Introduction to Fiber Optics*: Cambridge University Press, 1998.
[19] M. J. Adams, *Introduction to Optical Waveguide*: John Wiley & Sons, 1981.
[20] A. W. Snyder and J. D. Love, *Optical Waveguide Theory*. New York: Chapman and Hall, 1983.
[21] P. Yeh, *Optical Waves in Layered Media* John Wiley & Sons, 2005.
[22] K. Iizuka, *Elements of Photonics. Volume 2: Fiber and Integrated Optics*: Wiley-Interscience, 2002.
[23] G. Lifante, *Integrated Photonics: Fundamentals*: Wiley and Sons, 2003.
[24] T. Tamir, *Guided-wave optoelectronics*. New York : Springer-Verlag, 1990.
[25] E. Wolf, *Progress in Optics. Volume 32*: Elsevier, 1993.
[26] L. B. Soldano, "Multimode interference couplers," vol. PhD thesis. Delft, the Netherlands: Delft University of Technology, 1994.
[27] K. A. Latunde-Dada and F. P. Payne, "Theory and Design of Adiabatically Tapered Multimode Interference Couplers," *IEEE Journal of Lightwave Technology*, vol. 25, pp. 834-839 2007.
[28] M. R. Paiam, "Applications of multimode interference couplers in wavelength-division multiplexing ", vol. PhD thesis. Ottawa, Canada: University of Alberta, 1998.

[29] M. Bachmann, P. A. Besse, and H. Melchior, "Overlapping-image multimode interference couplers with a reduced number of self-images for uniform and nonuniform power splitting," *Applied Optics*, vol. 34, pp. 6898-6910, 1995.
[30] J. M. Heaton and R. M. Jenkins, " General matrix theory of self-imaging in multimode interference(MMI) couplers," *IEEE Photonics Technology Letters*, vol. 11, pp. 212-214, 1999.
[31] N. S. Lagali, "The general Mach-Zehnder interferometer using multimode interference coupler for optical communication network," vol. PhD thesis: University of Alberta, Canada, 2000.
[32] M. Shirakawa and J. Ohtsubo, "Optical Digital Fast Fourier Transform System," *Optical Review*, vol. 6, pp. 424-432, 1999.
[33] L. Pavesi and D. J. Lockwood, *Silicon Photonics*: Springer, 2004.
[34] G. T. Reed and A. P. Knights, *Silicon Photonics: An Introduction*: John Wiley and Sons, March 2004.
[35] T. Tsuchizawa, K. Yamada, and H. F. e. al., "Microphotonics Devices Based on Silicon Micro-Fabrication Technology," *IEEE Journal of Selected Topics in Quantum Electrononics*, vol. 11, pp. 232-, 2005.
[36] S. Janz, P. Cheben, and D. D. e. al., "Microphotonic elements for integration on the silicon-on-insulator waveguide platform," *IEEE Journal of Selected Topics in Quantum Electrononics*, vol. 12, pp. 1402-1415, Dec. 2006.
[37] R. Soref, "The past, present, and future of silicon photonics," *IEEE Journal of Selected Topics in Quantum Electrononics*, vol. 12, pp. 1678-1687, Dec. 2006.
[38] B. R. Koch, A. W. Fang, O. Cohen, and J. E. Bowers, "Mode-locked silicon evanescent lasers," *Optics Express*, vol. 15, pp. 11225-11233, August 2007.
[39] M. Paniccia, A. Liu, and H. R. e. al., "Optical Amplification and Lasing by Stimulated Raman Scattering in Silicon Waveguides," *IEEE Journal of Lightwave Technology*, vol. 24, pp. 1440-1455, March 2006.
[40] H. Rong, R. Jones, and A. L. e. al., "A continuous-wave Raman silicon laser," *Nature*, vol. 433, pp. 725-728, 2005.
[41] L. Liao, D. Samara-Rubio, and M. M. e. al., "High speed silicon Mach-Zehnder modulator," *Optics Express*, vol. 13, pp. 3119-2135, 2005.
[42] W. M. J. Green, M. J. Rooks, L. Sekaric, and Y. A. Vlasov, "Ultra-compact, low RF power, 10 Gb/s silicon Mach-Zehnder modulator," *Optics Express*, vol. 15, pp. 17106, 2007.
[43] S. Manipatruni, Q. Xu, and M. Lipson, "PINIP based high-speed high-extinction ratio micron-size silicon electrooptic modulator," *Optics Express*, vol. 15, 01 Oct. 2007.
[44] D. Yang, Y. Wang, and L. M. e. al, "A 2×2 SOI Mach-Zehnder thermo-optical switch based on strongly guided paired multimode interference couplers," *Optoelectronics Letters*, vol. 3, pp. 334-336, 09/2007.

[45] M. Harjanne, M. Kapulainen, T. Aalto, and P. Heimala, "Sub-μs switching time in silicon-on-insulator Mach–Zehnder thermooptic switch," *IEEE Photonics Technology Letters*, vol. 16, pp. 2039-2041, 2004.

[46] W. Ye, "Stress Engineering for Polarization Control in Silicon on Insulator Waveguides and Its Applications in Novel Passive Polarization Splitters/Filters," vol. PhD thesis: Carleton University, Ottawa, Canada, 2006.

[47] B. Luyssaert, "Compact Planar Waveguide Spot-Size Converters in Silicon-on-Insulator," vol. PhD thesis: INTEC, Ghent University, Belgium, 8/2005.

[48] R. Espinola, J. Dadap, R. Osgood, S. J. McNab, and Y. A. Vlasov, "C-band wavelength conversion in silicon photonic wire waveguides," *Optics Express*, vol. 13, pp. 4341 2005.

[49] J. T. Robinson, K. Preson, O. Painter, and M. Lipson, "First-principle derivation of gain in high-indexcontrast waveguides," *Optics Express*, vol. 16, 13 Oct. 2008.

[50] M. A. Foster, R. Salem, and D. F. G. e. al., "Silicon-chip-based ultrafast optical oscilloscope," *Nature*, vol. 456, pp. 81-84, 06 Nov. 2008.

[51] E. Dulkeith, F. Xia, and L. S. e. al., "Group index and group velocity dispersion in silicon-on-insulator photonic wires," *Optics Express*, vol. 14, pp. 3853-3863, 2006.

[52] J. I. Dadap, N. C. Panoiu, and X. C. e. al., "Nonlinear-optical phase modification in dispersion-engineered Si photonic wires," *Optics Express*, vol. 16, pp. 1280-1299, 2008.

[53] W. P. Huang, C. L. Xu, and S. K. Chaudhuri, "A finite-difference vector beam propagation method for three-dimensional waveguide structures," *IEEE Photonics Technology Letters*, vol. 4, pp. 148 - 151, 1992.

[54] F. Liu, Q. Li, and Z. Z. e. al., "Optically tunable delay line in silicon microring resonator based on thermal nonlinear effect," *IEEE Journal of Selected Topics in Quantum Electronics*, vol. 14, pp. 706 - 712, 2008.

[55] T. T. Le, L. W. Cahill, and D. Elton, "The Design of 2x2 SOI MMI couplers with arbitrary power coupling ratios," *Electronics Letters*, vol. 45, pp. 1118-1119, 2009.

Chapter 22.

A Low Noise Self-Cascode Configuration Operational Transconductance Amplifier (OTA) for Bandgap Reference Voltage Circuit

Leila Koushaeian[1], Bahram Ghafari[1], Farhad Goodarzy[1], Rob Evans[1], and Amir Zjajo[2]

1Department of Electrical and Electronic Engineering, University Of Melbourne, Victoria, Australia,
2Delft University of Technology, Delft, Netherlands
Leilak@unimelb.edu.au

This chapter provides an overview of a novel self-cascode configuration (SCC) operational transconductance amplifier (OTA) for use in a bandgap reference voltage (BGR) circuit. In this chapter we review BGR and self-cascode amplifier circuits. A small-signal model is used to derive an analytical noise expression for both the conventional Miller-amplifier and the SCC-amplifier employed in a bandgap reference voltage circuit.

22.1 INTRODUCTION

High precision CMOS operational amplifier circuits often a demanding and critical building block in many analog circuits; including analog-to-digital converters, digital-to-analog converters, and bandgap reference (BGR) voltage circuits. The bandgap reference noise performance depends strongly on the performance of the OTA employed. A temperature-independent bandgap reference signal can be derived utilizing the exponential characteristics of bipolar

4G Wireless Communication Networks: Design, Planning and Applications, 471-486,

devices for both negative and positive temperature coefficient [1]. For constant collector current, the base-emitter voltage V_{be} of bipolar transistors has negative temperature dependence at room temperature. This negative temperature dependence is cancelled by a proportional-to-absolute temperature dependence of the amplified difference between two base-emitter junctions. These junctions are biased at fixed but unequal current densities resulting in a relationship directly proportional to absolute temperature. This proportionality is, however, rather small (0.1-0.25 *mV/°C*) and needs to be amplified to allow further signal processing. Both processing and noise variation affect BGR accuracy. If BGR noise is not taken into account in the design, increased trimming, calibration, and die area are required [2]. Practically, the input-referred noise of the low noise amplifier should be kept below 5 μV_{rms} for 5-10 *kHz* bandwidth [3]. Previous publications on standard CMOS amplifier circuits have reported flicker noise corner frequencies between 7 *kHz* and 159 *kHz*, and noise voltage of 95 $nV/\sqrt{Hz}$ with the power of 80 μW at 1 *kHz* to 1.7 $\mu V/\sqrt{Hz}$ with the power of 7 *mW* at 1 *kHz* [4-6].

To reduce the effect of noise in analog circuits, few solutions have been proposed. Bipolar devices have a lower *1/f* noise level than CMOS devices. The noise level in CMOS device is usually 10 to 50 times larger than the noise level in bipolar device in a given device area. Therefore, a BiCMOS process using bipolar devices seems an appropriate choice for low-noise applications, similar to the one suggested in [7]. However, the integration of bipolar devices on the same substrate as CMOS devices is complex and costly. As a result, it is highly desirable to use standard CMOS technology to fabricate both analog and digital circuits.

Another method for reducing the output noise spectral density is to increase the gate area since flicker noise in CMOS devices is inversely proportional to the square root of the gate area. The inevitable penalties for doing this are increased power consumption, and a larger die area, and therefore a large input capacitance [7].

Several solutions have been proposed to reduce the effect of noise in BGR circuit. In [6], an external capacitor at the output has been applied to filter undesired noise, but this has the drawback of demanding a larger capacitance and as a result increased die area. Similarly, a chopper stabilizing technique proposed in [8] makes the circuit design more complex and as a consequence increases the die area and power consumption.

The main challenge for the design of the self-cascode configuration operational transconductance amplifier for bandgap reference voltage (SCC-BGR) is to overcome the noise-power trade-off, since the power of the amplifier is inversely related to the input-referred noise power of the amplifier v_{ni}^2 [3]. This means that reducing the power consumption will result in higher noise level. Also, through a modification of the OTA topology, the noise-power trade-off can be improved.

In this chapter, a novel self-cascode configuration (SCC) amplifier is employed in a bandgap reference voltage circuit. Compared to previously published work the circuits proposed here exhibits a lower temperature coefficient and lower power consumption without the need for trimming. The SCC-BGR does not require any external devices such as a filter or use of a chopper circuit to reduce the noise.

The chapter is organized as follows: Section 22.2 describes the principle of the bandgap reference circuit and system level requirements. Section 22.3 explains the Malcovati BGR, while Section 22.4 focuses on the design of a SCC-OTA amplifier. In Section 22.5 an analysis of the noise contribution of the proposed SCC-BGR is discussed. In Section 22.6, we compare the SCC-BGR with the Miller-BGR and demonstrate how the SCC-BGR can effectively reduce the noise performance of the BGR circuit. Section 22.7 concludes the chapter by summarising the SCC-OTA noise performance.

22.2 BASIC PRINCIPLE OF THE BANDGAP REFERENCE CIRCUIT AND SYSTEM LEVEL ACCURACY

As previously mentioned, this study uses a low-power bandgap reference voltage circuit to generate a stable DC voltage that is ideally independent of temperature. This is achieved by adding a voltage, proportional-to-absolute-temperature (PTAT) to a base–emitter voltage of the p-n devices in order to achieve a low-output voltage variation over a wide temperature range. The name “bandgap reference voltage” comes from the traditional bandgap circuit that is derived from the bandgap energy of silicon (E_g), which is approximately E_g=1.12 *eV* at room temperature, i.e., 25 °*C* [10]. However, the increasing demand for low-voltage, low-power wireless systems, and portable devices, forces the technology to reduce the supply voltage to less than 1 V. As a consequence, the traditional bandgap circuit can no longer produce 1.2 V [10]. Some techniques used to lower power consumption have been proposed in the literature. These techniques include resistive subdivision, BiCMOS technology, low-threshold, native devices, subthreshold operation technologies, and dynamic threshold MOS transistors (DTMOST) [11]. This chapter focuses on standard CMOS technology and seeks to design the circuit in the weak inversion region. The accuracy of the temperature sensor critically depends upon the analog-to-digital converter (ADC), which is highly dependent upon the accuracy of the bandgap reference voltage. Temperature coefficient (TC) is one of the key parameters in designing the BGR. Thus, TC represents the amount of the change in output reference voltage over the operating temperature range (0 °*C* to +100 °*C*) and is expressed in *ppm/°C*

$$TC\left(ppm/^{o}C\right)=\left(\frac{V_{ref\,\max}-V_{ref\,\min}}{V_{\text{nominal}}\times\left(T_{\max}-T_{\min}\right)}\right)\times 10^{6} \tag{22.1}$$

V_{refmax} and T_{max} are the maximum reference voltage and maximum temperature, respectively [12]. The BGR literature has not focused on optimising the maximum allowable temperature coefficient needed to consider ADC accuracy. If the inaccuracy is not taken into account, a concomitant increase could occur in the trimming and calibration time, and in the die area [12].

For example, an 8-bit ADC with a 1-V full-scale voltage reference, and a worse case maximum allowable conversion error of ½ LSB or 0.195 *mV*, requires a TC of 55.7 *ppm/°C @50°C*. For an *N*-bit ADC, the minimum allowable TC, at its high critical temperature, can be calculated using the following expressions:

$$TC = \left(\frac{V_{initial-Accuracy\%}}{2^{N+1} \times (T_{max} - T_{min})} \right) \quad (22.2)$$

where $T_{nominal}$ is the temperature at 25°C assuming a zero-temperature coefficient. $V_{initial\text{-}Accuracy\%}$ is the BGR initial-accuracy percentage [12]. Here, we aim to explore the TC of the maximum temperature boundary. For example, if the initial accuracy is given as 0.1 %, then the acceptable temperature coefficient will be 27.1 *ppm/°C @50°C* for an *8*-bit ADC. Therefore, the BGR temperature coefficient should be equal to or less than 27.1 *ppm/°C @50°C* in order to meet the required accuracy.

22.3 MALCOVATI CMOS BGR FOR SUB-NANO TECHNOLOGY

As noted, a bandgap reference circuit consists of two main voltage generators. In order to generate independent voltage reference versus temperature, the generated PTAT voltage should be eliminated by adding it to the complementary-to-absolute-temperature (CTAT) voltage, which has a negative temperature coefficient characteristic. The scaling factor needs to multiply the PTAT since its value is smaller than the CTAT [9].

The PTAT voltage is generated by using the base-emitter voltage difference of two parasitic vertical bipolar junction transistors (BJT) Q_1 and Q_2 (Figure 22.1) with different emitter sizes. One emitter size is (*m*) times larger than the other [1, 4]. This difference is related to absolute temperature based on equation (22.3).

$$\Delta V_Q = V_T . \ln\left(\frac{I_{S2}}{I_{S1}} \right) = V_T . \ln(m) \quad (22.3)$$

The value of the variation of the thermal voltage versus the temperature is approximately $\partial V_T / \partial T = 0.085$ *mV/°C* [9], where $I_{s1,2}$ is the p-n junction diode's scale current, and

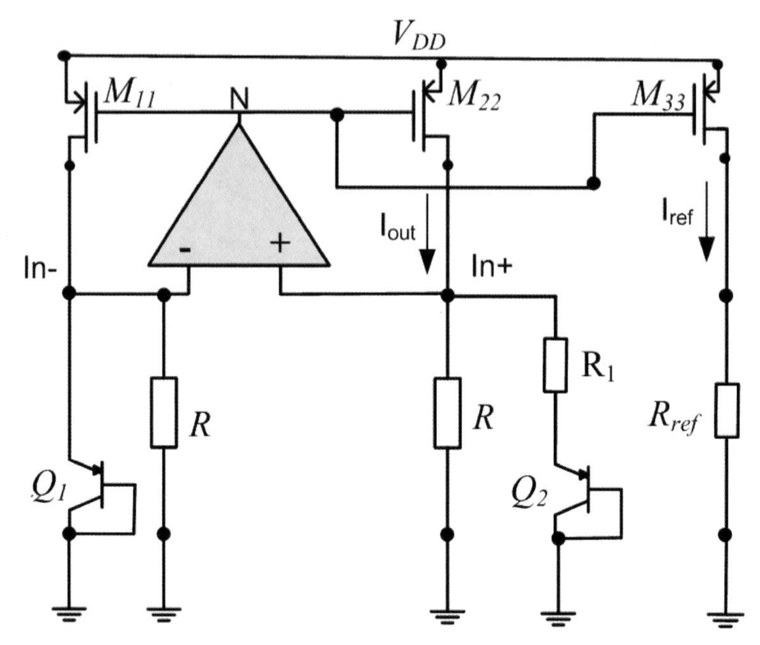

Figure 22.1 Basic Malcovati bandgap reference voltage.

$$V_T = \left(\frac{K.T}{q}\right) \tag{22.4}$$

where K is the Boltzmann constant (1.38 X 10^{-23} J/K), q is the electron charge (1.6 X 10^{-19} C), T is the room temperature (25 °C), and ΔV_Q is the base-emitter voltage difference. The generated reference voltage is given by

$$V_{ref} = \frac{R_{ref}}{R}.V_{Q1} + n.\frac{R_{ref}}{R}.\left(\frac{K.T}{q}\right).\ln(m) \tag{22.5}$$

where n is the scaling factor [1]. The temperature variation is calculated from

$$\frac{\partial V_{ref}}{\partial T} = \frac{R_{ref}}{R}.\frac{\partial V_{Q1}}{\partial T} + n.\frac{R_{ref}}{R}.\frac{\partial\left(\left(\frac{K.T}{q}\right).\ln(m)\right)}{\partial T} \tag{22.6}$$

The CTAT voltage is directly obtained from the base-emitter voltage drop of the vertical transistors Q_1 and Q_2 on the resistors R and R_1 (Figure 22.1) for which its variation versus temperature was found to be approximately equal to $\partial V_{Q1}/\partial T$= -1.6 $mV/°C$ [9]. The PTAT and CTAT currents were equal to

$$I_{PTAT} = \frac{\left(n.\left(\frac{K.T}{q} \right).\ln(m) \right)}{R_1} \quad and \quad I_{CTAT} = \frac{V_Q}{R} \tag{22.7}$$

respectively and their sum is equal to the total current through the reference resistor, i.e., R_{ref}, which was used to create the reference voltage of the bandgap circuit. Choosing the values of the resistors and the scaling ratio between R and R_1 enabled us to determine the zero temperature coefficients. To insure the temperature compensation, the R_1/R ratio should satisfy the following equation

$$\frac{R.\ln(m)}{R_1} = 22 \tag{22.8}$$

The operational amplifier senses any voltage variation in the opamp inputs. The voltage variation at these nodes is amplified by the high gain amplifier circuit. Consequently, the gate and the source of each side of the BGR current source (M_{11}-M_{22}) are maintained at the same voltage. The implementation of the operational amplifier and the compensation will be explained in the following section.

22.4 SELF-CASCODE OPERATIONAL AMPLIFIER

The self-cascode technique has been implemented in current mirror applications [13]. In [14], multi-threshold CMOS (MTCMOS) devices are used to make sure that the bottom transistor is working in the strong inversion region. In [15], a different approach is taken using the short channel effect to give a similar result. The combination of both previous approaches presented in [16], [13] forces the SCC-current mirror to bias the self-cascode devices into the strong inversion region. However, employing MTCMOS incurs additional process cost. Additionally, in a low-voltage design not enough voltage headroom is available to bias all transistors into the strong inversion region. For this reason, in this chapter we propose a different biasing approach to decrease the required headroom voltage.

We utilize the fact that the SCC-configuration employs transistors operating in the linear region and the weak inversion region. Since the voltage between the source and drain terminals of the transistor is negligible, the V_{DSsat} in a SCC has higher headroom voltage in comparison with a cascode configuration. This also means that the output swing voltage is maximized as shown in Figure 22.2.

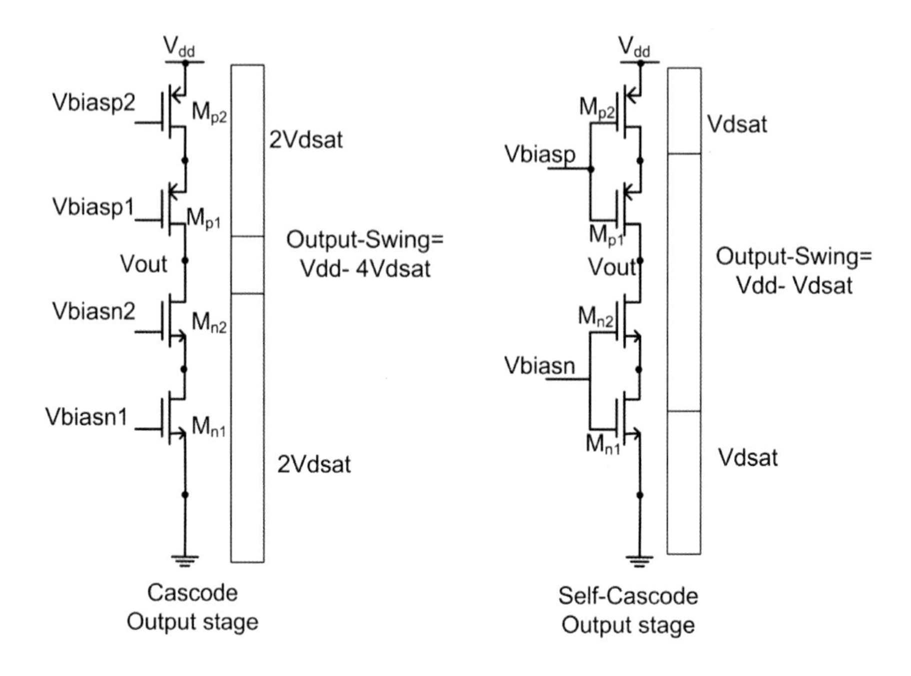

Figure 22.2 Output-Signal Swing Comparison.

The SCC-OTA is identical to a two-stage Miller opamp, with the addition of five self-cascoding transistors (Figure 22.3). The first stage is composed of the input differential self-cascode n-channel transistors (M_{1n} - M_{4n}) with self-cascode p-channel current mirror loads (M_{1p} -M_{4p}). The second stage is composed of the push-pull circuit. The gate of the input differential pair transistors (M_{1n} - M_{4n}) is connected to the node In+ of the BGR. Similarly, transistors (M_{3n} - M_{4n}) sense any voltage variation in the node *In-*. The voltage variation at nodes *In+* and *In-* is amplified by the amplifier gain. The minimum input common-mode voltage is given by:

$$V_{in,\min} = V_{GS} + V_{DSsat} \tag{22.9}$$

where the term V_{DSsat} is the minimum voltage required across the n-channel devices M_{in}. Furthermore, the input common-mode voltage is limited by the vertical BJT device which in a 65-nm process is around 728 mV. The maximum input common-mode voltage is determined through the drain voltage of input transistor V_{DD}-V_{GS} of the p-channel devices. Therefore, we have

$$\begin{aligned} V_{DS} &> V_{GS} - V_{thn}, \\ V_D &> V_G - V_{thn} \end{aligned} \tag{22.10}$$

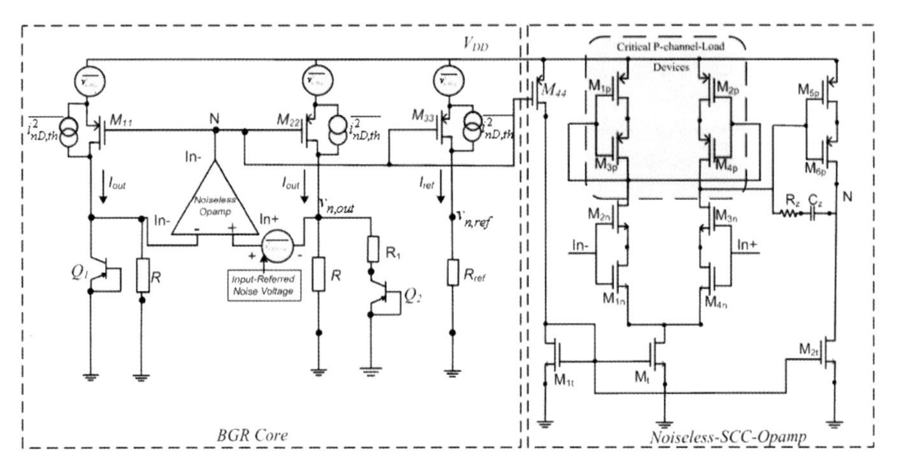

Figure 22.3 Modified SCC- Malcovati bandgap reference voltage (note that the start-up circuit has not been shown).

Hence

$$V_{in,\max} = V_{DD} - V_{GS} + V_{thn} \tag{22.11}$$

The effective transconductance of the self-cascode transistor in a SCC-OTA (e.g $M_{1n} - M_{2n}$) is approximately equal to the transconductance of a single transistor in a Miller-OTA M_{1n}. As a consequence, the equivalent output impedance of a self-cascode transistor ($M_{1n} - M_{2n}$) is ($m_{n,p}$ - 1) times larger than that of the output impedance of the single transistor ($r_{o,2n}$) [16], where ($m_{n,p}$) is width ratio of the self-cascode n-channel and p-channel devices, respectively. The SCC-OTA dc gain is given by

$$\begin{aligned} &|A_V|_{SCC-OTA} \\ &= \left[g_{m,1n,input} \cdot \left((m_n - 1).r_{o,2n,input}\right) \| \left((m_n - 1).r_{o,p,input}\right)\right]\left[g_{m,5p,output} \cdot r_{o,6p,output} \cdot \left(m_{p,out} - 1\right)\right] \\ &= \left(m_n . m_p^2\right)|A_V|_{Miller-OTA} \end{aligned} \tag{22.12}$$

where $g_{m,n,p,input}$ is the input transistor transconductance and $r_{o,n,p,output}$ is the output resistance for the n-channel and the p-channel device, respectively.

22.5 SCC-OTA NOISE ANALYSIS

We start by considering a small-signal circuit including the noise sources for individual transistors. The transfer function from each noise source to the output is calculated using the superposition principle. The noisy opamp model of the SCC-BGR circuit is illustrated in Figure 22.3 using a noiseless amplifier, with amplifier voltage noise in series of the input $V^2_{n,\,SCC\text{-}OTA}$. The thermal and flicker noise of the transistor is presented by a parallel noise current source ($i^2_{n,D,th}$) and series voltage source $V^2_{noise1/f\!,Mi}$ respectively. The input-referred noise of the each device can be expressed as

$$V^2_{noise,M,total}(f) = \sum_{i=1}^{M} |H_i|^2 .V^2_{noise,M_i}(f) \tag{22.13}$$

where M is the total number of device noise sources, $|H_i|$ is the transfer function from the i_{th} noise source to the output, $V^2_{noise,Mi}(f)$ is the device noise, and $V^2_{noise,M,total}(f)$ is the total input-referred noise voltage of the transistors.

To obtain the total noise equation, we also consider the noise contribution of the current sources M_{11} and M_{22}. As shown in Figure 22.3, by applying Kirchhoff Voltage Laws (*KVL*) starting at node N, the nominal dc voltage drop across this node is equal to the output current I_{out} times the impedance looking into node N:

$$V_N = I_{out}\cdot\frac{1}{g_{mP}} = \frac{V_{n,out}}{\left(\left(R_1 + g_{mQ}^{-1}\right)\right)}\cdot\frac{1}{g_{mP}} \tag{22.14}$$

where the small-signal equivalent of the BJT devices is modelled by resistance (g_{mQ}^{-1}). Considering the voltage noise of the M_{11} we can determine the noise at the output node:

$$V^2_{n,out}(f) = \left|H_{M_{11}}\right|^2 .V^2_{noise,M_{11}}(f),$$

$$\left|H_{M_{11}}\right|^2 = \left[\frac{1}{\left(\left(R_1 + g_{mQ}^{-1}\right)\|R\right)^{-1}.g_{mP}^{-1}\cdot\left(\frac{1}{|A_V|_{SCC-OTA}}\left(\left(\frac{g_{mQ}}{g_{mP}}\right)-1\right)+\frac{g_{mQ}}{g_{mP}}\right)}\right]^2 \tag{22.15}$$

The mismatch between the transistors is attenuated by a factor of (1+ (g_{mP} + g_{mbP}).R + R/r_o). Thus, the transconductance of both M_{11} and M_{22} is equal to g_{mP}. The contribution of the noise voltage source of transistor M_{11} to the input is the multiplication of the value of the transfer function $|H_{M11}|$ and the device-noise voltage. Similarly, the noise contribution of M_{22} to the output noise is

$$V_{n,out}^2(f)=\left|H_{M_{22}}\right|^2 .V_{noise,M_{22}}^2(f),$$

$$\left|H_{M_{22}}\right|^2=\left[\frac{1-\dfrac{g_{mQ}}{g_{mP}}}{\left(\left(R_1+g_{mQ}^{-1}\right)\|R\right)^{-1}.g_{mP}^{-1}.\left(\dfrac{1}{|A_V|_{SCC-OTA}}\left(\left(\dfrac{g_{mQ}}{g_{mP}}\right)-1\right)+\dfrac{g_{mQ}}{g_{mP}}\right)}\right]^2 \tag{22.16}$$

Note that M_{22} has an additional resistor when compared to M_{11}. Accordingly, the total noise is the combination of the opamp noise (M_{44} is part of the SCC-OTA) and the noise from the BGR current sources. For simplicity, the noise generated by resistors $V^2_{n,Resistori}$ *(f)* and BJT devices $V^2_{n,BJTi}$ *(f)* has been neglected since these noise sources are small compared with noise generated by the opamp. Similarly, the current of the current source of the BGR i.e. M_{22} is mirrored in the transistor M_{33}. By properly matching these two devices, their drain currents are approximately equal. As a consequence, $R_{ref}= ((R_1 + g_{mQ}^{-1}) \parallel R)$ which results in $V_{n,out}= V_{n,Ref}>>1$. To provide sufficient matching (less than 2 %) the BGR poly-resistors are series connected unit-matched resistors. Additionally, careful layout techniques such as common-centroid have been employed. The total noise as seen at the output is given by

$$\begin{aligned}\overline{V}_{n,out,total}^2(f)&=\frac{\overline{V}_{n,SCC-OTA}^2(f)}{B^2}+\sum_{i=1}^{M}\overline{V}_{noise,M_{ii}}^2(f)+\sum_{i=1}^{M}\left[\overline{V}_{n,\mathrm{Re}sistor_i}^2(f)+\overline{V}_{n,BJT_i}^2(f)\right]\\&=\frac{\overline{V}_{n,SCC-OTA}^2(f)}{B^2}+\overline{V}_{n,out}^2(f).\left[\sum_{i=1}^{M}\frac{1}{\left|H_{M_{ii}}\right|^2}\right]\end{aligned} \tag{22.17}$$

Where

$$B=\left[\frac{1}{\left(\left(R_1+g_{mQ}^{-1}\right)\|R\right)^{-1}.\left(g_{mQ}\|R\right)^{-1}-\left(|A_V|_{SCC-OTA}.g_{mP}\right)^{-1}-1}\right] \tag{22.18}$$

Since the OTA noise dominates the output noise [10], the first term in (22.17) dominates. Therefore, it is important to increase the gain of the amplifier to achieve the lower noise performance. In comparison to the classic Miller BGR, the noise in the SCC-OTA is reduced by the factor of $(m_n .m_p^2)$.

The analytical approach shows that the SCC-configuration also provides less low-frequency *(1/f)* noise since the device area has been increased by implementing the SCC-topology. The dependency of the device area and flicker noise is

$$\overline{V}^2_{noise,M_i}(f)=\sum_{i=1}^{M}\left[\overline{V}^2_{thermal-noise,M_i}(f)+\overline{V}^2_{flicker-noise,M_i}(f)\right] \\ =\sum_{i=1}^{M}\left[4KT.\frac{\gamma.g_{mi}}{g^2_{dsi}}+\frac{k_f}{Cox_i.W_i.L_i}.\frac{1}{f}\right] \tag{22.19}$$

where γ is equal to *2/3* for strong inversion operation and *1/2* for weak inversion operation, and k_f is the flicker noise coefficient which is the process dependant constant on the order of 10^{-25} V^2F. The g_{mi} and $1/g_{dsi}$ are transistor transconductance and the output resistance respectively. The major source of flicker noise in the opamp is due to the load transistor i.e., $M_{1p\text{-}4p}$ (Figure 22.3) [16]. For this design, the load transistors in the op-amp are p-channel devices that intrinsically have less flicker noise than the similarly sized n-channel devices.

Similarly, power supply rejection ration (PSRR) expression for SCC-BGR is found as:

$$PSRR=\left[\frac{s+\omega_P}{s+\omega_t}\right].\frac{1}{|A_V|_{SCC-OTA}.\beta} \tag{22.20}$$

where ω_p indicate the dominate pole and ω_t unity gain bandwidth respectively. The factor of $|A_V|_{SCC\text{-}OTA}.\beta$ is the feedback loop gain of the SCC-OTA. Therefore, the PSRR in a SCC-OTA is improved approximately by the factor of $(m_n.m_p)^2$.

22.6 RESULTS

The CMOS version of a Miller BGR with a Miller opamp and the proposed BGR have been designed and simulated using a 65-nm IBM bulk CMOS process. The open loop frequency response of the self-cascode operational amplifier has a DC gain of 88 *dB*, a unity gain bandwidth (UGB) of 356 *kHz*, and a phase margin of 61° from a 1.2 *V* supply voltage (Table 22.1). The corner simulation has been

Table 22.1 Simulated reference voltage @65-nm CMOS

	Classic-BGR	This Work
V_{ref}(V) @ room temp	777mV	700mV
TC (ppm/°C)	50.3 ppm/°C	16.4ppm/°C
±10% Supply Voltage Variation	1.6mV	180μV
Power Consumption	126 μW	23μW
Power Supply Voltage	1.2V	1.2V

carried out for each individual corner by applying ±10% variation to supply voltage, bias current for extreme temperature points i.e. 0 °C and 100°C (Table 22.2). The DC gain and UGB for Miller-OTA is equal to 55 *dB* and 10 *MHz*, respectively. The SCC-BGR generates a reference of 700 *mV* and consumes 23 *μW* at room temperature. The temperature coefficient for an output voltage of 700 *mV* is 16.4 *ppm/°C* without trimming, for a temperature ranging from 0°C to 100°C. The PSRR is 42.83 *dB* at low-frequency (10 *kHz*) (Figure 22.4), which significantly lower than PSRR performance of the Miller-BGR. Table 22.3 shows the transistor dimension and resistor value. All the resistors used in the BGR are P+ poly resistors.

Table 22.2 Corner simulation Opamp @65-nm CMOS

Parameter	Min	Max
Unity Gain Bandwidth (kHz)	163.9	220.1
Phase Margin	53.25°	64.19°
Open Loop Gain(dB)	65.8	84.27
Input Offset (μV)	214.1	19.88

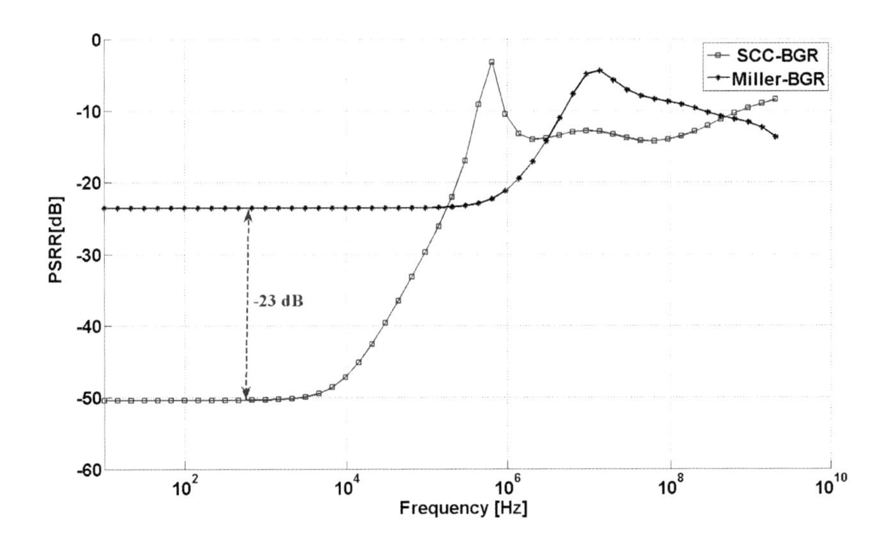

Figure 22.4 PSRR comparison between SCC-BGR and Miller BGR.

Table 22.3 Transistor dimension and resistance value used in the SCC-BGR

Device	Parameter
$M_{11,22,33}$	12μm/4.8μm
$Q_{1,2}$	3.2μm/3.2μm , m=8(Q_2)
$R, R_{1,2}$	24KΩ 270 KΩ
$M_{1n,4n}$	2μm/3μm
$M_{2n,3n}$	10X (2μm/3μm)
$M_{1p,2p}$	900nm/3μm
$M_{3p,4p}$	10X(900nm/3μm)
$M_{5p,6p}$	3.6μm/3μm,10X (3.6μm/3μm)
M_t	1μm/3μm
M_{2t}	2μm/3μm
M_{44}	260nm/4.8μm

Table 22.4 Critical devices- noise contribution for SCC-BGR

Device	Flicker Noise	Thermal Noise
BGR-Current source	48.8%	1.26%
BGR-output Branch	30.70%	0.81%
SCC-BGR-Current source	0.18%	1.75%
SCC-BGR-output Branch	0.09%	0.50%

The noise spectra of both the Miller-BGR and the SCC-BGR are shown in Figure 22.5. For comparison, the noise summary of the critical device sources is presented in Table 22.4. The results verify the analytical approach for noise performance predicted when the self-cascode topology is applied to the Miller amplifier. Compared to previously published work on bandgap reference-voltage circuits, the circuits proposed here exhibit a lower temperature coefficient and lower power consumption without the need for trimming (Table 22.5).

22.7 CONCLUSION

This chapter describes a small-signal noise analysis for a novel self-cascode configuration (SCC) amplifier employed in a bandgap reference voltage circuit. Compared to previously published work using standard CMOS technology on

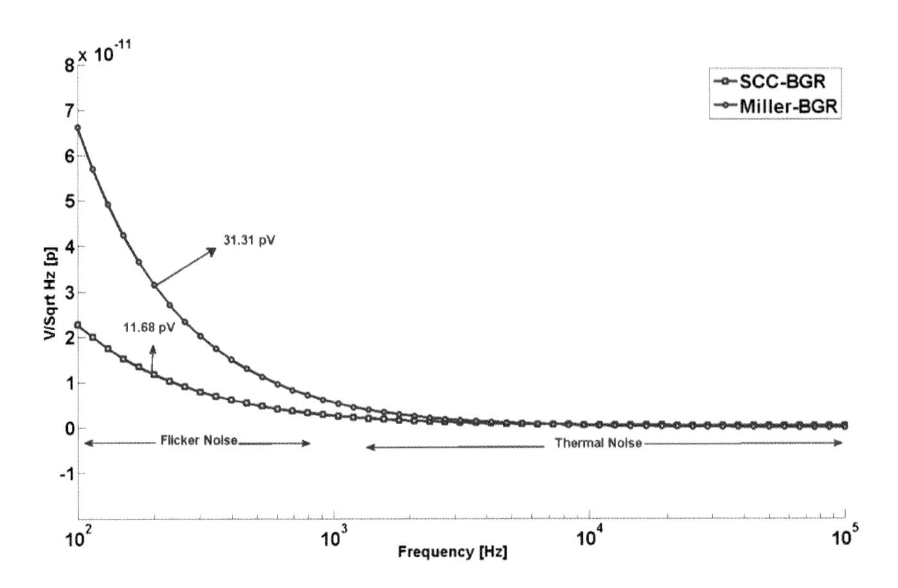

Figure22.5 Noise spectra comparison between SCC-BGR and Miller BGR.

Table 22.5 Comparison with state-of-the art

BGR Topology	Technology Process	TC	V_{DD}	V_{ref}	Power Consumption	PSRR
[17]	BiCMOS 0.8-μm	7.5 ppm/°K	1 V	540mV	92 μW	-
[18]	CMOS 0.13-μm	93 ppm/°C	0.6 V	400mV	923 μW	-
[19]	CMOS 0.35-μm	16.9 ppm/°C	(1 to 4) V	190.1mV	0.25@1V 0.56@4V	V_{DD}=1V @100Hz -41 dB @10MHz-17dB
[20]	CMOS 90-nm	28.3 ppm/°C	1.2 V	497.2mV	276.6 μW	-
Classic-BGR	CMOS 65-nm	50.3 ppm/°C	1.2 V	777mV	126 μW	@10Hz&10KHz −24 dB
This Work	CMOS 65-nm	16.4ppm/°C	1.2 V	700mV	23μW	@10Hz&10KHz −42.83 dB

bandgap reference-voltage circuits [17-20], the circuits proposed here using standard CMOS technology exhibits a lower temperature coefficient and lower power consumption without the need for trimming. The analytical approach and simulation results indicate that the SCC-configuration provides higher output impedance while also decreasing low-frequency (flicker) noise. The low-noise features of the modified structure make it a desirable choice for low power, high accuracy application.

References

[1] Giustolisi, G., and Palumbo, G.," A detailed analysis of power-supply noise attenuation in bandgap voltage references," IEEE Transactions on Circuits and Systems I: Fundamental Theory and Applications, Feb 2003, vol. 50, no. 2, pp. 185-197.

[2] Colombo, D., Wirth, G., Bampi, S., and Fayomi, C.,"Impact of noise on trim circuits for bandgap voltage references," 14th IEEE International Conference on Electronics Circuits and Systems, Dec 2007, pp. 775-778.

[3] Wattanpanitch, W., Fee, M., and Sarpeshkar, R.," An energy-efficient micropower neural recording amplifier," IEEE Transactions on Biomedical Circuits and Systems, June 2007, vol. 1, no. 2, pp. 136-147.

[4] Holman, W.T., Connelly, J.A., "A compact low noise operational amplifier for a 1.2-μm digital CMOS technology," IEEE Journal of Solid-State Circuits, Jun 1995, vol.30, no.6, pp.710-714.

[5] Maeding, D.G., "A CMOS operational amplifier with low impedance drive capability," IEEE Journal of Solid-State Circuits , Apr 1983, vol.18, no.2, pp. 227- 229.

[6] Ka Nang Leung, Mok, P.K.T., "A CMOS voltage reference based on weighted ΔV_{GS} for CMOS low-dropout linear regulators," IEEE Journal of Solid-State Circuits, Jan 2003, vol.38, no.1, pp. 146- 150.

[7] Sanborn, K., Dongsheng Ma, Ivanov, V., "A Sub-1-V Low-Noise Bandgap Voltage Reference," IEEE Journal of Solid-State Circuits, Nov. 2007, vol.42, no.11, pp.2466-2481.

[8] Yueming Jiang, Lee, E.K.F., "A low voltage low 1/f noise CMOS bandgap reference," IEEE International Symposium on Circuits and Systems (ISCAS), May 2005, vol.4, pp. 3877- 3880.

[9] Baker, R., "CMOS, circuit design, layout, and simulation, " Wiley-IEEE Press, New Jersy, 2001.

[10] Razavi, B.," Design of Analog CMOS Integrated Circuits," Boston-McGraw-Hill, 2001.

[11] Fayomi, J.C., Wirth, G., Facpong, H., and Matsuzawa, A.," Sub 1-V CMOS bandgap reference design techniques: a survey," *Analog Integrated Circuits Signal Process,* February 2010, vol.62, issue.2, pp.141-157.

[12] Maxim.Semiconductor,"Calculating Temperature coefficient (Tempco) and initial Accuracy for Voltage Reference," Application note 3998.

[13] Gerosa, A., Neviani, A., "Enhancing output voltage swing in low-voltage micro-power OTA using self-cascode," Electronics Letters, April 2003, vol.39, no.8, pp. 638- 639.

[14] Castello, R., Grassi, A.G., and Donati, S.," A 500-nA sixth-order bandpass SC filter," IEEE Journal of Solid-State Circuits, 1990, vol.25, no.3, pp.669-676.

[15] Fujimori, I.; Sugimoto, T. , "A 1.5 V, 4.1 mW dual-channel audio delta-sigma D/A converter," IEEE Journal of Solid-State Circuits, Dec 1998, vol.33, no.12, pp.1863-1870.
[16] Sanchez-Snecio, E.," Low voltage analog circuit design techniques," IEICE Transaction of Electronics: A tutorial, 2000, pp. 179-196.
[17] Malcovati, P., Maloberti, F., Fioncchi, C., and Pruzzi, M.," Curvature compensated BiCMOS bandgap with 1-V supply voltage," IEEE J. Solid-State Circuits, 2001, vol. 36, no. 7, pp. 1076 -1081.
[18] Ytterdal, T.," Bandgap voltage reference circuit for supply voltages down to 0.6-V," Electronics Letters, October 2003, vol. 39, no. 20, pp. 1427- 1428.
[19] Hsieh, C., Huang, H., and Chen, K.," 1-V, 16.9 ppm/° C, 250 nA switched-capacitor CMOS voltage reference," IEEE Transactions on Very Large Scale Integration (VLSI) Systems, April 2011, vol. 19, no. 4, pp. 659-667.
[20] Lee, S., Lee, H., Woo, J., and Kim, S.," Low-voltage bandgap reference with output regulated current mirror in 90-nm CMOS," Electronics Letters, July 2010, vol. 46, no. 14, pp. 976-977.

Acronyms

1G	First Generation
2G	Second Generation
3G	Third Generation
3GPP	Third Generation Partnership Program
4G	Fourth Generation
AAA	Authentication, Authorisation and Accounting
AAS	Adaptive Antenna System
AC	Admission Control
AC	Alternating Current
ACI	Adjacent Channel Interference
ADC	Analog to Digital Converter
AF	Assured Forwarding
AF	Amplify and Forward
AFFR	Adaptive Fractional Frequency Reuse
AJ	Anti-Jamming
AMC	Adaptive Modulation and Coding
ANC	Analogue Network Coding
ARQ	Automatic Repeat Request
ASN	Access Service Network
ASN-GW	Access Service Network - Gateway
ASP	Application Service Provider
ATM	Asynchronous Transfer Mode
AWGN	Additive White Gaussian Noise
BC	Best Channel Scheduler
BCH	Broadcast Control Channel
BE	Best Effort
BER	Bit Error Rate
BLER	Block Error Rate
BPSK	Binary Phase Shift Keying
BS	Base Station
BSC	Base Station Controller
BTS	Base Transceiver Station
CB	Coherence Bandwidth
CBR	Call Blocking Ratio
CC	Convolution Coding
CC	Congestion Control

CCPCH	Common Control Physical Channel
CDF	Cummulative Distribution Function
CDMA	Code Division Multiple Access
CEP	Components of Error Probability
CHF	Characteristic Function
CINR	Carrier to Interference Noise Ratio
CLT	Central Limit Theorem
CN	Core Network
CP	Cyclic Prefix
CPCH	Common Packet Channel
CPICH	Common Pilot Channel
CQI	Channel Quality Indicator
CQICH	Channel Quality Information CHannel
CR	Coherence Ratio
CRC	Cyclic Redundancy Check
CSI	Channel State Information
CSN	Core Service Network
CTC	Convolution Turbo Coding
DAC	Digital to Analog Converter
DAS	Distributed Antenna System
dB	Decibel
DC	Direct Current
DCH	Dedicated Channel
DCM	Digital Clock Manager
DCT	Discrete Cosine Transform
DDS	Digital Direct Synthesiser
DF	Decode and Forward
DFDMA	Distributed Frequency Division Multiple Access
DFT	Discrete Fourier Transform
DHCP	Dynamic Host Configuration Protocol
DL	Down Link
DLFP	Down Link Frame Prefix
DL-MAP	Down Link Media Access Protocol
DPCCH	Dedicated Physical Control Channel
DPCH	Dedicated Physical Channel
DPDCH	Dedicated Physical Data Channel
DS-CDMA	Direct Sequence - Code Division Multiple Access
DVB-H	Dynamic Video Broadcast – Handheld
DWT	Discrete Wavelet Transform
ECG	Electrocardiography
EDGE	Enhanced Data rate for GSM Evolution
EEG	Electroencephalography
EF	Expedited Forwarding

EIRP	Effective Isotropic Radiated Power
EMC	Electromagnetic Compatibility
EMC	Electromagnetic Compatibility
EMF	Electromagnetic Field
EMF	Electromagnetic Field
EMG	Electromyography
EMI	Electromagnetic Interference
EMI	Electromagnetic Interference
eNB	Evolved Node B
EOG	Electrooculography
EPC	Evolved Packet Core
ERP	Effective Radiated Power
E-UTRAN	Evolve UMTS Terrestrial Access Network
FCH	Frame Control Header
FDD	Frequency Division Duplex
FDMA	Frequency Domain Multiple Access
FDTD	Finite Difference Time Domain
FEC	Forward Equivalence Class
FEC	Forward Error Correction
FFR	Fractional Frequency Reuse
FFT	Fast Fourier Transform
FH	Frequency Hopping
FH-CDMA	Frequency Hopping - Code Division Multiple Access
FPGA	Field Programmable Gate Array
FSH	Frame Synchronisation Header
FUSC	Fully Used Sub-Channelisation
GF	Galois Field
GoS	Grade of Service
GPRS	Generalised Packet Radio Service
GSM	Global System for Mobile Communications
H-ARQ	Hybrid Automatic Repeat reQuest
HO	HandOver
HPEM	High-Power Electromagnetic
HSDPA	High Speed Downlink Packet Access
IBI	Inter-Block Interference
IC	Interference Cancellation
ICI	Inter-Carrier Interference
ICIC	Inter Cell Interference Coordination
IDFT	Inverse Discrete Fourier Transform
IEEE	Institute of Electrical and Electronic Engineers
IFDMA	Interleaved Frequency Division Multiple Access
IFFT	Inverse Fast Fourier Transform
IP	Internet Protocol

IPTV	Internet Protocol Television
IS-95	Interim Standard-95
ISI	Inter-Symbol Interference
LDF-BC	Large Delay First to Round Robin Scheduler
LFDMA	Localised Frequency Division Multiple Access
LFSR	Linear Feedback Shift Register
LOS	Line-of-Sight
LPI	Low Probability of Intercept
LSB	Lowest Significant Bit
LTE	Long Term Evolution
LTE-Advance	Long Term Evolution Advance
MAC	Medium Access Control
MAI	Multiple Access Intereference
MAN-SC	Meltropolitan Area Network – Signal Carrier
MCS	Modulation Coding Sequence
MCS	Modulation and Coding Scheme
ME	Mobile Equipment
MFG	Magnetic Field Generators
MGF	Multiple Gamma Functions
MIMO	Multiple Input Multiple Output
MISO	Multiple Input Single Output
MMI	Multimode Interference
MMSE	Minimum Mean Square Error
MPEG	Motion Picture Expert Group
MRC	Maximum Ratio Combining
MSB	Most Significant Bit
MSS	Maximum SNR Scheme
MT	Mobile Terminal
NAP	Network Access Provider
NAS	Non Access Stratum
NLOS	Non-Line-of-Sight
NMT	Nordic Mobile Telephone
NSP	Network Service Provider
NWG	Network Working Group
OCXO	Oven Controlled Crystal Oscillator
OFDM	Orthogonal Frequency Division Multiplexing
OFDMA	Orthogonal Frequency Division Multiple Access
OFUSC	Optional Fully Used Sub-Channelisation
OSR	Over Subscription Ratio
OVSF	Orthogonal Variable Spreading Factor (codes)
OWR	One Way Rclaying
PAD	Personal Alarm Device
PAPR	Peak to Average Power Ratio

PBP	Power Bandwidth Profile
PCB	Printed Circuit Board
PCS	Personal Communications Service
PDF	Probability Density Function
PDP	Power Delay Profile
PDSCH	Physical Dedicated Shared Channel
PF	Probability of Intercept
PFS	Proportional Fair Scheduler
PFS	Proportional Fairness Scheme
PHY	Physical Layer
PICH	Page Indication Channel
PIR	Peak Information Rate
PLL	Phase-Locked Loop
PMP	Point-to-Multi-Point
PN	Pseudo Noise
PRACH	Physical Random Access Channel
PRB	Physical Resource Block
PS	Packet Service (also Polling Service)
PS	Physical Slot (WiMAX)
PSC	Primary Synchronisation Code
PSD	Power Spectral Density
PUSC	Partially Used Sub-Channelisation
QAM	Quadrature Amplitude Modulation
QoS	Quality of Service
QPSK	Quadrature Phase Shift Keying
RACH	Random Access Channel
RF	Radio Frequency
RFI	Radio Frequency Interference
RFID	Radio Frequency IDentification
RNC	Radio Network Controller
RR-BC	Round Robin Best Channel Scheduler
RRM	Radio Resource Management
RSRP	Minimum Reference Signal Received Power
RT	Relay Terminal
RTG	Receive-Transmit Transition Gap
RTLS	Real Time Location System
RX	Receiver
SC-FDE	Sub-Carrier Frequency Division Equalisation
SC-FDMA	Sub-Carrier Frequency Division Multiplexing
SCH	Synchronisation Channel
SDMA	Space Division Multiple Access
SF	Spreading Factor
SFBC	Space Frequency Block Coding

SGSN	Serving GPRS Support Nod
SIMO	Single Input Multiple Output
SINR	Signal to Interference Noise Ratio
SIP	Session Initiation Protocol
SISO	Single Input Single Output
SME	Small-to-Medium Enterprise
SNR	Signal-to-Noise Ratio
SOFDMA	Scalable Orthogonal Frequency Division Multiple Access
SOI	Silicon-On-Insulator
SS	Spread Spectrum
SS	Subscriber Station (WiMAX)
SSC	Secondary Synchronisation Code
SS-CDMA	Spread Spectrum - Code Division Multiple Access
SUI	Stanford University Interim
SuperDARN	Super Dual Auroral Radar Network
TCP/IP	Transmission Control Protocol / Internet Protocol
TDBC	Time Delay Broadcast
TDD	Time Domain Duplex
TDMA	Time Domain Multiple Access
TE	Transverse Electric
THS	Timing Hardware Server
TIGER	Tasman International Geospace Environment Radar
TM	Transverse Magnetic
TTG	Transmit-Receive Transition Gap
TTI	Transmission Time Indicator
TUSC	Total Usage of Sub-Channels
TWR	Two Way Relaying
TX	Transmitter
UE	User Equipment
UGS	Unsolicited Grant Service
UL	Up Link
UL-MAP	Up Link Media Access Protocol
UMTS	Universal Mobile Telecommunications System
USART	Universal Synchronous/Asynchronous Communication Interface
UTRA	UMTS Terrestrial Radio Access
UTRAN	UMTS Terrestrial Radio Access Network
VBR	Variable Bit Rate
VoIP	Voice Over IP
WBAN	Wireless Body Area Network
WCDMA	Wideband CDMA
WiBRO	Wireless Broadband

WiMAX	Worldwide Interoperability for Microwave Access
WLAN	Wireless Local Area Network
WSS	Wide Sense Stationary
WSSUS	Wide Sense Stationary Uncorrelated Scattering
XOR	Exclusive OR
ZF-FDE	Zero Forcing Frequency Division Equalisation

Index

Z

About The Editors

Johnson Agbinya receivedPhD from La Trobe University in 1994 in microwave radar remote sensing (MSc Research University of Strathclyde, Glasgow, Scotland (1982) in microprocessor techniques in digital control systems) and BSc (Electronic/Electrical Engineering, Obafemi Awolowo University, Ife, Nigeria (1977). He is currently Associate Professor (remote sensing systems engineering) in the department of electronic engineering, Honorary Professor in Computer Science at the University of Witwatersrand in Johannesburg, South Africa and Extraordinary Professor in telecommunications at Tshwane University of Technology/French South African Technical Institute in Pretoria, South Africa. He was Senior Research Scientist at CSIRO Telecommunications and Industrial Physics (1994 – 2000) in biometrics and remote sensing and Principal Engineer Research at Vodafone Australia research from 2000 to 2003. He is the author of six technical books in electronic communications including Principles of Inductive Near Field Communications for Internet of Things (River Publishers, Aalborg Postkontor, Denmark, 2011); IP Communications and Services for NGN (Auerbach Publications, Taylor & Francis Group, USA, 2010) and Planning and Optimisation of 3G and 4G Wireless Networks (River Publishers, Aalborg Postkontor, Denmark, 2009).

Dr Agbinya is Consulting Editor for River Publishers Denmark on emerging areas in Telecommunications and Science and also the founder and editor-in-chief of the African Journal of ICT (AJICT) and founder of the International Conference on Broadband Communications and Biomedical Applications (IB2COM). His current research interests include remote and short range sensing, inductive communications and wireless power transfer, Machine to Machine communications (M2), Internet of Things, wireless and mobile communications and biometric systems.

Dr. Agbinya is a member of IEEE, ACS and African Institute of Mathematics (AIMS). He has published extensively on broadband wireless communications, remote sensing, inductive communications, biometrics, vehicular networks, video and speech compression and coding, contributing to the development of voice over IP, intelligent multimedia sub-system and design and optimisation of 3G networks. He was recipient of several research and best paper awards and has held several advisory roles including the Nigerian National ICT Policy initiative.

Mari Carmen Aguayo-Torres received the Ph.D. degree in Telecommunication Engineering from the University of Malaga, Spain, in 2001 (M.S. 1994) with a thesis on Adaptive OFDM. She is currently an Associate Professor in the department of Communications Engineering Department at University of Malaga. Her main research interests include adaptive modulation and coding for fading channels, generalized MIMO, cooperative communications, OFDM and SC-FDMA, crosslayer design, and probabilistic QoS guarantees (often applied to machine to machine traffic) within cellular, satellite and vehicular networks.

Dr. Aguayo-Torres is a member of IEEE and has been an organizer for several international conferences, such as IB2Com 2010. She is involved in a number of public and private funded projects and actively collaborates with industry, mainly in the field of wireless communications (LTE, LTE-Advanced, WiMax).

Ryszard Klempous holds PhD in Control & Automation, Institute of Engineering Cybernetics and MS in Automation, from Faculty of Electronics, Wroclaw University of Technology, Poland. He is Senior Member of IEEE and NYAS. He has published over 150 papers inmainly high ranked journals including Journal of Computational and Applied Mathematics (Elsevier), IEEE, European Journal of Operational Research as well as conferences proceedings. He is the main coordinator of international teams within the main topics: Virtual Student Exchange: Lessons learned in virtual International Teaming in Interdisciplinary Design Education, Assessment of the quality of teaching and learning based on data driven evaluation methods, 24/7 Software development in Virtual Student Exchange groups: redefining the work and study week. In addition, he is the Project leader from Polish side on: Expanding the 24-Hour Workplace in cooperation with UTS, Sydney, TRAF - Tool for Creation Realistic Animation of Human-Like Figures. His main research goals are in Multi-model approach to human motion designing; Feasibility analysis of human motion identification using motion capture; Models and methods for biometric motion identification based on motion capture. He has been Program Committee member of over 40 top ranked conferences under IEEE, Springer Verlag, IFAC, and CHAOS auspices and has beenreviewer of papers submitted to these conferences as well as for the top ranked journals: (e.g. Journal of Computational and Applied Mathematics (Elsevier), International Journal of Adaptive Control and Signal Processing (John Wiley), International Journal of Computing Anticipatory Systems, LNCS). He is co-editor (with Janos Fodor, Jan Nikodem, Witold Jacak and Carmen Paz Suarez Araujo) for Springer Verlag Series Studies in Computational Intelligence. Essential outcomes: Methods of sensors localization in WSN, Data driven methods and data analysis of a distributed solar collector field; Hierarchical Control of a distributed solar collector field; Methodology and technology in distributed intelligent building systems.

Jan Nikodem is assistant professor in the Institute of Computer Engineering, Automatics and Robotics at the Faculty of Electronics, Wroclaw University of Technology, Wroclaw, Poland. He completed an MSc in Electrical Engineering and Automation, Wroclaw University of Technology in 1978 and PhD in Computer Science, Wroclaw University of Technology. His research interests include the study of methodologies, techniques and tools for analysis and design of goal oriented distributed systems. Distributed components in such systems self-organize, collaborate and evolve over time to optimise survivability, adaptability, flexibility, efficiency and effectiveness in the domain of communications. He has published extensively on Wireless Sensor Networks: using relations for communication activity, neighbours cooperation, collective decisions making, spatial routing. Important applications areas: Data driven methods and data analysis of a distributed solar collector field; Hierarchical Control of a distributed solar collector field; 24/7 Software development in Virtual Student Exchange groups. He is also involved in research into new methods of teaching, learning and creating of software development in large groups (Expanding the 24-Hour Workspace in cooperation with UTS, Sydney).